Essential Algebra

Table of Contents

Part 1

Part 2

Chapter 1 Variables and Equations

Lesson 1.1 Variables

- Use letters to represent variables
- Make a table of values for an expression
- Find a pattern and write a rule
- Plot points from a table of values
- Read values from a graph

Lesson 1.2 Algebraic Expressions

- Write expressions
- Evaluate expressions
- Plot points from a table of values
- Use formulas to write and evaluate expressions

Lesson 1.3 Equations and Graphs

- Graph equations
- Solve equations graphically

Lesson 1.4 Solving Equations

- Solve linear equations with one operation
- Solve problems using formulas
- Write an equation to solve a problem

Lesson 1.5 Order of Operations

- Use the associative laws for addition and multiplication
- Follow the order of operations to simplify expressions

1.1 Variables

What Is a Variable?

A **variable** is a numerical quantity that changes over time or in different situations.

We can show the values of a variable in a table or a graph.

Example 1 Life expectancy is the average age to which people will live. The table at right shows that your life expectancy depends on the year of your birth. Thus, life expectancy is a variable.

a. To what age could people born in 1940 expect to live?

b. In what year did people's life expectancy reach 70 years of age?

Year Born	Life Expectancy
1900	49
1910	51
1920	58
1930	59
1940	63
1950	68
1960	70
1970	71
1980	74
1990	75

Solutions a. People born in 1940 lived to 63 years of age on average.

b. Life expectancy reached 70 years of age in 1960.

Example 2 The graph at right shows the average annual salaries of NFL football players, starting in 1940. The average salary of NFL football players is a variable.

a. Use the graph to estimate the average annual salary of NFL football players in 1975.

b. By how much did salaries increase from 1975 to 1980?

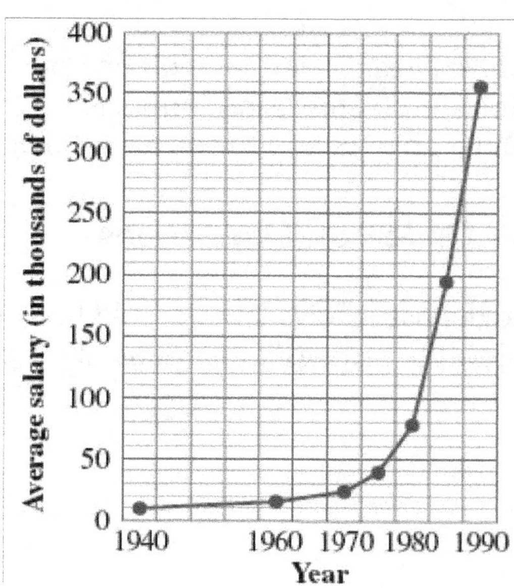

Solutions a. In 1975, the average NFL player's salary was $40,000.

b. In 1980, the average salary was $80,000, so salaries had increased from $40,000 to $80,000, or by

$$\$80{,}000 - \$40{,}000 = \$40{,}000$$

RQ1. A numerical quantity that changes over time or in different situations is called a __________ .

RQ2. We can show the values of a variable in a __________ or a __________ .

Look Closer: By displaying the values of a variable in a table or a graph, we sometimes see trends or patterns in those values. For example, we see that life expectancy has been increasing slowly over time, and that salaries of NFL football players have been increasing ever more rapidly since 1940.

Using Letters to Represent Variables

Imagine that you are traveling by airplane to another city. The table below shows the altitude of the airplane, in thousands of feet, at ten-minute intervals after take-off. The two quantities shown in the table, time and altitude, are both variables.

Time (min)	0	10	20	30	40	50	60	70	80	90	100	110
Altitude (1000 ft)	0	8	20	21	23	28	29	30	25	13	5	0

We often use a letter as a kind of short-hand to represent a variable. For this example, we use the following letters:

t stands for the **time** elapsed after take-off,
h stands for the plane's **altitude** at that time

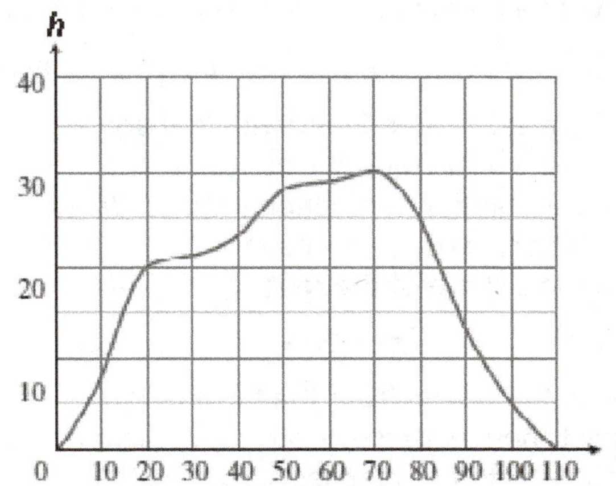

We can get a better feel for the variables by plotting them on a graph. In the figure at right, the values of t are displayed on the horizontal scale, or **axis**, and the values of h are shown on the vertical axis. The graph shows how the values of h are related to the values of t.

Example 3a. What is the value of h when $t = 85$? What do these values mean in this situation?

b. When is the plane descending?

Solutions a. The graph shows that h is approximately 20 when $t = 85$. This means that at 85 minutes into the flight, the plane's altitude is about 20,000 feet.

b. The plane is descending when its altitude is decreasing. On the graph, h begins to decrease at $t = 70$ and continues decreasing until $t = 110$. Thus, the plane is descending from the 70th minute until the end of the flight at 110 minutes.

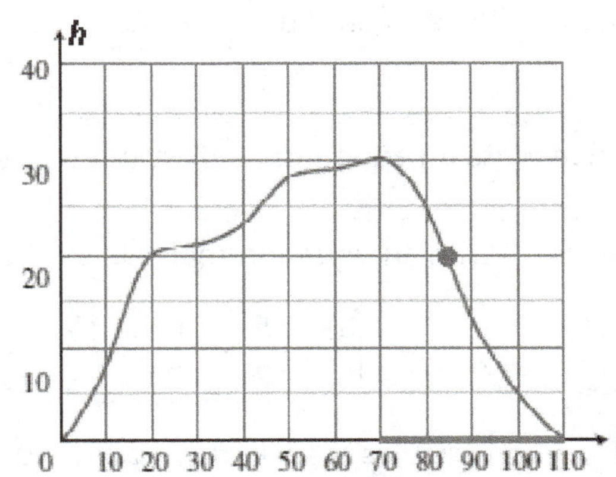

RQ3. We often use __________ as a shorthand to represent variable quantities.
RQ4. Which variable is displayed on the vertical axis in Example 3?

Look Ahead: It would be useful if we could discover a formula or rule to help us *predict* the values of an interesting variable.

Connections Between Variables

Sometimes there is a simple mathematical relationship between the values of two variables.

Example 4 Fernando plans to share an apartment with three other students next year. The table shows his share of the rent for apartments of various prices.

Rent	280	300	340	360	400	460	500
Fernando's share	70	75	85	90	100	**?**	**?**

a. Fill in the blanks in the table. Describe in words how you found the unknown values.

Divide the rent by 4

b. Write a mathematical sentence that gives Fernando's share of the total rent.

Fernando's share = **Rent ÷ 4**

c. The total rent and Fernando's share are variables. Let R stand for the total rent, and let F stand for Fernando's share. Using these letters, write a formula for Fernando's share of the rent.

$$\boldsymbol{F = R \div 4}$$

d. Plot the points from the table and connect them with a smooth curve. Extend your line so that it reaches across the entire grid, not just the points that you plotted.

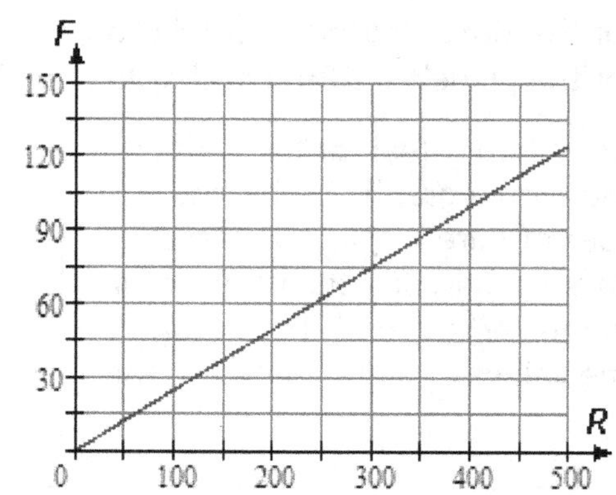

A formula relating two variables is a type of **equation**, and an equation is just a statement that two quantities are equal.

By studying the values in a table, we may be able to find a relationship between the values, and then write an equation relating the two variables.

Example 5 Write an equation that relates the two variables in the table.

x	5	10	15	20	25
y	2	7	12	17	22

Solution The two variables are x and y. Each column of the table shows a related pair of values. Notice that 2 is 3 less than 5, 7 is 3 less than 10, and so on. That is:

$$2 = 5 - 3, \quad 7 = 10 - 3,$$
$$12 = 15 - 3, \quad 17 = 20 - 3,$$

and so on. In each column, we can find y by subtracting 3 from x. Thus, $y = x - 3$.

RQ5. A statement that two quantities are equal is called an ______________.

Skills Warm-Up

Choose the correct arithmetic operation (addition, subtraction, multiplication, or division) and explain how to answer the question.

1. An air conditioner keeps the inside temperature 16° cooler than the outside temperature. If the outside temperature is 90°, how can you find the inside temperature?

2. Tom's recipe for punch calls for three times as much fruit juice as soda. If he has half a gallon of soda, how can he find the amount of fruit juice he needs?

3. A clothesline should be 2 feet longer than the distance between the supporting poles so that it can be tied at each end. If the poles are 20 feet apart, how can you find the length of the clothesline?

4. The weight of a bridge is supported equally by 8 pillars. If the bridge weighs one million tons, how can you find the weight each pillar must support?

5. Katrin has 4 hours to complete both her math and geography homework. If her math assignment takes $2\frac{1}{2}$ hours, how can she calculate how long she has for geography?

6. The cost of leasing a compact car is 63% the cost of a luxury car. If the lease on the luxury car is $500 per month, how can you find the cost of leasing the compact car?

Answers: **1.** Subtract 16 from the outside temperature. **2.** Multiply the amount of soda by 3. **3.** Add 2 feet to the distance between the poles. **4.** Divide the weight of the bridge by 8. **5.** Subtract the time for math from 4 hours. **6.** Multiply the cost of the luxury car by 0.63.

Homework 1.1

Skills Practice

For Problems 1-4,

a. Find the pattern and fill in the table.

b. Write an equation for the second variable in terms of the first variable.

1.

m	g
2	5
3	6
5	8
10	13
12	
16	
18	
m	

2.

t	w
0	20
2	18
4	16
5	15
6	
10	
12	
t	

3.

b	x
0	0
2	1
4	2
5	2.5
6	
8	
9	
b	

4.

z	3	6	8	12	15	18	20	z
r	2	4	$\frac{16}{3}$	8				

Hint: What fraction can you multiply z by to get r?

In Problems 5-6, make your own table. Choose values for the first variable, and use the rule to find the values of the second variable.

5. $W = 1.2 \times n$

n					
W					

6. $M = \frac{3}{2} \times x$

x					
M					

For Problems 7-8, fill in the tables. What do you notice? Explain.

7a. $y = \frac{x}{8}$

x	y
4	
8	
10	
16	

b. $y = \frac{1}{8} \times x$

x	y
4	
8	
10	
16	

c. $y = 0.125 \times x$

x	y
4	
8	
10	
16	

8a. $y = \frac{3 \times x}{5}$

x	y
5	
10	
12	
1	

b. $y = \frac{3}{5} \times x$

x	y
5	
10	
12	
1	

c. $y = 0.6 \times x$

x	y
5	
10	
12	
1	

Applications

9. The temperature in Sunnyvale is usually 15° hotter than it is in Ridgecrest, which is at a higher elevation. Fill in the table.

Temperature in Ridgecrest	70	75	82	86	90	R
Calculation						
Temperature in Sunnyvale						

a. Explain in words how to find the temperature in Sunnyvale.

b. Write your explanation as a mathematical sentence:

Temp in Sunnyvale =

c. Let R stand for the temperature in Ridgecrest and S for the temperature in Sunnyvale. Write an equation for S in terms of R.

d. Plot the points from the table and connect them with a smooth curve.

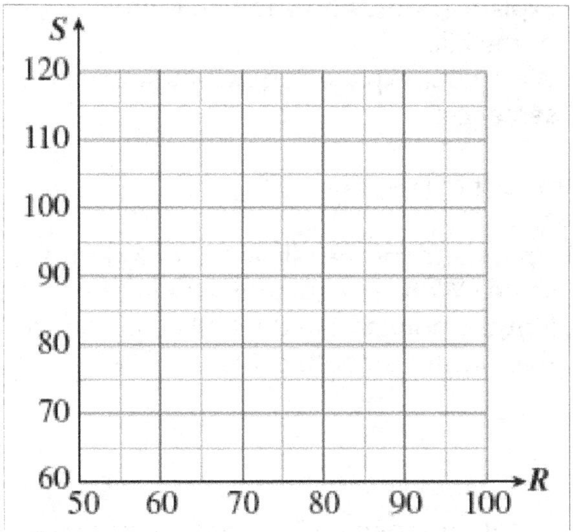

10. Jerome is driving from his home in White Falls to his parents' home in Castle Heights, 200 miles away. Fill in the table.

Miles driven	40	60	90	120	140	170	d
Calculation							
Miles remaining							

a. Explain in words how to find the number of miles Jerome has left to drive.

b. Write your explanation as a mathematical sentence:

Miles remaining =

c. Let d stand for the number of miles Jerome has driven and r for the number of miles that remain. Write an equation for r in terms of d.

d. Plot the points from the table and connect them with a smooth curve.

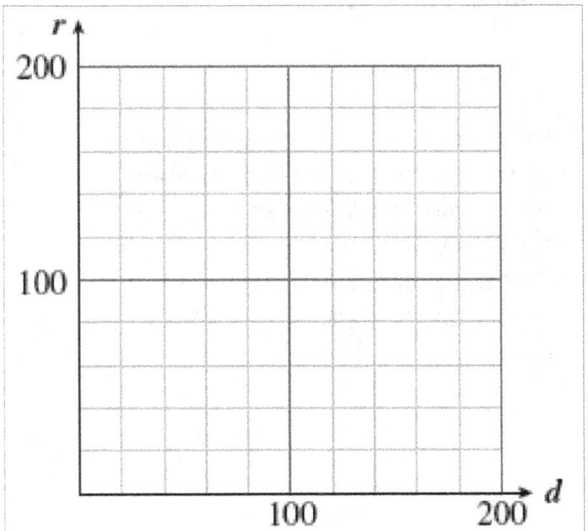

11. Milton goes to a restaurant with two friends, and they agree to split the bill equally. Fill in the table.

Total bill	15	30	45	60	75	81	b
Calculation							
Milton's share							

a. Explain in words how to find Milton's share of the bill.

b. Write your explanation as a mathematical sentence:

Milton's share =

c. Let b stand for the bill and s for Milton's share. Write an equation for s in terms of b.

d. Plot the points from the table and connect them with a smooth curve.

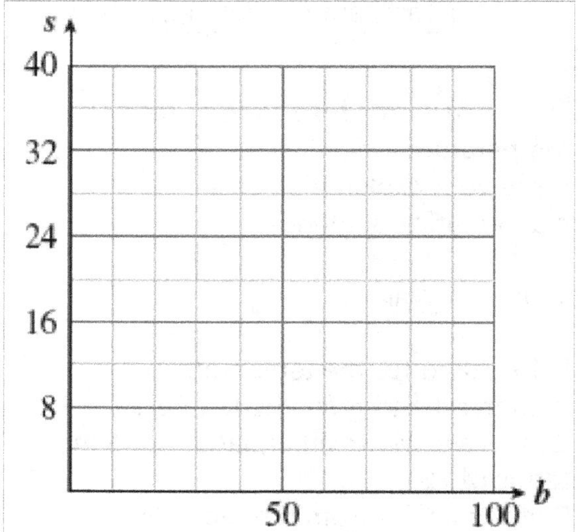

12. Nutrition experts tell us that no more than 30% of the calories we consume should come from fat. Fill in the table with the fat calories allowed daily for various calorie levels.

Total calories	1000	1500	2000	2500	3000	C
Calculation						
Fat calories						

a. Explain in words how to find the number of fat calories allowed.

b. Write your explanation as a mathematical sentence:

Fat calories =

c. Let C stand for the total number of calories and F for the number of fat calories. Write an equation for F in terms of C.

d. Plot the points from the table and connect them with a smooth curve.

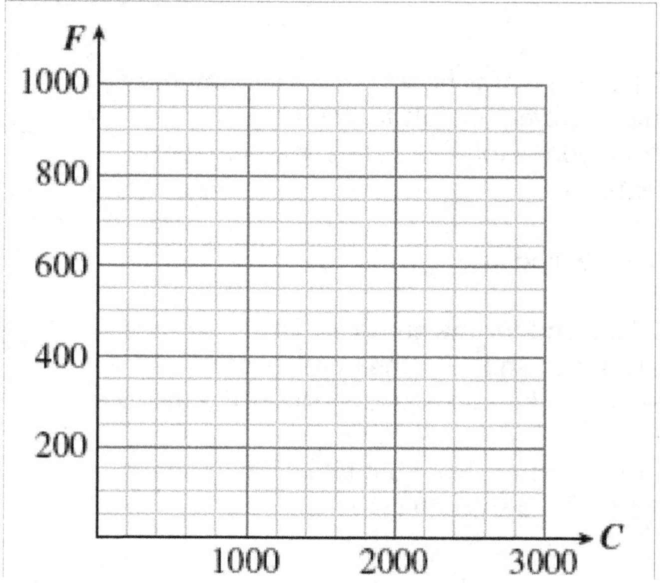

Use the graphs to answer the questions in Problems 13 and 14. You may have to estimate some of your answers.

13. Suppose you invest $2000 in a retirement account that pays 8% interest compounded continuously. The amount of money in the account each year is shown by the graph at left below.

a. What variable is displayed on the horizontal axis? The vertical axis?
b. How much money will be in the account 5 years from now?
c. When will the amount of money in the account exceed $6000?
d. How much will the account grow between the second and third years?
e. How much will the account grow between the twelfth and thirteenth years?

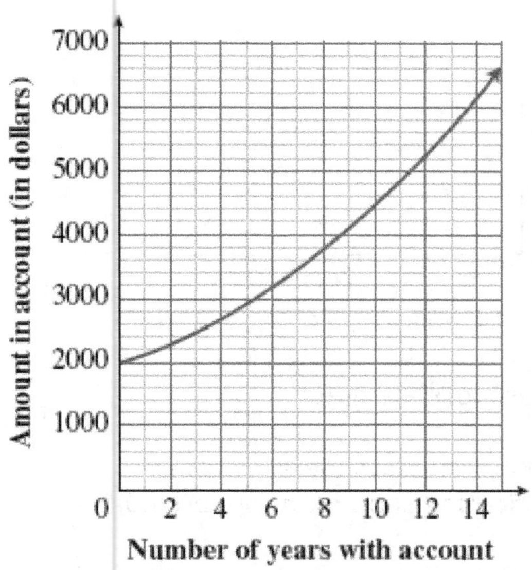

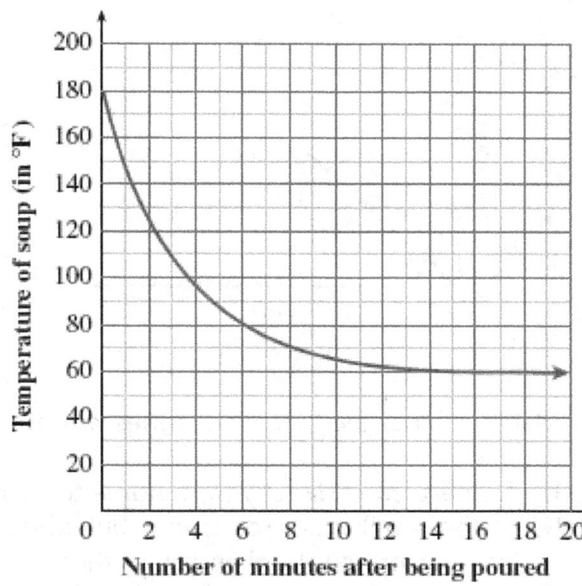

14. Wendy brings a Thermos of soup for her lunch on a hike in the mountains. When she pours the soup into a bowl, it begins to cool off. The graph above right shows the temperature of the soup after it is served.

a. What variable is displayed on the horizontal axis? The vertical axis?
b. How long does it take the soup to cool below 100°?
c. What is the temperature of the soup after 8 minutes?
d. How many degrees does the soup cool in the first 2 minutes?
e. After a while, the temperature of the soup levels off at the same as the outside temperature. What was the outside temperature that day?

15. A cantaloupe is dropped from a tall building. Its speed increases until it hits the ground. Which of the four graphs below best represents the speed of the cantaloupe? Explain why your choice is the best one.

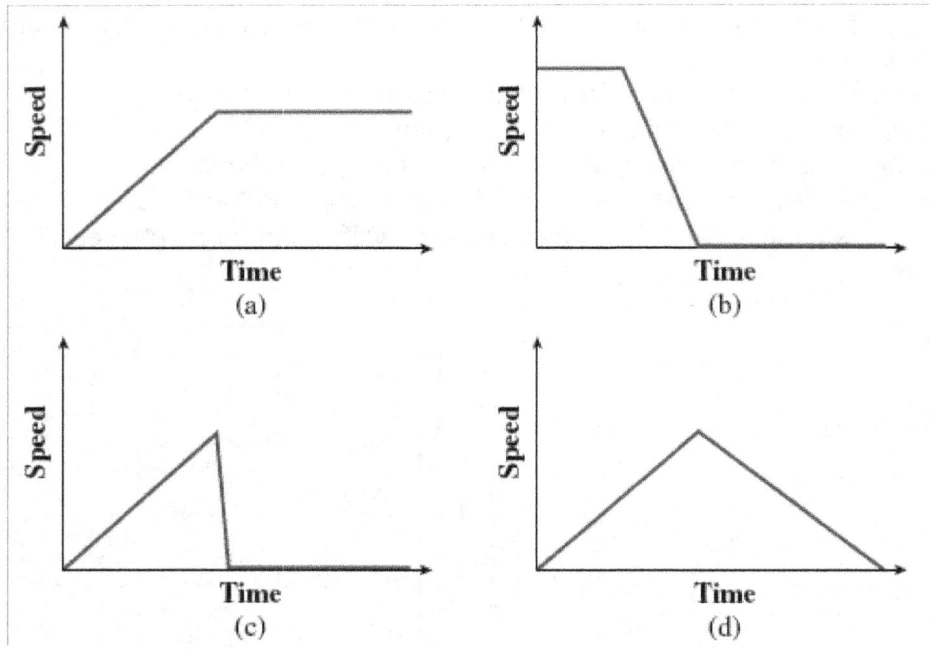

16. a. Match each of the following stories with the appropriate graph.

- **I.** Delbert walks directly from home to school, then stays there.
- **II.** Delbert walks towards school, but stops at coffee shop for a cappuccino, then he goes on to school and stays there.
- **III.** Delbert walks toward school, but decides to return home, where he stays.

b. Write your own story for the remaining graph.

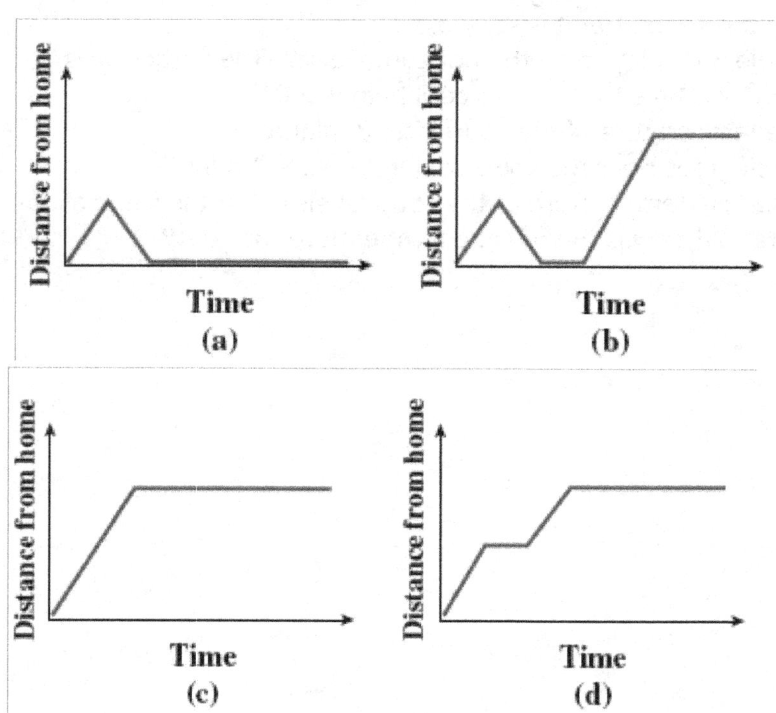

17. Plot each pair of values on the grid provided. Connect the points with a smooth curve. Which graph appears to be a straight line?

a.

t	0	1	2	5	6	8
v	16	14	12	6	4	0

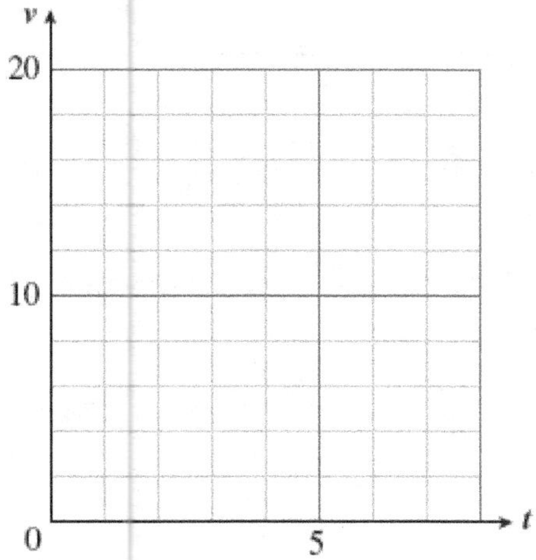

b.

w	0	1	2	4	6	10
Q	12	8	6	4	3	2

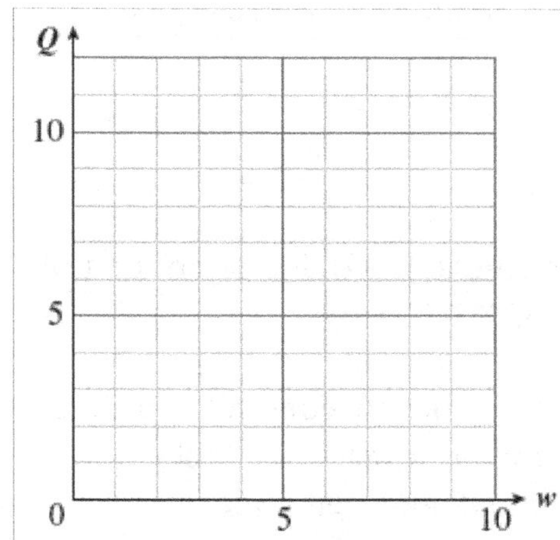

18. a. Read the graph at right to fill in the table.

x	0	10	30	40	60	70
y						

b. Find a formula expressing y in terms of x.

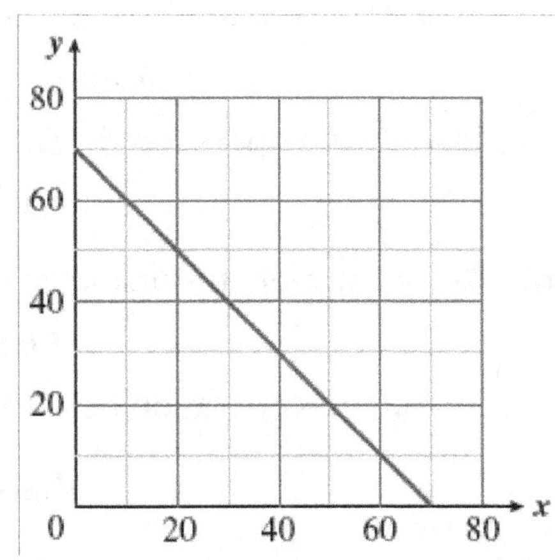

For Problems 19-20,

a. Write an equation for the second variable in terms of the first variable.

b. Fill in the table. Plot the points and connect them with a smooth curve.

19.

x	0	0.2	0.3	0.5	0.6	0.7
y	2	1.8	1.7			

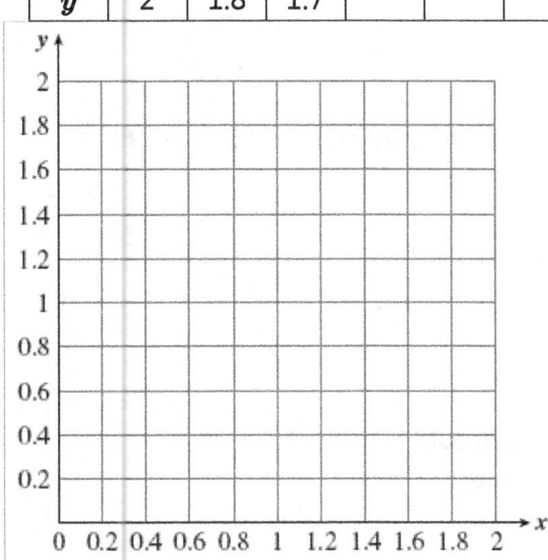

20.

x	1	2	3	4	6	8
y	120	60	40			

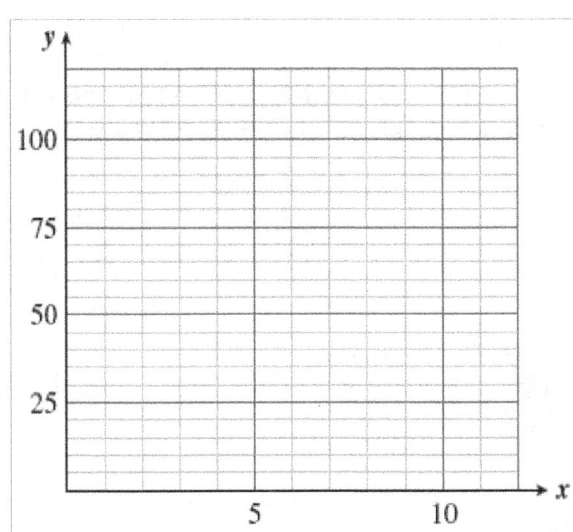

1.2 Algebraic Expressions

Writing an Algebraic Expression

> An **algebraic expression**, or simply an **expression**, is any meaningful combination of numbers, variables, and operation symbols.

For example,

$$4 \times g, \quad 3 \times c + p, \quad \text{and} \quad \frac{n-7}{w}$$

are algebraic expressions. An important part of algebra involves translating word phrases into algebraic expressions.

Example 1 Write an algebraic expression for each quantity.

a. 30% of the money invested in stocks

b. The cost of the dinner split three ways

Solutions a. The amount invested is unknown, so we choose a variable to represent it.

Amount invested in stocks: s

Next, we identify the operation described: "30% of" means 0.30 times:

The expression is $\mathbf{0.30\,s}$

b. The cost of the dinner is unknown, so we choose a variable to represent it.

Cost of dinner: C

Next, we identify the operation described: "Split" means divided:

The expression is $\mathbf{\frac{C}{3}}$ ■

> **RQ1.** A meaningful combination of numbers, variables and operation symbols is called an ______.

Sums

When we add two numbers a and b, the result is called the **sum** of a and b. We call the numbers a and b the **terms** of the sum.

Example 2 Write sums to represent the following phrases.

a. Six more than x

b. Fifteen greater than r

Solutions a. $6 + x$ or $x + 6$

b. $15 + r$ or $r + 15$ ■

The terms can be added in either order. We call this fact the **commutative law** for addition.

> **Commutative Law for Addition**
> If a and b are any numbers, then
> $$a + b = b + a$$

Products

When we multiply two numbers a and b, the result is called the **product** of a and b. The numbers a and b are the **factors** of the product.

Look Closer: In arithmetic we use the symbol $\times$ to denote multiplication. However, in algebra, $\times$ may be confused with the variable x. So, instead, we use a dot or parentheses for multiplication, like this:

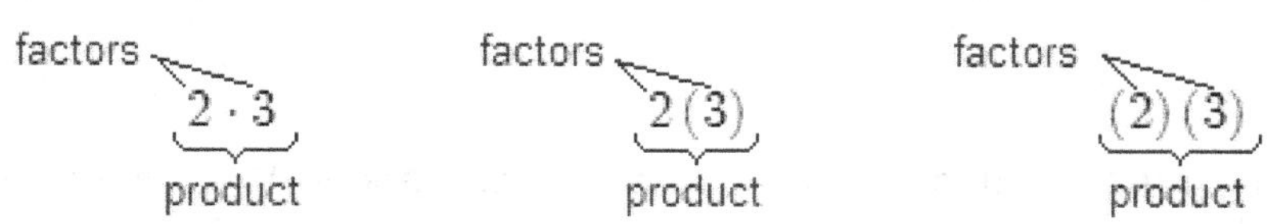

If one of the factors is a variable, we can just write the factors side by side. For example,

ab	means	the product of a times b
$3x$	means	the product of 3 times x

Example 3 Write products to represent the following phrases.

a. Ten times z **b.** Two-thirds of y **c.** The product of x and y

Solutions a. $10z$ **b.** $\frac{2}{3}y$ **c.** xy or yx

Look Closer: In Example 3a, $10z = 10 \cdot z = z \cdot 10$, because two numbers can be multiplied in either order to give the same answer. This is the **commutative law** for multiplication.

> **Commutative Law for Multiplication**
> If a and b are any numbers, then
> $$a \cdot b = b \cdot a$$

In algebra, we usually write products with the numeral first. Thus, we write $10z$ for "10 times z". Each factor in a product is called a **coefficient** for the other factors. Most often, we use the word "coefficient" for the numerical factor of a product. For example, we say that 10 is the **numerical coefficient** of the expression $10z$.

> **RQ2.** When we add two numbers, the result is called the _______ and the two numbers are called the _________ .
> **RQ3.** When we multiply two numbers, the result is called the _______ and the two numbers are called the _________ .

Differences

When we subtract b from a the result is called the **difference** of a and b. As with addition, a and b are called **terms**.

$$\underbrace{a - b}_{\text{difference}} \qquad \underbrace{8 - c}_{\text{difference}} \qquad \underbrace{5 - 2}_{\text{difference}}$$

(In each, the arrows labelled "terms" point to the two terms.)

> **Caution!** The difference $5 - 2$ is *not* the same as $2 - 5$. When we translate "a subtracted from b," the order of the terms is important.
>
> $b - a$ means a subtracted from b
> $x - 7$ means 7 subtracted from x
>
> The operation of subtraction is *not* commutative.

Quotients

When we divide a by b the result is called the **quotient** of a and b. We call a the **dividend** and b the **divisor**. In algebra we indicate division by using the division symbol, $\div$, or a fraction bar.

$$\text{dividend} \rightarrow \underbrace{a \div b}_{\text{quotient}} \leftarrow \text{divisor} \qquad \text{fraction bar} \rightarrow \left.\frac{a}{b}\right\} \leftarrow \text{quotient}$$

(division symbol: $\div$; dividend: a; divisor: b)

Look Closer: The operation of division is **not** commutative. The order of the numbers in a quotient makes a difference. For example, $12 \div 3 = 4$ but $3 \div 12 = \frac{1}{4}$.

Example 4 Write an algebraic expression for each phrase.

a. z subtracted from 13 **b.** 25 divided by R

Solutions a. $13 - z$ **b.** $\dfrac{25}{R}$

> **RQ4.** Which operations obey the **commutative laws**? What do these laws say?
> **RQ5.** When we subtract two numbers, the result is called the ________ .
> **RQ6.** When we divide two numbers, the result is called the ________ .

Evaluating an Algebraic Expression

An algebraic expression is a pattern or rule for different versions of the situation it describes. We can replace the variable by specific numbers to fit a particular situation.

> Substituting a specific value into an expression and calculating the result is called **evaluating** the expression.

Example 5 Fernando has three roommates, and his share of the rent is $\frac{R}{4}$, where R is a variable representing the total rent on the apartment. If Fernando and his roommates find an apartment whose rent is \$540, we can find Fernando's share of the rent by replacing R by 540.

$$\frac{R}{4} = 540 \div 4 = 135$$

Fernando's share of the rent is \$135.

In Example 5, we evaluated the expression $\frac{R}{4}$ for $R = 540$.

Some Useful Algebraic Formulas

Here are some examples of common formulas. Formulas may involve several different variables.

1. To find the **distance** traveled by an object moving at a constant speed or rate for a specified time, we multiply the rate by the time.

   distance = rate × time $\qquad d = rt$

Example How far will a train moving at 80 miles per hour travel in three hours?

Solution Evaluate the formula with $r = 80$ and $t = 3$ to find

$$\begin{aligned} d &= rt \\ &= (80)(3) = 240 \end{aligned}$$

The train will travel 240 miles.

2. To find the **profit** earned by a company we subtract the costs from the revenue.

   profit = revenue − cost $\qquad P = R - C$

Example The owner of a sandwich shop spent \$800 last week for labor and supplies. She received \$1150 in revenue. What was her profit?

Solution Evaluate the formula with $R = 1150$ and $C = 800$ to find

$$\begin{aligned} P &= R - C \\ &= 1150 - 800 = 350 \end{aligned}$$

The owner's profit was \$350.

3. To find the **interest** earned on an investment we multiply the amount invested (the principal) by the interest rate and the length of the investment.

   interest = principal × interest rate × time $\qquad I = Prt$

Example How much interest will be earned on \$800 invested for 3 years in a savings account that pays $5\frac{1}{2}\%$ interest?

Solution Evaluate the formula with $P = 800$, $r = 0.055$, and $t = 3$ to find

$$\begin{aligned} I &= Prt \\ &= 800 \times 0.055 \times 3 = 132 \end{aligned}$$

The account will earn \$132 in interest. ■

4. To find a percentage or **part** of a given amount, we multiply the whole amount by the percentage rate.

part = percentage rate × whole $\quad P = rW$

Example How much ginger ale do you need to make 60 gallons of a fruit punch that is 20% ginger ale?

Solution Evaluate the formula with $r = 0.20$ and $W = 60$ to find

$$\begin{aligned} P &= rW \\ &= 0.20 \times 60 = 12 \end{aligned}$$

You would need 12 gallons of ginger ale. ■

5. To find the **average** of a collection of scores, we divide the sum of the scores by the number of scores.

$$\textbf{average score} = \frac{\textbf{sum of scores}}{\textbf{number of scores}} \qquad A = \frac{S}{n}$$

Example Bert's quiz scores in chemistry are 15, 16, 18, 18, and 12. What is his average score?

Solution Evaluate the formula with $S = 15 + 16 + 18 + 18 + 12$ and $n = 5$ to find

$$A = \frac{S}{n} = \frac{15 + 16 + 18 + 18 + 12}{5} = 15.8$$

Bert's average score is 15.8. ■

RQ7. Whhat does it mean to evaluate an expression?

RQ8. How do we change a percent to a decimal fraction?

Look Ahead: You will also need to know the formulas for the area and perimeter of a rectangle. See the Skills Warm-Up to review these formulas. •

Skills Warm-Up

Recall the formulas for the **area** and **perimeter** of a rectangle:

$$A = lw \quad \text{and} \quad P = 2l + 2w$$

1. Delbert's living room is 20 feet long and 12 feet wide. How much oak baseboard does he need to border the floor? (Don't worry about doorways.)
2. How much wood parquet tiling must he buy to cover the floor?

Which of the following phrases describe a perimeter, and which describe an area?

3. The distance you jog around the shore of a small lake
4. The amount of Astroturf needed for the new football field
5. The amount of grated cheese needed to cover a pizza
6. The number of tulip bulbs needed to border a patio

Find the perimeter and area of each figure. All angles are right angles.

7. 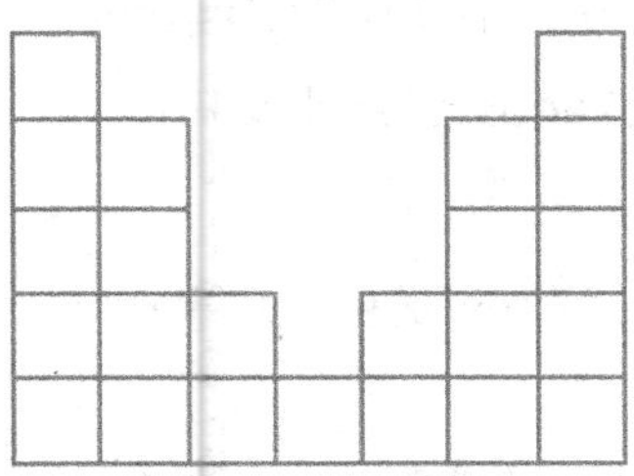

All squares are
1 centimeter on a side.

8. 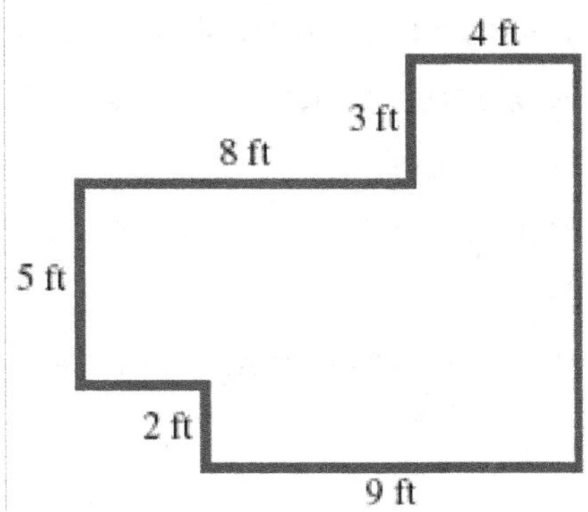

Answers: **1.** 64 ft **2.** 240 sq ft **3.** perimeter **4.** area **5.** area **6.** perimeter
7. 32 cm; 23 sq cm **8.** 44 ft; 90 sq ft

Homework 1.2

Skills Practice

For Problems 1-10, write the word phrase as an algebraic expression.

1. Product of 4 and y
2. Twice b
3. 115% of g
4. t decreased by 5
5. The quotient of 7 and w
6. 4 divided into B
7. 20 more than T
8. 16 reduced by p
9. The ratio of 15 to M
10. The difference of R and 3.5

For Problems 11-14, choose the correct algebraic expression.

$n+6$	$n-6$	$6-n$
$6n$	$\frac{n}{6}$	$\frac{6}{n}$

11. Rashad is 6 years younger than Shelley. If Shelley is n years old, how old is Rashad?
12. Each package of sodas contains 6 cans. If Antoine bought n packages of sodas, how many cans did he buy?
13. Lizette and Patrick together own 6 cats. If Lizette owns n cats, how many cats does Patrick own?
14. Mitra divided 6 cupcakes among n children. How much of a cupcake did each child get?

For Problems 15-17, write an algebraic expression for the area or perimeter of the figure. Include units in your answers. The dimensions are given in inches.

15. 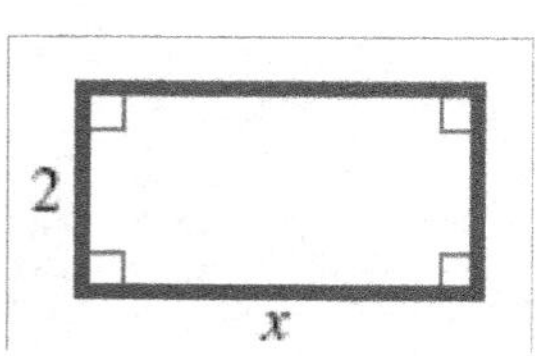

area =

16. 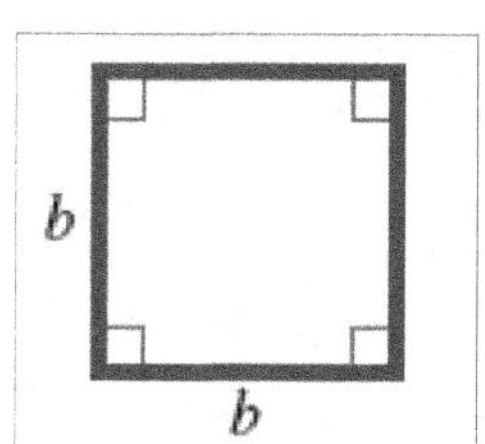

perimeter =

17. 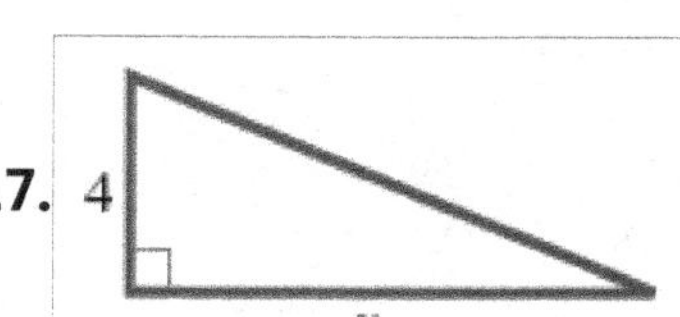

area =

18. **a.** Which of these expressions says "one-sixth of x" ?

$$\frac{x}{6}, \quad \frac{6}{x}, \quad \frac{1}{6}x, \quad \frac{1}{6x}$$

b. Which of these expressions says "6% of x" ?

$$0.6x, \quad 0.06x, \quad \frac{6}{100}x, \quad \frac{1}{6}x$$

19. Francine is saving up to buy a car, and deposits $\frac{1}{5}$ of her spending money each month into a savings account.

a. If m stands for Francine's spending money, write an expression for the amount she saves.

b. Evaluate the expression to complete the table.

Spending money (dollars)	m	20	25	60	80
Amount saved (dollars)					

20. Delbert wants to enlarge his class photograph. Its height is $\frac{1}{4}$ of its width. The enlarged photo should have the same shape as the original.

a. If W stands for the width of the enlargement, write an expression for its height.

b. Evaluate the expression to complete the table.

Width of photo (inches)	W	6	10	24	30
Height of photo (inches)					

21. If the governor vetoes a bill passed by the State Assembly, $\frac{2}{3}$ of the members present must vote for the bill in order to overturn the veto.

a. If p stand for the number of Assembly members present, write an expression for the number of votes needed to overturn a veto.

b. Evaluate the expression to complete the table.

Members present	p	90	96	120	129
Votes needed					

22. Marla's investment club buys some stock. She will get $\frac{3}{5}$ of the dividends.

a. If D stands for the stock dividends, write an expression for Marla's share.

b. Evaluate the expression to complete the table.

Stock dividend	D	40	85	115	170
Marla's share					

Applications

■ For Problems 23-30, choose a variable for the unknown quantity and translate the phrase into an algebraic expression. Use the three steps you learned in the Lesson.

23. The product of the ticket price and \$15
24. Three times the cost of a light bulb
25. Three-fifths of the savings account balance
26. The price of the pizza divided by 6
27. The weight of the copper in ounces divided by 16
28. 9% of the school buses
29. \$16 less than the cost of the vaccine
30. The total cost of 32 identical computers

■ For Problems 31-32, write algebraic expressions in terms of x.

31. a. Daniel and Lara together made \$480. If Daniel made x dollars, how much did Lara make?

b. Alix spent \$500 on tuition and books. If she spent x dollars on books, how much was her tuition?

c. Thirty children signed up for summer camp. If x boys signed up, how many girls signed up?

32. a. Rona spent \$15 less than her sister on shoes. If Rona's sister spent x dollars, how much did Rona spend?

b. Phoenix had 12 fewer rain days than Boston last year. If Boston had x rain days, how many rain days did Phoenix have?

c. Jared scored 18 points lower on his second test than he scored on his first test. If he scored x points on the first test, what was his score on the second test?

■ For Problems 33-36, name the variable and write an algebraic expression.

33. Eggnog is 70% milk. Write an expression for the amount of milk in a container of eggnog.

34. Errol has saved \$1200 for his vacation this year. Write an expression for the average amount he can spend on each day of his vacation.

35. Garth received 432 fewer votes than his opponent in the election. Write an expression for the number of votes Garth received.

36. The cost of the conference was \$2000 over budget. Write an expression for the cost of the conference.

■ For Problems 37-40, use the formulas in this Lesson.

37. a. Write an equation for the distance d traveled in t hours by a small plane flying at 180 miles per hour.

b. How far will the plane fly in 2 hours? In $3\frac{1}{2}$ hours? In half a day?

38. a. A certain pesticide contains 0.02% by volume of a dangerous chemical by volume. Write an equation for the amount of chemical C that enters the environment in terms of the number of gallons g of pesticide used.

b. How much of the chemical enters the environment if 400 gallons of the pesticide are used? 5000 gallons? 50,000 gallons?

39. a. BioTech budgets 8.5% of its revenue for research. Write an equation for the research budget B in terms of BioTech's revenue R.

b. What is the research budget if BioTech's revenue is \$100,000? \$500,000? \$2,000,000?

40. a. Hugo's Auto Shop paid \$4000 in expenses this month. Write an equation for their profit P in terms of their revenue R.

b. What was their profit if their revenue was \$10,000? \$6500? \$2500?

1.3 Equations and Graphs

Anatomy of a Graph

We use equations to express the relationship between two variabes. A graph is a way of visualizing an equation.

> A graph has two **axes**, horizontal and vertical, and the values for the variables are displayed along the axes. The first, or **input,** variable is displayed on the horizontal axis. The second, or **output,** variable is displayed on the vertical axis.

The graph itself shows how the two variables are related.

Example 1 The graph at right shows the relationship between Delbert's age, D, and Francine's age, F. Write an equation for Francine's age in terms of Delbert's age.

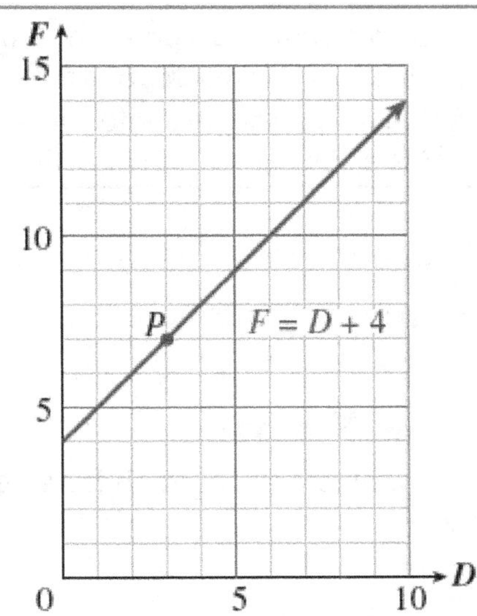

Solution In this graph, D is the input variable and F is the output variable. We make a table of values by reading points on the graph.

D	0	2	3	4	6	9
F	4	6	7	8	10	13

From the table, we see that the value of F is always 4 more than the value of D. Thus, Francine is exactly four years older than Delbert, or $F = D + 4$.

RQ1. A ________ is a way of visualizing an algebraic equation.

RQ2. The values of the variables are displayed on the ___________ .

RQ3. The input variable is located on the __________ axis.

Each point on a graph has two **coordinates,** which designate the position of the point. For example, the point labeled P in Example 1 has horizontal coordinate 3 and vertical coordinate 7.

> In Example 1 above, we write the coordinates of point P inside parentheses as an **ordered pair**: $(3, 7)$. The order of the coordinates makes a difference. *We always list the horizontal coordinate first, then the vertical coordinate.*

RQ4. The position of a point on the graph is given by its _________ .

RQ5. The notation (x, y) is called ______________ .

RQ6. In an ordered pair, we always list the _________ coordinate first.

Look Closer: How does the graph illustrate the equation $F = D + 4$? The coordinates of each point on the graph are values for D and F that make the equation true. The coordinates of the point P, namely $D = 3$ and $F = 7$, represent the fact that when Delbert was 3 years old, Francine was 7 years old. If we substitute these values into our equation we get

$$F = D + 4$$
$$7 = 3 + 4$$

which is true.

An ordered pair that makes an equation true is called a **solution** of the equation. Each point on the graph represents a solution of the equation.

Exercise a. Locate on the graph each of the ordered pairs listed in the table above, and make a dot there. Label each point with its coordinates
b. By substituting its coordinates into the equation, verify that each point you labeled in part (a) represents a solution of the equation $F = D + 4$.

RQ7. An ordered pair whose coordinates make the equation true is called a __________ of the equation.

Graphing an Equation

The simplest way to draw a graph for an equation is to make a table of values and plot the points.

Example 2 Graph the equation $y = 8 - x$.

Solution In this equation, x is the input variable and y is the output variable. We choose several values for x and use the equation to find the corresponding values for y.

x	Calculation	y	(x, y)
0	$y = 8 - \mathbf{0} = 8$	8	$(0, 8)$
2	$y = 8 - \mathbf{2} = 6$	6	$(2, 6)$
5	$y = 8 - \mathbf{5} = 3$	3	$(5, 3)$
8	$y = 8 - \mathbf{8} = 0$	0	$(8, 0)$

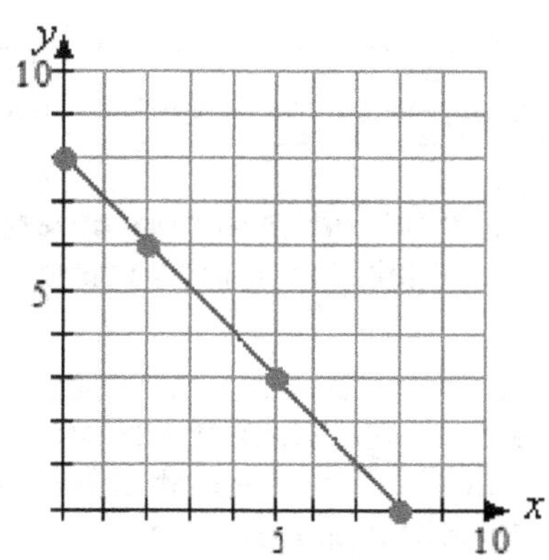

Each ordered pair (x, y) represents a point on the graph of the equation. We plot the points on the grid and connect them with a smooth curve, as shown at right.

Choosing Scales for the Axes

If the variables in an equation have very large (or very small) values, we must choose scales for the axes that fit these values.

Example 3 Graph the equation $y = 250 + x$.

Solution To graph this equation, we choose multiples of 50 for the x-values.

x	Calculation	y	(x, y)
0	$y = 250 + \mathbf{0} = 250$	250	(0, 250)
50	$y = 250 + \mathbf{50} = 300$	300	(50, 300)
100	$y = 250 + \mathbf{100} = 350$	350	(100, 350)
200	$y = 250 + \mathbf{200} = 450$	450	(200, 450)

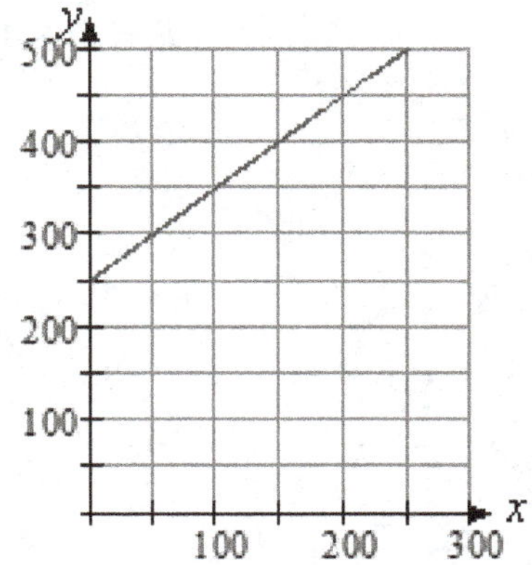

The largest y-value in the table is 450, so we scale the axis to a little larger than 450, say, 500. We plot the ordered pairs to obtain the graph shown at right.

Look Closer: How do we know what scales to use on the axes? For most graphs, it is best to have between ten and twenty tick marks on each axis, or the graph will be hard to read. We choose intervals of convenient size for the particular problem, such as 5, 10, 25, 100, or 1000. It is not necessary to use the same scale on both axes.

Caution! It is important that the scales on the axes be evenly spaced. Each tick mark must represent the same interval. The scales on the graph at right are incorrectly labeled.

Incorrect! →

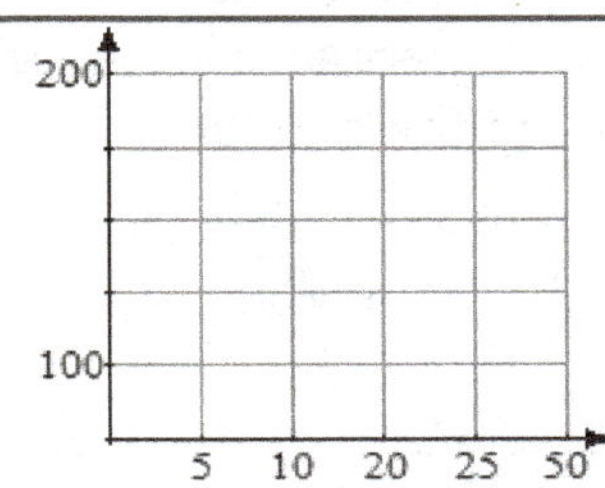

Solving Equations with Graphs

In Example 1 we showed a graph of the equation

$$F = D + 4$$

which gives Francine's age, F, in terms of Delbert's age, D. We can use the graph to answer two types of questions about the equation $F = D + 4$:

1. Given a value of D, find the corresponding value of F.
2. Given a value of F, find the corresponding value of D.

The first of these tasks is another way of **evaluating** the algebraic expression $D + 4$, and the second task is another way of **solving** an equation.

Example 4a. Use the graph of the equation $F = D + 4$ to evaluate the expression $D + 4$ for $D = 7$. Verify your answer algebraically.
b. Use the graph of the equation $F = D + 4$ to solve the equation $13 = D + 4$. Verify your answer algebraically.

Solutions a. We locate $D = 7$ on the horizontal axis, as shown at left below. Then we move vertically to point A on the graph with D-coordinate 7. Finally, we move horizontally from point A to the vertical axis to find the F-coordinate. The coordinates of point A are $(7, 11)$, which tells us that when $D = 7$, $F = 11$. To verify the answer algebraically, we **evaluate** the expression for $D = 7$: we substitute 7 for D into the equation:

$$F = D + 4 = 7 + 4 = 11$$

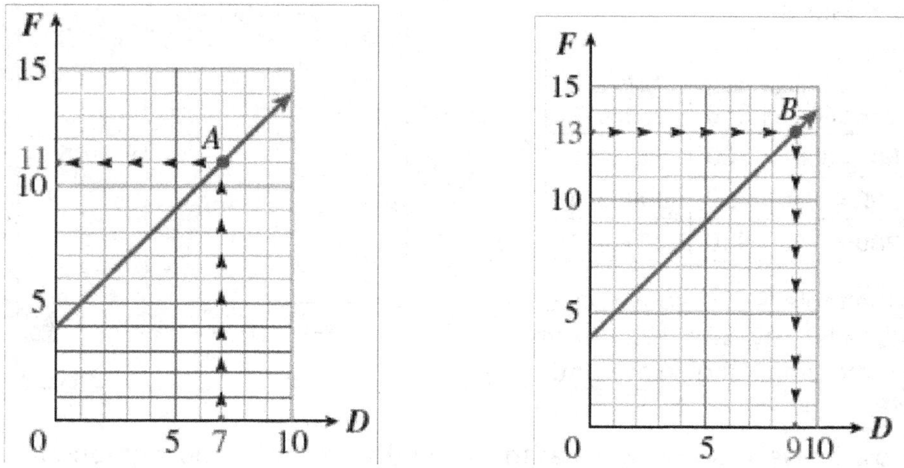

a. b.

b. We locate $F = 13$ on the vertical axis. We move horizontally to point B on the graph, with F-coordinate 13. From point B, we move vertically to the horizontal axis to find the D-coordinate. The coordinates of point B are $(9, 13)$, which tells us that when $F = 13$, $D = 9$. To verify the answer algebraically, we **solve** the equation when $F = 13$: we subtract 4 from both sides of the equation:

$$\begin{aligned} 13 &= D + 4 \quad \text{Subtract 4 from both sides.} \\ \underline{-4} & \quad \underline{\quad -4} \\ 9 &= D \end{aligned}$$

Example 5 Here is a graph of

$$y = 1.5x$$

Use the graph to solve the equation $1.5x = 3.75$.

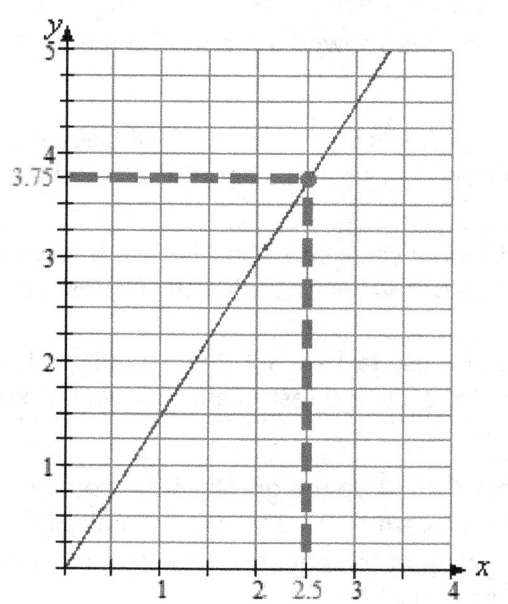

Solution By comparing the equation of the graph with the equation we want to solve, we see that y has been replaced by 3.75.

$$\begin{gathered} \mathbf{y} = 1.5x \\ \downarrow \\ \mathbf{3.75} = 1.5x \end{gathered}$$

We locate 3.75 on the y-axis, then find the point on the graph with y-coordinate 3.75. The x-coordinate of this point is 2.5, so the ordered pair $(2.5, 3.75)$ is a solution of the equation $y = 1.5x$, and $x = 2.5$ is the solution of the equation $1.5x = 3.75$.

RQ8. The graph of an equation is a picture of its ___________.
RQ9. To solve an equation using a graph, we start on the _______ axis.

Skills Warm-Up

In Exercises 1 and 2, what value corresponds to each labeled point?

1. 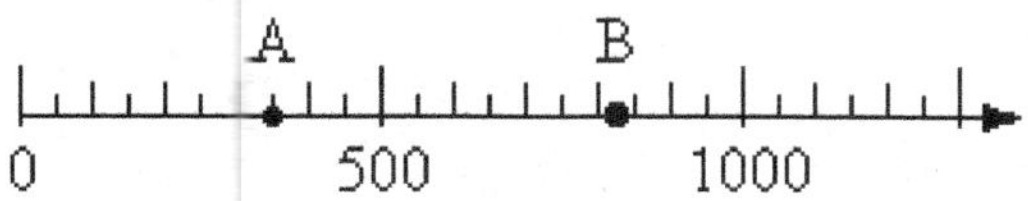

2. 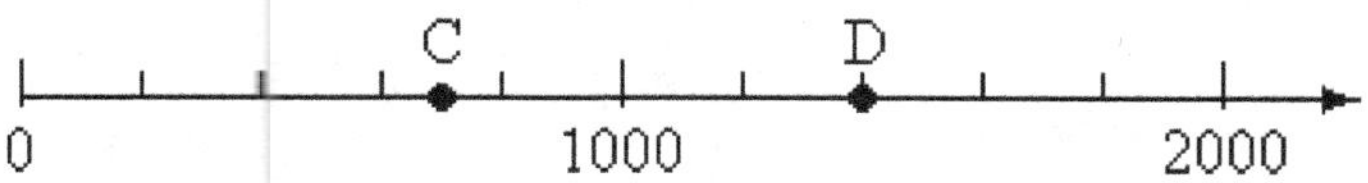

3. a. Label the axis. 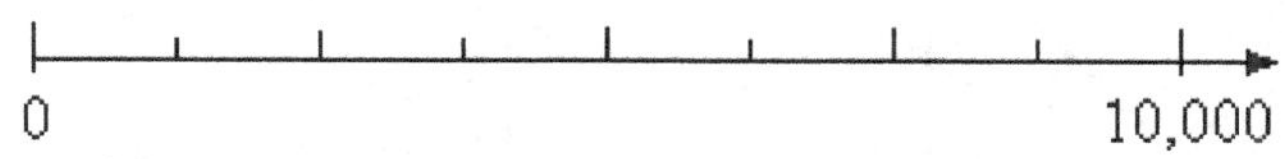

b. On the axis in part (a), plot 1400 and 8350.

4. a. Label the axis below with 16 tick marks (not counting zero), the highest being 800.

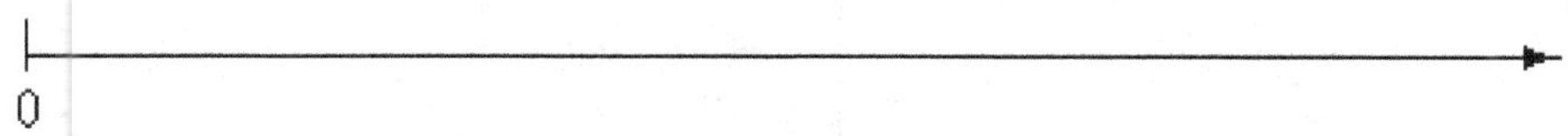

b. On the axis in part (a), plot 132 and 614.

5. a. Label an axis in increments of 40,000 from 0 to 600,000.

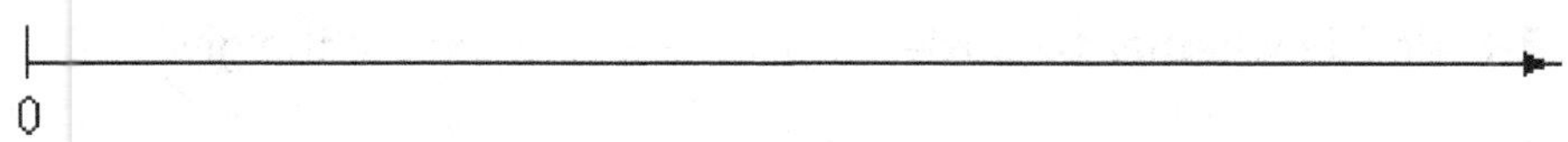

b. On your axis, plot 250,000 and 472,600.

6. a. Label an axis in increments of 0.5 from 0 to 5.

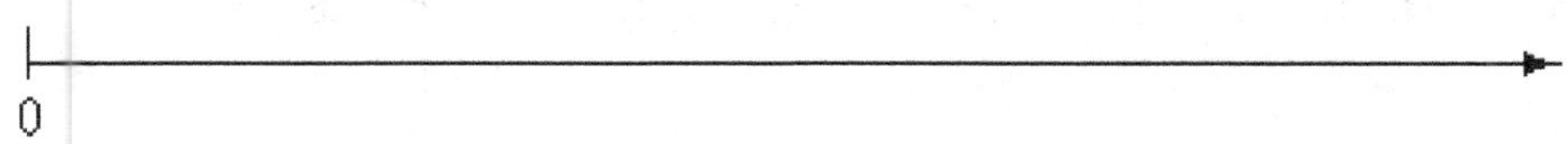

b. On your axis, plot 1.3 and 3.77.

Answers: 1. A : 350; B : 825 **2.** 700; 1400

3. 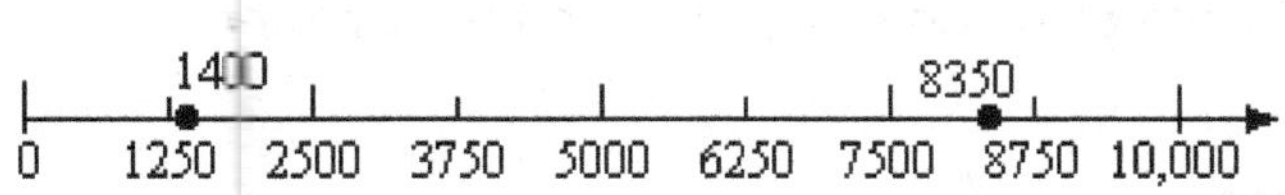

4. 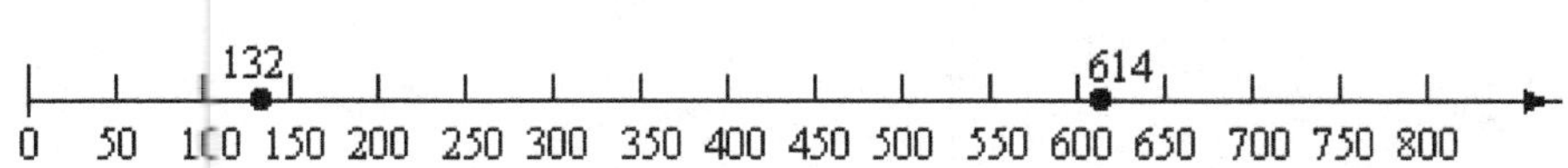

5. 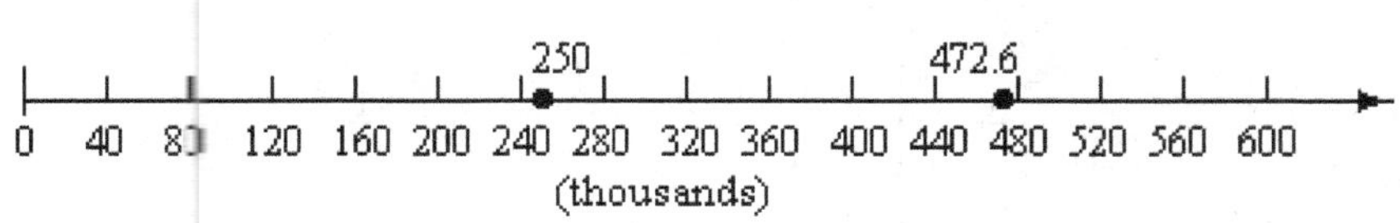

6. 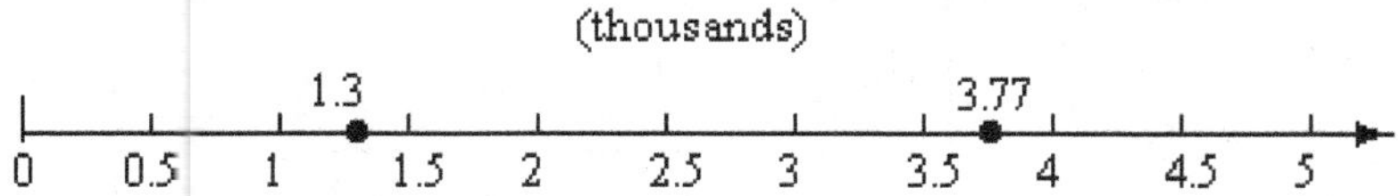

Homework 1.3

Skills Practice

For Problems 1-2, decide whether the ordered pairs are solutions of the equation whose graph is shown.

1. **a.** $(6.5, 3)$ **b.** $(0, 3.5)$ **c.** $(8, 2)$ **d.** $(4.5, 1)$

2. **a.** $(2, 6)$ **b.** $(4, 2)$ **c.** $(10, 0)$ **d.** $(11, 7)$

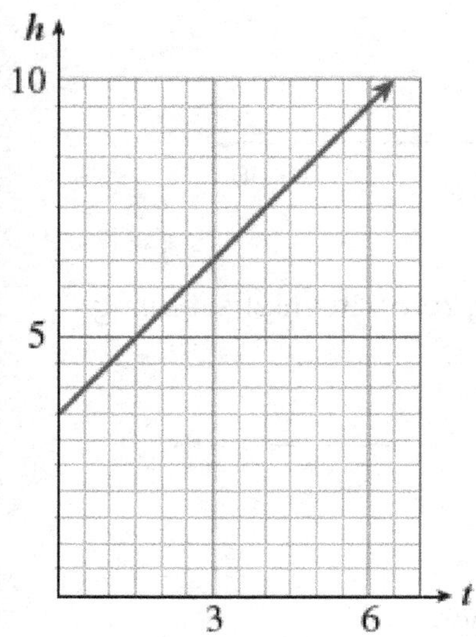

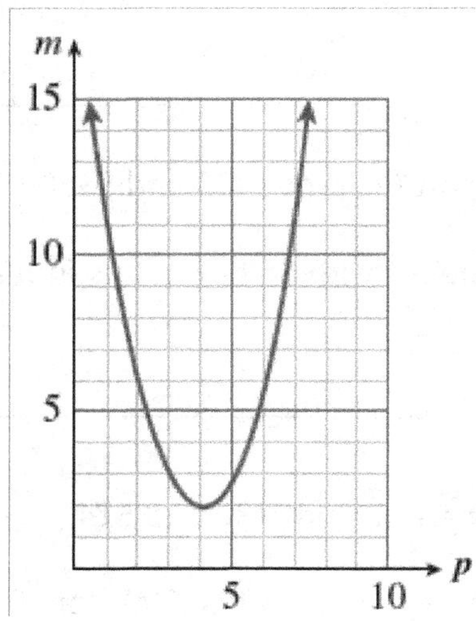

For Problems 3-6, decide whether the ordered pairs are solutions of the given equation.

3. $y = \frac{3}{4}x$

a. $(8, 6)$ **b.** $(12, 16)$ **c.** $(2, 3)$ **d.** $(6, \frac{9}{2})$

4. $y = \frac{x}{2.5}$

a. $(1, 2.5)$ **b.** $(25, 10)$ **c.** $(5, 2)$ **d.** $(8, 20)$

5. $w = z - 1.8$

a. $(10, 8.8)$ **b.** $(6, 7.8)$ **c.** $(2, \frac{1}{5})$ **d.** $(9.2, 7.4)$

6. $w = 120 - z$

a. $(0, 120)$ **b.** $(65, 55)$ **c.** $(150, 30)$ **d.** $(9.6, 2.4)$

For Problems 7-8, state the interval that each grid line represents on the horizontal and vertical axes.

7.

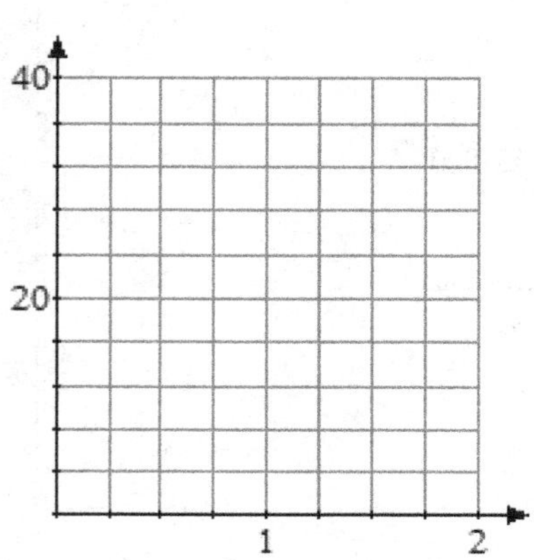

8.

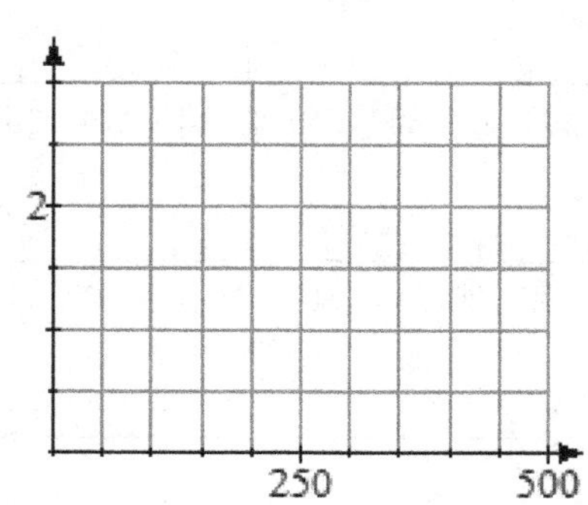

For the graphs in Problems 9-10:

a. Fill in the table with the correct coordinates. Choose your own points to complete the table.

b. Look for a pattern in your table, and write an equation for the second variable in terms of the first variable.

9.

x		2	5		10		
y	16			10			

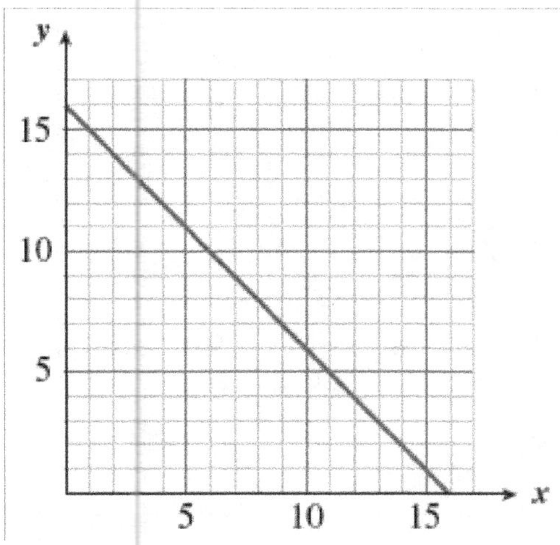

10.

x	0	1			7		
y			16	20			

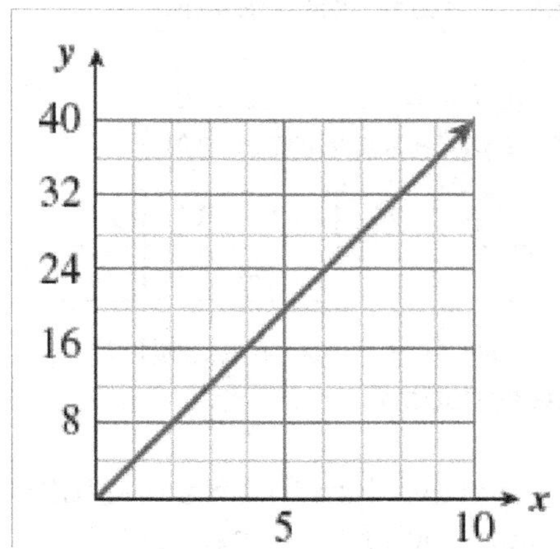

For Problems 11-14, choose the equation that describes the graph. (A graph might have more than one equation.)

a. $y = 150 - x$	**b.** $y = 15 - x$	**c.** $y = 0.2x$	**d.** $y = \dfrac{x}{60}$
e. $y = \dfrac{x}{5}$	**f.** $y = 15 + x$	**g.** $y = \dfrac{60}{x}$	**h.** $y = 5x$

11.

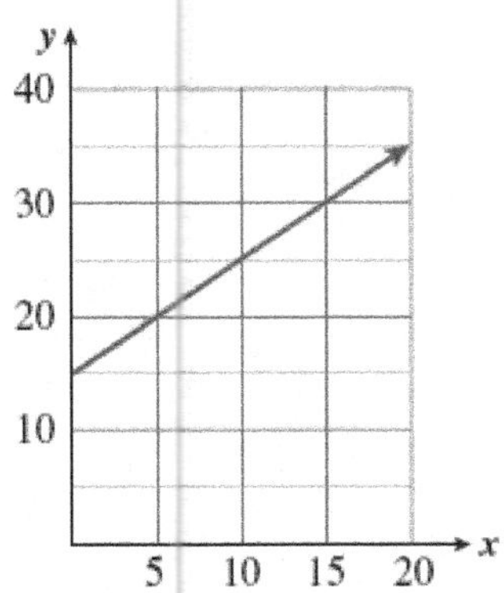

12.

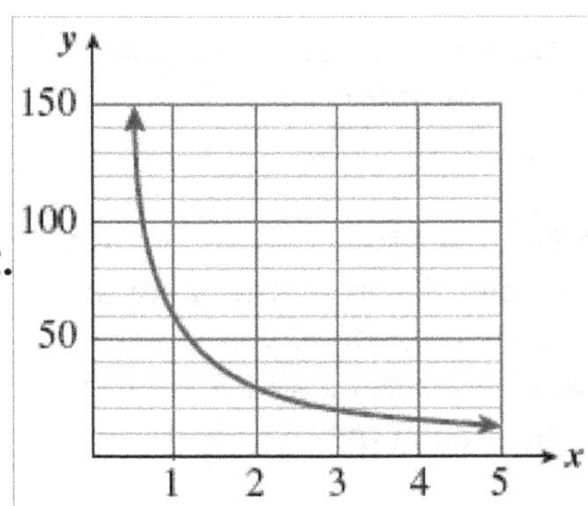

13.

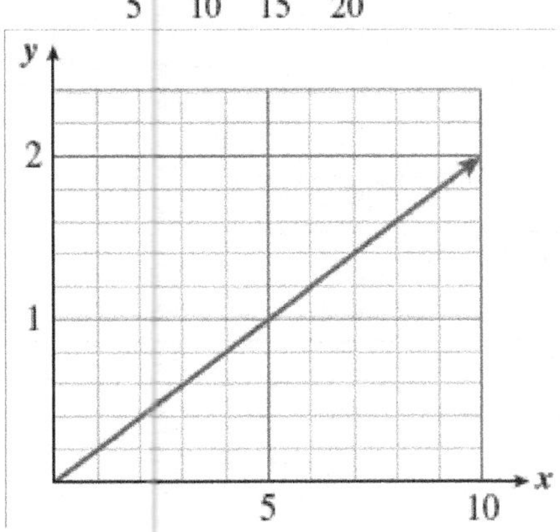

14.

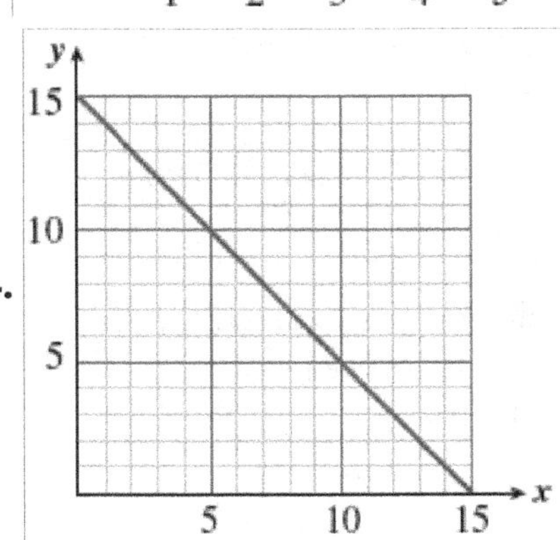

Applications

■ Use the three steps in the Lesson to make the graphs in Problems 15-17.

15. The college Bookstore charges a 25% markup over wholesale prices. If the wholesale price of a book is w dollars, the bookstore's price p is given by

$$p = 1.25\,w$$

a. How much does the bookstore charge for a book whose wholesale price is \$28?

b. Evaluate the formula to find the bookstore's price for various books.

w	12	16	20	30	36	40
p						

c. Plot the values from the table and graph the equation.

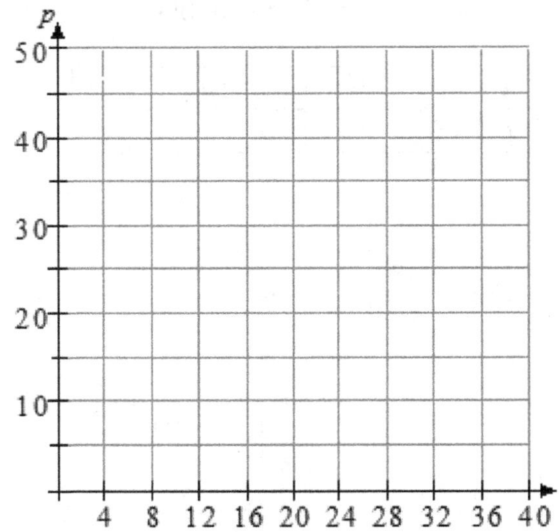

16. Uncle Ray's diet allows him to eat a total of 1000 calories for lunch and dinner. Let L stand for the number of calories in Uncle Ray's lunch, and D for the number of calories in his dinner.

a. Write an equation for D in terms of L.

b. Complete the table.

L	250	300	450	600
D				

c. Label the scales on the axes and graph the equation. Use intervals of 100 on both axes.

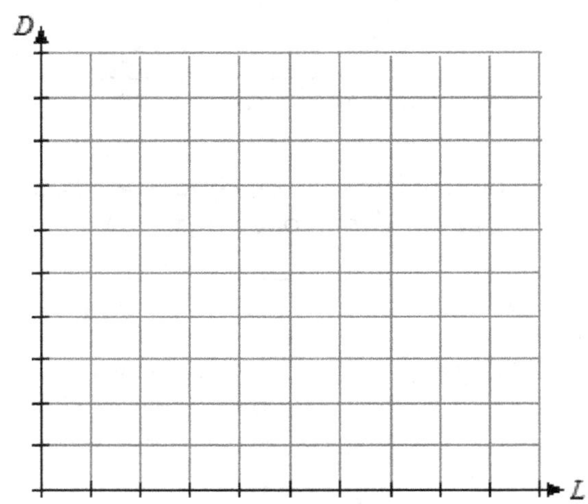

17. The Green Co-op plans to divide its profit of \$3600 from the sale of environmentally safe products among its members. Let m stand for the number of members, and S for the share of the profits each member will receive.

a. Write an equation for S in terms of m.

b. Complete the table.

m	20	40	60	100
S				

c. Label the scales on the axes and graph the equation. Use intervals of 20 on the vertical axis.

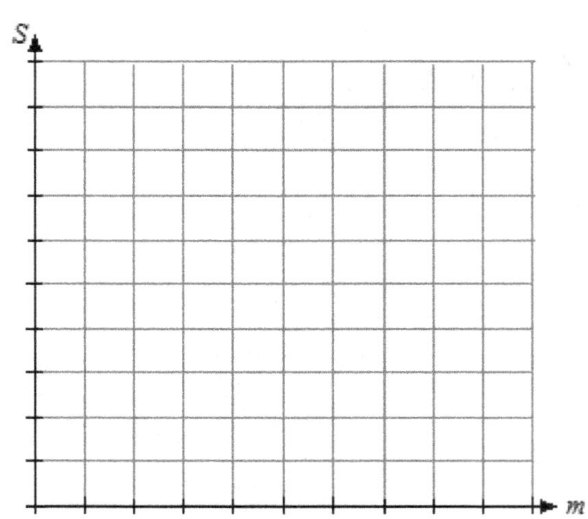

For Problems 18-19, solve the equation graphically by drawing on the graph. (You may have to estimate some of your solutions.) Then verify your answer by solving algebraically.

18. The graph of $k = p - 20$ is shown below. Use the graph to solve the equations:

a. $90 = p - 20$ **b.** $p - 20 = 40$

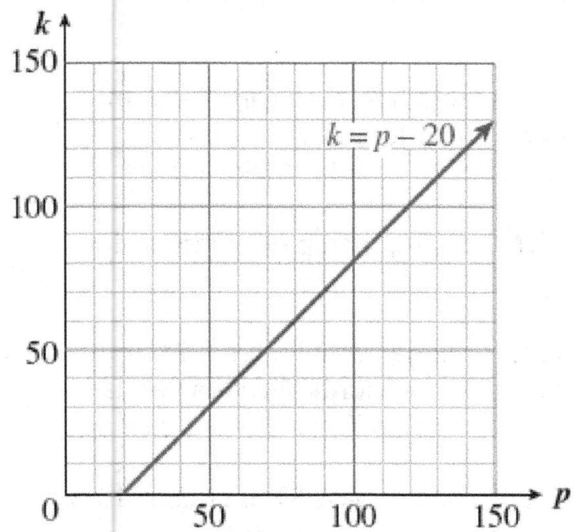

19. The graph of $d = 18g$ is shown below. Use the graph to solve the equations:

a. $250 = 18g$ **b.** $18g = 210$

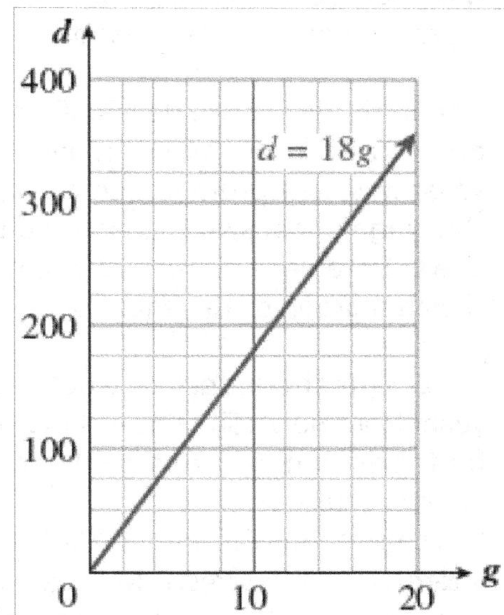

For Problems 20-21,

a. Give the coordinates of the indicated point.

b. Explain what those coordinates tell us about the problem situation.

20. The graph at left below gives the sales tax, T, in terms of the price p of an item, where the amounts are in dollars.

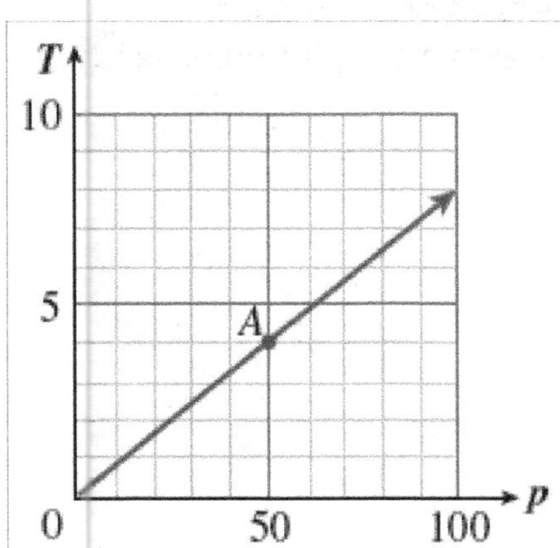

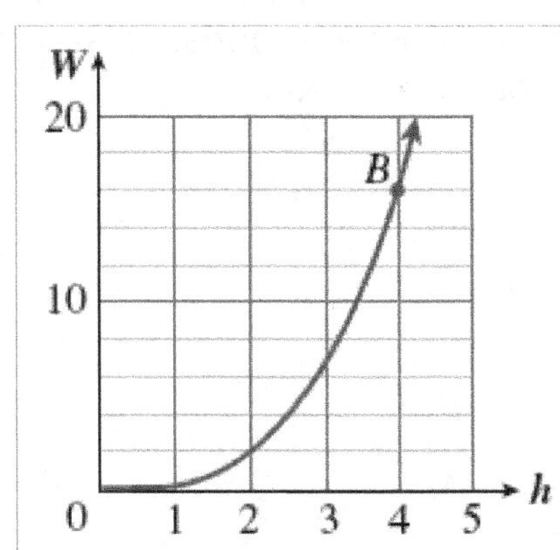

21. The graph at right above gives the weight, W, in ounces of a chocolate bunny that is h inches tall.

For Problems 22-25, use the formulas in Section 1.2.

22. a. Suppose your father loans you \$2000 to be repaid with 4% annual interest when you finish school. Write an equation for the amount of interest I you will owe after t years.

b. How much interest will you owe if you finish school in 2 years? In 4 years? In 7 years? Make a table showing your answers as ordered pairs.

c. Use your table to choose appropriate scales for the axes, then plot points and sketch a graph of the equation.

23. a. The Earth Alliance made \$6000 in revenue from selling tickets to Earth Day, an event for children. Write an equation for its profit P in terms of its costs.

b. What was their profit if their costs were \$800? \$1000? \$2500? Make a table showing your answers as ordered pairs.

c. Use your table to choose appropriate scales for the axes, then plot points and sketch a graph of the equation.

24. a. Edgar's great-aunt plans to put \$5000 in a trust for Edgar until he turns 21, three years from now. She has a choice of several different accounts. Write an equation for the amount of interest I the money will earn in 3 years if the account pays interest rate r.

b. How much interest will the \$5000 earn in an account that pays 8% interest? $10\frac{1}{4}$% interest? 12% interest? Make a table showing your answers as ordered pairs.

c. Use your table to choose appropriate scales for the axes, then plot points and sketch a graph of the equation.

25. a. Delbert keeps track of the total number of points he earns on homework assignments, each of which is worth 30 points. At the end of the semester he has 540 points. Write an equation for Delbert's average homework score S in terms of the number of assignments, n.

b. What is Delbert's average score if there were 20 assignments? 25 assignments? 30 assignments? Make a table showing your answers as ordered pairs.

c. Use your table to choose appropriate scales for the axes, then plot points and sketch a graph of the equation.

1.4 Solving Equations

What is an Equation?

An **equation** is a statement that two expressions are equal. It may involve one or more variables.

■ **Example 1** Liz makes \$6 an hour as a tutor in the Math Lab. Her weekly earnings, w, depend on the number of hours she works, h. The equation relating these two variables is

$$w = 6h$$

If we know the value of h, we can evaluate the expression $6h$ to find Liz's earnings. For example, if $h = 7$, then

$$w = 6(7) = 42$$

Liz makes \$42 for 7 hours of work. ■

In Example 1 we used evaluation to find the value of w when we knew the value of h. What if we know the value of w and want to find h? Suppose Liz earned \$54 last week. How many hours did she work? To answer this question we substitute 54 for w, so that the equation becomes

$$54 = 6h$$

Now we have an equation in just one variable. We would like to find the value of h that makes this equation true.

A value of the variable that makes an equation true is called a **solution** of the equation, and the process of finding this value is called **solving** the equation.

Trial and Error

Sometimes we can solve an equation by trial and error. By trying different values of h, you will quickly see that the solution of the equation $54 = 6h$ is 9. We can use a table to help us solve an equation by trial and error.

Exercise One of Aunt Esther's Chocolate Dream cookies contains 42 calories, so d cookies contain c calories, and $c = 42d$. If Albert consumed 546 calories, how many cookies did he eat? Use trial and error to help you solve the equation $546 = 42d$.

d	5	6	7	8	9	10	11	12	13	14	15
c											

Look Ahead: Trial and error can be time-consuming, especially for complicated equations. In this section we investigate some algebraic methods for solving equations. ●

RQ1. What is an equation?
RQ2. What is a solution of an equation?

Inverse Operations

Instead of relying on trial and error to solve equations, we will use **inverse operations**. Try the Exercise below.

Exercise a. Choose any number for x: $x =$ ______________.
Multiply your number by 5: $5x =$ ______________.
Divide the result by 5: $\frac{5x}{5} =$ ______________.

Did you end up with your original number?
Try multiplying and dividing your number by another number besides 5. Do you still end up with your original number?

Multiplication and division are opposite or **inverse operations**, because each operation undoes the effects of the other.

Exercise b. Choose any number for x: $x =$ ______________.
Add 4 to your number: $x + 4 =$ ______________.
Subtract 4 from the result: $x + 4 - 4 =$ ______________.

Did you end up with your original number?
Try adding and subtracting another number besides 4. Do you still end up with your original number?

Addition and subtraction are opposite or **inverse operations**, because each operation undoes the effects of the other.

We see that the opposite operation for multiplication is division, and the opposite operation for addition is subtraction. Now we can solve the equation in Example 1.

■ **Example 2** Solve the equation $54 = 6h$.

Solution We will transform this equation into a new one of the form $h = k$ (or $k = h$), where the number k is the solution. To do this, we must isolate the variable h on one side of the equals sign. Because h is multiplied by 6 in this equation, we divide both sides by 6, like this:

$$\frac{54}{\mathbf{6}} = \frac{6h}{\mathbf{6}} \qquad \textbf{Divide both sides by 6.}$$
$$9 = h$$

The solution is 9. Liz worked for 9 hours. ■

In Example 2, we transformed the equation into a simpler one (namely, $h = 9$) which tells us the solution. We can use the following operations to transform an equation:

To solve an equation:
1. We can add or subtract the same number from both sides.
2. We can multiply or divide both sides by the same number, as long as that number is not zero.

Look Closer: Our method for solving equations works because of the **properties of equality**, one for each of the four arithmetic operations. They are listed below. •

> **Addition and Subtraction Properties of Equality**
> If the same quantity is added to or subtracted from both sides of an equation, the solution is unchanged. In symbols,
>
> **If** $a = b$, **then** $a + c = b + c$ **and** $a - c = b - c$
>
> **Multiplication and Division Properties of Equality**
> If both sides of an equation are multiplied or divided by the same nonzero quantity, the solution is unchanged. In symbols,
>
> **If** $a = b$, **then** $ac = bc$ **and** $\dfrac{a}{c} = \dfrac{b}{c}$, $c \neq 0$

Applying any one of these properties to an equation produces an **equivalent equation**: one with the same solutions as the original equation.

Solving Equations

Now we can solve equations algebraically. We'll start with the simple Example from Section 1.3 about Delbert's and Francine's ages.

■ **Example 3** Francine is exactly four years older than Delbert, so an equation relating their ages is $F = D + 4$. How old is Delbert when Francine is 19?

Solution We must solve the equation

$$19 = D + 4$$

We see that 4 has been added to the variable, D. To isolate D on one side of the equals sign, we subtract 4 from both sides of the equation, like this:

$$\begin{array}{rcl} 19 & = & D + 4 \\ \underline{-4} & & \underline{\quad -4} \\ 15 & = & D \end{array} \qquad \text{Note that } D + 4 - 4 = D.$$

The solution to this equation is 15. You can check that substituting 15 for D does make the equation true.

Check: $19 = D + 4$ Substitute 15 for D.
$19 = 15 + 4$ True: the solution checks. ■

The last example is easy enough to solve without algebra, but the method will help us with harder problems. Here is a general strategy for solving such equations:

> **To solve an equation algebraically:**
> **Step 1** Ask yourself what operation has been performed on the variable.
> **Step 2** Perform the opposite operation on *both* sides of the equation in order to isolate the variable.

RQ3. What is the opposite operation for subtraction? For multiplication?
RQ4. What are equivalent equations?

Problem Solving with Equations

Some problems can be solved by applying one of our familiar formulas.

Example 4 A saving's account at Al's Bank earns 5% annual interest. Jan's account earned \$42.50 in interest last year. What was her balance at the beginning of the year?

Solution First, we choose the appropriate formula. This is a problem about interest, so we use the interest formula: $I = Prt$

Next, we list the values of the variables. (Which variable is unknown?)

$$I = 42.50 \qquad r = 0.05$$
$$P = \text{unknown} \qquad t = 1$$

We substitute the values into the formula: $42.50 = P(0.05)(1)$

Then we solve the equation:

$$\frac{42.50}{0.05} = \frac{P(0.05)}{0.05} \qquad \textbf{Divide both sides by 0.05.}$$
$$850 = P$$

Finally, we answer the question in a sentence:

Jan's account balance was \$850 at the beginning of the year.

Many practical problems can be solved by first writing an equation that describes or **models** the problem, and then solving the equation. We will use three steps to solve applied problems.

Steps for solving an applied problem:

Step 1 Identify the unknown quantity and choose a variable to represent it.

Step 2 Find some quantity that can be described in two different ways, and write an equation using the variable to model the problem situation.

Step 3 Solve the equation and answer the question in the problem.

Example 5 Jerry needs an additional \$35 for airfare to New York. The ticket to New York costs \$293. How much money does Jerry have?

Solution We follow the three steps.

Step 1 The amount of money Jerry has is unknown.

Amount of money Jerry has: m

Step 2 The airfare to New York can be described in two different ways:

m	$+$	35	$=$	293
amount Jerry has		amount he needs		airfare

Step 3 We solve the equation and answer the question.

$$\begin{aligned} m + 35 &= 293 \\ -35 \quad &\;\; -35 \\ m &= 258 \end{aligned}$$

Subtract 35 from both sides.

Jerry has $258.

Caution! In Example 5, each side of the equation represents the airfare to New York, **not** the amount of money Jerry has, which is what we want to find. This is typical of applied problems: the equation does not usually represent the unknown quantity itself, but some related quantity.

Look Ahead: Although you may be able to solve the problems in this lesson with arithmetic, *the important thing is to learn to write the algebraic equation, or model,* for the problem. This skill will help you to solve problems that are too difficult to solve with arithmetic alone.

RQ5. What is the first step in solving an applied problem?
RQ6. True or false: One side of the equation that models a problem will be the unknown variable.

Skills Warm-Up

Choose the correct algebraic expression.

$m + 15$ $\qquad$ $m - 15$ $\qquad$ $15 - m$

1. Carol weighs 15 pounds less than Garth. If Garth weighs m pounds, how much does Carol weigh?
2. Amber and Beryl together planted 15 trees. If Amber planted m trees, how many trees did Beryl plant?
3. Fred earned $15 more this week than last week. If he earned m dollars last week, how much did he earn this week?
4. Meg bicycled 15 miles farther than Kwan. If Kwan rode m miles, how far did Meg ride?
5. There are 15 children in Amy's swim class. If there are m girls, how many are boys?
6. The sale price of a sweater is $15 less than the regular price. If the regular price is m dollars, what is the sale price?

Choose the correct algebraic expression.

$12p$ $\qquad$ $\dfrac{12}{p}$ $\qquad$ $\dfrac{p}{12}$

7. Julian earns $12 an hour. If he works for p hours, how much will he make?
8. Farmer Brown collected p eggs this morning. How many dozen is that?
9. Melissa bought 12 colored markers. If their total cost was p dollars, how much did each marker cost?

10. Rosalind is baby-sitting for p children. If she brings 12 puzzles, how many will each child get?

11. Hector has to read 12 chapters in his history text. If he has p days to complete the assignment, how many chapters should he read per day?

12. Roma swims 12 laps per day. After p days, how many laps has she swum?

Answers: **1.** $m-15$ **2.** $15-m$ **3.** $m+15$ **4.** $m+15$ **5.** $15-m$ **6.** $m-15$
7. $12p$ **8.** $\frac{p}{12}$ **9.** $\frac{p}{12}$ **10.** $\frac{12}{p}$ **11.** $\frac{12}{p}$ **12.** $12p$

Homework 1.4

Skills Practice

For Problems 1-2, fill in the table for each equation. Explain how you found the unknown values.

1. $q = 9 + t$

t	2	4			21	
q			15	18		39

2. $p = 5n$

n	0	2		5		
p			20		35	55

3. Decide whether the given value for the variable is a solution of the equation.
   a. $x - 4 = 6$; $x = 10$ **b.** $4y = 28$; $y = 24$ **c.** $\frac{0}{z} = 0$; $z = 19$

For Problems 4-12, solve, and check your solution.

4. $x - 3 = 11$
5. $10.6 = 7.8 + y$
6. $3y = 108$
7. $42 = 3.5b$
8. $2.6 = \dfrac{a}{1.5}$
9. $x - 4 = 0$
10. $34x = 212$
11. $6z = 20$
12. $9 = k + 9$

For Problems 13-16,
a. Choose the appropriate formula to write an equation.
b. Solve the equation and answer the question.

13. Clive loaned his brother some money to buy a new truck, and his brother agreed to repay the loan in 1 year with 3% interest. Clive earned $75 interest on the loan. How much did Clive loan his brother?

14. Andy's average homework score on eight assignments was 38.25. How many homework points did Andy earn altogether?

15. How long will it take a cyclist traveling at 13 miles an hour to cover 234 miles?

16. A roll of carpet contains 400 square feet of carpet. If the roll is 16 feet wide, how long is the piece of carpet?

For Problems 17-22, choose the appropriate equation.

$x + 7 = 26$	$7x = 26$	$\dfrac{x}{7} = 26$
$x - 7 = 26$	$\dfrac{x}{26} = 7$	$\dfrac{26}{x} = 7$

17. Sarah drove 7 miles farther to her high school reunion than Jenni drove. If Sarah drove 26 miles, how far did Jenni drive?

18. Lurline and Rozik live 26 miles apart. They meet at a theme park between their homes. If Lurline drove 7 miles to the park, how far did Rozik drive?

19. Doris is training for a triathlon. This week she averaged 26 miles per day on her bicycle. If she rode every day, what was her total mileage?

20. Glynnis jogged the same route every day this week for a total of 26 miles. how long is her route?

21. Astrid divided her supply of colored pencils among the 26 children in her class, and each child got 7 pencils. How many pencils does she have?

22. Ariel lost 7 of the beads on her necklace, and now there are 26. How many were there originally?

Applications

■ Follow the steps to solve Problems 23-26.

23. Lupe spent $24 at the Craft Fair. She now has $39 left. How much did she have before the Craft Fair?
    a. What are we asked to find? Choose a variable to represent it.
    b. Find two ways to express the amount of money Lupe had after the Craft Fair, and write an equation.
    c. Solve the equation and answer the question in the problem.

24. Danny weighs 32 pounds more than Brenda. If Danny weighs 157 pounds, how much does Brenda weigh?
    a. What are we asked to find? Choose a variable to represent it.
    b. Find two ways to express Danny's weight, and write an equation.
    c. Solve the equation and answer the question in the problem.

25. Miranda worked 20 hours this week and made $136. What is Miranda's hourly wage?
    a. What are we asked to find? Choose a variable to represent it.
    b. Find two ways to express Miranda's total earnings, and write an equation.
    c. Solve the equation and answer the question in the problem.

26. Struggling Students Gardening Service splits their profit equally among their eight members. If each member made $64 last week, what was the total profit?
    a. What are we asked to find? Choose a variable to represent it.
    b. Find two ways to express each member's share, and write an equation.
    c. Solve the equation and answer the question in the problem.

■ Write algebraic equations to solve Problems 27-29. Follow the three steps in the Lesson.

27. Martha paid $26 less for a suit at a discount store than her mother paid at a boutique for the same suit. If Martha paid $89 for the suit, how much did her mother pay?

28. Emily spends 40% of her monthly income on rent. If her rent is $360 a month, how much does Emily make?

29. After she wrote a check for $2378, Avril's bank account shows a balance of $1978. How much money was in Avril's account before the check cleared?

■ In Problems 30-32, we compare evaluating an expression and solving an equation.

30. A used car costs $3400 less than the new version of the same model.

a. Choose two variables and write an equation for the cost of the used car in terms of the cost of the new car.

b. If the new car costs $14,500, how much does the used car cost?

c. If the used car costs $9200, how much does the new car cost?

31. The Dodgers won 60% of their games last season.

a. Choose two variables and write an equation for the number of games the Dodgers won in terms of the number of games they played.

b. If the Dodgers played 120 games, how many did they win?

c. If the Dodgers won 96 games, how many did they play?

32. Sunshine Industries manufactures beach umbrellas. Their profit on each umbrella is 18% of the selling price.

a. Choose two variables and write an equation for the profit on each umbrella in terms of its selling price.

b. If a beach umbrella sells for $60, what is the profit?

c. If the profit on one umbrella is $7.20, what is the selling price?

1.5 Order of Operations

Addition and Multiplication

If you add together three or more numbers, such as

$$2+5+8$$

it doesn't matter which addition you do first; you will get the same answer either way.

Example 1a. A sum of three or more terms can be added in any order. In the sums below, the parentheses tell us which part of the expression to simplify first.

$$(\mathbf{2+5})+8 \quad \text{and} \quad 2+(\mathbf{5+8})$$
$$=\mathbf{7}+8=15 \qquad\qquad =2+\mathbf{13}=15$$

b. Similarly, a product of three factors can be multiplied in any order. Thus

$$(\mathbf{3\cdot 2})\cdot 4 \quad \text{and} \quad 3\cdot(\mathbf{2\cdot 4})$$
$$=\mathbf{6}\cdot 4=24 \qquad\qquad =3\cdot\mathbf{8}=24$$

These two facts illustrate the **associative laws** for addition and multiplication.

Associative Law for Addition

If a, b, and c are any numbers, then

$$(a+b)+c=a+(b+c)$$

Associative Law for Multiplication

If a, b, and c are any numbers, then

$$(a\cdot b)\cdot c=a\cdot(b\cdot c)$$

Subtraction and Division

What about a string of subtractions or a string of divisions, such as

$$20-8-5 \quad \text{or} \quad 36\div 6\div 2$$

In these calculations, we get different answers, depending on which operations we perform first, as you can see in Example 2.

Example 2a. Subtraction:

$$(\mathbf{20-8})-5=\mathbf{12}-5=7 \quad \text{but} \quad 20-(\mathbf{8-5})=20-\mathbf{3}=17$$

b. Division:

$$(\mathbf{36\div 6})\div 2=\mathbf{6}\div 2=3 \quad \text{but} \quad 36\div(\mathbf{6\div 2})=36\div\mathbf{3}=12$$

The associative laws do **not** hold for subtraction or division.

So, if there are no parentheses in the expression, how do we know which operations to perform first?

1. In a string of additions and subtractions, we perform the operations **in order from left to right**.
2. Similarly, we perform multiplications and divisions **in order from left to right**.

Example 3 Simplify each expression.

a. $20 - 8 - 5$ **b.** $36 \div 6 \div 2$

Solutions Perform the operations in order from left to right.

a. $\mathbf{20 - 8} - 5 = 12 - 5 = 17$ **b.** $\mathbf{36 \div 6} \div 2 = 6 \div 2 = 3$

RQ1. What do the associative laws tell us?
RQ2. Which two operations are not associative?
RQ3. In what order should we perform a string of additions and subtractions?

Combined Operations

How should we simplify the expression $4 + 6 \cdot 2$?

If we do the addition first, we get

$$(\mathbf{4+6}) \cdot 2 = \mathbf{10} \cdot 2 = 20$$

If we do the multiplication first, we get

$$4 + (\mathbf{6 \cdot 2}) = 4 + \mathbf{12} = 16$$

Which one is correct? In order to avoid confusion, we make the following agreement.

Always perform multiplications and divisions before additions and subtractions.

Example 4 The correct way to simplify the expression $4 + 6 \cdot 2$ is

$$4 + \mathbf{6 \cdot 2} \quad \text{Multiply first.}$$
$$= 4 + \mathbf{12} \quad \text{Then add.}$$
$$= 16$$

Look Closer: In longer expressions, it can be helpful to group the expression into its terms before beginning. **Terms** are expressions separated by addition or subtraction symbols. We simplify each term before combining them.

Example 5 Simplify $6 + 2 \cdot 5 - 12 \div 3 \cdot 2$.

Solution We start by underlining each term of the expression separately. Then we simplify each term.

$\underline{6} + \underline{2 \cdot 5} - \underline{12 \div 3 \cdot 2}$ Perform multiplications and divisions from left to right.

$= \mathbf{6 + 10} - 8$ Perform additions and subtractions from left to right.

$= \mathbf{16} - 8 = 8$

Caution! 1. In Example 5, we do **not** start by adding $6+2$. We must perform multiplications before additions.
2. We simplify $12 \div 3 \cdot 2$ by performing the operations **from left to right**

$$\mathbf{12 \div 3} \cdot 2 = 4 \cdot 2 = 8$$

Parentheses

What if the addition should come first in a particular calculation? In that case, we use parentheses to enclose the sum, like this:

$$(4+6) \cdot 2$$

Perform any operations inside parentheses first.

Example 6 Simplify.

a. $2+5(\mathbf{7-3})$ — Subtract inside parentheses.
$= 2+\mathbf{5(4)}$ — Multiply.
$= 2+20 = 22$ — Add.

b. $6(10-\mathbf{2 \cdot 4}) \div 4$ — Multiply inside parentheses.
$= 6(\mathbf{10-8}) \div 4$ — Subtract inside parentheses.
$= \mathbf{6(2)} \div 4$ — Multiply and divide in order from left to right.
$= 12 \div 4 = 3$

RQ4. Is it true that multiplications should be performed before divisions? Why or why not?
RQ5. What do parentheses tell us?
RQ6. If there are no parentheses, in what order should we perform multiplications and divisions?

Fraction Bars

Like parentheses, a fraction bar is a **grouping device**.

Expressions that appear above or below a fraction bar should be simplified first.

Example 7 Simplify $\dfrac{24+6}{12+6}$

Solution We begin by computing the sums above and below the fraction bar. Then we can reduce the fraction. Thus,

$$\frac{24+6}{12+6} = \frac{30}{18} = \frac{5}{3}$$

Caution! Do not be tempted to "cancel" the terms in Example 7. For example, it would be incorrect to write

$$\frac{24+6}{12+6} = \frac{\not{2}4+6}{\not{1}2+6} = \frac{2+6}{1+6} \quad \leftarrow \textbf{Incorrect!}$$

According to the order of operations, we must simplify the numerator and denominator first, before dividing. ■

RQ7. Besides parentheses, what other symbol serves as a grouping device?
RQ8. When adding fractions, which should you do first: reduce each fraction, or find an LCD?

Summary

Combining all our guidelines for simplifying expressions, we state the rules for the order of operations.

Order of Operations

1. First, perform any operations that appear inside parentheses, or above or below a fraction bar.
2. Next, perform all multiplications and divisions in order from left to right.
3. Finally, perform all additions and subtractions in order from left to right.

Algebraic Expressions

The order of operations applies to variables as well as constants. For example, the expression

$$5+2x$$

tells us to multiply x by 2, then add 5 to the result. Thus, to evaluate the expression for $x=8$, we write

For $x=8$, $\quad 5+2x = 5+2(8)$ **Multiply first.**
$\quad = 5+16 = 21$

In Example 8, note how parentheses change the meaning of an expression.

■ **Example 8** Write algebraic expressions for each of the following phrases. Then evaluate each phrase for $x=2$ and $y=6$.

a. Three times the sum of x and y

b. The sum of $3x$ and y

Solutions

a. $3(x+y)$
$3(2+6) = 3(8) = 24$

b. $3x+y$
$3(2)+6 = 6+6 = 12$ ■

Skills Warm-Up

Perform the operations on fractions as indicated.

1a. $\frac{5}{4}+\frac{3}{4}$ **b.** $\frac{5}{4}-\frac{3}{4}$ **c.** $\frac{5}{4}\cdot\frac{3}{4}$ **d.** $\frac{5}{4}\div\frac{3}{4}$

2a. $\frac{1}{2}+\frac{1}{6}$ **b.** $\frac{1}{2}-\frac{1}{6}$ **c.** $\frac{1}{2}\cdot\frac{1}{6}$ **d.** $\frac{1}{2}\div\frac{1}{6}$

3a. $\frac{2}{3}+\frac{1}{4}$ **b.** $\frac{2}{3}-\frac{1}{4}$ **c.** $\frac{2}{3}\cdot\frac{1}{4}$ **d.** $\frac{2}{3}\div\frac{1}{4}$

Answers: 1a. 2 **b.** $\frac{1}{2}$ **c.** $\frac{15}{16}$ **d.** $\frac{5}{3}$ **2a.** $\frac{2}{3}$ **b.** $\frac{1}{3}$ **c.** $\frac{1}{12}$ **d.** 3
3a. $\frac{11}{12}$ **b.** $\frac{5}{12}$ **c.** $\frac{1}{6}$ **d.** $\frac{8}{3}$

Homework 1.5

Skills Practice

For Problems 1-12, simplify each expression by following the order of operations.

1. $2+4(3)$

2. $15-\frac{3}{4}(16)$

3. $6\div\frac{1}{4}\cdot 3$

4. $3+3(2+3)$

5. $\frac{1}{3}\cdot 12-3\left(\frac{5}{6}\right)$

6. $2+3\cdot 8-6+3$

7. $\frac{3(8)}{12}-\frac{6+4}{5}$

8. $28+6\div 2-2(5+3\cdot 2)$

9. $3[3(3+2)-8]-17$

10. $\frac{3(3)+5}{6-2(2)}+\frac{2(5)-4}{9-4-2}$

11. $7[15-24\div 2]-9[5(4+2)-4(6+1)]$

12. $20-4(9-3\cdot 2)+8-5\cdot 3$

For Problems 13-16, use your calculator to simplify each expression. Round your answers to three decimal places if necessary.

13. $\frac{6.4+3.5}{3.6(3.2)}$

14. $\frac{14.6-6.8}{7.3+3.4}$

15. $\frac{26.2-9.1}{8.4\div 7.7}+5.1(6.9-1.6)$

16. $\frac{1728(847-603)}{216(98-38)}+6(876-514)$

For Problems 17-22, evaluate each pair of expressions mentally.

17a. $8+2\cdot 5$ **b.** $(8+2)\cdot 5$

18a. $\frac{24}{2+6}$ **b.** $\frac{24}{2}+6$

19a. $(9-4)-3$ **b.** $9-(4-3)$

20a. $6\cdot 8-6$ **b.** $6(8-6)$

21a. $\frac{36}{6(3)}$ **b.** $\frac{36}{6}(3)$

22a. $(30-5)(5)$ **b.** $30-(5)(5)$

For Problems 23-24, fill in the table to evaluate the expression in two steps.

23. $5z-3$

z	$5z$	$5z-3$
2		
4		
5		

24. $2(12+Q)$

Q	$12+Q$	$2(12+Q)$
0		
4		
8		

For Problems 25-30, evaluate for the given values.

25. $2y+x$ for $x=8$ and $y=9$

26. $4a+3b$ for $a=8$ and $b=7$

27. $\frac{a}{b}-\frac{b}{a}$ for $a=8$ and $b=6$

28. $\frac{24-2x}{2+y}-\frac{4x+1}{3y}$ for $x=4$ and $y=6$

29. $\frac{a}{1-r}$ for $a=6$ and $r=0.2$

30. $mx + b$ for $m = \frac{3}{5}$, $x = \frac{2}{3}$, and $b = \frac{9}{10}$

Applications

31. Consider the expression $20 - 2 \cdot 8 + 1$. How would you change the expression if you wanted:

a. The subtraction performed first?
b. The addition performed first?
c. The multiplication performed first?

32. Write an algebraic expression for the following instructions:
"Multiply the sum of 5 and 7 by 4, then divide by the difference of 10 and 8."

33. Write two algebraic expressions for "the difference of 20 and 8, divided by the sum of 6 ane 4 and 10":

a. Using a fraction bar
b. Using a division symbol

34. Without performing the calculations, write down the steps you would use to simplify the expression: $825 - 32(12) \div 4 + 2$

For problems 35-38, choose the correct algebraic expression for each English phrase.

$\frac{m}{12} - 3$	$\frac{m-3}{12}$
$\frac{12}{m-3}$	$3 - \frac{12}{m}$

35. Twelve divided by 3 less than m.
36. Three less than the quotient of m divided by 12.
37. The quotient of 3 less than m divided by 12.
38. Subtract from 3 the quotient of 12 and m.

39. Find and correct the error in the calculation:

$$\begin{aligned}&(5+4) - 3(8 - 3 \cdot 2)\\&= 9 - 3(8-6)\\&= 6(2) = 12\end{aligned}$$

40. Think of the region shown at right as a rectangle with a smaller rectangle removed, and write an expression for its area. Then simplify your expression to find the area. (The measurements given are in centimeters.)

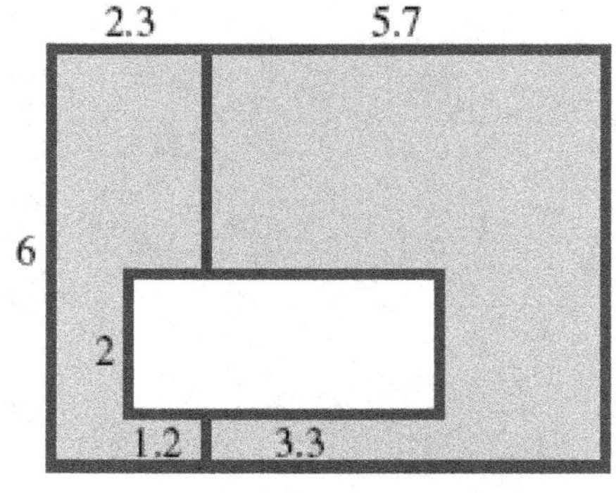

Chapter 1 Summary and Review

Lesson 1.1 Variables

- A **variable** is a numerical quantity that changes over time or in different situations.
- We can use a table, a graph, or an equation to compare the values of two related variables.
- A formula relating two variables is a type of **equation**, and an equation is just a statement that two quantities are equal.

Lesson 1.2 Algebraic Expressions

- An **algebraic expression** is any meaningful combination of numbers, variables, and operation symbols.

> **Commutative Laws**
>
> **Addition:** If a and b are any numbers, then
>
> $$a + b = b + a$$
>
> **Multiplication:** If a and b are any numbers, then
>
> $$a \cdot b = b \cdot a$$

- Subtraction and division are not commutative; the order of the numbers matters in a difference or a quotient.

> **To write an algebraic expression:**
>
> **Step 1** Identify the unknown quantity and write a short phrase to describe it.
>
> **Step 2** Choose a variable to represent the unknown quantity.
>
> **Step 3** Use mathematical symbols to represent the relationship.

- The process of substituting a specific value for the variable into an expression is called **evaluating** the expression.
- An **equation** is a mathematical statement that two algebraic expressions are equal. A **formula** is an equation that relates two or more variables.

Some Useful Algebraic Formulas

1.	distance = rate × time	$d = rt$
2.	profit = revenue − cost	$P = R - C$
3.	interest = principal × interest rate × time	$I = Prt$
4.	part = percentage rate × whole	$P = rW$
5.	average score = $\dfrac{\text{sum of scores}}{\text{number of scores}}$	$A = \dfrac{S}{n}$

Lesson 1.3 Equations and Graphs

- A graph is a way of visualizing an algebraic equation in two variables. The first or **input** variable is displayed on the horizontal axis and the second or **output** variable is displayed on the vertical axis.
- Each point on a graph has two **coordinates,** which designate the position of the point.

- An ordered pair of numbers that make an equation true is called a **solution** of the equation. Each point on the graph represents a solution of the equation.

> **To graph an equation:**
> 1. Make a table of values.
> 2. Choose scales for the axes.
> 3. Plot the points and connect them with a smooth curve.

- If the variables in an equation have very large (or very small) values, we must choose scales for the axes that fit these values.
- We can use a graph to evaluate an algebraic expression or to solve an equation.

Lesson 1.4 Solving Equations

- To solve an equation algebraically, we use **inverse**, or opposite, operations.

> **To solve an equation algebraically:**
> **Step 1** Ask yourself what operation has been performed on the variable.
> **Step 2** Perform the opposite operation on both sides of the equation. This should isolate the variable.

> **Operations for solving an equation:**
> 1. We can add or subtract the same number from both sides.
> 2. We can multiply or divide both sides by the same number, as long as that number is not zero.

> **Addition and Subtraction Properties of Equality**
> If the same quantity is added to or subtracted from both sides of an equation, the solution is unchanged. In symbols,
>
> $$\textbf{If}\quad a=b,\quad \textbf{then}\quad a+c=b+c\quad \textbf{and}\quad a-c=b-c$$
>
> **Multiplication and Division Properties of Equality**
> If both sides of an equation are multiplied or divided by the same nonzero quantity, the solution is unchanged. In symbols,
>
> $$\textbf{If}\quad a=b,\quad \textbf{then}\quad ac=bc\quad \textbf{and}\quad \frac{a}{c}=\frac{b}{c},\quad c\neq 0$$

- Many practical problems can be solved by first writing an equation that describes or **models** the problem, and then solving the equation.

> **Steps for solving an applied problem:**
> **Step 1** Identify the unknown quantity and choose a variable to represent it.
> **Step 2** Find some quantity that can be described in two different ways, and write an equation using the variable to model the problem situation.
> **Step 3** Solve the equation and answer the question in the problem.

Lesson 1.5 Order of Operations

Associative Laws

Addition: If a, b, and c are any numbers, then

$$(a+b)+c=a+(b+c)$$

Multiplication: If a, b, and c are any numbers, then

$$(a \cdot b) \cdot c=a \cdot (b \cdot c)$$

- The associative laws do *not* hold for subtraction or division.

Order of Operations

1. First, perform any operations that appear inside parentheses, or above or below a fraction bar.
2. Next, perform all multiplications and divisions in order from left to right.
3. Finally, perform all additions and subtractions in order from left to right.

- When using a calculator, you must enclose in parentheses any expression that appears above or below a fraction bar.

Review Questions

For Problems 1-10, use complete sentences to answer the questions.

1. What is a variable? Give an example.
2. What is the difference between an algebraic expression and an equation?
3. What is the difference between factors and terms?
4. **a.** Write as an algebraic expression "p subtracted from w."
   b. Evaluate your expression for $p=5.6$ and $w=56$.
5. **a.** Write as an algebraic expression "m divided by g."
   b. Evaluate your expression for $m=12$ and $g=192$.
6. State a formula from this chapter for each of the variables below.
   a. Distance: $d=$ **b.** Profit: $P=$
   c. Interest: $I=$ **d.** Percentage or part: $P=$
   e. Average value: $A=$
7. State formulas for the perimeter and area of a rectangle, and for the area and circumference of a circle.
8. When we construct the graph of an equation, which variable is displayed on which axis?
9. Write a paragraph explaining how the following terms are related to an equation in two variables: solution, coordinates, ordered pair.
10. State the order of operations for simplifying algebraic expressions.

Review Problems

1. Martha can travel 22 miles on a gallon of gasoline. Fill in the table with the distance Martha can drive on various amounts of gasoline.

Gallons of gas	3	5	8	10	11	12
Calculation						
Miles driven						

   a. Explain in words how to find the distance Martha can drive.
   b. Write your explanation as a mathematical sentence:
   Miles driven =
   c. Let g stand for the number of gallons of gas in Martha's car and m for the number of miles she can drive. Write an equation for m in terms of g.
   d. Use the grid in below left to plot the values from your table. Connect the data points with a smooth curve.

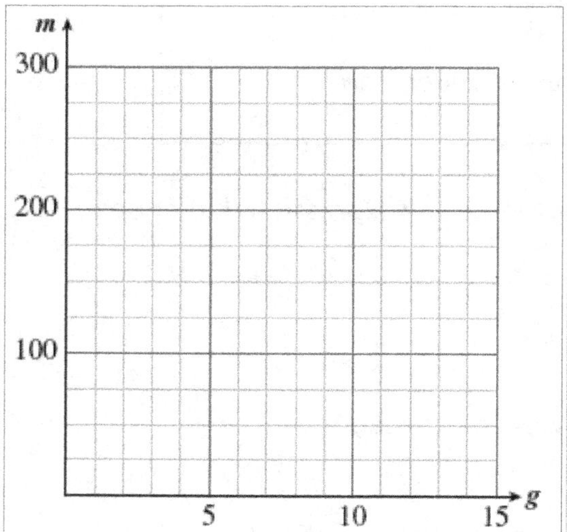

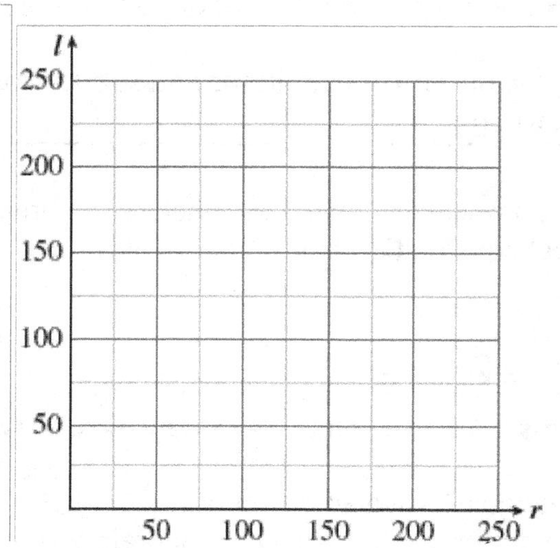

2. Mariel has a 200-page history assignment to read. Fill in the table with the number of pages Mariel has left to read after completing a certain number of pages

Pages read	20	50	85	110	135	180
Calculation						
Pages left						

   a. Explain in words how to find the number of pages left to read.
   b. Write your explanation as a mathematical sentence:
   Pages left =
   c. Let r stand for the number of pages Mariel has already read, and l for the number of pages she has left to read. Write an equation for l in terms of r.
   d. Use the grid above right to plot the values from your table. Connect the data points with a smooth curve.

3. Find the pattern and fill in the table. Then write an equation for the second variable in terms of the first.

x	0.5	1.0	1.5	2.0	4.0	6.0	7.5
y	0.125	0.25	0.375	0.5			

4. Use the graph to fill in the table. Then write an equation for the second variable in terms of the first.

x	y

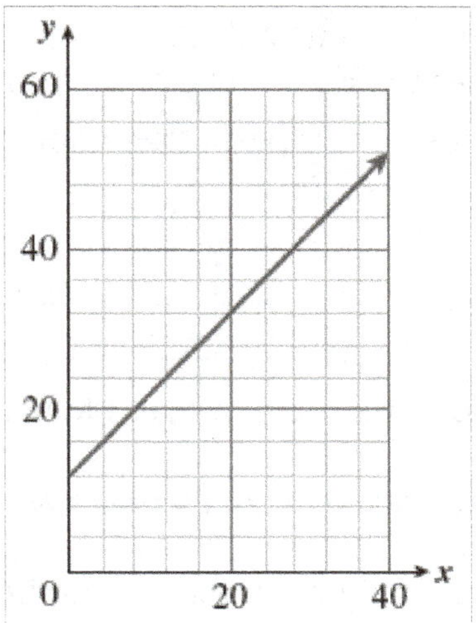

For Problems 5-12, write an algebraic expression for the phrase.

5. Five greater than z
6. 28% of t
7. Sixty dollars less than first-class air fare
8. The quotient of the volume of the sphere and 6
9. Six degrees hotter than yesterday's temperature
10. The cost of the gasoline divided three ways
11. 8% of the total bill
12. Five inches less than the height of the triangle

13. a. Write an expression for the distance traveled in t seconds by a car moving at 88 feet per second.
b. How far will the car travel in 5 seconds? In half a second? In half a minute?

14. a. Write an expression for the distance you can travel in 3 hours at a speed of r miles per hour.
b. How far can you travel in 3 hours on a bicycle at a speed of 6 miles per hour? On a motorboat at a speed of 20 miles per hour?

15. a. Write an expression for the amount of interest earned in t years by \$500 deposited in an account that pays 7% annual interest.
b. How much interest will the account earn after 1 year? After 2 years? After 5 years?

16. a. Write an expression for the amount of interest earned in two years by \$1500 deposited in an account that pays interest rate r.
b. How much interest will the money earn at $6\frac{1}{2}$% interest? At 8.3% interest?

17. Rachel has saved \$60 to help pay for her books this semester. Let b stand for the price of her books, and n for the amount she still needs.
a. Write an equation for n in terms of b.
b. Fill in the table.

b	100	120	150	180	200
n					

c. Choose appropriate scales for the axes and graph your equation.

18. Matt wants to travel 360 miles to visit a friend over spring break. He is deciding whether to ride his bike or drive. If he travels at an average speed of r miles per hour, then the trip will take t hours, where $t = \dfrac{360}{r}$.

a. Fill in the table.

r	10	20	30	40	60	80	90
t							

b. Choose appropriate scales for the axes and graph your equation.

19. On a typical evening, there are twice as many non-smokers as smokers in a popular restaurant.

a. Write an equation for the number of non-smokers, n, in terms of the number of smokers, s.

b. Make a table of values and graph your equation.

20. Match each equation with its graph.

a. $y = 8 - x$ **b.** $y = x + 8$ **c.** $y = 8x$

I

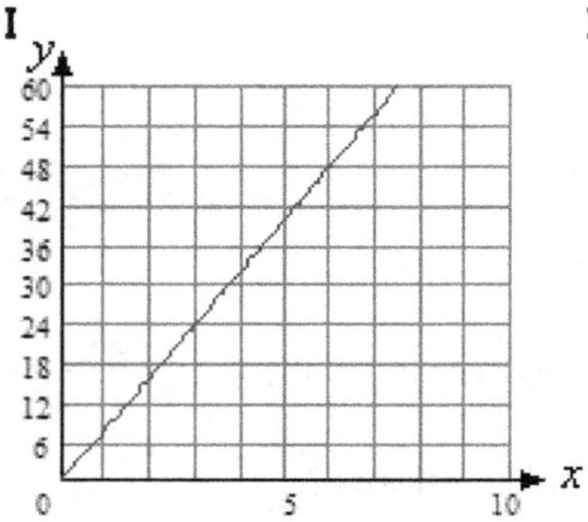

II

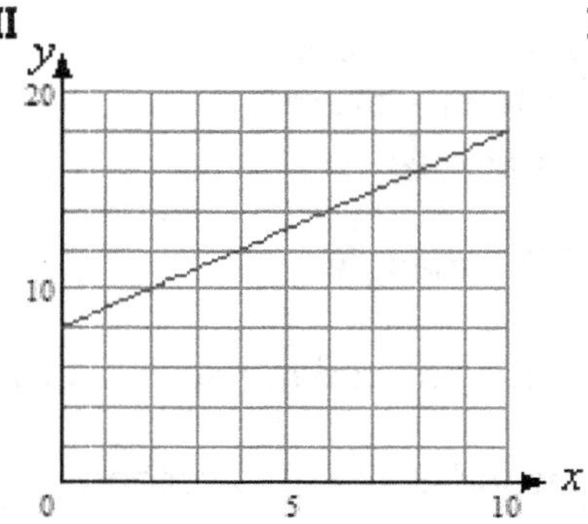

III

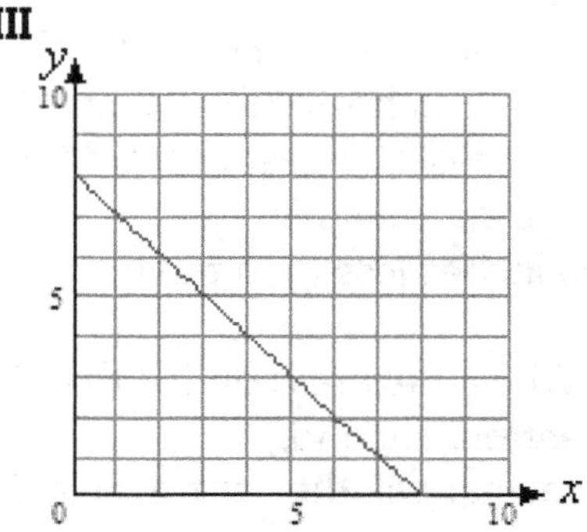

21. Decide which ordered pairs are solutions of the equation $y = \dfrac{5}{2}x$.

a. $(4, 10)$ **b.** $(5, 2)$
c. $(1.8, 4.5)$ **d.** $(3, 7.5)$

22. Decide which ordered pairs are solutions of the equation whose graph is shown.

a. $(5, 0)$ **b.** $(2, 8)$
c. $(8, 3)$ **d.** $(6, 8)$

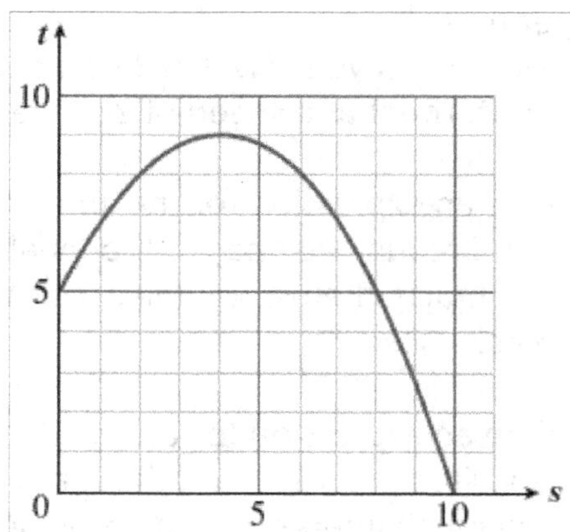

For Problems 23-24, decide whether the given values are solutions of the equation.

23. $\dfrac{t}{6} = 3$

a. $t = 18$ **b.** $t = 2$

24. $8z = 36.8$

a. $z = 28.8$ **b.** $z = 4.6$

For Problems 25-30, solve the equation.

25. $12 = 3 + y$ **26.** $2.7 = t - 1.8$ **27.** $12.2a = 4.88$

28. $8 = \frac{y}{4}$ **29.** $\frac{w}{2} = 0$ **30.** $9.5 = \frac{b}{0.6}$

For problems 31-32, solve the equation graphically.

31. The graph of $y = 1.25x$ is shown below left. Use the graph to solve each equation.
a. $1.25x = 40$ **b.** $30 = 1.25x$

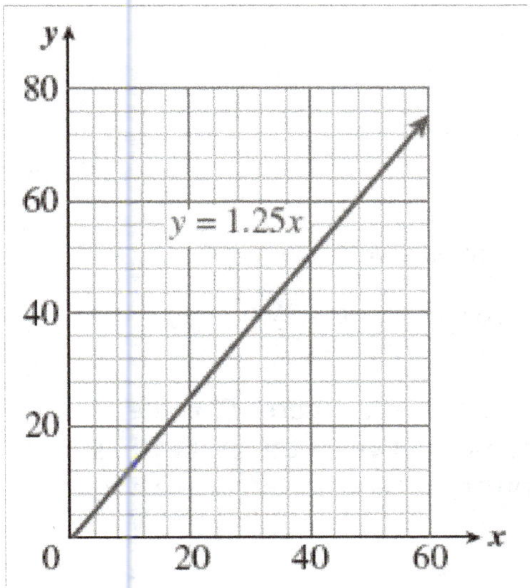

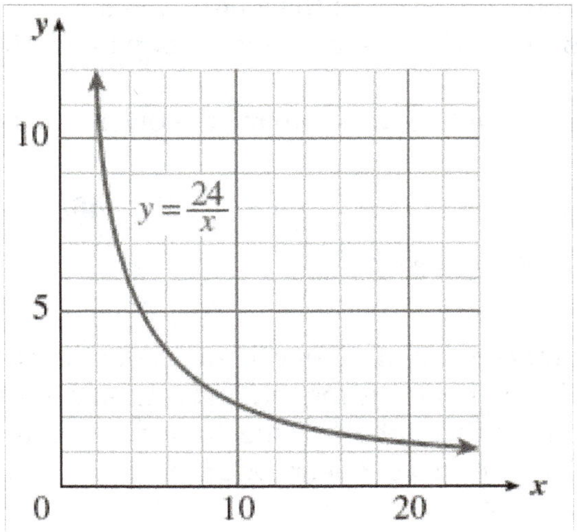

32. The graph of $y = \frac{24}{x}$ is shown above right. Use the graph to solve each equation.
a. $\frac{24}{x} = 6$ **b.** $1.5 = \frac{24}{x}$

For Problems 33-38, write and solve an equation to answer the question.

33. Staci invested some money in a T-bill account that pays $9\frac{1}{2}\%$ interest, and 1 year later the account had earned \$171 interest. How much did Staci deposit in the T-bill?

34. A jet flew 2800 miles at a speed of 560 miles per hour. How long did the trip take?

35. Fifty-three percent of the electorate voted for candidate Phil I. Buster. If candidate Buster received 106,000 votes, how many people voted?

36. Salim's test average was 3.2 points below the minimum grade for an A. If Salim's average was 89.3, what grade was required for an A?

37. Ivan has a new puppy whose adult weight will be about 85 pounds. When the ratio of the puppy's weight to its adult weight is 0.7, Ivan should take the puppy to the vet for shots. How much will the puppy weight when it is ready for shots?

38. If a rectangular garden plot is 12 feet wide, how long must it be to provide 180 square feet of gardening space?

■ For Problems 39-42, simplify by following the order of operations.

39. $6 + 18 \div 6 - 3 \cdot 2$

40. $\frac{2}{3} + \frac{1}{3} \cdot 4$

41. $36 \div (9 - 3 \cdot 2) \cdot 2 - 24 \div 4 \cdot 3$

42. $\frac{1.2(7.7)}{4.3 - 2.3}$

■ For Problems 43-36, write an algebraic expression for the phrase.

43. Four pages longer than half the length of the first draft
44. \$100 short of twice last year's price
45. Three times the sum of the length and 5.6
46. 73% of the difference of Robin's weight and 100 pounds

■ For Problems 47-50, evaluate for the given values.

47. $mx + b$ for $m = \frac{1}{2},\ x = 3,\ b = \frac{5}{2}$

48. $\frac{1}{m} - \frac{1}{n}$ for $m = 4,\ n = 6$

49. $\frac{3w + z}{z}$ for $w = 8,\ z = 6$

50. $2(l + w)$ for $l = \frac{1}{3},\ w = \frac{1}{6}$

51. The total amount of money in an account is given by $P(1 + rt)$, where P is the initial investment, r is the interest rate, and t is the number of years the money has been in the account. How much money is in an account that earns 10% interest if \$1000 was invested 3 years ago?

52. If an object is thrown downwards with initial speed v, then after t seconds it will fall a distance given in feet by $t(16t + v)$. How far will a penny fall in 2 seconds if it is thrown at a speed of 10 feet per second?

Chapter 2 Linear Equations

Lesson 2.1 Signed Numbers

- Perform arithmetic operations on signed numbers
- Write a subtraction fact as an equivalent addition fact
- Simplify quotients involving zero
- Use operations on signed numbers in applications

Lesson 2.2 Expressions and Equations

- Write an expression with two operations
- Evaluate expressions using the order of operations
- Apply the order of operations to expressions involving signed numbers

Lesson 2.3 Graphs of Linear Equations

- Plot points on a Cartesian coordinate system
- Graph a linear equation in four quadrants
- Graph a linear model
- Use a graph to evaluate an expression

Lesson 2.4 Linear Equations and Inequalities

- Solve equations with two or more operations
- Write an equation to model an applied problem
- Solve an inequality

Lesson 2.5 Like Terms

- Decide whether two expressions are equivalent
- Combine like terms
- Write and simplify expressions using like terms
- Write an equation to model a situation
- Solve equations that involve like terms

2.1 Signed Numbers

Types of Numbers

Numbers greater than zero are called **positive numbers**, and numbers less than zero are **negative numbers**. We use a **number line** to illustrate relationships among numbers.

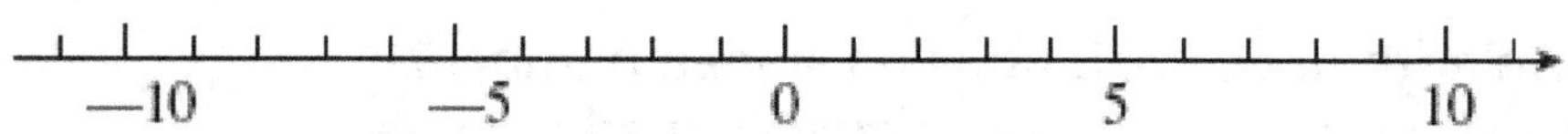

A Number Line

To the right of zero we mark the numbers 1, 2, 3, 4, ... at evenly spaced intervals. These numbers are called the **natural** or **counting numbers**. The **whole numbers** include the natural numbers and zero: 0, 1, 2, 3, ...

> The natural numbers, zero, and the negatives of the natural numbers are called the **integers**:
>
> $$\ldots, -3, -2, -1, 0, 1, 2, 3, \ldots$$

Fractions, such as $\frac{2}{3}$, $-\frac{5}{4}$, and 3.6, lie between the integers on the number line.

> **RQ1.** What are integers?

As we move from left to right on a number line, the numbers increase. In the figure below, the graph of −6 lies to the left of the graph of −2. Therefore, −6 is less than −2, or, equivalently, −2 is greater than −6.

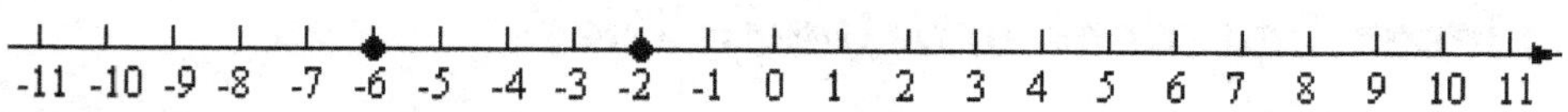

> We use special symbols to indicate order:
>
> $<$ means **is less than**
> $>$ means **is greater than**

For example,

$-6 < -2$ means −6 is less than −2
$-2 > -6$ means −2 is greater than −6

The small ends of the symbols $<$ and $>$ always point to the smaller number.

> ■ **Example 1a.** Which is the lower altitude, −81 feet or −94 feet?
> **b.** Express the relationship using one of the order symbols.
>
> **Solutions a.** Negative altitudes correspond to feet below sea level, and 94 feet is farther below sea level than 81 feet. Therefore, −91 feet is the lower altitude.
> **b.** $-94 < -81$, or $-81 > -94$. ■

Adding Two Numbers with the Same Sign

We'll use number lines to review operations on signed numbers.

Case 1: The sum of two positive numbers

Illustrate the sum $5+3$ on a number line.

1. Graph the first number, $+5$, as shown below.
2. Move 3 units in the positive direction, or to the right.
3. This brings us to the sum, which is $+8$. Thus, $5+3=8$.

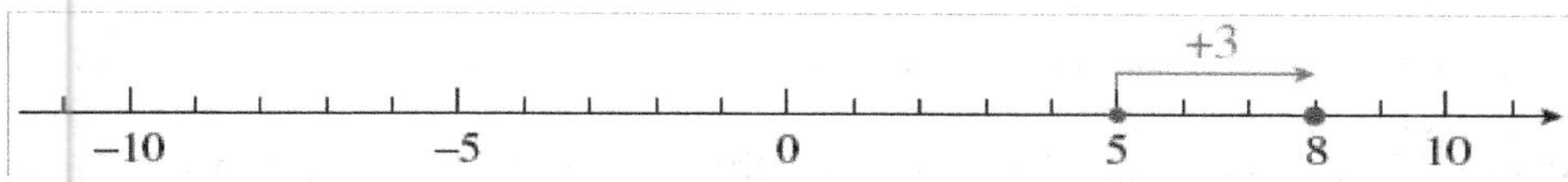

Case 2: The sum of two negative numbers

Think of negative numbers as representing debts. The sum of two debts is an even bigger debt, so we would expect the sum of two negative numbers to be a negative number. Thus, we guess that the sum $(-5)+(-3)$ is –8.

Illustrate the sum $(-5)+(-3)$ on a number line.

1. Graph −5, as shown below.
2. Move 3 units in the negative direction, or to the left.
3. This brings us to the sum, −8, as expected. Thus, $(-5)+(-3)=-8$.

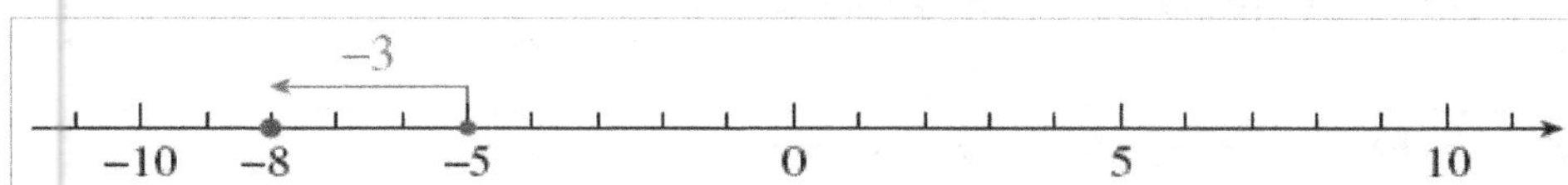

Look Closer: In the examples above, we see that:

The sum of two positive numbers is positive.
The sum of two negative numbers is negative.

Thus, the *sign* of the sum is the same as the sign of the two terms. To find the *value* of the sum, we add the **absolute values**, or unsigned parts, of the two terms.

Example 2 Add $(-4)+(-7)$

Solution The sum is negative. We add the absolute values of −4 and −7 (the numbers 4 and 7 without their signs) to get 11, then make the sum negative: $(-4)+(-7)=-11$ ■

Adding Two Numbers with Opposite Signs

Suppose you have debts of \$9, and assets of \$6. (Your "assets" are your possessions.) What is your net worth? We can model this situation by the sum

$$-9+(+6)$$

where debts are represented by negative numbers and assets by positive numbers. If you use your \$6 to pay off part of your debt, you will still owe \$3, so your net worth is $-\$3$. It makes sense that if your debts are greater than your assets, your net worth is negative.

Illustrate the sum $-9+(+6)$ on a number line.

1. Graph -9, as shown below.
2. Move 6 units in the positive direction, or to the right.
3. This brings us to the sum, -3.

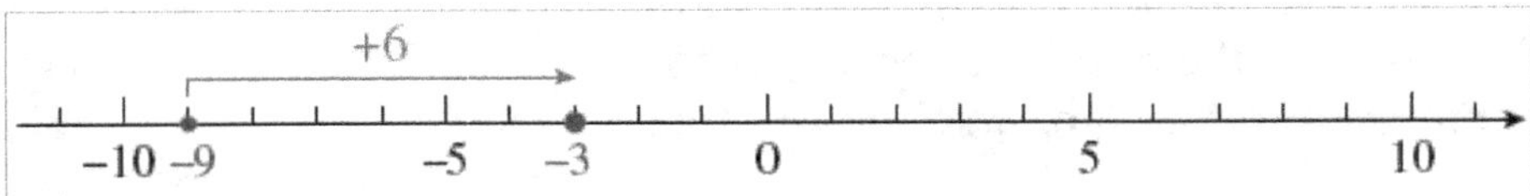

Look Closer: In the example above, we see that the sum has the same sign as the number with the larger absolute value. (Can you explain this in terms of debts and assets?) When the two numbers have opposite signs, we find the value of the sum by **subtracting the absolute values of the two numbers**.

Example 3a. Add $(-5)+(+9)$ **b.** Add $(+5)+(-9)$

Solutions a. The two numbers have opposite signs, so we subtract 5 from 9 to get 4, then make the answer positive, because $+9$ has a larger absolute value than -5. Thus, $(-5)+(+9)=4$.

b. We subtract 5 from 9 to get 4, then make the answer negative, because -9 has a larger absolute value than $+5$. Thus, $(+5)+(-9)=-4$.

RQ2. What is the absolute value of a number?
RQ3. When we add two numbers with opposite signs, we __________ their absolute values.

We now have two rules for adding signed numbers.

Rules for Adding Integers

1. To add two numbers with the **same** sign, add their absolute values. The sum has the same sign as the numbers.
2. To add two numbers with **opposite** signs, subtract their absolute values. The sum has the same sign as the number with the larger absolute value.

Subtracting a Positive Number

Let's compare the two problems:

$$9+(-5) \quad \text{and} \quad 9-(+5)$$

Illustrate the sum $9+(-5)$ on a number line.

1. Plot 9 on the number line.
2. Move 5 units in the negative direction, or to the left.
3. The sum is 4.

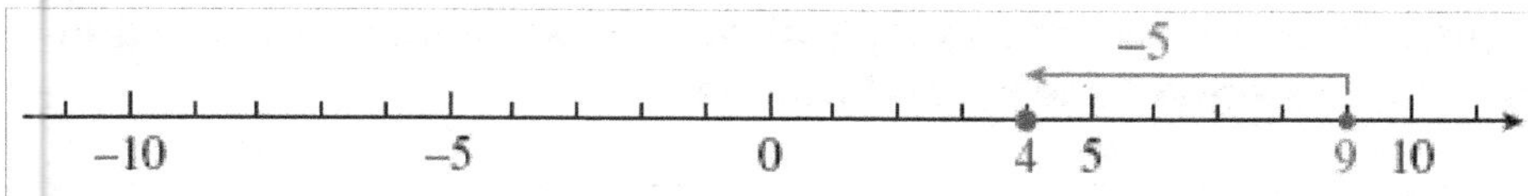

Illustrate the subtraction problem $9-(+5)$ on a number line.

1. Plot 9 on the number line
2. Move 5 units to the left, to indicate subtraction.
3. The result is 4. The picture is identical to the graph above.

Look Closer: We see that subtracting a positive number has the same result as adding the negative number with the same absolute value. Thus, **to subtract a positive number, we add its opposite**. •

Subtracting a Negative Number

When we *add* a negative number we move to the *left* on the number line. There are only two directions we can move on a number line, so when we *subtract* a negative number we must move to the *right*.

Compare the graphs for adding –3 and subtracting –3.

Addition: $5+(-3)$ Move 3 units to the **left,** to get 2.

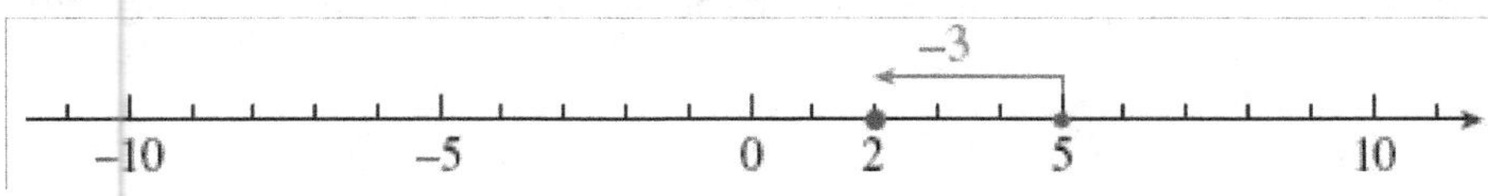

Subtraction: $5-(-3)$ Move 3 units to the **right,** to get 8.

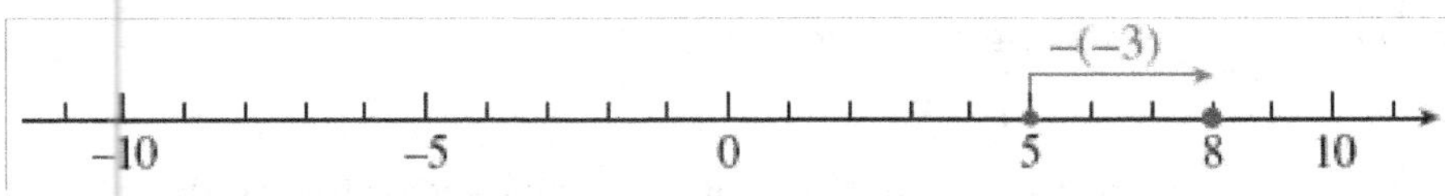

Look Closer: Notice that

$$5-(-3)=8 \quad \text{and} \quad 5+(+3)=8$$

Subtracting a negative number has the same result as adding the positive number with the same absolute value. Thus, **to subtract a negative number, we add its opposite**. •

RQ4. True or False: $-x$ always represents a negative number. Give an example to support your answer.

RQ5. Subtracting a negative number is the same as ______________ .

From the examples above, we see that to subtract any number, positive or negative, we add its opposite. We can do this in steps as follows.

Rule for Subtracting Integers

To subtract b from a:

1. Change the sign of b.
2. Change the subtraction to addition.
3. Proceed as in addition.

This rule tells us that **we can rewrite every subtraction problem as an addition problem by changing the sign of the second number**.

Example 4 Rewrite each subtraction problem as an addition, then compute the answer.

change to addition

a. $3-(+10)=3+(-10)=-7$

change sign

change to addition

b. $-6-(+3)=-6+(-3)=-9$

change sign

change to addition

c. $2-(-6)=2+(+6)=8$

change sign

change to addition

d. $-7-(-4)=-7+(+4)=-3$

change sign

Look Closer: At first it may be difficult to understand why subtracting a negative number is the same as adding a positive number. Try thinking in terms of debts and assets. For example, suppose you have \$25 in your wallet, and you have \$5 in your hand to pay off a debt you owe a friend. If your friend now cancels (or subtracts) the debt, you are actually \$5 richer:

$$25-(-5)=30$$

Caution! For simple subtractions we don't really need the subtraction rule. For example, we can subtract $11-5$ just as we did in arithmetic:

$$11-5=6$$

However, if the subtraction involves negative numbers, it is usually easier to use the subtraction rule. Thus,

$$11-(-5)=11+(+5)=16$$

Look Ahead: We can interpret the expression $7-9$ either as a subtraction or as an addition:

$$7-(+9) \quad \text{or} \quad 7+(-9)$$

Both give the same result, -2. Because addition is easier than subtraction, it is better to think of a long expression such as

$$7-9+12-6$$

as a *sum* of signed numbers. To do this, we add all the numbers together, and the $+$ or $-$ sign in front of each number tells us whether that number positive or negative.

Remember that additions should be performed **in order from left to right**.

Example 5 Simplify $7 - 9 + 12 - 6$

Solution We add the signed numbers 7 and –9 and 12 and –6, in order from left to right.

$$7 - 9 + 12 - 6 = \mathbf{-2 + 12} - 6 \qquad \text{The sum of 7 and } -9 \text{ is } -2;$$
$$= \mathbf{10} - 6 = 4 \qquad \text{the sum of } -2 \text{ and 12 is 10.}$$

Caution! The following calculation is incorrect:

$$16 - 9 - \mathbf{4 + 3} = 16 - \mathbf{5} + 3 = 11 + 3 = 14 \qquad \leftarrow \textbf{Incorrect!}$$

We must perform additions and subtractions in order from left to right:

$$\mathbf{16 - 9} - 4 + 3 = \mathbf{7} - 4 + 3 = 3 + 3 = 6$$

Multiplying Signed Numbers

Case 1: The product of numbers with opposite signs

We can think of multiplication as repeated addition. For example,

$$3(2) \quad \text{means} \quad 2 + 2 + 2 \qquad \text{The sum of three 2's}$$

Similarly,

$$3(-2) \quad \text{means} \quad -2 + (-2) + (-2) \qquad \text{The sum of three } -2\text{'s}$$

which is –6, so $3(-2) = -6$. Think in terms of money: if you owe three different people 2 dollars each, then you are 6 dollars in debt.

Because multiplication is commutative, it is also true that $-2(3) = -6$. This example illustrates the following fact.

The product of a positive number and a negative number is a negative number.

Example 6 You can use your calculator to verify each of the following products.

a. $(-4)(7) = -28$ **b.** $5(-2.6) = -13$

Case 2: The product of numbers with the same sign

You know that the product of two positive numbers is positive. It is also true that the product of two negative numbers is positive.

To understand why the product of two negative numbers is positive, you might think of canceling, or subtracting, several debts of the same amount. Or, you could observe the pattern in the lists of products below. As we move down the list, the product increases by equal amounts.

$3(-2) = -6$	$3(-5) = -15$	
$2(-2) = -4$	$2(-5) = -10$	
$1(-2) = -2$	$1(-5) = -5$	← Positive times negative is negative.
$0(-2) = 0$	$0(-5) = 0$	
$-1(-2) = 2$	$-1(-5) = 5$	← Negative times negative is positive.
$-2(-2) = 4$	$-2(-5) = 10$	
$-3(-2) = 6$	$-3(-5) = 15$	

Based on the pattern in these calculations, the following rule seems reasonable.

The product of two negative numbers is a positive number.

Example 7 You can use your calculator to verify the following products.

a. $(-6)(-3) = 18$ **b.** $(-2.1)(-3.4) = 7.14$

Now we have a pair of rules about the products of signed numbers.

Products of Signed Numbers

1. The product of two numbers with opposite signs is a negative number.
2. The product of two numbers with the same sign is a positive number.

Dividing Signed Numbers

Division is the inverse operation for multiplication. For example, we know that

$$\frac{12}{3} = \boxed{4} \qquad \text{because} \qquad 3 \cdot \boxed{4} = 12$$

Thus, every division fact can be converted into an equivalent multiplication fact. For example,

$$\text{The division fact } \frac{\mathbf{400}}{16} = 25 \text{ is equivalent to } 16 \cdot 25 = \mathbf{400}$$

Note that the *numerator* in the division fact becomes the *product* in the multiplication fact. The same relationship between multiplication and division holds for negative numbers as well.

Example 8 Rewrite each division fact as a multiplication fact.

a. $\dfrac{-144}{64} = -2.25$ **b.** $-36 \div \dfrac{-3}{8} = 96$

Solutions The numerator in each division fact becomes the product in the multiplication fact.

a. $-144 = -2.25(64)$ **b.** $-36 = \dfrac{-3}{8}(96)$

Example 8 illustrates the rules for division of signed numbers.

Quotients of Signed Numbers

1. The quotient of two numbers with opposite signs is a negative number.
2. The quotient of two numbers with the same sign is a positive number.

RQ6. Which of these is a product?

$3 + (-4) \qquad -3 - 4 \qquad -3(-4) \qquad 3 - (-4)$

RQ7. Rewrite the division fact $\dfrac{48}{D} = Q$ as a multiplication fact.

Remember that a fraction can be thought of as a division. For example, $\frac{3}{4}$ means $3 \div 4$, or 0.75. What about negative fractions? Because the quotient of two numbers with unlike signs is negative, the fractions $\frac{-3}{4}$ and $\frac{3}{-4}$ both represent the negative number $-\frac{3}{4}$. On the other hand, $\frac{-3}{-4}$ is equal to the positive number $\frac{3}{4}$.

For many of the calculations in algebra, it is more convenient to write a negative fraction with the minus sign in the numerator, like this: $\frac{-3}{4}$.

> A negative fraction is written in **standard form** when the minus sign is in the numerator: $\frac{-a}{b}$.
>
> $$\frac{-a}{b} = \frac{a}{-b} = -\frac{a}{b}$$
>
> The standard form for a positive fraction has positive numerator and denominator.
>
> $$\frac{a}{b} = \frac{-a}{-b}$$

Quotients Involving Zero

What is the meaning of an expression such as $\frac{0}{2}$? We can rewrite this quotient as an equivalent multiplication fact.

$$\frac{0}{2} = \boxed{?} \quad \text{is equivalent to} \quad 2 \cdot \boxed{?} = 0$$

What number can we substitute for the question mark to make $2 \cdot \boxed{?} = 0$ a true statement? The only solution for this equation is 0. Thus, replacing the question marks by 0, we have

$$\frac{0}{2} = 0 \quad \text{because} \quad 2 \cdot 0 = 0$$

In general, **0 divided by any (non-zero) number is 0**.

Now consider a quotient with zero in the denominator, such as $\frac{2}{0}$.

$$\frac{2}{0} = \boxed{?} \quad \text{is equivalent to} \quad 0 \cdot \boxed{?} = 2$$

What number can we substitute for the question mark to make $0 \cdot \boxed{?} = 2$ a true statement? This equation has no solution, because 0 times any number is 0, not 2. Thus, there is no answer for the division problem $\frac{2}{0}$. We say that the quotient $\frac{2}{0}$ is **undefined**. In general, **we cannot divide any number by zero**.

> **Quotients Involving Zero**
>
> If a is any nonzero number, then
>
> $$\frac{0}{a} = 0 \quad \text{and} \quad \frac{a}{0} \text{ is undefined.}$$

Example 9a. $6 \div 0$ is undefined. **b.** $0 \div 8 = 0$

RQ8. Which of these is undefined? $\frac{-6}{0}$ or $\frac{0}{-6}$

Skills Warm-Up

1. Graph the numbers on the number line: $0,\ -3,\ 3,\ \frac{1}{2},\ \frac{-5}{3},\ 5\frac{1}{4}$

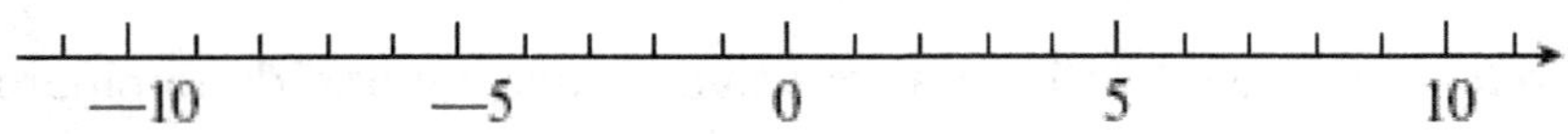

2. Graph the numbers on the number line: $-4.5,\ -2,\ -1.4,\ 0.6,\ 2$

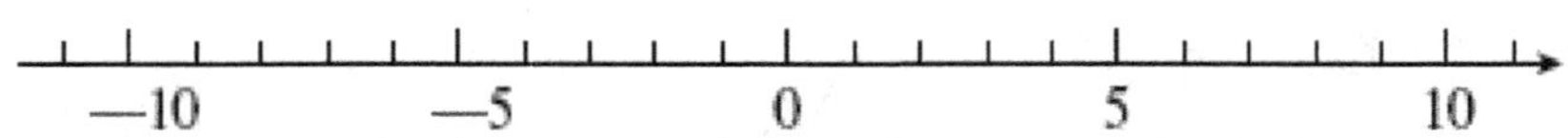

Replace the comma in each pair by the proper symbol, $<$, $>$, or $=$.

3. a. $0, -4$ **b.** $-5, -9$ **c.** $-2\frac{5}{8}, -2\frac{1}{8}$ **d.** $13.6, 13.66$

4. a. $3, -7$ **b.** $-1, -6$ **c.** $-3, -3\frac{3}{4}$ **d.** $-18.4, -19.6$

Answers: 3.a. $>$ **b.** $>$ **c.** $<$ **d.** $<$ **4.a.** $>$ **b.** $>$ **c.** $>$ **d.** $>$

Homework 2.1

Skills Practice

For Problems 1-4, add.

1. a. $5+(-3)$ b. $-5+3$ c. $-5+(-3)$
2. a. $-15+(-20)$ b. $15+(-20)$ c. $-15+20$
3. a. $-47+22$ b. $6.8+(-2.7)$ c. $-\frac{5}{6}+\frac{2}{3}$
4. a. $-13+(-36)$ b. $-2.5+4.9$ c. $-\frac{3}{4}+\left(-\frac{3}{8}\right)$

5. Rewrite each subtraction problem as an addition problem, and give the answer.
   a. $4-8$ b. $3-(-9)$ c. $-8-(-6)$ d. $-6-5$

For Problems 6-8, add or subtract as indicated.

6. a. $12+(-6)$ b. $6-(-4)$ c. $-2-8$ d. $-7+9$
7. a. $-14-(-3)$ b. $-5+(-4)$ c. $-6-(-6)$ d. $-4-4$
8. a. $24-(-10)$ b. $18+(-12)$ c. $-16+14$ d. $-25-(-15)$

For Problems 9-10, compute each sum or difference in parts (a) and (b), and decide which is easier. Then decide how to simplify the expression in part (c).

9. a. $15-(+5)$ b. $15+(-5)$ c. $15-5$
10. a. $-6-(+2)$ b. $-6+(-2)$ c. $-6-2$

For Problems 11-12, compute mentally.

11. a. $2-7$ b. $-2-7$ c. $-2+7$ d. $2-(-7)$ e. $-2-(-7)$
12. a. $13-5$ b. $-13-5$ c. $-13+5$ d. $13-(-5)$ e. $-13-(-5)$

For Problems 13-14, add.

13. a. $-5+3+(-4)$ b. $-4+(-7)+(-7)$
14. a. $6+(-14)+12+(-17)$ b. $-35+(-5)+28+(-21)+13+(-14)$

15. Simplify by adding the signed numbers.
    a. $2+5-8-1$ b. $-23+28-14+21$ c. $-34-52+68-21$

16. Simplify by following the rules for addition and subtraction.
    a. $-6+5+(-3)-(-8)$ b. $-11-2-(-4)-(-3)$ c. $-14-(-16)-4+(-7)$

For problems 17-20, multiply or divide.

17. a. $(-8)(-4)$ b. $\frac{12}{-4}$ c. $-20\div(-5)$
18. a. $-6(-1)(3)$ b. $\frac{-8}{0}$ c. $(-5)(0)(6)$
19. a. $0\left(\frac{-7}{15}\right)$ b. $(-2)(-2)(-2)$ c. $-30\div 0$

20. a. $\left(\frac{1}{2}\right)\left(\frac{-3}{4}\right)$ **b.** $-0.1(-26)$ **c.** $\frac{-1}{2} \div \left(\frac{-3}{4}\right)$

■ For Problems 21-22, perform the indicated operations.

21. a. $-12-4$
b. $-12(-4)$
c. $-12-(-4)$
d. $-12+(-4)$
e. $-12 \div 4$

22. a. $-3-3$
b. $-3(-3)$
c. $-3-(-3)$
d. $-3+(-3)$
e. $-3 \div (-3)$

23. a. Find two numbers whose sum equals zero.
b. Find two numbers whose difference equals zero.
c. Find two numbers whose product equals zero.
d. Find two numbers whose quotient equals zero.

24. a. Find two numbers whose product is 1.
b. Find two numbers whose product is –1.
c. Find two numbers whose quotient is 1.
d. Find two numbers whose quotient is –1.

■ For Problems 25-30, solve.

25. $x-9=-4$ **26.** $-9z=12$ **27.** $\frac{-a}{4}=8$

28. $9-x=3$ **29.** $t+5=-8$ **30.** $-6b=-27$

31. a. What is the solution of the equation $-x=-3$?
b. What is the solution of the equation $-x=6$?

32. Find and correct the error in the following calculation:

$$18-6+2-5=18-8+5=15 \quad \textbf{(Incorrect!)}$$

Applications

■ For problems 33-40, write and simplify an expression using negative numbers to answer the question.

33. Thelma's failing company is worth –\$1000. She decides to merge with Louise's company, whose net worth is \$1500. What is the combined worth of the two companies?

34. Emily is at an elevation of –87 feet when she begins climbing a mountain. What is her elevation after ascending 127 feet?

35. The temperature at noon was 2° and dropped 8° over the next 12 hours. What was the temperature at midnight?

36. Rocky lost \$450 betting on the horses this afternoon and then dropped \$245 at a poker game in the evening. What was the net change in his financial status for the day?

37. A tourist in California can travel from Death Valley, at an elevation of −282 feet, to Mt. Whitney, at 14,494 feet. What is his net change in elevation?

38. Despite being overdrawn by $24.20, Nelson writes a check for $11.20. What is his new balance?

39. Whitney is climbing down a sheer cliff. She has pitons spaced vertically every 6 meters. What is her net elevation change after descending to the eighth piton from the top?

40. The temperature on the ice planet Hoth dropped 115° in just 4 hours. What was the average temperature change per hour?

■ Choose the correct equation to model Problems 41-44.

$-12 + n = -5$	$-12 + n = 5$
$n + 12 = -5$	$n + 12 = 5$

41. Last night the temperature rose 12° and this morning the temperature is 5°. What was the temperature yesterday?

42. Marta's score was −12, but after the last hand her score rose to −5. How many points did she make on the last hand?

43. Bradley rode the elevator down 12 floors and emerged five floors below ground. What floor did he start on?

44. Orrin was $12 in debt, but he did some yard work this weekend and after paying his debt he has $5. How much did he make?

2.2 Expressions and Equations

Writing Algebraic Expressions

When we write an algebraic expression, we use the same operations on a variable that we would use to calculate with a specific number. Writing down the expression for a specific numerical value can help us write an algebraic expression.

Example 1 Alida keeps \$100 in cash from her weekly paycheck, and deposits 40% of the remainder in her savings account. If Alida's paycheck is p, write an expression for the amount she deposits in savings.

Solution How would we calculate Alida's deposit if we knew her paycheck? Suppose Alida's paycheck is \$500. First she subtracts \$100 from that amount to get $500 - 100$, and then she takes 40% of the remainder for savings:

$$0.40\,(500 - 100)$$ **Perform operations inside parentheses first.**

If her paycheck is p dollars, we perform the same operations on p instead of on 500. The expression is thus

$$0.40\,(p - 100)$$

RQ1. We write an expression with variables using the same operations we would use to calculate with a ________________ .

RQ2. In Example 1, how does Alida calculate the amount to deposit in savings? How do we write that in an algebraic expression?

Evaluating Algebraic Expressions

When we evaluate an algebraic expression, we must follow the order of operations.

Example 2 Four students bought concert tickets with their \$20 student-discount coupons.

a. If the regular price of a ticket is t dollars, write an expression for the total amount the four students paid.

b. How much did the students pay if the regular price of a ticket is \$38?

Solutions a. The discount price of one ticket is $t - 20$ dollars. So four tickets cost

$$4(t - 20) \text{ dollars}$$

b. We evaluate the expression in part (a) for $t = 38$.

$$\begin{aligned} 4(t - 20) &= 4(38 - 20) \\ &= 4(18) = 72 \end{aligned}$$ **Simplify inside parentheses first.**

The students paid \$72 for the tickets.

RQ3. Explain why we need parentheses in Example 2.

RQ4. How should we write "3 times the sum of x and 12"?

Negative Numbers

The order of operations applies to signed numbers. Operations inside parentheses or other grouping devices should be performed first.

Example 3 Simplify $4-3-[-6+(-5)-(-2)]$

Solution Perform the operations inside brackets first. Simplify each step by rewriting subtractions as equivalent additions.

$4-3-[-6+(-5)-(-2)]$	$-5-(-2)=-5+2$
$=4-3-[-6-5+2]$	$-6-5=-11;\ -11+2=-9$
$=4-3-[-9]$	Rewrite as an addition.
$=4-3+9=10$	Add from left to right.

RQ5. Which operation is performed first in the expression $6-4x$?
RQ6. What is the first step in evaluating the expression $2(18-x)$?

Caution! When using negative numbers, we must be careful to distinguish between products and sums. In Example 4, note how the parentheses and minus signs are used in each expression.

Example 4 Simplify each expression.

a. $3(-8)$ **b.** $3-(-8)$ **c.** $3-8$ **d.** $-3-8$

Solutions a. This expression is a product: $3(-8)=-24$

b. This is a subtraction. We follow the rule for subtraction by changing the sign of the second number and then adding: $3-(-8)=3+8=11$

c. This is an addition; the negative sign in front of 8 tells us that we are adding -8 to 3. Thus, $3-8=3+(-8)=-5$

d. This is also an addition: $-3-8=-3+(-8)=-11$

RQ7. What is wrong with this calculation: $4-3+9=4-12=-8$

Look Closer: When we evaluate an algebraic expression at a negative number, we enclose the negative numbers in parentheses. This will help prevent us from confusing multiplication with subtraction.

Example 5 Evaluate $2x-3xy$ for $x=-5$ and $y=-2$.

Solution We substitute -5 for x and -2 for y, then follow the order of operations.

$$\begin{aligned}2x-3xy &= 2(-5)-3(-5)(-2) \quad \text{Do multiplications first.}\\ &= -10-(-15)(-2)\\ &= -10-30=-40\end{aligned}$$

RQ8. When we evaluate an expression at a negative number, we should ________ the negative number in ________________.

Skills Warm-Up

Choose the correct algebraic expression for each of the following situations.

$n+12$	$12n$	$\frac{n}{12}$
$\frac{12}{n}$	$12-n$	$n-12$

1. Helen bought n packages of tulip bulbs. If each package contains 12 bulbs, how many bulbs did she buy?

2. Henry bought a package of n gladiolus bulbs, then bought 12 loose bulbs. How many bulbs did he buy?

3. Together Karen and Dave sold 12 tickets to the spring concert. If Karen sold n tickets, how many did Dave sell?

4. Together Karl and Diana collected n used books for the book sale. If Karl collected 12 books, how many did Diana collect?

5. Greta made n dollars last week. If she worked for 12 hours, how much did she make per hour?

6. Gert jogged for 12 minutes. If she jogged n miles, how many minutes does it take her to jog 1 mile?

Answers: **1.** $12n$ **2.** $n+12$ **3.** $12-n$ **4.** $n-12$ **5.** $\frac{n}{12}$ **6.** $\frac{12}{n}$

Homework 2.2

Skills Practice

1. The perimeter of a rectangle of length l and width w is given by $P = 2l + 2w$. Find the perimeter of a rectangular meeting hall with dimensions 8.5 meters by 6.4 meters.

2. The area of a trapezoid with bases B and b and height h is given by $A = \frac{1}{2}(B + b)h$. Find the area of a trapezoid whose bases are 9 centimeters and 7 centimeters and whose height is 3 centimeters.

3. If a company sells n items at a cost of c dollars each and sells then at a price of p dollars each, the company's profit is given by $P = n(p - c)$. Find the profit earned by a manufacturer of bicycle equipment by selling 300 bicycle helmets that cost \$32 each to produce and sell for \$50 apiece.

4. The temperature in degrees Celsius (°C) is given by $C = \frac{5}{9}(F - 32)$, where F stands for the temperature in degrees Fahrenheit (°F). Find normal body temperature in degrees Celsius if normal temperature is 98.6° Fahrenheit.

■ In Problems 5-8, simplify by following the order of operations.

5. $-18 - [8 - 12 - (-4)]$

6. $3 - (-6 + 2) + (-1 - 4)$

7. $-7 + [-8 - (-2)] - [6 + (-4)]$

8. $0 - [5 - (-1)] + [-6 - 3]$

9. Find each product.
   a. $(-2)(3)(4)$
   b. $(-2)(-3)(4)$
   c. $(-2)(-3)(-4)$
   d. $(-2)(-3)(4)(2)$
   e. $(-2)(-3)(-4)(2)$
   f. $(-2)(-3)(-4)(-2)$

10. Use your results from Problem 9 to complete the statements:
   a. The product of an odd number of negative numbers is ________ .
   b. The product of an even number of negative numbers is ________ .

■ In Problems 11-14, use the order of operations to simplify each expression.

11a. $-2(-3) - 4$ **b.** $5(-4) - 3(-6)$

12a. $(-4 - 3)(-4 + 3)$ **b.** $-3(8) - 6(-2) - 5(2)$

13a. $\dfrac{15}{-3} - \dfrac{4-8}{8-12}$ **b.** $\dfrac{4 - 2(-5)}{-4 + 3(-1)}$

14a. $12.6 - 0.32(0.25)(4.2 - 8.7)$ **b.** $(5.8 - 2.6)(-2.5)(-0.6 + 3)$

15. Simplify each expression.
   a. $-3(-4)(-5)$
   b. $-3(-4) - 5$
   c. $-3(-4 - 5)$
   d. $-3 - (-4 - 5)$
   e. $-3 - (-4)(-5)$
   f. $(-3 - 4)(-5)$

16. Simplify each expression.
   a. $24 \div 6 - 2$
   b. $24 \div (-6 - 2)$
   c. $24 \div (-6) - 2$
   d. $24 - 6 \div (-2)$
   e. $24 \div (-6) \div (-2)$
   f. $24 \div (-6 \div 2)$

In Problems 17-22, evaluate for the given values.

17. $15 - x - y$ for $x = -6,\ y = 8$

18. $p - (4 - m)$ for $p = -2,\ m = -6$

19. $12x - 3xy$ for $x = -3,\ y = 2$

20. $\dfrac{y-3}{x-4}$ for $x = -9,\ y = 2$

21. $\dfrac{1}{2}t(t-1)$ for $t = \dfrac{2}{3}$

22. $\dfrac{x-m}{s}$ for $x = 4,\ m = \dfrac{9}{4},\ s = \dfrac{3}{2}$

23. a. Evaluate each expression for $x = 5$. What do you notice?

(i) $\dfrac{-3}{4}x$ **(ii)** $\dfrac{-3x}{4}$ **(iii)** $-0.75\,x$

b. Does $\dfrac{-8}{5}x = \dfrac{-8}{5x}$? Support your answer with examples.

24. Which of the following are equivalent to $-\dfrac{4}{9}x$?

a. $\dfrac{4}{9}(-x)$ **b.** $\dfrac{-4x}{-9}$ **c.** $\dfrac{-4}{9x}$ **d.** $\dfrac{-4x}{9}$ **e.** $\dfrac{-4}{9}(-x)$

Write an algebraic expression for area or perimeter.

25.

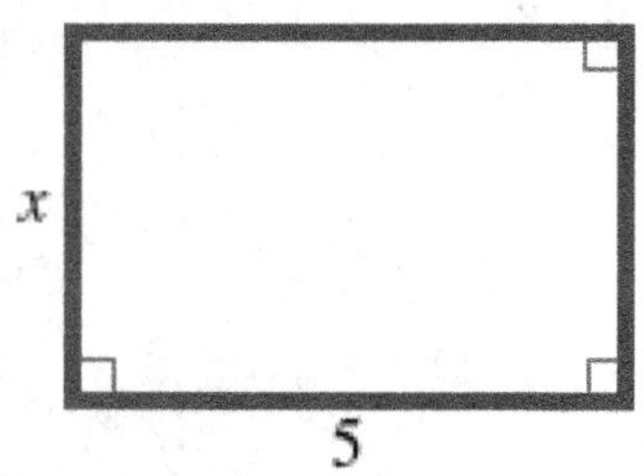

Perimeter =

26.

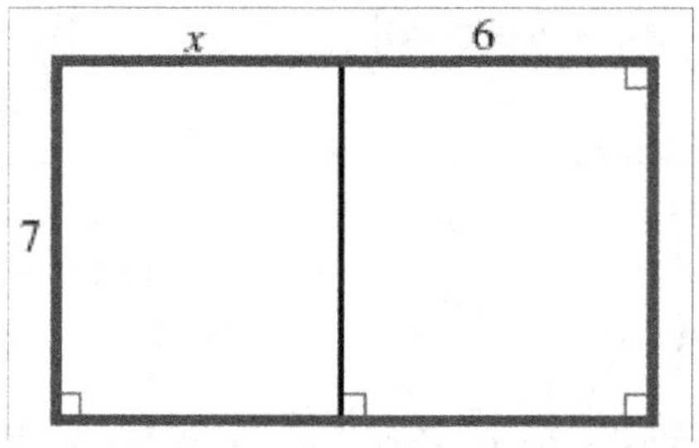

Area =

Applications

27. Salewa saved \$5000 to live on while going to school full time. She spends \$200 per week on living expenses.

a. How much of Salewa's savings will be left after 3 weeks? After 12 weeks?

b. Describe in words how you calculated your answers to part (a).

c. Fill in the table below.

Number of weeks	2	4	5	6	10	15	20
Calculation							
Savings left							

d. Write an algebraic expression for Salewa's savings after w weeks.

e. Write an equation that gives Salewa's savings, S, in terms of w.

28. To calculate how much state income tax she owes, Francine subtracts \$2000 from her income, and then takes 12% of the result.

a. What is Francine's state income tax if her income is \$8000? If her income is \$10,000?

b. Describe in words how you calculated your answers to part (a).

c. Fill in the table below.

Income	5000	7000	12,000	15,000	20,000	24,000	30,000
Calculation							
State Tax							

d. Write an algebraic expression for Francine's state income tax if her income is I.

e. Write an equation that gives Francine's state income tax, T, in terms of I.

For Problems 29-32,

a. Choose a variable for the unknown quantity and write an algebraic expression.

b. Evaluate the expression for the given values.

29. a. Three inches less than twice the width

b. The width is 13 inches.

30. a. Twenty dollars more than 40% of the principal

b. The principal is \$500.

31. a. \$8 times the number of children's tickets subtracted from 150

b. There are 83 children's tickets

32. a. One-third of \$50 less than the profit

b. The profit is \$500

In Problems 33-40, write an algebraic expression.

33. The oven temperature started at 65° and is rising at 30° per minute. Write an expression for the oven temperature after t minutes.

34. Luisa's parents have agreed to pay her tuition (\$800 per year) plus half her annual living expenses while she is in school. Write an expression for the amount her parents will pay if Luisa's annual expenses are a dollars.

35. Mildred canned 80 pints of tomatoes. She kept some for herslf and divided the rest equally among her four daughters. If Mildred kept M pints, write an expression for the number of pints she gave each daughter.

36. Moira's income is \$50 more than one-third of her mother's income. Write an expression for Moira's income if her mother's income is I.

37. Otis buys 200 pounds of dog food and uses 15 pounds a week for his dog Ralph. Write an expression for the amount of dog food Otis has left after w weeks.

38. Renee receives $600 for appearing in a cola commercial, plus a residual of $80 each time the commercial is aired. Write an expression for Renee's earnings if the commercial plays t times.

39. Digby bought scuba diving gear by making a $50 down payment and arranging to pay the balance in ten equal installments. If the total cost of the gear is C dollars, write an expression for the amount of each installment.

40. Each passengers and three crew members on a small airplane is allowed 35 pounds of luggage. Write an expression for the weight of the luggage if there are x passengers.

■ In Problems 41-42,

a. Write an algebraic expression.

b. Make a table for the expression showing at least two positive values for the variable and two negative values.

41. The ratio of a number to 4 more than the number.

42. The sum of a and 5, times the difference of a and 5.

2.3 Graphs of Linear Equations

Graphing Equations

> An equation of the form $y = ax + b$, where a and b are constants, is called a **linear equation** because its graph is a straight line.

We can graph a linear equation by evaluating the expression $ax + b$ at several values of x and then plotting points.

Example 1 At 6 am the temperature was 50°, and it has been rising by 4° every hour.

a. Write an equation for the temperature, T, after h hours.

b. Graph your equation.

Solutions a. We set $h = 0$ at 6 am. At that time, $T = 50$. Because the temperature is rising by 4° every hour, we multiply the number of hours by 4 and add the result to 50 to get the new temperature. Thus, our equation is $T = 50 + 4h$.

b. Step 1 We make a table of values for the equation. We choose some values for h, as shown in the first column of the table below. For each value of h, we evaluate the formula for T, and we record these values in the second column of the table.

h	T		Ordered pairs
0	50	$T = 50 + 4(\mathbf{0}) = 50 + 0 = 50$	$(0, 50)$
1	54	$T = 50 + 4(\mathbf{1}) = 50 + 4 = 54$	$(1, 54)$
3	62	$T = 50 + 4(\mathbf{3}) = 50 + 12 = 62$	$(3, 62)$
5	70	$T = 50 + 4(\mathbf{5}) = 50 + 20 = 70$	$(5, 70)$

Step 2 We use the values in the table to help us choose scales for the axes. For this graph, the x-values are less than 10, but the largest y-value is 70. We use increments of 1 on the horizontal or h-axis and increments of 10 on the vertical or T-axis.

Step 3 Each ordered pair gives us a point on the graph. We plot the points on the grid and connect them with a smooth curve. All the points should lie on a straight line.

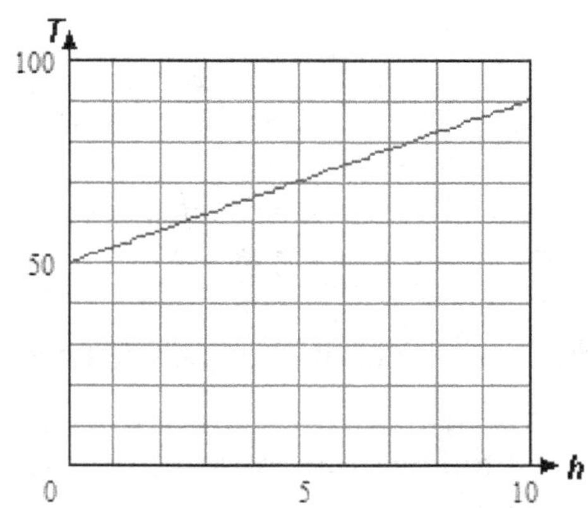

> **RQ1.** A linear equation has the form ______________.
> **RQ2.** We use ____________ to help us choose scales for the axes.

Cartesian Coordinate System

Many graphs include negative values as well as positive values. To make such a graph, we construct a **Cartesian coordinate system**, named after the French mathematician and philosopher René Descartes.

We draw two perpendicular number lines for the horizontal and vertical axes, as shown below left. The two axes divide the plane into four **quadrants**, or regions, as shown at right. We often use x for the input variable and y for the output variable, so the horizontal axis is called the **x-axis** and the vertical axis is the **y-axis**. The axes intersect at the zero point of each number line. The coordinates of this intersection point, called the **origin**, are $(0, 0)$.

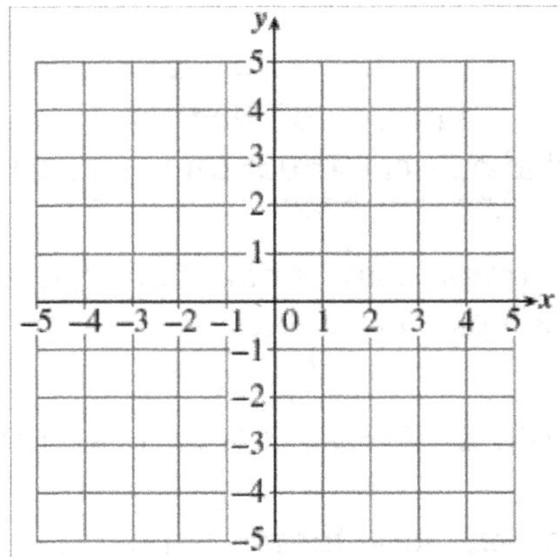

RQ3. What are the coordinates of the origin?
RQ4. True or False: The x-coordinate of any point on the y-axis is zero.

Example 2 The figure at right shows how to plot points on a Cartesian grid. Points in the upper right quadrant have both coordinates positive. Points with negative x-coordinates are plotted to the left of the y-axis, and points with negative y-coordinates are plotted below the x-axis. For example, the point $(-2, 1)$ is plotted 2 units to the left of the y-axis and 1 unit above the x-axis.

In which quadrant are both coordinates negative?

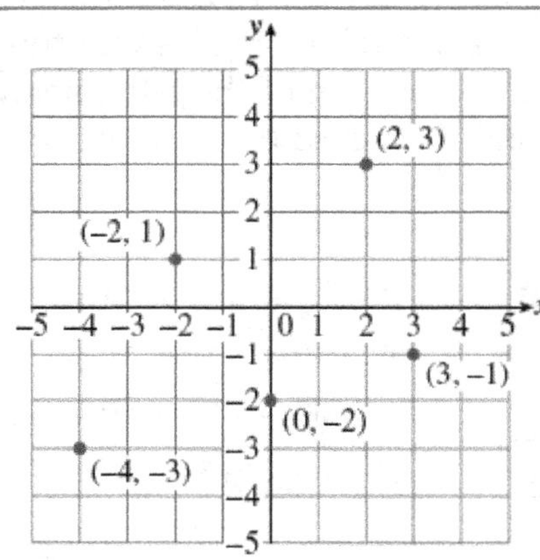

RQ5. True or False: The notation $(5, 3)$ represents two points, one on the x-axis and one on the y-axis.
RQ6. True or False: If both coordinates of a point are negative, the point is located in the lower left quadrant.

Now we can graph an equation that includes negative values.

Example 3 The temperature in Nome was -12°F at noon. It has been rising at a rate of 2°F per hour all day.

a. Write an equation for the temperature, T, after h hours.

b. Fill in the table and graph your equation. (Negative values of h represent hours before noon.)

h	-3	-2	-1	0	1	2	3
T							

Use your graph to answer the following questions:

c. What was the temperature 8 hours before noon?

d. When will the temperature reach $-4°$F?

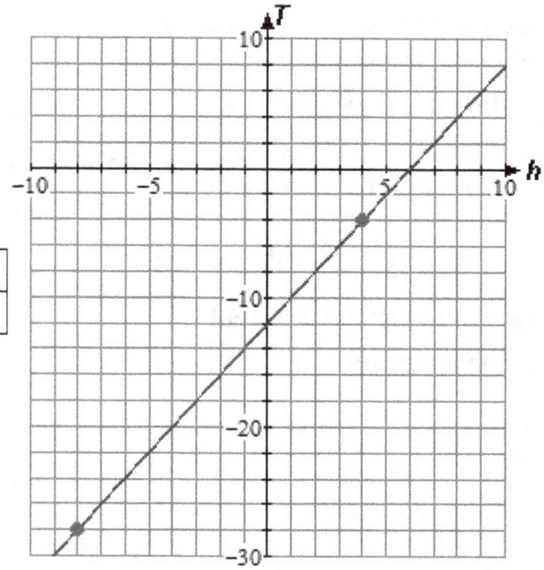

Solutions a. The temperature at h hours after noon is given by $T = -12 + 2h$.

b. We evaluate the expression for T at each of the h-values in the table.

h	−3	−2	−1	0	1	2	3
T	−18	−16	−14	−12	−10	−8	−6

c. We find the point on the graph with h-coordinate -8. Its T-coordinate is -28, so the temperature was $-28°$.

d. We find the point on the graph with T- coordinate -4. Its h-coordinate is 4, so it will be $-4°$ at 4 pm.

Look Closer: We can also answer parts (c) and (d) of Example 3 using algebra. For part (c), we evaluate the expression for T, namely $-12 + 2h$, at $h = -8$. For part (d), we solve the equation $T = -12 + 2h$ with $T = -4$. Make sure that you understand both methods, graphical and algebraic.

Graphing Tip #1

If the coefficient of x is a fraction, we can make our work easier by choosing multiples of the denominator for the x-values. That way we won't have to work with fractions to find the y-values. In Example 4, the coefficient of x is $\frac{-2}{3}$, so we choose multiples of 3 for the x-values.

Example 4a. Graph $y = \dfrac{-2}{3}x - 4$

b. Use the graph to solve the equation $\dfrac{-2}{3}x - 4 = -2$.

Solutions a. We choose x-values that are multiples of 3 and make a table of values.

x	y	
−9	2	$y = -\frac{2}{3}(\mathbf{-9}) - 4 = 2$
−6	0	$y = -\frac{2}{3}(\mathbf{-6}) - 4 = 0$
0	−4	$y = -\frac{2}{3}(\mathbf{0}) - 4 = -4$
3	−6	$y = -\frac{2}{3}(\mathbf{3}) - 4 = -6$
6	−8	$y = -\frac{2}{3}(\mathbf{6}) - 4 = -8$

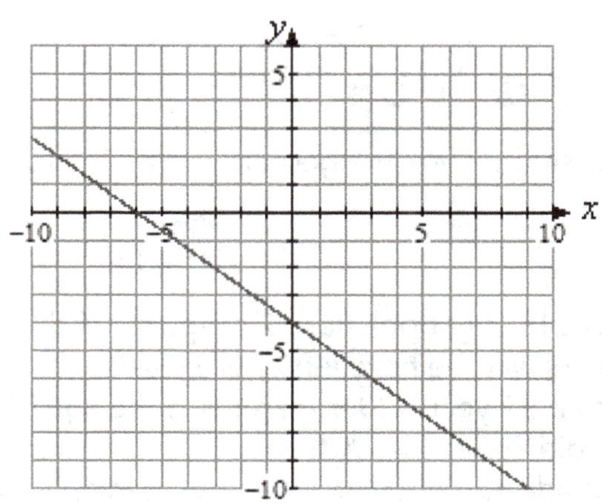

Then we plot the points and connect them with a straight line.

b. We see that y has been replaced by -2 in the equation of the graph. So we look for the point on the graph that has y-coordinate -2. This point, labeled P on the graph at right, has x-coordinate -3. Because it lies on the graph, the point $P(-3, -2)$ is a solution of the equation $y = \frac{-2}{3}x - 4$. When we substitute $x = -3$ and $y = -2$ into the equation, we get a true statement:

$$-2 = \frac{-2}{3}(-3) - 4$$

But this statement also tells us that -3 is a solution of the equation $-2 = \frac{-2}{3}x - 4$.

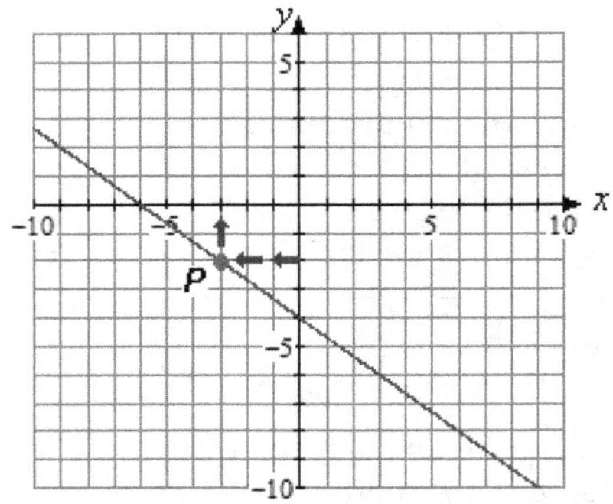

RQ7. What is the first step in graphing an equation?
RQ8. Which x-values would you choose to graph $y = \frac{5}{4}x - 3$?

Graphing Tip #2

When you graph a linear equation, you should extend the line far enough in both directions so that it will cross both the x-axis and the y-axis. The points where the graph crosses the axes are important for applications.

Look Closer: The graph of an equation is a picture of the solutions of the equation. Each point on the graph represents a solution. If a point lies on the graph, its coordinates make the equation true.

For example, the point $(-3, -2)$ lies on the graph in Example 4. When we substitute $x = -3$ and $y = -2$ into the equation $y = \frac{-2}{3}x - 4$, we get a true statement:

$$y = \frac{-2}{3}x - 4 \qquad \text{Substitute } x = -3 \text{ and } y = -2.$$
$$-2 = \frac{-2}{3}(-3) - 4$$
$$-2 = 2 - 4 \qquad \text{True.}$$

Skills Warm-Up

Choose the correct algebraic expression for each situation.

$2t + 12$	$12 - 2t$
$2(t + 12)$	$2t - 12$

1. Janine's history book has 12 chapters. If she studies 2 chapters a week, how many chapters will she have left after t weeks?
2. Arturo is 12 years older than twice the age of his nephew. If Arturo's nephew is t years old, how old is Arturo?
3. Rick made 12 fewer than twice as many phone calls as his roommate made this month. If Rick's roommate made t phone calls, how many calls did Rick make?
4. Every winter, the Civic Society knits mittens for the children of the county orphanage. This year there are 12 more children than last year. If there were t children last year, how many mittens will they need this year?

Answers: 1. $12 - 2t$ **2.** $2t + 12$ **3.** $2t - 12$ **4.** $2(t + 12)$

Homework 2.3

Skills Practice

■ In Problems 1-2, the figure shown is the graph of an equation. Decide which of the given points are solutions of the equation.

1. **a.** $(2, 0)$ **b.** $(3, 2)$ **c.** $(-1, -2)$ **d.** $(-3, -6)$

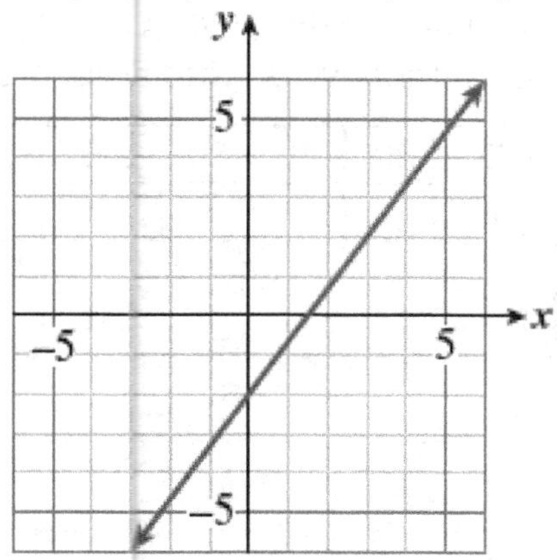

2. **a.** $(4, -2)$ **b.** $(4, 4)$ **c.** $(0, 4)$ **d.** $(-1, -4)$

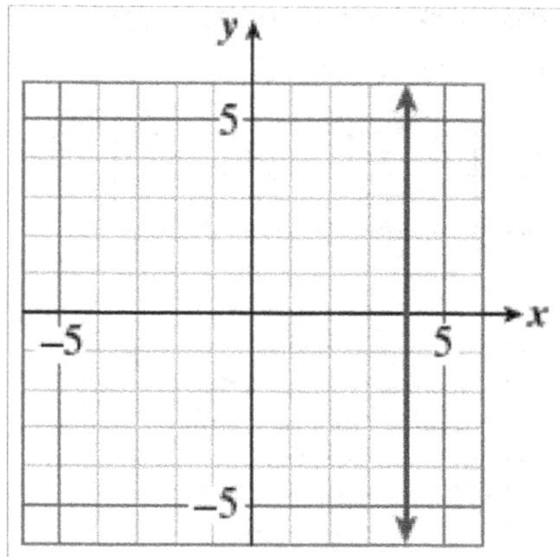

■ For Problems 3-5, give the coordinates of the labeled points.

3.

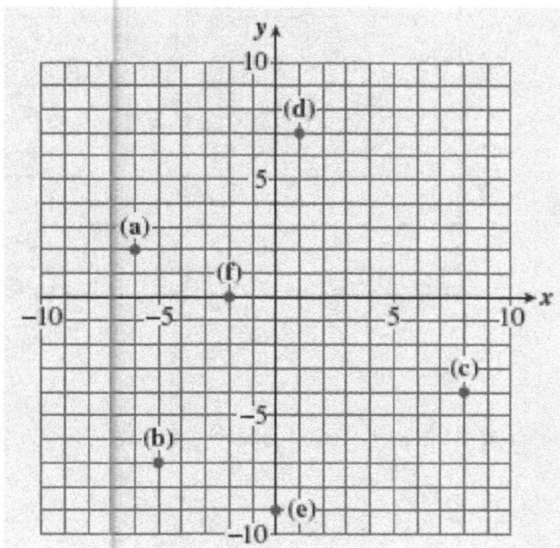

4.

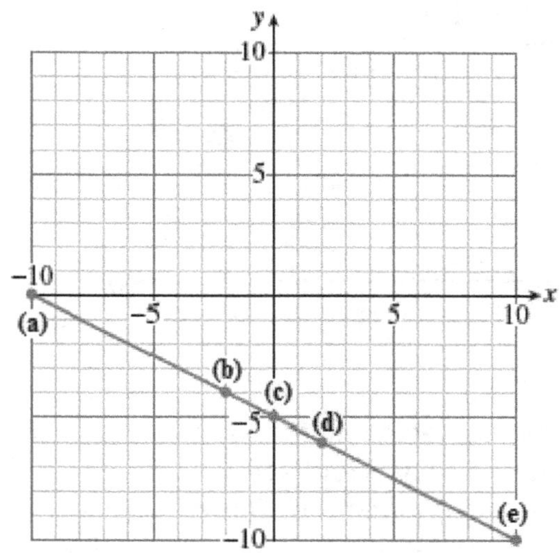

5.

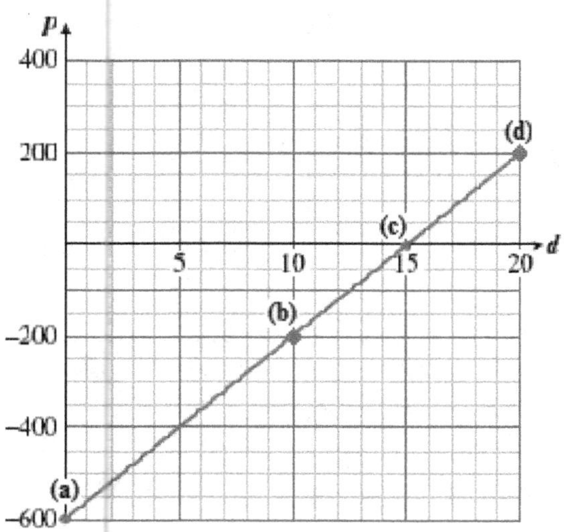

Applications

In Problems 6-7,

a. Write an equation relating the variables.
b. Complete the table of values.
c. Graph your equation on the grid.

6. Greta's math notebook has 100 pages, and she uses on average 6 pages per day for notes and homework. How many pages, P, will she have left after d days?

d	0	2	5	10	15
P					

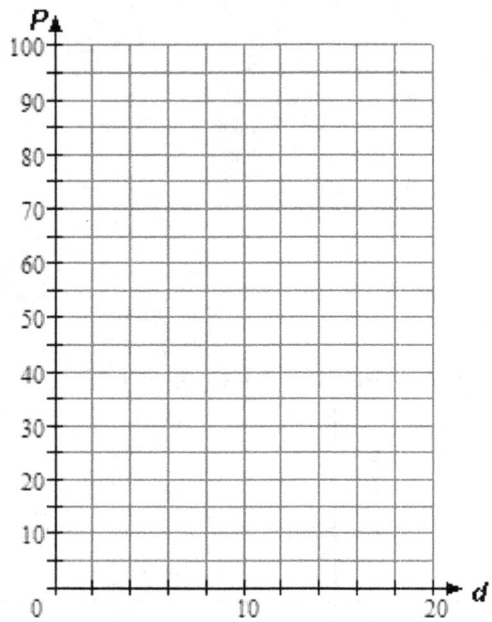

7. Delbert shares a house with four roommates. He pays $200 rent per month, plus his share of the utilities. If the utilities cost U dollars, how much money, M, does Delbert owe?

U	20	40	80	100	200
M					

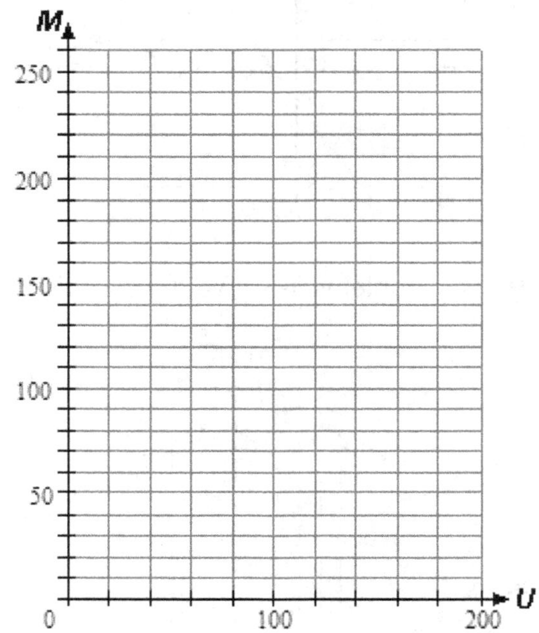

8. Darren inherited some money and has been spending it at the rate of $100 per day. Right now he has $2000 left.

a. Write an equation for Darren's balance, B, after d days.
b. Fill in the table. Negative values of d represent days in the past.

d	−15	−5	0	5	10	15	20
B							

c. Graph your equation, using the values in the table.
Use your graph to answer the following questions:
d. What was Darren's balance 15 days ago?
e. When will Darren's balance reach $500?

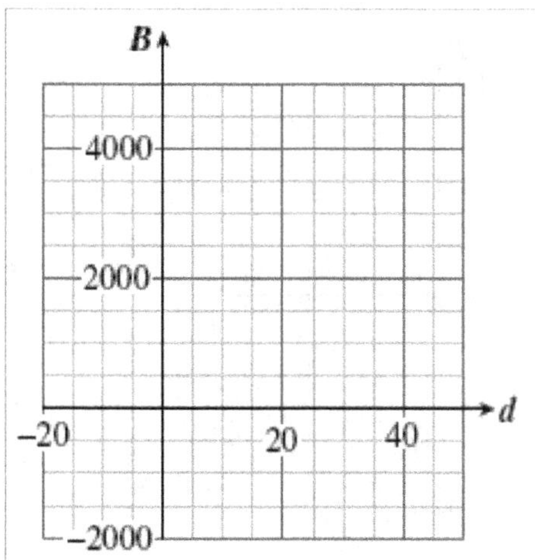

9. Gregory purchased stereo equipment on a monthly installment plan. After m months, Gregory still owes a balance of B dollars.

a. What was the price of the stereo equipment?

b. How much does Gregory pay each month?

c. How much does Gregory owe after 6 months?

d. How many monthly payments will Gregory make?

e. Write an algebraic equation for B in terms of m.

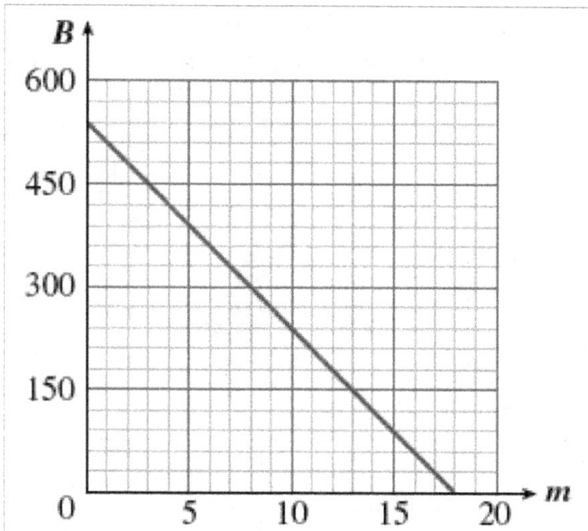

For part (c) of Problems 10-11, write an equation and use your graph to solve it.

10. On a 100-point test, Lori loses 5 points for each wrong answer.

a. Write an equation for Lori's score, s, if she gives x wrong answers.

b. Complete the table of values and graph your equation.

x	2	5	6	12
s				

c. If Lori's score is 65, how many wrong answers did she give?

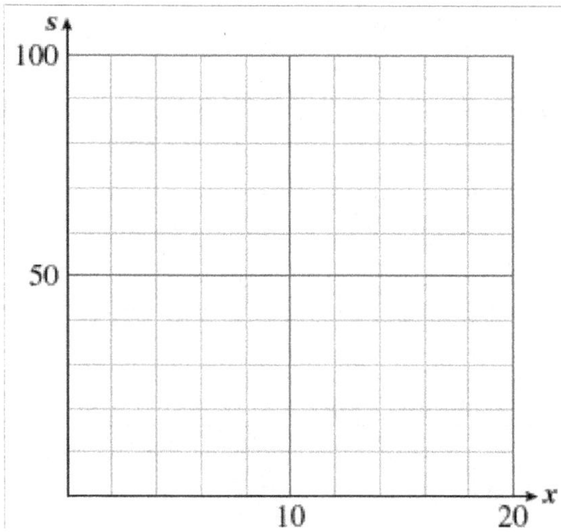

11. The water in Silver Pond is 10 feet deep, but the water level is dropping at a rate of $\frac{1}{2}$ inch per week.

a. Write an equation for the depth, d, of the pond after w weeks.

b. Complete the table of values and graph your equation.

w	24	36	84	120
d				

c. How long will it take until the depth of the pond is 8 feet? (**Hint:** Convert all the units to inches.)

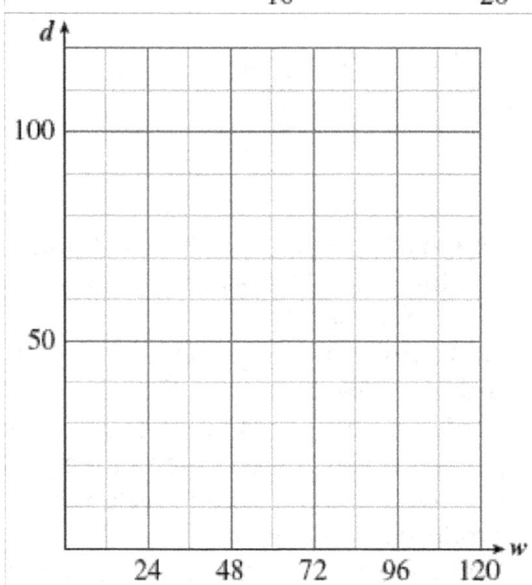

In Problems 12-15, choose four x-values and make a table of values, then graph the equation. Extend your line far enough that it crosses both axes.

12. $y = x + 3$

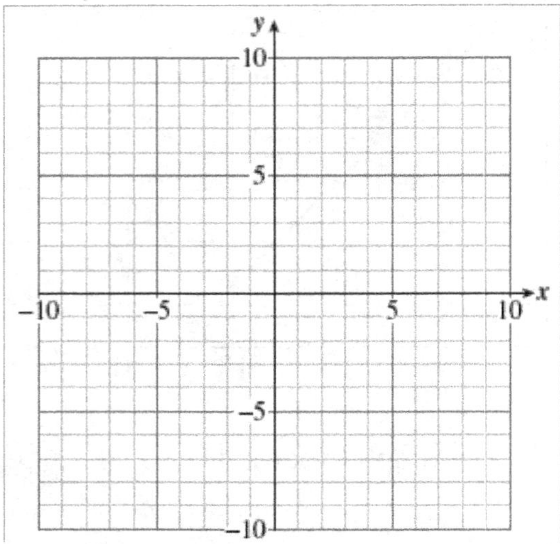

13. $y = 2x + 1$

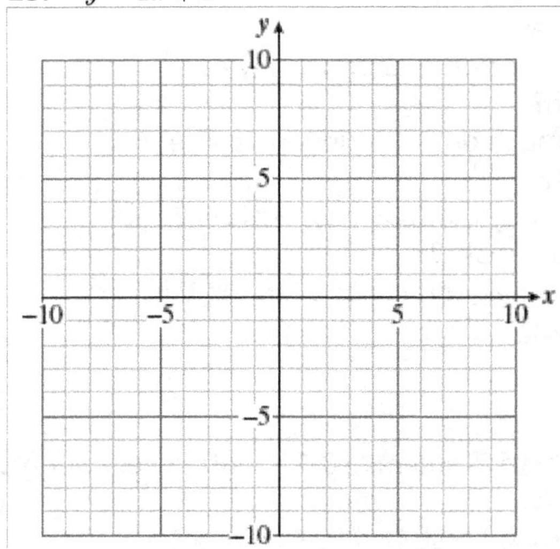

14. $y = -\frac{1}{2}x - 5$

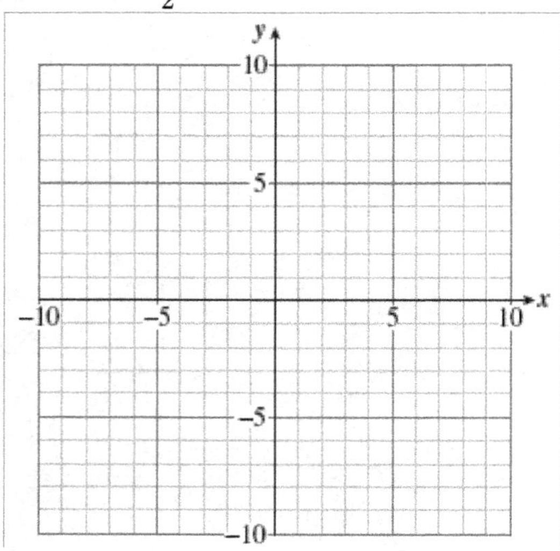

15. $y = \frac{5}{4}x - 4$

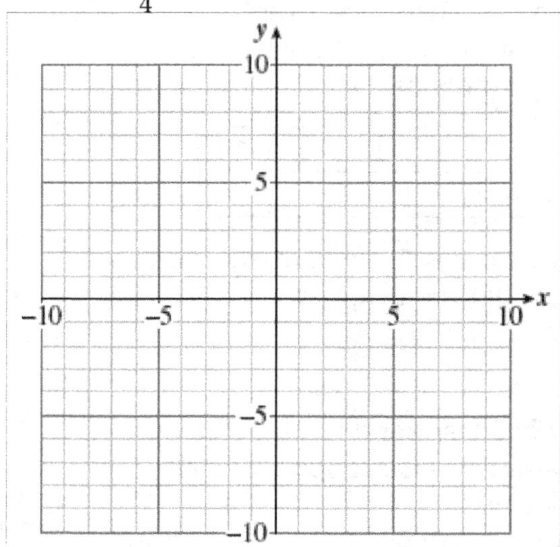

16. The graph of $y = 2x + 6$ is shown at right.

a. Use the graph to evaluate the expression $2x + 6$ for $x = -5$.

b. Find a point on the graph with $y = -4$. What is the x-value of the point?

c. Verify that the coordinates of your point in part (b) satisfy the equation of the graph.

d. Use the graph to find an x-value that produces a y-value of 8.

e. Find two points on the graph for which $y > -4$. What are their x-values?

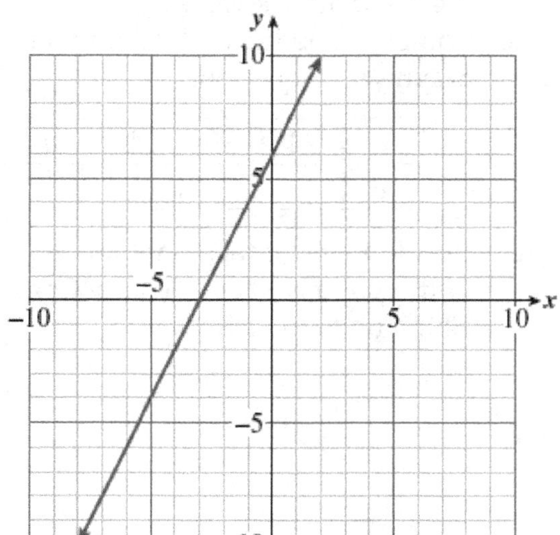

In Problems 17-18, use the graph to solve the equations. Show your solutions on the graph. Then check your solutions algebraically.

17.

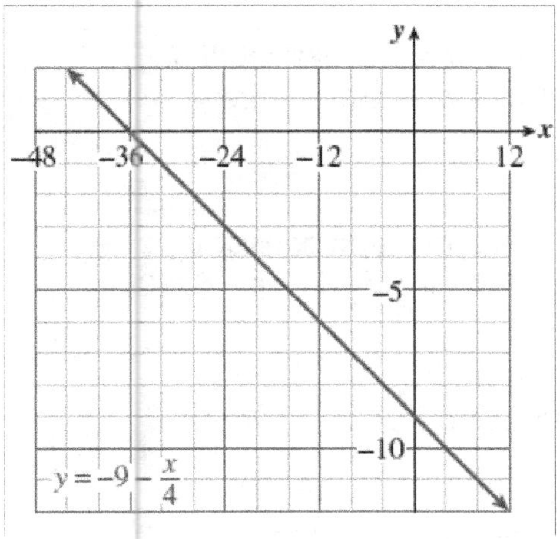

a. $-9-\dfrac{x}{4}=-5$

b. $-9-\dfrac{x}{4}=1$

18.

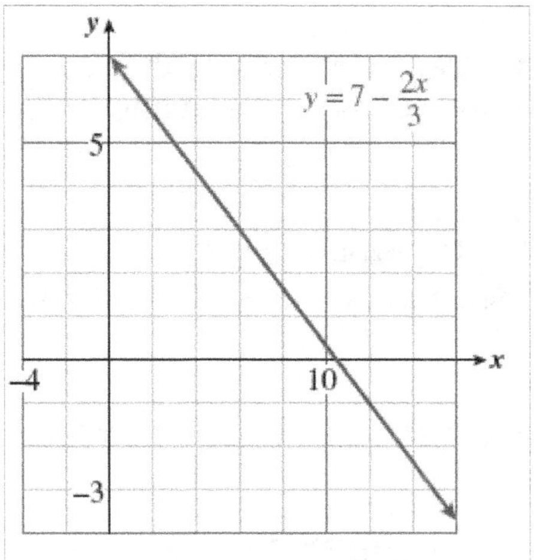

a. $7-\dfrac{2x}{3}=3$

b. $7-\dfrac{2x}{3}=-1$

In Problems 19-20, use the graph to estimate the solution of the equation. Then use a calculator to help you verify the solution algebraically. Do you think your estimate was too big or too small?

19. $37.21-8.4t=24.61$

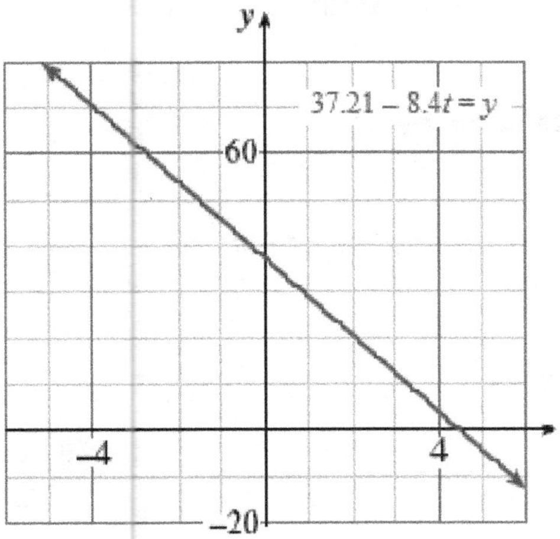

20. $-26.4=-3.65+9.1x$

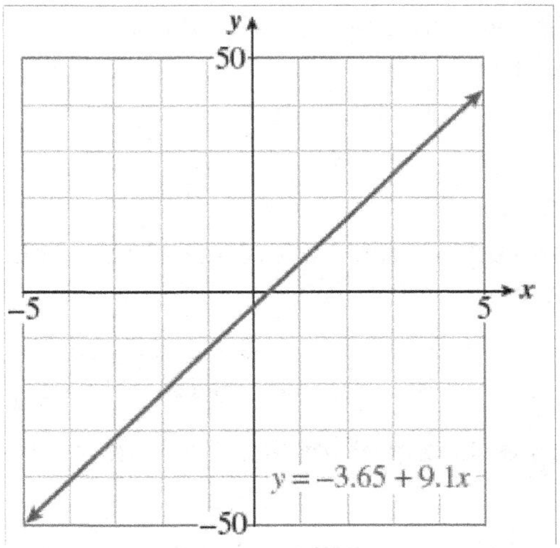

2.4 Linear Equations and Inequalities

Solving Equations

We solve an equation by isolating the variable on one side of the equation. If an equation involves two or more operations, we must undo those operations in reverse order.

Example 1 The expression $T = 50 + 15u$ describes college tuition consisting of a \$50 registration fee plus \$15 per unit. How many units can you take if you have \$290 saved for tuition?

Solution We can answer the question by solving the equation

$$\mathbf{290} = 50 + 15u$$

where we have substituted \$290 for the tuition, T. Think about the expression $50 + 15u$. How would you evaluate this expression if you were given a value for u? Following the order of operations, you would

1. first multiply by 15
2. then add 50

In order to solve the equation, we must *reverse these two steps* to undo the operations and isolate the variable. We first subtract 50 from both sides of the equation:

$$\begin{array}{rr} 290 = & 50 + 15u \\ \underline{-\mathbf{50}} & \underline{-\mathbf{50}\quad\quad} \\ 240 = & 15u \end{array}$$ **Subtract 50 from both sides.**

This isolates the term that contains the variable, $15u$. Then we divide both sides of the equation by 15.

$$\frac{240}{\mathbf{15}} = \frac{15u}{\mathbf{15}}$$ **Divide both sides by 15.**

$$16 = u$$

You can enroll in 16 units. We can check the solution by substituting 16 for u in the original equation.

Check:

$$50 + 15u = 290$$

$$50 + 15(\mathbf{16}) = 50 + 240 = 290$$ **True.**

Because a true statement results, the solution checks.

Look Closer: In Example 1, notice how we reversed the operations used in the equation.

Operations performed on u	*Steps for solution*
1. Multiply by 15	1. Subtract 50
2. Add 50	2. Divide by 15

Strategy for solving equations.

To solve an equation that involves two or more operations, we undo those operations *in reverse order.*

Example 2 Solve $8 - 3x = -10$

Solution The left side of the equation has two terms: 8 and $-3x$. We want to isolate the term containing the variable, so we subtract 8 from both sides.

$$\begin{array}{rcl} 8 - 3x &=& -10 \\ \underline{-8\quad} & & \underline{-8} \\ -3x &=& -18 \end{array}$$ **Subtract 8 from both sides.**

Next, we divide both sides by -3 to get

$$\frac{-3x}{-3} = \frac{-18}{-3}$$ **Divide both sides by -3.**

$$x = 6$$

Thus, the solution is 6.

Check: $8 - 3(6) = 8 - 18 = -10$

Caution! In Example 1, the term $-3x$ means "-3 times x," so we **divide** both sides by -3. Do not try to add 3 to both sides.

RQ1. If an equation involves more than one operation, how must we undo those operations?

RQ2. Delbert says he solved the equation $-5x = 15$ by adding 5 to both sides. What is wrong with his method?

Applied Problems

Problem solving often involves signed numbers, either in the equation that models the problem or in its solution, or both.

Example 3 The trout population in Clear Lake is decreasing by approximately 60 fish per year, and this year there are about 430 trout in the lake. If the population drops to 100, the Park Service will have to restock the lake.

a. Write an equation for the population P of trout x years from now.

b. When will the Park Service have to restock the lake?

c. Graph your equation for P, and illustrate your answer to part (b) on the graph.

Solutions a. The population starts this year at 430, and decreases by 60 for each following year. Thus, $P = 430 - 60x$.

b. We would like to find the value of x when $P = 100$. We substitute 100 for P, and solve the equation for x.

$$\begin{array}{rcl} 100 &=& 430 - 60x \\ \underline{-430} & & \underline{-430\quad} \\ -330 &=& -60x \end{array}$$ **Subtract 430 from both sides.**

$$\frac{-330}{-60} = \frac{-60x}{-60}$$ **Divide both sides by -60.**

$$5.5 = x$$

The Park Service will have to restock the lake in five and a half years, if the population continues to decline at the current rate.

c. The figure shows the graph of the equation $P = 430 - 60x$. To solve

$$100 = 430 - 60x$$

we locate the point on the graph with $P = 100$, and read its x-coordinate, at about 5.5.

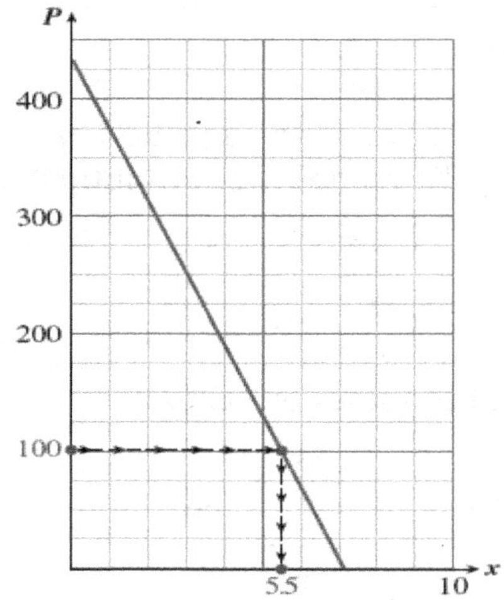

Solving Inequalities

> A statement that uses one of the symbols $<$ or $>$ is called an **inequality**.

Examples of inequalities are

$$-1 > -3 \quad \text{and} \quad x < 2$$

Unlike the equations we have studied, which have at most one solution, an inequality can have infinitely many solutions. A solution of an equation or inequality is said to **satisfy** the equation or inequality.

Example 4 Solve the inequality $x < 2$.

Solution The solutions include 1, 0, −1, −2, and all the other negative integers, as well as fractions less than 2, such as $1\frac{3}{5}$, $\frac{2}{3}$, and $-\frac{17}{8}$. All of these values satisfy the inequality $x < 2$. In fact, *all* the numbers to the left of 2 on the number line are solutions of $x < 2$. Because we cannot list all these solutions, we often graph them on a number line as shown below.

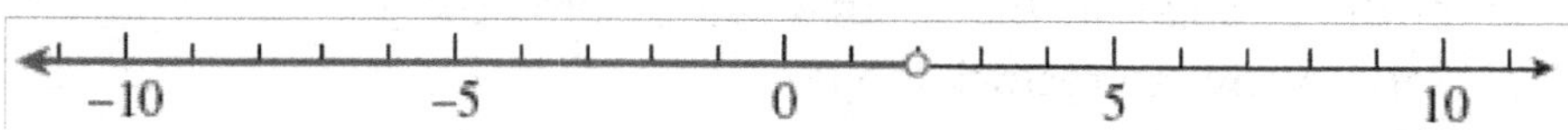

The open circle at 2 shows that 2 itself is not a solution (because 2 is not less than 2.)

Look Closer: An inequality that uses the symbol for less than, $<$, or greater than, $>$, is called a **strict inequality**. A **nonstrict inequality** uses one of the following symbols.

> $\geq$ means **greater than or equal to**
> $\leq$ means **less than or equal to**

For example, the graph of all solutions to the inequality

$$x \geq -2$$

is shown on the next page. We use a solid dot at −2 to show that $x = -2$ is included in the solutions.

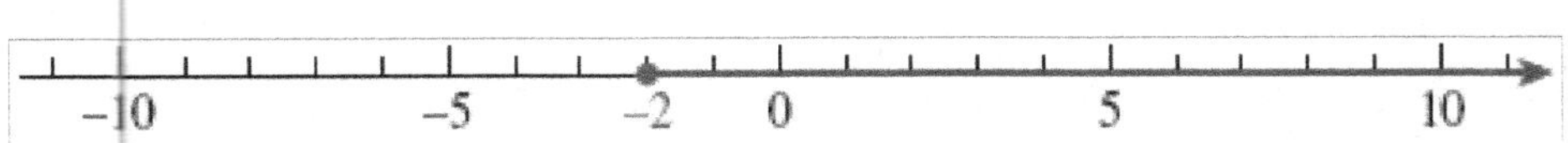

RQ3. A solution of an inequality is said to __________ the inequality.
RQ4. What is the difference between a strict and a nonstrict inequality?

The rules for solving inequalities are very similar to the rules for solving equations, with one important difference. In the Activities we will develop the following strategies for solving inequalities.

To Solve an Inequality:

1. We can add or subtract the same quantity on both sides.
2. We can multiply or divide both sides by the same positive number.
3. If we multiply or divide both sides by a **negative** number, we must **reverse** the direction of the inequality.

Example 5 Solve $-3x+1>7$, and graph the solutions on a number line.

Solution We isolate x on one side of the inequality.

$$-3x+1-1>7-1 \qquad \text{Subtract 1 from both sides.}$$
$$-3x>6$$
$$\frac{-3x}{-3}<\frac{6}{-3} \qquad \text{Divide both sides by } -3\text{; reverse the direction of the inequality.}$$
$$x<-2$$

The graph of the solutions is shown below.

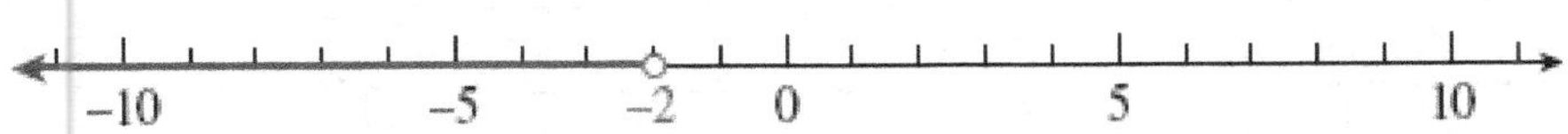

Compound Inequalities

An inequality in which the variable expression is bounded from above and from below is called a **compound inequality**. For example,

$$-3<2x-5\leq 6$$

is a compound inequality. To solve a compound inequality, we must perform the steps needed to isolate x on all three sides of the inequality.

Example 6 Solve $-3<2x-5\leq 6$

Solution To solve for x, we first add 5 on each side of the inequality symbols.

$$\begin{array}{rcccl} -3 & < & 2x-5 & \leq & 6 \\ +5 & & +5 & & +5 \\ \hline 2 & < & 2x & \leq & 11 \end{array}$$

Next, to solve $2<2x\leq 11$, we divide each side by 2.

$$\frac{2}{2} < \frac{2x}{2} \leq \frac{11}{2}$$
$$1 < x \leq \frac{11}{2}$$

The solution consists of all numbers greater than 1 but less than or equal to $\frac{11}{2}$. The graph of the solutions is shown below.

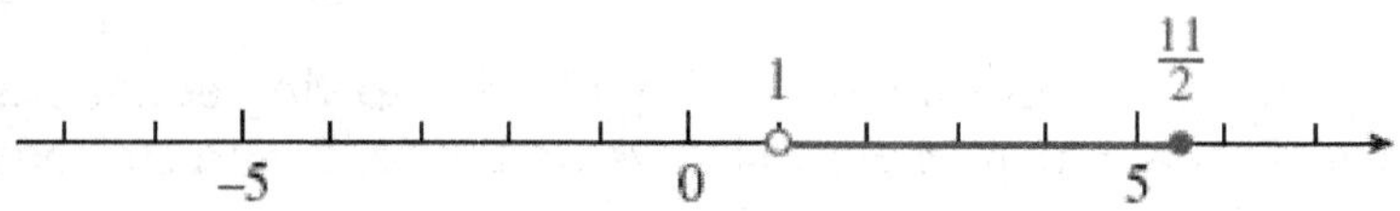

RQ5. When do we need to reverse the direction of an inequality?
RQ6. What is a compound inequality?

Skills Warm-Up

Solve each equation. Try to do so mentally (without using pencil and paper.)

1. $\frac{u}{3} = 6$ **2.** $7 = 3 + s$ **3.** $a - \frac{1}{3} = \frac{2}{3}$

4. $20 = 5m$ **5.** $\frac{1}{4}p = 8$ **6.** $7t = 5$

Fill in the tables. Then analyze the order of operations in your calculations:

Table 1

n	$3n$	$3n - 5$
2		
5		
		7
		22

Table 2

m	$\frac{m}{4}$	$\frac{m}{4} + 1$
8		
12		
		6
		2

7. Consider the equation $3n - 5 = p$. Look at Table 1 to help you answer the questions:
  a. Let $n = 2$. Explain how to find p in two steps.
  b. Let $p = 7$. Explain how to find n in two steps.

8. Consider the equation $\frac{m}{4} + 1 = h$. Look at Table 2 to help you answer the questions:
  a. Let $m = 8$. Explain how to find h in two steps.
  b. Let $h = 6$. Explain how to find m in two steps.

9. a. If you put on socks and then put on shoes, what operations are needed to reverse the process?
  b. You leave home and bicycle north for 3 miles and then east for 2 miles. Suddenly you notice that you have dropped your wallet. How should you retrace your steps?

Answers: **1.** 18 **2.** 4 **3.** 1 **4.** 4 **5.** 32 **6.** $\frac{5}{7}$

Homework 2.4

Skills Practice

Fill in the table in Problems 1-2. Use two steps for each row: Fill in the middle column first. Then use the table to help you answer the questions.

1.

x	$2x$	$2x+4$
3		
6		
		14
		20

Consider the equation $y = 2x + 4$.

a. Let $x = 3$. Explain how to find y in two steps.

b. Let $y = 14$. Explain how to find x in two steps.

2.

q	$q-3$	$5(q-3)$
3		
		10
4		
		20

Consider the equation $5(q-3) = R$.

a. Let $q = 3$. Explain how to find R in two steps.

b. Let $R = 10$. Explain how to find q in two steps.

For Problems 3-19, solve.

3. $6x - 13 = 5$

4. $\dfrac{2a}{5} = 8$

5. $\dfrac{x}{4} + 2 = 3$

6. $6x + 5 = 5$

7. $24 = 4(p - 7)$

8. $0 = \dfrac{5z}{7}$

9. $\dfrac{k+4}{5} = 9$

10. $\dfrac{2x}{3} - 5 = 7$

11. $7 = \dfrac{4b-3}{3}$

12. $3c - 7 = -13$

13. $-5 = -2 - 3t$

14. $-3(2h - 6) = -12$

15. $1 - \dfrac{b}{3} = -5$

16. $\dfrac{3y}{5} + 2 = -4$

17. $\dfrac{5x}{2} + 10 = 0$

18. $11.8w - 37.8 = 120.32$

19. $9.7(2.6 + v) = 58.2$

In Problems 20-28,

a. Solve each inequality algebraically.

b. Graph your solutions on a number line.

c. Give at least one value of the variable that is a solution, and one value that is not a solution.

20. $x + 10 \le -5$

21. $-3y < 15$

22. $\dfrac{x}{3} \le 4$

23. $2x + 3 > 7$

24. $-3x + 2 \le 11$

25. $-3 > \dfrac{2x}{3} + 1$

26. $-3 \le 3z \le 12$

27. $23 > 9 - 2b \ge 13$

28. $-8 \le \dfrac{5w+3}{4} < -3$

Applications

29. Delbert is thinking of a number. If he multiplies the number by 5 and then subtracts 6, the result is 29. What is the number?

■ In Problems 30-31, find the error in the "solution," then write a corrected solution.

30.
$$6 - 3x = -12$$
$$-3x = -6$$
$$x = 2$$

31.
$$-2 + \frac{2}{3}x = -4$$
$$-2 + 2x = -12$$
$$2x = -10$$
$$x = -5$$

■ In Problems 32-33, use the graphs to answer the questions.

32. Francine's new puppy weighed 3.8 pounds when she brought it home, and it should gain approximately 0.6 pound per week. The graph shows the puppy's weight, W, after t weeks.

a. Estimate the puppy's weight after 8 weeks.

b. Estimate how long it will take for the puppy to reach 26 pounds.

c. Write an algebraic equation for W in terms of t.

d. Write and solve an equation to verify your answer to part (b).

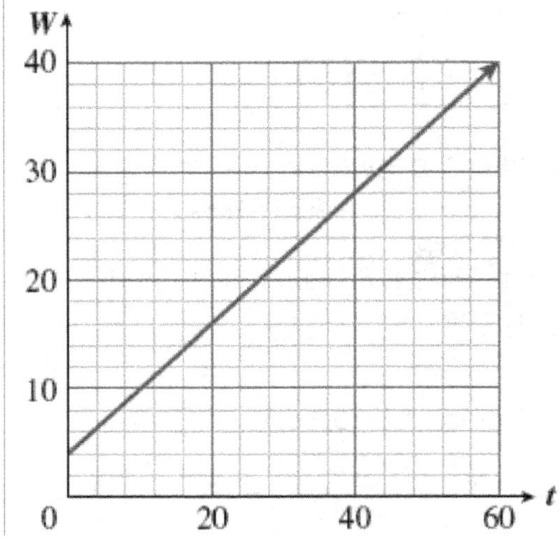

33. Marvin's telephone company charges \$13 a month plus \$0.15 per minute. The graph shows Marvin's phone bill, P, if he talks for m minutes.

a. Estimate Marvin's phone bill if he talks for 45 minutes.

b. Estimate the number of minutes Marvin talked if his phone bill is \$25.

c. Write an algebraic equation for P in terms of m.

d. Write and solve an equation to verify your answer to part (b).

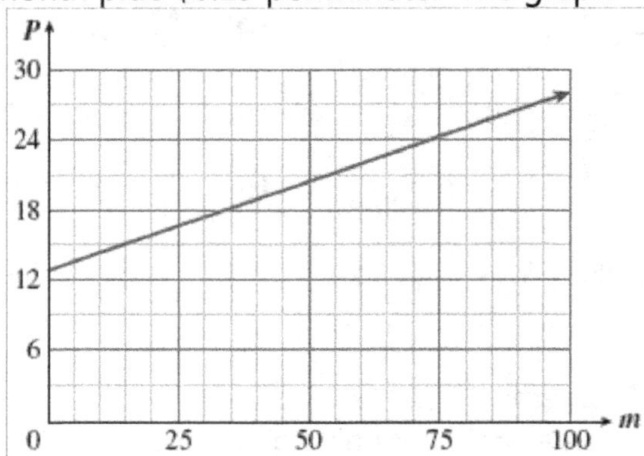

■ In Problems 34-38, solve in three steps:

a. Choose a variable for the unknown quantity.

b. Write an equation that involves your variable.

c. Solve your equation, and answer the question.

34. Simona bought a car for \$10,200. She paid \$1200 down and will pay the rest in 36 monthly installments. How much is each installment?
(**Hint:** Write an equation about the total amount Simona pays.)

35. Every school day, each member of the cross-country team runs a certain distance d, which is assigned on the basis of ability, plus a half mile around the track. Greta runs 22.5 miles per week. What is her assigned distance?
(**Hint:** Write an equation about the total distance Greta runs in one school week.)

36. The Tree People planted new trees in an area that was burned by brush fires. 60% of the seeds sprouted, but gophers ate 38 of the new sprouts. That left 112 new saplings. How many seeds were planted?
(**Hint:** Write an equation about the number of seeds that survived.)

37. During a four-day warming trend, the temperature rose from $-6°$ to $26°$. What was the average change in temperature per day?

38. Eric is on a diet to reduce his current weight of 196 pounds to 162 pounds. If he loses 4 pounds per week, how long will it take him to reach his desired weight?

■ Use one of the formulas below to solve Problems 39-40.

Perimeter of a rectangle $P = 2l + 2w$

Area of a triangle $A = \frac{bh}{2}$

39. A farmer has 500 yards of fencing material to enclose a rectangular pasture. He would like the pasture to be 75 yards wide. How long will it be?

40. A triangular sail requires 12 square meters of fabric. If the base of the sail measures 4 meters, how tall is the sail?

■ In Problems 41-42, write and solve an equation to find the value of x.

41.

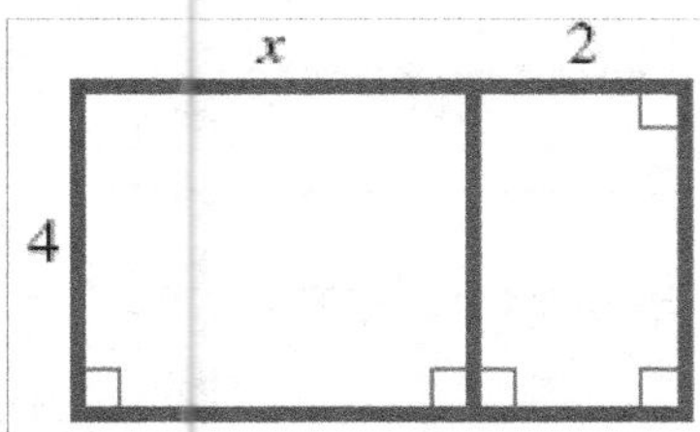

Area $= 36$

42.

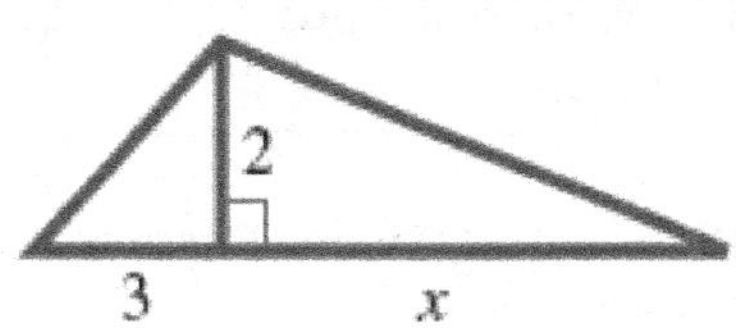

Area $= 20$

2.5 Like Terms

Equivalent Expressions

When we solve an equation such as

$$2 + 3x = 17$$

we find a value for the variable that makes the equation true. The solution to this equation is 5, because the two expressions $2 + 3x$ and 17 are equal if $x = 5$. If we use any other value for x, say $x = 9$ or $x = -1$, then $2 + 3x$ does not equal 17.

It is possible for two algebraic expressions to be equal for **all** values of the variable. For instance, the expression $2x + 3x$ and $5x$ will always give the same value when the same number is substituted for x in each expression. For example,

if $x = 3$, then $2x + 3x = 2(3) + 3(3) = 6 + 9 = 15$

and $5x = 5(3) = 15$

Both expressions equal 15 when $x = 3$.

Exercise Verify that $2x + 3x$ and $5x$ are equal when $x = -4$.

$2x + 3x =$ ____________

and $5x =$ ____________

Two algebraic expressions are **equivalent** if they name the same number for all values of the variable.

Thus, $2x + 3x$ and $5x$ are equivalent expressions, and we can replace one by the other if it will simplify our work.

Caution! Although two expressions may name the same number for *some* values of the variable, they are not necessarily equivalent. For example, to show that the expressions $2 + 3x$ and $5x$ are *not* equivalent, we only have to find a single value of x for which they are not equal. If we evaluate each expression for $x = 6$, we find:

$$2 + 3x = 2 + 3(6) = 20$$
$$5x = 5(6) = 30$$

Because they are not equal for $x = 6$, the expressions $2 + 3x$ and $5x$ are not equivalent. ■

RQ1. When are two expressions equivalent?

Like Terms

We saw above that $2x + 3x$ and $5x$ are equivalent expressions. This makes sense when we consider what these expressions mean: $2x$ means two x's added together, and $3x$ means three x's added together. Then

$$2x + 3x = (x + x) + (x + x + x) = 5x$$

We can add two x's to three x's and get five x's, just as we can add two pencils to three pencils, or two dollars to three dollars. In algebra this is called **combining like terms**.

Like terms are any terms that are exactly alike in their variable factors.

Example 1a. $2x$ and $3x$ are like terms
$-4a$ and $7a$ are like terms

because their variable factors are identical.

b.

$2x$ and $3y$ are not like terms
$2x$ and 3 are not like terms

because their variable factors are different.

The numerical factor in a term is called the **numerical coefficient**, or just the **coefficient** of the term.

Example 2a. In the expression $3xy$, the number 3 is the numerical coefficient.

b. In a term such as xy or b, the numerical coefficient is 1.

c. If a variable is preceded by a negative sign, the coefficient of the term is -1. For example,

$-x$ means $-1 \cdot x$
$-a$ means $-1 \cdot a$

Adding and Subtracting Like Terms

To add like terms:

Add the numerical coefficients of the terms. Do not change the variable factors of the terms.

Example 3 Add like terms.

a. $5n + 3n$ **b.** $-4y - 3y$ **c.** $-6st + 9st$

Solutions In each calculation, we do not change the variables we are adding; only the coefficient changes.

a. $5n + 3n = (5 + 3)n = 8n$ **Add the coefficients, 5 and 3.**

b. $-4y - 3y = (-4 - 3)y = -7y$ **Add the coefficients, -4 and -3.**

c. $-6st + 9st = (-6 + 9)st = 3st$ **Add the coefficients, -6 and 9.**

Look Ahead: In Example 3, we replaced each expression by a simpler but equivalent expression. In other words, $-7y$ is equivalent to $-4y - 3y$, and $3st$ is equivalent to $-6st + 9st$.

Replacing an expression by a simpler equivalent one is called **simplifying** the expression.

We subtract like terms in the same way that we add like terms.

To subtract like terms:
Subtract the numerical coefficients of the terms. Do not change the variable factors of the terms.

■ **Example 4** $6a - (-8a) = [6 - (-8)]a = 14a$ **Subtract the coefficients.** ■

RQ2. What are like terms?
RQ3. How do we add or subtract like terms?

Caution! We cannot add or subtract unlike terms. Thus,

$2x + 3y$ cannot be simplified
$-6x + 4xy$ cannot be simplified
$6 - 2x$ cannot be simplified ■

Many expressions contain both like and unlike terms. In such expressions we can combine only the like terms.

■ **Example 5** Simplify $-4a + 2 - 5 + 8a + 5a$

Solution We combine all the a-terms, and all the constant terms, like this:

$$-4a + 8a + 5a = 9a \quad \text{and} \quad 2 - 5 = -3$$

so

$$-4a + 2 - 5 + 8a + 5a = 9a - 3$$ ■

Caution! In Example 5, $9a - 3 \neq 6a$, because $9a$ and -3 are not like terms. ■

Removing Parentheses

When we add signed numbers, we can remove parentheses that follow an addition symbol. For instance,

$$5 + (-3) = 5 - 3 = 2$$

We can also remove parentheses when we add like terms. Thus,

$$5x + (-3x) = 5x - 3x = 2x$$

Removing parentheses that follow a plus sign:
Parentheses preceded by a plus sign may be omitted, and each term within parentheses keeps its original sign.

Example 6 Simplify $(2x+3)+(5x-7)$

Solution Each set of parentheses is preceded by a plus sign, so we may remove the parentheses.

$$\begin{aligned}(2x+3)+(5x-7) &= 2x+3+5x-7 && \textbf{Combine like terms.}\\ &= 7x-4\end{aligned}$$

We must be careful when removing parentheses preceded by a **minus** sign. In the expression

$$(5x-2)-(7x-4)$$

the minus sign preceding $(7x-4)$ applies to *each term* inside the parentheses. Therefore, when we remove the parentheses we must change the sign of $7x$ from $+$ to $-$ *and* we must change the sign of –4 from $-$ to $+$. Thus,

$$\begin{aligned}(5x-2)-(7x-4) &= 5x-2-7x+4 && \textbf{Change each sign inside parentheses.}\\ &= 5x-7x-2+4 && \textbf{Combine like terms.}\\ &= -2x+2\end{aligned}$$

Removing parentheses that follow a minus sign:
If an expression inside parentheses is preceded by a minus sign, we **change the sign of each term within parentheses** and then omit the parentheses.

Example 7 Simplify $-5b-(3-2b)+(4b-6)$

Solution Before combining like terms, we must remove the parentheses. We change the signs of any terms inside parentheses preceded by a minus sign; we do *not* change the signs of terms inside parentheses preceded by a plus sign.

$$\begin{aligned}-5b-(3-2b)+(4b-6) &= -5b-3+2b+4b-6\\ &= -5b+2b+4b-3-6\\ &= b-9\end{aligned}$$

(signs changed: $-3+2b$; signs not changed: $+4b-6$)

RQ4. We can remove parentheses that follow ________________.
RQ5. What must you do if you remove parentheses that follow a minus sign?

Solving Equations and Inequalities

We can now solve equations in which two or more terms contain the variable.

Example 8 Solve the equation $2x+7=4x-3$

Solution We first get all terms containing the variable on one side of the equation. For this example, we will subtract $2x$ from both sides of the equation
to get

$$\begin{array}{rrl} 2x+7= & 4x-3 & \quad \textbf{Subtract } 2x \textbf{ from both sides.} \\ \underline{-\mathbf{2x}\quad} & \underline{-\mathbf{2x}\quad} & \\ 7= & 2x-3 & \end{array}$$

Now the equation looks like ones we already know how to solve, and we proceed as usual to isolate the variable.

$$\begin{array}{rrl} 7= & 2x-3 & \quad \textbf{Add 3 to both sides.} \\ \underline{+\mathbf{3}} & \underline{\quad +\mathbf{3}} & \\ 10= & 2x & \quad \textbf{Divide both sides by 2.} \\ \frac{10}{\mathbf{2}}= & \frac{2x}{\mathbf{2}} & \\ 5= & x & \end{array}$$

The solution is 5.

Check: Does $2(\mathbf{5})+7=4(\mathbf{5})-3$? **True.** ■

If one side of an equation or inequality contains like terms, we should combine them before beginning to solve.

■ **Example 9** Solve $2x-4x+14<3+5x-10$

Solution Combine like terms on each side of the inequality, to get

$$\begin{array}{ll} \mathbf{2x}-\mathbf{4x}+14<\mathbf{3}+5x-\mathbf{10} & \quad \mathbf{2x-4x=-2x} \textbf{ and } \mathbf{3-10=-7} \\ -2x+14<5x-7 & \end{array}$$

Now continue solving as usual: We want to get all the terms containing x on one side of the inequality, and all the constant terms on the other side.

$$\begin{array}{rrl} -2x+14< & 5x-7 & \quad \textbf{Subtract } 5x \textbf{ from both sides.} \\ \underline{-\mathbf{5x}\qquad} & \underline{-\mathbf{5x}\quad} & \\ -7x+14< & -7 & \\ \underline{-\mathbf{14}} & \underline{-\mathbf{14}} & \quad \textbf{Subtract 14 from both sides.} \\ -7x< & -21 & \\ \frac{-7x}{\mathbf{-7}}> & \frac{-21}{\mathbf{-7}} & \quad \textbf{Divide both sides by } -7\textbf{;} \\ & & \quad \textbf{reverse the direction of the inequality.} \\ x>3 & & \end{array}$$

The solution is all x-values greater than 3. You can check the solution by substituting one x-value greater than 3 and one value less than 3 into the original inequality, for instance:

Check: $x=4:$ $2(\mathbf{4})-4(\mathbf{4})+14<3+5(\mathbf{4})-10$? **True:** $6<13$

$x=2:$ $2(\mathbf{2})-4(\mathbf{2})+14<3+5(\mathbf{2})-10$? **False:** $10 \not< 3$ ■

RQ6. What should you do if one side of an equation or inequality contains like terms?

Here are our guidelines for solving linear equations.

Steps for Solving Linear Equations

1. Combine like terms on each side of the equation.
2. By adding or subtracting the same quantity on both sides of the equation, get all the variable terms on one side and all the constant terms on the other.
3. Divide both sides by the coefficient of the variable to obtain an equation of the form $x = a$.

Skills Warm-Up

Write algebraic expressions.

1. Getaway Tours offers a Caribbean cruise for \$2000 per person if 12 people sign up. For each additional person who signs up, the price per person is reduced by \$60. How much will you pay for a cruise if p additional people sign up?
2. Anita figures her taxes by subtracting a \$1200 deductible from her income, then taking 6% of the result. What is her tax bill if her income is I?
3. The area of a trapezoid is one-half the product of its height h and the sum of the two bases, a and b.
4. The ratio of a number k to 5 less than k.

Evaluate.

5. $3n(n-k)(n+k)$ for $n=-4$ and $k=-6$
6. $(t-1)(2t+3)$ for $t=\dfrac{2}{3}$

Answers: **1.** $2000-60p$ **2.** $0.06(I-1200)$ **3.** $\dfrac{1}{2}h(a+b)$ **4.** $\dfrac{k}{k-5}$ **5.** 240 **6.** $\dfrac{-13}{9}$

Homework 2.5

Skills Practice

■ In Problems 1-4, add or subtract like terms.

1. $-6x+2x$
2. $-7.6a-5.2a$
3. $3t-4t+2t$
4. $3bc-(-4bc)-8bc$

■ In Problems 5-6, combine like terms.

5. $3+4y-(-8y)-7$
6. $-2st-2+5s-6st-(-4s)$

■ In Problems 7-10, simplify and combine like terms.

7. $4x+(5x-2)$
8. $22y-34-(16y-24)$
9. $6a-5-2a-(2a-5)$
10. $(-2-3b)-(2b-4)$

■ In Problems 11-16, solve each equation or inequality.

11. $4m-3=2m+5$
12. $15-9t=33-5t$
13. $-6s=3s$
14. $3x+5>2x+3$
15. $-8g+35=9g-13+g$
16. $-15y+5-2y-4\geq -12y+21$
17. $-2-3w+5w\leq 4w-44$
18. $-5p-4+3p=2+7p+3$

19. a. Evaluate the expression $-3y+2+7y-6y-4y-8$ for $y=2.5$.
    b. Simplify the expression in part (a).
    c. Evaluate your answer to part (b) for $y=2.5$. Check that you got the same answer for part (a).

20. Evaluate the expression for the given values of the variables.
    $4-3a+6ab-(-8a)-10+9ab+2a$ for $a=-2,\ b=5$

Applications

■ In Problems 21-22, choose a value for the variable and show that the following pairs of expressions are **not** equivalent.

21. $2+7x$ and $9x$
22. $-(a-3)$ and $-a-3$

■ Write and simplify an algebraic expression for the perimeter of the figure in Problems 23-24.

23.

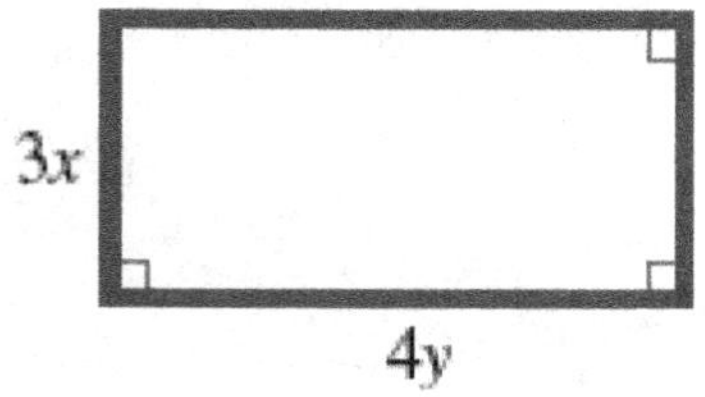

24.

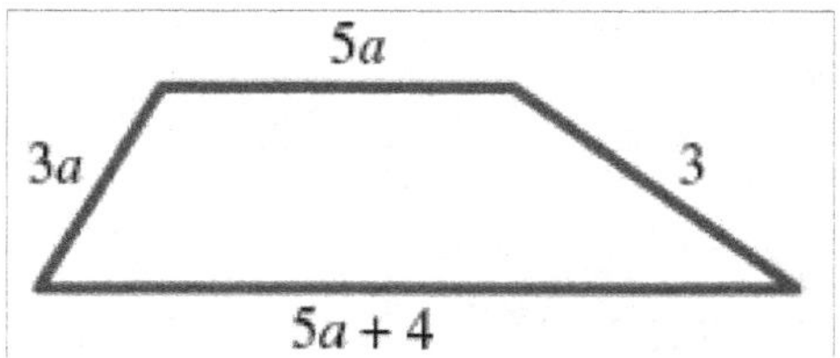

For Problems 25-29, write an algebraic expression or an equation to answer each question.

25. There are 8 more cats than dogs at the county shelter. Write expressions in terms of the number of dogs, d.

a. The number of cats

b. The total number of cats and dogs

c. If there are 40 dogs and cats at the shelter, how many dogs are there?

26. For every smoker in the restaurant there are three nonsmokers. Write expressions in terms of x, the number of smokers.

a. The number of nonsmokers

b. The total number of people in the restaurant

c. If there are 36 people in the restaurant, how many are smokers?

27. A tortoise and a hare are having a race. After t seconds, the tortoise has traveled $0.2t$ feet and the hare has traveled $10t$ feet.

a. Express the distance between them in terms of t.

b. How far apart are the tortoise and the hare after 10 seconds?

c. When will they be 147 feet apart?

28. Delbert bought a high definition TV at a price of x dollars. The sales tax in his state is 8%.

a. Write an expression for the tax on Delbert's TV.

b. Write an expression for the price Delbert paid, including tax.

c. Delbert paid $928.80 for his TV. What was the price before tax?

29. The cost of producing m stereos is $251m + 1355$ dollars, and each stereo sells for $847.

a. What profit is made from producing and selling m stereos?

b. How many stereos must be produced to make a profit of $16,525?

For Problems 30-34, write and solve an equation.

30. Last summer Kevin made 8 batches of barbecue sauce using one bottle of catsup. This summer he made only 5 batches and had 12 ounces of catsup left over. How much catsup does Kevin use in a batch of barbecue sauce?

31. Arch has two wooden beams of equal length. He makes 3 posts for mailboxes with one beam, and has 33 inches left. He makes another mailbox post with the other beam, and has 131 inches left. How long is each mailbox post?

32. Quentin bought 12 cans of cat food and a half-gallon of milk that cost $1.80. Tilda bought 8 cans of cat food and a $5 bag of kibble. They paid the same price at the check-out stand. How much does one can of cat food cost?

33. If Sarah drinks three glasses of milk a day she will exceed her minimum daily requirement for calcium by 70 milligrams. If she drinks only two glasses of milk she will still need another 220 milligrams of calcium. How many milligrams of calcium does a glass of milk contain?

34. On her last history test, Parisa answered 9 short-answer questions correctly and earned 15 points on the essay. Seana answered 10 short-answer questions correctly and got 23 points on the essay. Seana's score on the test was 14 points higher than Parisa's. How many points was each short-answer question worth?

Chapter 2 Summary and Review

Section 2.1 Signed Numbers

- The natural numbers, zero, and the negatives of the natural numbers are called the **integers.**
- We use special symbols to indicate order:

$<$ means **is less than**
$>$ means **is greater than**

- The **absolute value** of a number is the unsigned part of the number. It is never negative.

Rules for Operations on Integers
To add two integers:
1. To add two numbers with the same sign, add their absolute values. The sum has the same sign as the numbers.
2. To add two numbers with opposite signs, subtract their absolute values. The sum has the same sign as the number with the larger absolute value.

To subtract b from a:
1. Change the sign of b.
2. Change the subtraction to addition.
3. Proceed as in addition.

To multiply or divide two integers:
1. The product or quotient of two numbers with opposite signs is a negative number.
2. The product or quotient of two numbers with the same sign is a positive number.

Quotients Involving Zero
If a is any nonzero number, then

$$\frac{0}{a} = 0 \qquad \text{and} \qquad \frac{a}{0} \textbf{ is undefined}$$

Section 2.2 Expressions and Equations

- When we evaluate an algebraic expression, we follow the order of operations.
- When we evaluate an algebraic expression at a negative number, we enclose the negative numbers in parentheses.

Section 2.3 Graphs of Linear Equations

- An equation of the form $y = ax + b$, where a and b are constants, is called a **linear equation** because its graph is a straight line.
- We can graph a linear equation by evaluating the expression $ax + b$ at several values of x and then plotting points.
- We use a **Cartesian coordinate system** to make a graph that includes negative values of the variables.

- The graph of an equation is a picture of the solutions of the equation. Each point on the graph represents a solution.

Section 2.4 Linear Equations and Inequalities

- To solve an equation that involves two or more operations, we undo those operations *in reverse order.*
- A statement that uses one of the symbols $<$ or $>$ is called an **inequality**.

> **Rules for Solving Inequalities**
> 1. We can add or subtract the same quantity on both sides.
> 2. We can multiply or divide both sides by the same positive number.
> 3. If we multiply or divide both sides by a **negative** number, we must **reverse** the direction of the inequality.

- Each point on the graph of an equation represents a solution of the equation. To solve the equation for a particular y-value, we locate the corresponding point on the graph.
- To solve a **compound inequality**, we perform the steps needed to isolate x on all three sides of the inequality.

Section 2.5 Like Terms

- Two algebraic expressions are **equivalent** if they name the same number for *all* values of the variable.
- **Like terms** are any terms that are *exactly alike* in their variable factors. By adding or subtracting like terms we can replace one algebraic expression by a shorter or simpler one.
- The numerical factor in a term is called the **numerical coefficient**, or just the **coefficient** of the term.
- Replacing an expression by a simpler equivalent one is called **simplifying** the expression.

> **To add or subtract like terms:**
> Add or subtract the numerical coefficients of the terms. Do not change the variable factors of the terms.

> **Removing parentheses**
> 1. Parentheses following a plus sign may be omitted; each term within parentheses keeps its original sign.
> 2. To remove parentheses that follow a minus sign, we change the sign of each term within parentheses and then omit the parentheses and the minus sign.

> **Steps for Solving Linear Equations**
> 1. Combine like terms on each side of the equation.
> 2. By adding or subtracting the same quantity on both sides of the equation, get all the variable terms on one side and all the constant terms on the other.
> 3. Divide both sides by the coefficient of the variable to obtain an equation of the form $x = a$.

Review Questions

Use complete sentences to answer the questions in Problems 1-10.

1. Explain the terms natural numbers, whole numbers, and integers.
2. A classmate claims that the opposite of a number and the absolute value of a number are the same, and uses $x = -3$ as an example. Do you agree? Give examples of your own.
3. You have probably heard people say that "two negatives make a positive." For which of the four arithmetic operations is this statement
   a. always true?
   b. always false?
   c. sometimes true and sometimes false?
   Make up an example for each case.
4. In a Cartesian coordinate system, the axes divide the plane into four __________. The point $(0, 0)$ is called the __________.
5. Explain how to add or subtract like terms.
6. What is wrong with this statement? $8x - (2x - 3) = 6x - 3$
7. When solving an inequality, we must remember to __________ if we __________ both sides by __________.
8. Explain how to use a graph to solve an equation.
9. Give an example of a quotient that is undefined.
10. Explain why $-\frac{a}{b}$ is not equal to $\frac{-a}{-b}$.

Review Problems

For Problems 1-6, choose the appropriate equation.

a. $5x - 8 = 30$	**b.** $\frac{x}{5} - 30 = 8$	**c.** $\frac{x - 30}{5} = 8$
d. $5x + 8 = 30$	**e.** $\frac{x}{5} - 8 = 30$	**f.** $\frac{x + 30}{5} = 8$

1. Ilciar has earned a total of 30 points on the first four quizzes in his biology class. What must he earn on the fifth quiz to end up with an average of 8?
2. Jocelyn ordered five exotic plants from a nursery. She paid a total of \$30, including an \$8 shipping fee. How much did she pay for each plant?
3. The five members of the chess team pitched in to buy new equipment. They used \$30 from their treasury, and each member donated \$8. How much did the new equipment cost?
4. Hemman bought 5 tapes on sale, and he cashed in a gift certificate for \$8. He then owed the clerk \$30. How much was each tape?
5. Nirusha and four other people won the office baseball pool. After spending \$30 of her share, Nirusha had \$8 left. What was the total amount in the pool?
6. One-fifth of the members at Sportslife Health Club signed up for a yoga class. Eight of them dropped out, leaving 30 in the class. How many members are in the Club?

For Problems 7-10, solve.

7. $3x - 4 = 1$
8. $1.2 + 0.4z = 3.2$
9. $\frac{7v}{8} - 3 = 4$
10. $13 = \frac{2}{7}x + 13$

For Problems 11-12, write an equation relating the variables, make a table of values, and graph your equation on the grid provided.

11. A new computer station for a graphics design firm costs \$2000 and depreciates in value \$200 every year. Write and graph an equation that gives the value, V, of the station after t years. Use the grid below left.

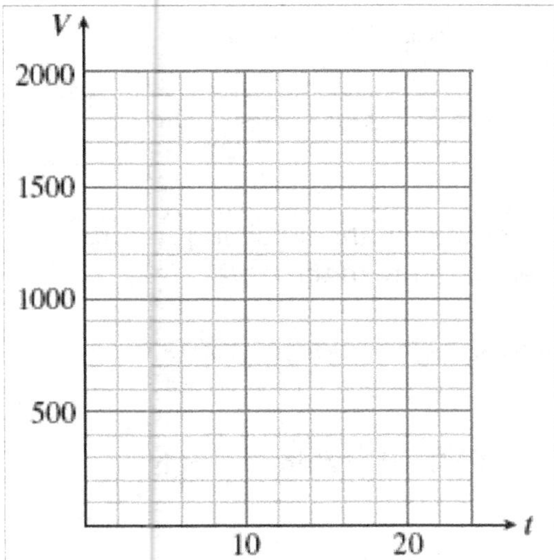

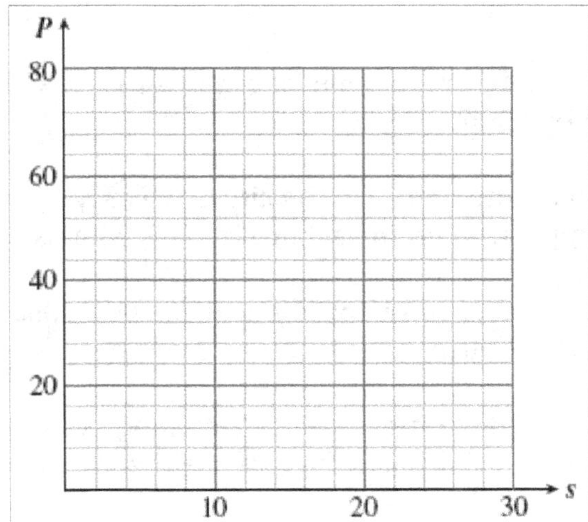

12. Ranwa teaches aerobic dance. Her gym pays her \$5 for an hour-long class, plus \$2 for each paying student in the class. Write and graph an equation that gives Ranwa's pay, P, if there are s students in the class. Use the grid above right.

For Problems 13-14, write and solve an equation to answer the question.

13. Customers at Sunny Orchard can pick their own fruit. There is a \$5.75 entrance fee, and the fruit costs \$4.50 a bushel. Bahn spent \$19.25 at the orchard. How many bushels of fruit did he pick?

14. The perimeter of a rectangle is the sum of twice its width and twice its length. If 150 meters of fence enclose a rectangular yard of length 45 meters, what is the width of the yard?

15. If $x = -3$, evaluate each expression.

a. $-x$ **b.** $x - 3$ **c.** $3 - x$ **d.** $-(-x)$

16. Graph each set of numbers on a separate number line.

a. $1.4,\ 2\frac{1}{2},\ -2\frac{1}{3},\ -3.5$ **b.** $-15, -5, 10, 25$

For Problems 17-20, replace the comma by the proper symbol; $>$, $<$, or $=$.

17. $-2, -3$

18. $-2.02, -2.1$

19. $2 - (-5), -7$

20. $-6(\frac{-1}{3}), -2$

For Problems 21-28, simplify.

21. $28 - 14 - 9 + 15$

22. $11 - 14 + (-24) - (-18)$

23. $12 - [6 - (-2) - 5]$

24. $-2 + [-3 - (-14) + 6]$

25. $5 - (-4)3 - 7(-2)$

26. $5 - (-4)(3 - 7)(-2)$

27. $\dfrac{6(-3)-8}{-4(-3-5)}$

28. $\dfrac{6(-2)-8(-9)}{5(2-7)}$

For Problems 29-32, evaluate.

29. $2-ab-3a,$ for $a=-5,\ b=-4$

30. $(8-6xy)xy,$ for $x=-2,\ y=2$

31. $\dfrac{-3-y}{4-x},$ for $x=-1,\ y=2$

32. $\dfrac{5}{9}(F-32)+273,$ for $F=-22$

33. If the overnight low in Lone Pine was –4°F, what was the temperature after it had warmed by 10°F?

34. The winter temperature in the city is typically 3°F warmer than in the adjoining suburb. If the temperature in the suburb is –7°F, what is the temperature in the city?

35. Jordan's clothing company is worth 280,000 dollars and Asher's is worth –180,000 dollars. How much more is Jordan's company worth than Asher's?

36. A certain arsenic compound has a melting point of –8.5°F. If the melting point is reduced by 1.1°F, what is the new melting point?

37. In his first football game, Bo rushes for three consecutive losses of 4 yards each. What is his total net yardage for the three plays?

38. In testing a military aircraft's handling at low altitude, the pilot runs a series of flights over a set course, each at an altitude 150 feet lower than the previous flight. What is the net change in altitude between the first and the sixth flights?

For Problems 39-42, find three solutions for the inequality.

39. $x-2<5$

40. $4x\geq -12$

41. $-9\leq -3x$

42. $-15>5+x$

43. The temperature in Maple Grove was 18°F at noon, and it has been dropping ever since at a rate of 3°F per hour.

a. Fill in the table. T stands for the temperature h hours after noon. Negative values of h represent hours before noon.

h	–4	–2	0	1	3	5	8
T							

b. Write an equation for the temperature, T, after h hours.

c. Graph your equation on the grid.

d. What was the temperature at 10 am?

e. When will the temperature reach –15°F?

f. How much did the temperature drop between 3 pm and 9 pm?

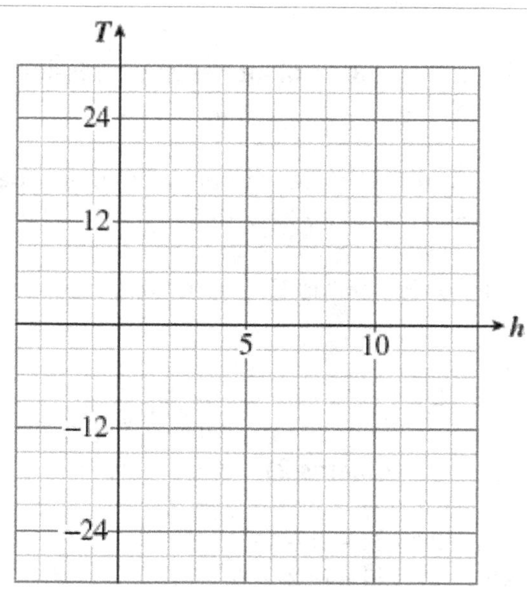

44. Graph each pair of points on the grid, then find the distance between them.

a. $A(-2,-4)$ and $B(-7,-4)$

b. $P(6,-2)$ and $Q(6,8)$

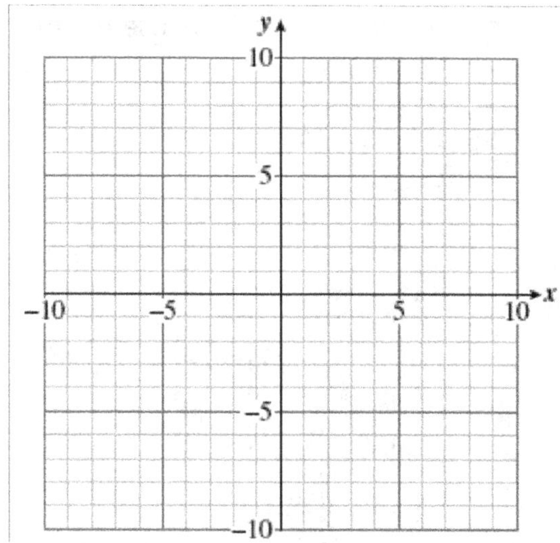

For Problems 45-46, make a table of values and graph each equation.

45. $y=-2x+7$

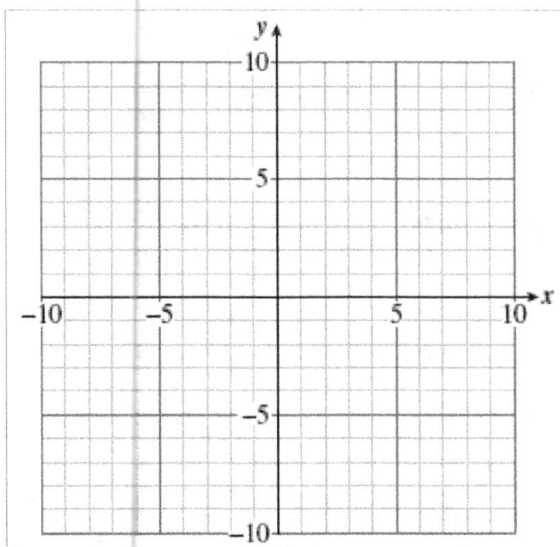

46. $y=\dfrac{4}{3}x-2$

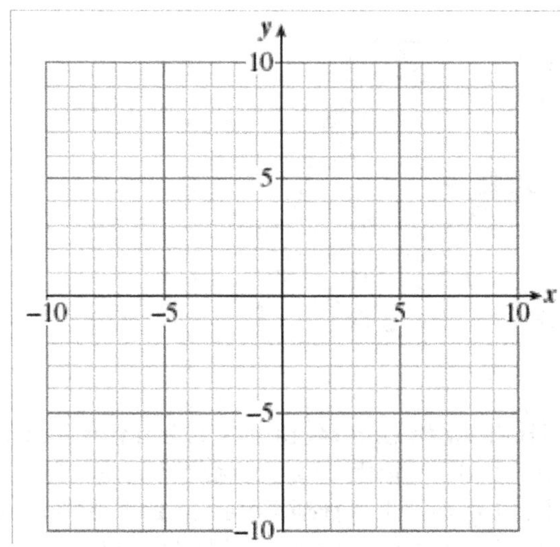

For Problems 47-48, simplify the expression.

47. $(4m+2n)-(2m-5n)$

48. $(-5c-6)+(-11c+15)$

For problems 49-54, solve.

49. $4z-6=-10$

50. $3-5x=-17$

51. $-1=\dfrac{5w}{3}+4$

52. $4-\dfrac{2z}{5}=8$

53. $3h-2=5h+10$

54. $7-9w=w+7$

For Problems 55-56, use the graph to solve the equation. Estimate your solutions if necessary.

55. a. $-3x + 9 = 24$
b. $-3x + 9 = 3$

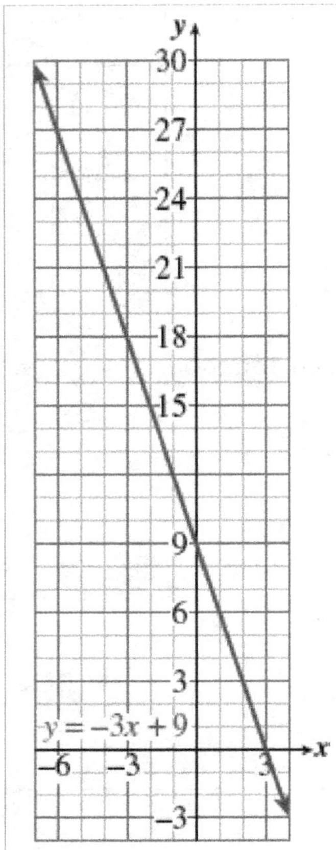

56. a. $24x - 1800 = -1250$
b. $24x - 1800 = 500$

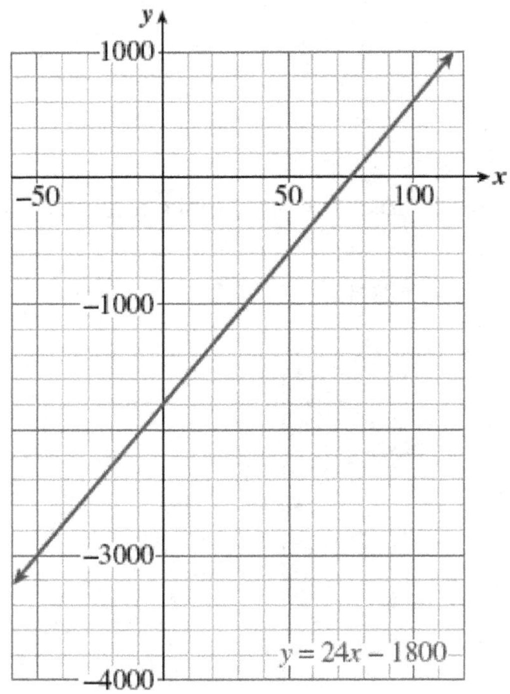

57. The length of a rectangle is 3 times its width.
a. If the width of a rectangle is x, what is its length?
b. Express the perimeter of the rectangle in terms of x.
c. Suppose the perimeter of the rectangle is 48 centimeters. Find the dimensions of the rectangle.

58. In a city council election the winner received 132 votes more than her opponent.
a. If the winner received y votes, how many did her opponent receive?
b. Write an expression for the total number of votes cast for the two candidates.
c. If 12,822 votes were cast, how many did each candidate receive?

For Problems 59-64, solve the inequality algebraically, and graph the solutions on a number line.

59. $2 - 3z \leq -7$

60. $\frac{t}{-3} - 1.7 > 2.8$

61. $3k - 13 < 5 + 6k$

62. $12a - 28 < -18 + 2a$

63. $-9 < 5 - 2n \leq -1$

64. $15 \geq -6 + 3m \geq -6$

Chapter 3 Graphs of Linear Equations

Lesson 3.1 Some Examples of Linear Models

- Write an equation for a linear model
- Find the intercepts of a line from its equation
- Graph a line by the intercept method
- Interpret the intercepts in context

Lesson 3.2 Using a Graphing Calculator

- Solve for y in terms of x
- Choose a graphing window
- Find coordinates using the Trace
- Solve equations and inequalities graphically

Lesson 3.3 Ratio and Proportion

- Compare two quantities with a ratio or a rate
- Solve a proportion
- Write a proportion to model a problem
- Recognize proportional variables

Lesson 3.4 Slope

- Compute the slope of a line
- Interpret the slope as a rate of change

Lesson 3.5 Slope-Intercept Form

- Write an equation in slope-intercept form
- Identify the slope and y-intercept of a line from its equation
- Graph a linear equation by the slope-intercept method
- Interpret the slope and y-intercept in context

Lesson 3.6 The Distributive Law

- Apply the distributive law
- Write and simplify expressions using the distributive law
- Write an equation to model a situation
- Solve equations that involve the distributive law

3.1 Some Examples of Linear Models

Tables, Graphs, and Equations

A **mathematical model** is a simplified representation of reality that helps us understand a process or phenomenon.

Look Ahead: We can use a model to analyze data, identify trends, and predict the effects of change. The first step in creating a model is to describe relationships between the variables involved. Starting from a verbal description of a situation, we can represent the relationship by a table of values, by a graph, or by an algebraic equation. Each of these mathematical tools is useful in a different way.

1. **A table of values** displays specific data points with precise numerical values.
2. **A graph** is a visual display of the data. It is easier to spot trends and describe the overall behavior of the variables from a graph.
3. **An algebraic equation** is a compact summary of the model. It can be used to analyze the model and to make predictions.

Example 1 In 1990, the median home price in the US was about $90,000. The median price increased by about $5000 per year over the next decade.

(Source: www.realestateabc.com)

a. Make a table of values showing the median price of a house in 1990, 1994, 1998, and 2000.
b. Choose suitable scales for the axes and plot the values you found in part (a) on a graph. Use t, the number of years since 1990, on the horizontal axis and the price of the house, P, on the vertical axis. Draw a curve through the points.
c. Write an equation that expresses P in terms of t.

Solutions a. In 1990 the median price was $90,000. Four years later, in 1994, the price had increased by $4(5000) = 20,000$ dollars, so

$$P = 90,000 + 4(5000) = 110,000$$

In 1998 the price had increased by $8(5000) = 40,000$ dollars, so

$$P = 90,000 + 8(5000) = 130,000$$

Year	Price of House	(t, P)
1990	90,000	$(0, 90,000)$
1994	110,000	$(4, 110,000)$
1998	130,000	$(8, 130,000)$
2000	140,000	$(10, 140,000)$

You can verify the price of the house in 2000 by a similar calculation.

b. Let t stand for the number of years since 1990, so that $t = 0$ in 1990, $t = 4$ in 1994, and so on. To choose scales for the axes, look at the values in the table. For this graph we scale the horizontal axis, or t-axis, in 2-year intervals and the vertical axis, or P-axis, up to \$200,000 in intervals of \$10,000. The points lie on a straight line.

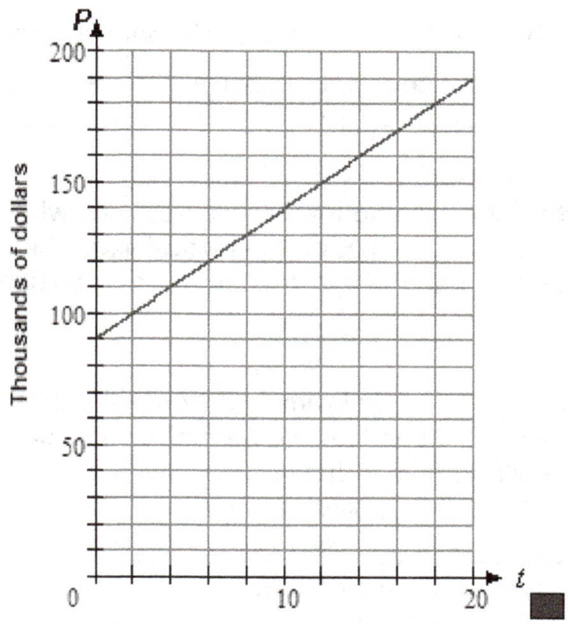

c. Look back at the calculations in part (a). The price of the house started at \$90,000 in 1990 and increased by $t \times 5000$ dollars after t years. Thus,

$$P = 90,000 + 5000t$$

RQ1. Name three ways of describing a mathematical model.

The equation we wrote in Example 1 is an example of a **linear model**, which describes a variable that increases or decreases at a constant rate.

Linear Model

A **linear model** describes a variable that increases or decreases at a constant rate. It has the form

$$y = \textbf{(starting value)} + \textbf{(rate)} \times t$$

Example 2 Use the equation from Example 1 to answer the following questions. Then show how to find the answers by using the graph.

a. According to the model, what was the median cost of a house in 1997?

b. When did the median cost of a house reach \$150,000?

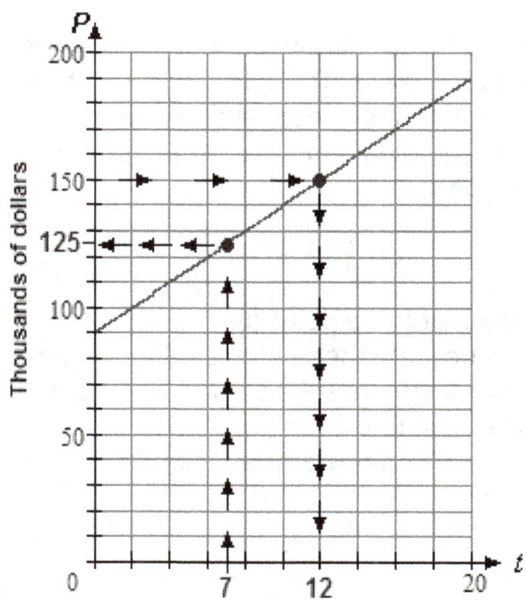

Solutions a. We substitute $t = 7$ into the equation to find

$$P = 90,000 + 5000(7) = 125,000$$

In 1997, the median cost of a house was \$125,000. To find the cost graphically, we locate $t = 7$ on the horizontal axis, then find point A on the graph whose t-coordinate is 7. The P-coordinate of this point is 125,000.

b. We substitute $P = 150,000$ into the equation and solve for t.

$$150,000 = 90,000 + 5000t \qquad \textbf{Subtract 90,0000 from both sides.}$$
$$60,000 = 5000t \qquad \textbf{Divide both sides by 5000.}$$
$$12 = t$$

The median price of a house reached $150,000 when $t = 12$, or in 2002. To use the graph, we first locate $P = 150,000$ on the vertical axis. Then we move horizontally to find the point B on the graph whose P-coordinate is 150,000. The t-coordinate of this point is 12. ■

Look Closer: In Example 2, notice the different algebraic techniques we used in parts (a) and (b). In part (a), we were given a value of t and we **evaluated the expression** $90,000 + 5000t$ to find P. In part (b) we were given a value of P and we **solved the equation** $P = 90,000 + 5000t$ to find t. •

RQ2. What is a linear model?
RQ3. What is the difference between evaluating an expression and solving an equation?

The graph in Examples 1 and 2 is an **increasing graph**. As we move along the graph from left to right (in the direction of increasing t), the second coordinate increases as well. Activity 2 in the Lesson illustrates a **decreasing graph**.

Intercepts of a Graph

The points at which a graph crosses the axes are called the **intercepts** of the graph.

The intercepts of the graph of $y = \frac{1}{2}x + 4$ are shown in the figure below. The coordinates of the intercepts are easy to find.

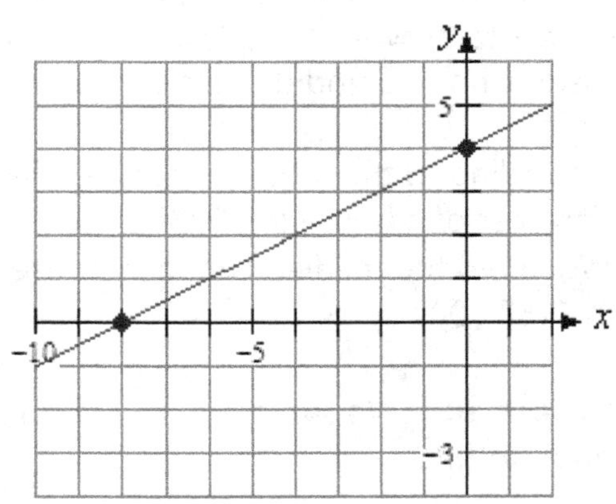

Look Closer: Because the y-intercept of a graph lies on the y-axis, its x-coordinate must be zero. Thus, to find the y-intercept we can substitute 0 for x in the equation and solve for y. For our example we have:

$$y = \frac{1}{2}(0) + 4 = 4$$

so the y-intercept is the point $(0, 4)$.

Similarly, because the x-intercept lies on the x-axis, its y-coordinate must be zero. To find the x-intercept we substitute 0 for y and solve the equation for x. For our example we find:

$$0 = \frac{1}{2}x + 4 \qquad \textbf{Subtract 4 from both sides.}$$
$$-4 = \frac{1}{2}x \qquad \textbf{Multiply both sides by 2.}$$
$$-8 = x$$

The x-intercept is $(-8, 0)$. •

Intercepts of a Graph

The points where a graph crosses the axes are called the **intercepts** of the graph.

1. To find the x-intercept, set $y = 0$ and solve for x.
2. To find the y-intercept, set $x = 0$ and solve for y.

Example 3 The temperature in Nome was $-12°$ at noon and has been rising at a rate of $2°$ per hour all day. A graph of the temperature is shown at right.

a. Write an equation for the temperature, T, at time h, where h is the number of hours after noon.

b. Find the intercepts of the graph and interpret their meaning for this situation.

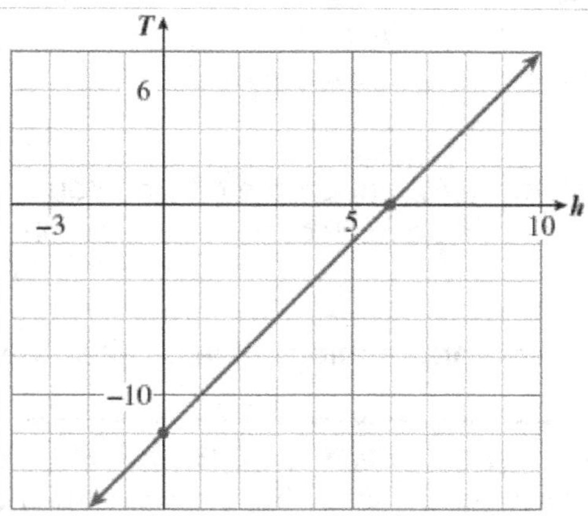

Solutions a. The starting value of the temperature was $-12°$, and it increased at a rate of $2°$ per hour. Using the form for a linear model, we have

$$T = \quad -12 \quad + \quad 2h$$
$$y = \textbf{(starting value)} + \textbf{(rate)} \times t$$

b. We find the T-intercept by substituting $h = 0$ to obtain

$$T = -12 + 2(0) = -12$$

When $h = 0$ (at noon), the temperature was $-12°$. To find the h-intercept, we substitute $T = 0$ and solve for h.

$$0 = -12 + 2h$$
$$12 = 2h$$
$$6 = h$$

The temperature reached $0°$ when $h = 6$, or at 6 pm.

RQ4. What are the intercepts of a graph?

Intercept Method of Graphing

We can use the x- and y-intercepts to draw the graph of a linear equation. Instead of choosing several different values of x to find points on the graph, we find the two intercepts. We set up a table as shown at right:
By finding the missing values for x and y we are finding the intercepts of the graph.

x	y
0	
	0

Example 4 Find the intercepts and graph the equation $3x - 2y = 12$.

Solution We set $x = 0$ to find the y-intercept, and set $y = 0$ to find the x-intercept. The completed table is:

$$3(0) - 2y = 12$$
$$-2y = 12$$
$$y = -6$$

x	y
0	−6
4	0

$$3x - 2(0) = 12$$
$$3x = 12$$
$$x = 4$$

We plot the two points $(0, -6)$ and $(4, 0)$ and draw a line through them to obtain the graph shown. It is still a good idea to find a third point as a check. If we choose $x = 2$ we find

$3(2) - 2y = 12$ **Simplify the left side.**
$6 - 2y = 12$ **Subtract 6.**
$-2y = 6$ **Divide by -2.**
$y = -3$

so $(2, -3)$ is another point on the graph. You can check that this point does lie on the graph.

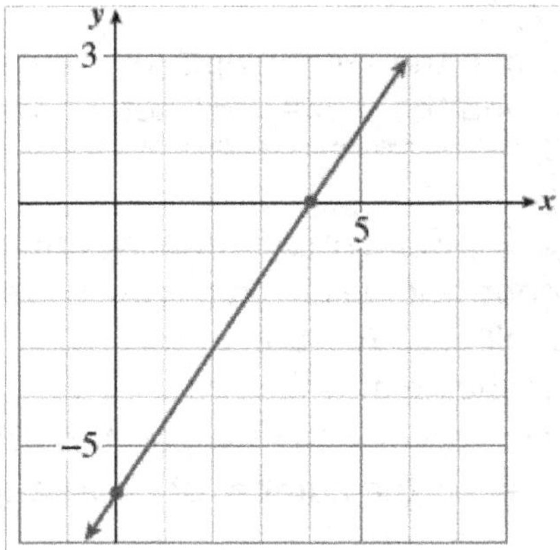

This method for graphing a line is called the **intercept method**.

To Graph a Linear Equation Using the Intercept Method:
1. Find the x- and y-intercepts of the graph.
2. Draw the line through the two intercepts.
3. Find a third point on the graph as a check. (Choose any convenient value for x and solve for y.)

General Form for a Linear Equation

The graphs of the equations above were all portions of straight lines. For this reason such equations are called **linear equations**. The order of the terms in the equation does not matter.

For example, the equation in Example 1,

$P = 90,000 + 5000t$ can be written equivalently as $-5000t + P = 90,000$

and the equation in Example 3,

$T = -12 + 2h,$ can be written as $T - 2h = -12$

This form of a linear equation, $Ax + By = C$, is called the **general form**.

General Linear Form
The **general form** for a linear equation is

$$\boldsymbol{Ax + By = C} \qquad (\boldsymbol{A} \text{ and } \boldsymbol{B} \text{ cannot both be } 0)$$

Example 5 The manager at Albert's Appliances has \$3000 to spend on advertising for the next fiscal quarter. A 30-second spot on television costs \$150 per broadcast, and a 30-second radio ad costs \$50.

a. The manager decides to buy x television ads and y radio ads. Write an equation relating x and y.
b. Use the intercept method to graph the equation.

c. What do the intercepts tell us about this situation?

Solutions a. Each television ad costs \$150, so x ads will cost \$$150x$. Similarly, y radio ads will cost \$$50y$. The manager has \$3000 to spend, so the sum of the costs must be \$3000. Thus,

$$150x + 50y = 3000$$

b. We find the intercepts of the graph by substituting first $x = 0$, and then $y = 0$. (Can you do this mentally?)

x	y
0	60
20	0

We plot the intercepts and connect them with a straight line, as shown in the figure.

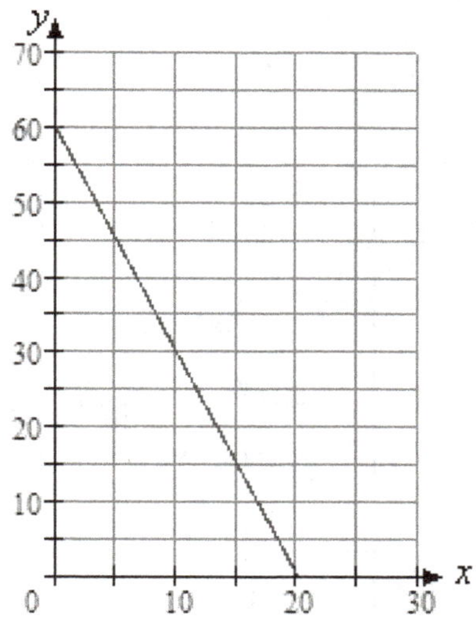

c. The y-intercept is $(0, 60)$. It tells us that if the manager buys no television ads, he can buy 60 radio ads. The x-intercept is $(20, 0)$. It tells us that if the manager buys no radio ads, he can buy 20 television ads.

RQ5. What is the general form for a linear equation?

Skills Warm-Up

Find each percent without pencil, paper, or calculator.

1. 25% of 80 **2.** 75% of 12 **3.** 50% of 28 **4.** 20% of 45
5. 30% of 50 **6.** 60% of 60 **7.** $66\frac{2}{3}$% of 36 **8.** $33\frac{1}{3}$% of 27

Answers: 1. 20 **2.** 9 **3.** 14 **4.** 9 **5.** 15 **6.** 36 **7.** 24 **8.** 9

Homework 3.1

Skills Practice

For Problems 1-6,

a. Find the x- and y-intercepts of each line.

b. Use the intercept method to graph the line.

1. $2x + 4y = 8$

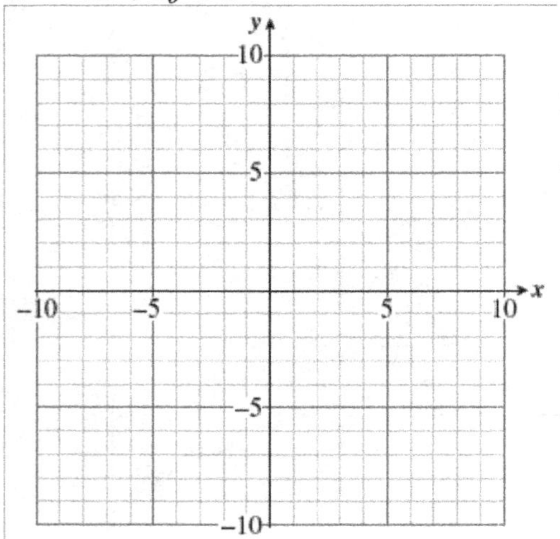

2. $x + 2y + 10 = 0$

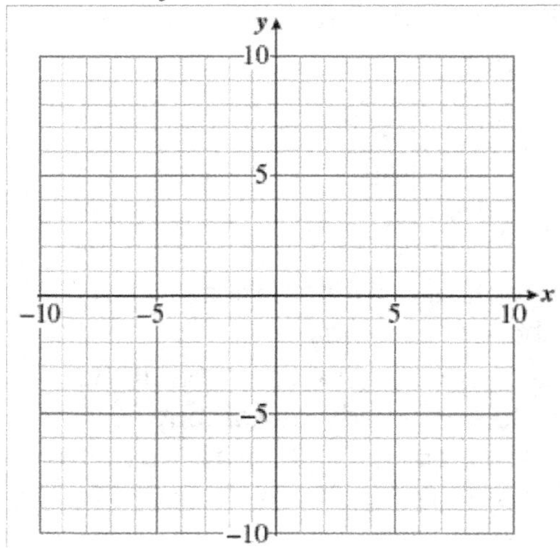

3. $2x = 14 + 7y$

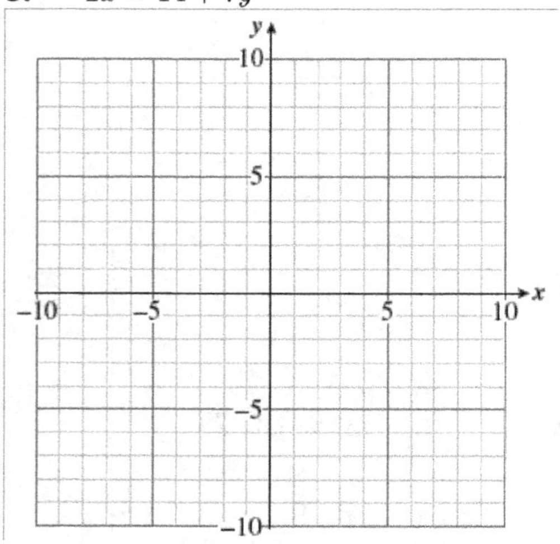

4. $y = -4x + 8$

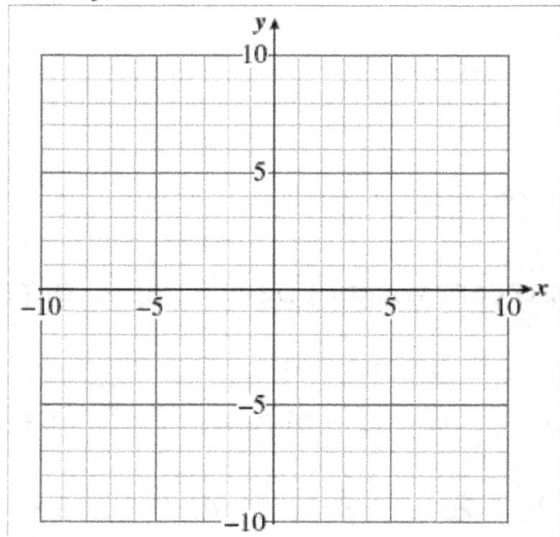

5. $\dfrac{x}{20} + \dfrac{y}{30} = 1$

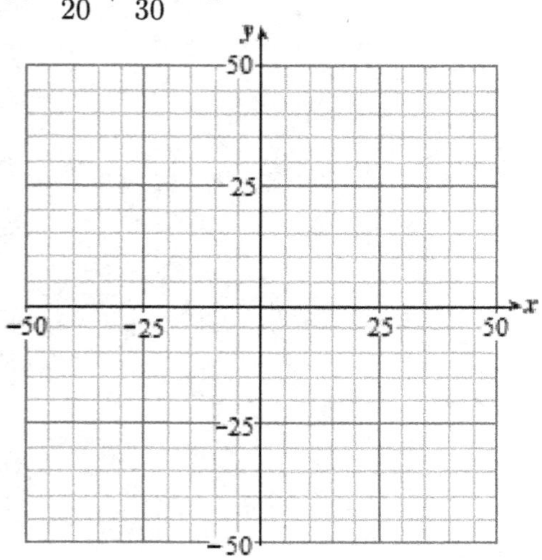

6. $3x - 2y = 120$

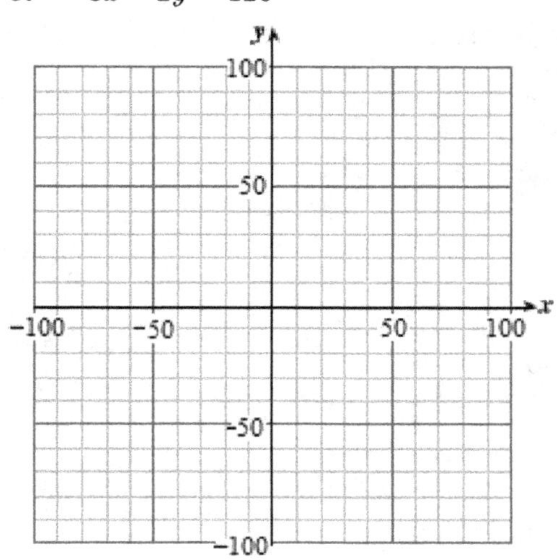

For Problems 7-12, match each equation with its graph. (More than one equation may describe the same graph.)

7. $2x + 3y = 12$ **8.** $2x - 3y = 12$ **9.** $3x - 2y = 12$

10. $-3y - 2x = 12$ **11.** $\dfrac{x}{6} - \dfrac{y}{4} = 1$ **12.** $\dfrac{x}{4} - \dfrac{y}{6} = 1$

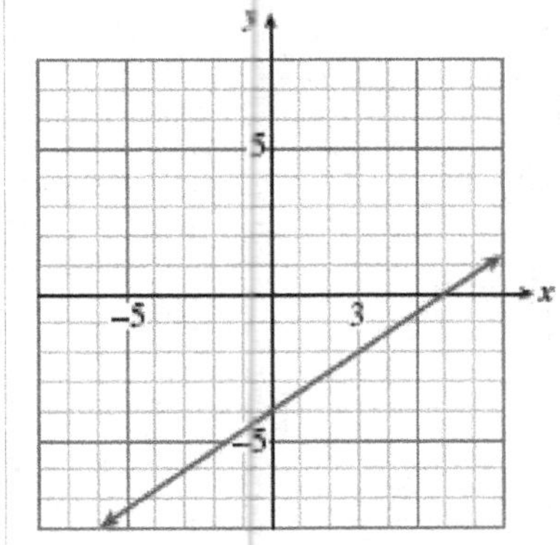

a.

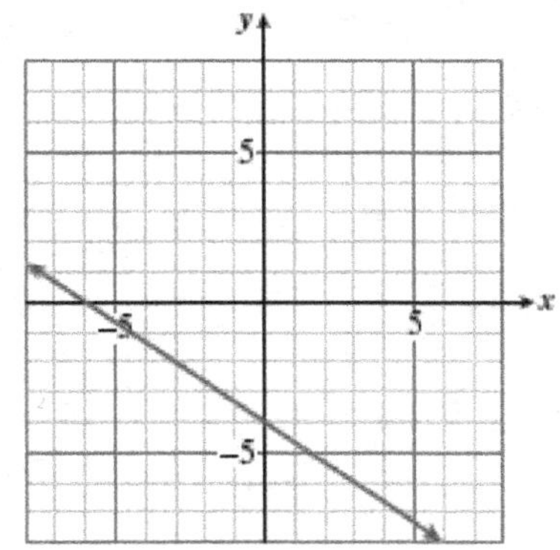

b.

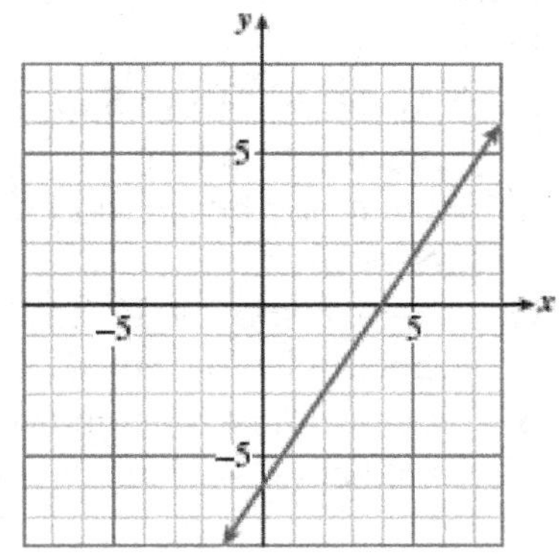

c.

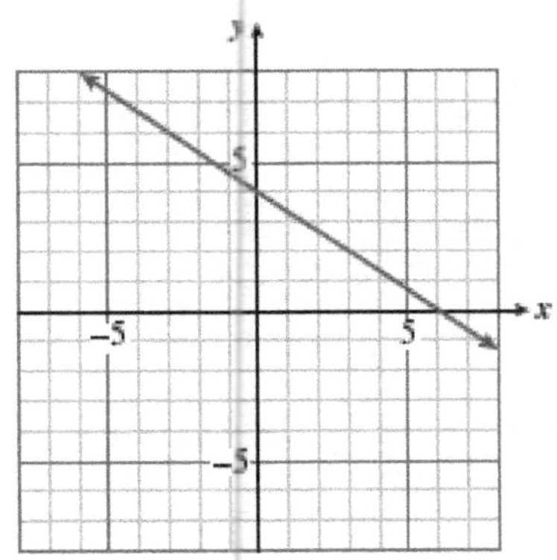

d.

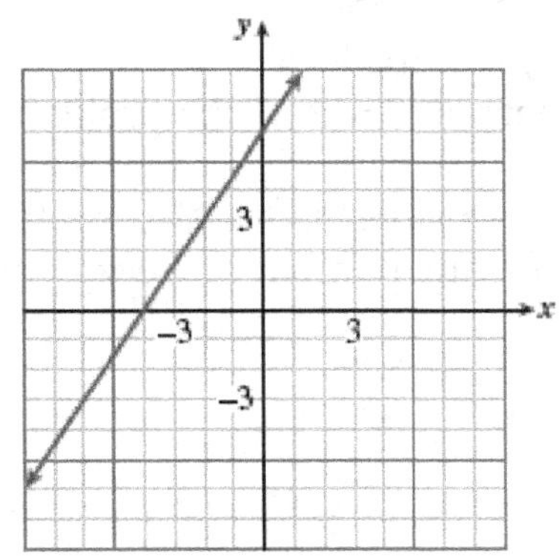

e.

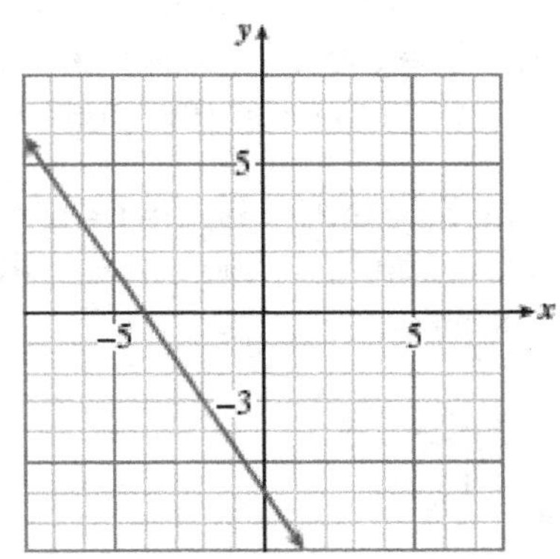

f.

For Problems 13-18, solve each pair of equations. In part (b) of each problem, your answer will involve the constant k.

13. a. $-2x = 6$
b. $-2x = k$

14. a. $x + 4 = 7$
b. $x + k = 7$

15. a. $x - 5 = 9$
b. $x - 5 = k$

16. a. $2x + 3 = 8$
b. $2x + k = 8$

17. a. $15 - 4x = 3$
b. $15 - 4x = k$

18. a. $9 + 3x = -1$
b. $9 + kx = -1$

Applications

■ Use the graph to answer the questions.

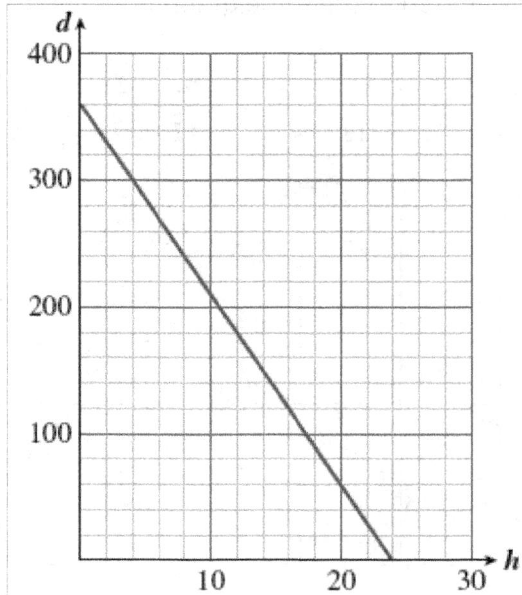

19. During spring break, Francine took the train to San Francisco and then bicycled home. The graph below left shows Francine's distance d from home, in miles, after cycling for h hours.

a. How far is it from San Francisco to Francine's home?

b. How many hours did Francine cycle to get home?

c. After cycling for 12 hours, how far is Francine from home?

d. How far did Francine cycle in the first four hours?

■ For Problems 20-22,

a. Find the intercepts of each linear equation.

b. Use the intercept method to graph the line.

c. Explain what the intercepts mean in terms of the problem situation.

20. The amount of home heating oil (in gallons) in the Olsons' tank is given by the equation $G = 200 - 15w$, where w is the number of weeks since they turned on the furnace.

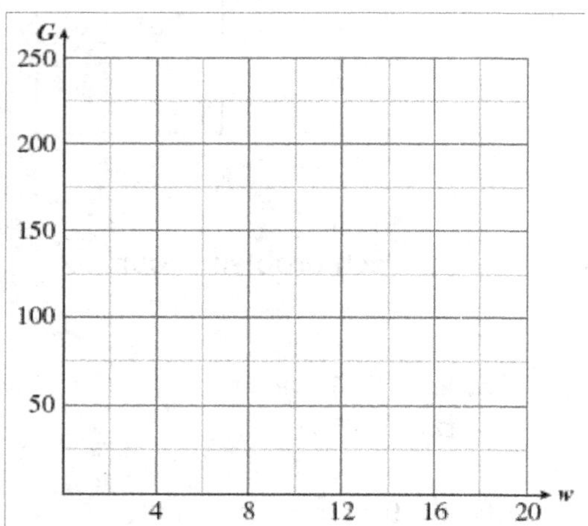

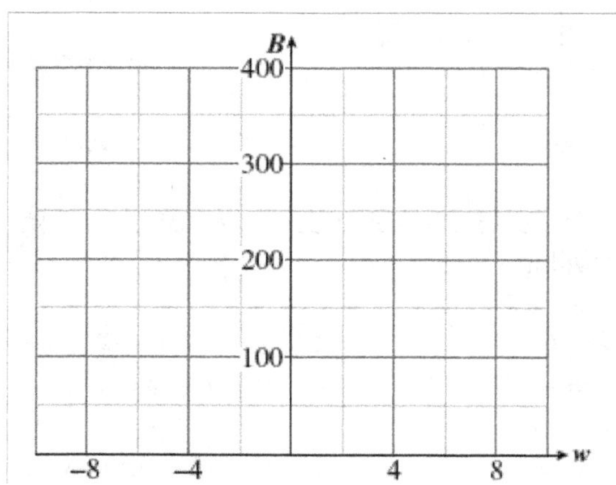

21. Dana joined a savings plan some weeks ago. Her bank balance is growing each week according to the formula $B = 225 + 25w$, where $w = 0$ represents this week.

22. Delbert bought some equipment and went into the dog-grooming business. His profit is increasing according to the equation $P = -600 + 40d$, where d is the number of dogs he has groomed.

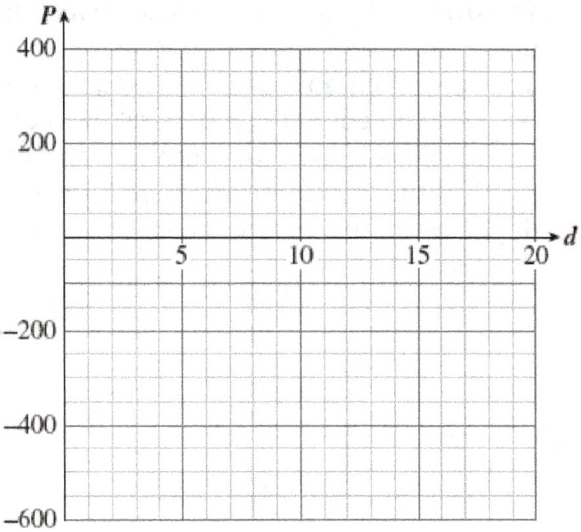

23. a. At what point does the graph of

$$2x - 3y = 25$$

cross the x-axis?

b. At what point does the graph of

$$1.4x + 3.6y = -18$$

cross the y-axis?

24. The x-intercept of a line is positive and its y-intercept is negative. Is the line increasing or decreasing? Sketch a possible example of such a line.

25. On October 31, Betty and Paul fill their 250-gallon oil tank for their heater. Beginning in November they use an average of 15 gallons of oil per week.

a. Write an equation that expresses the amount of oil, A, in the tank in terms of the number of weeks, w, since October 31.

b. Find the intercepts and graph the equation.

c. What do the intercepts tell us about this situation?

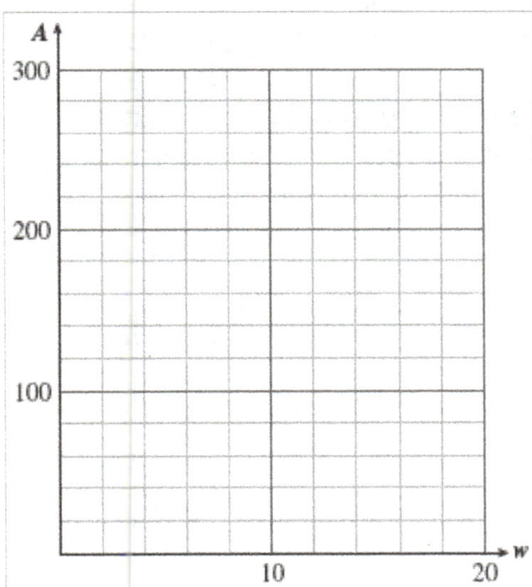

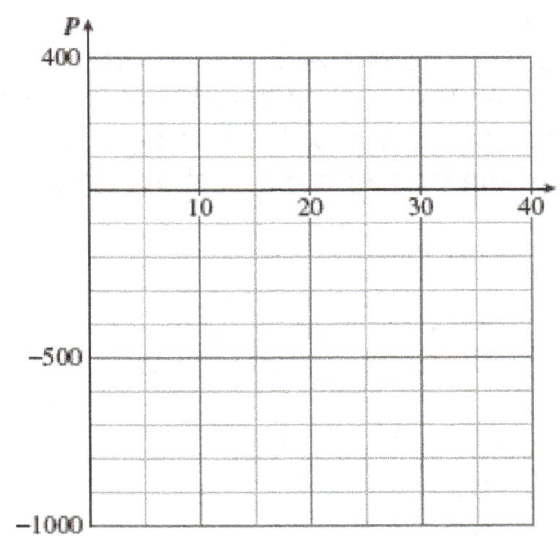

26. Phil and Ernie buy a used photocopier for \$800 and set up a copy service on their campus. For each hour that the copier runs, Phil and Ernie make \$40.

a. Write an equation that expresses Phil and Ernie's profit (or loss), P, in terms of the number of hours, t, they run the copier.

b. Find the intercepts and sketch the graph. (Suggestion: Scale the horizontal axis from 0 to 40 in increments of 5, and the vertical axis from -1000 to 400 in increments of 100.)

c. What do the intercepts tell us about Phil and Ernie's profit?

27. The owner of a gas station has $19,200 to spend on unleaded gas this month. Regular unleaded costs him $2.40 per gallon, and premium unleaded costs $3.20 per gallon.

a. How much do x gallons of regular cost? How much do y gallons of premium cost?

b. Write an equation in general form that relates the amount of regular unleaded gasoline, x, the owner can buy and the amount of premium unleaded, y.

c. Find the intercepts and sketch the graph.

d. What do the intercepts tell us about the amount of gasoline the owner can purchase?

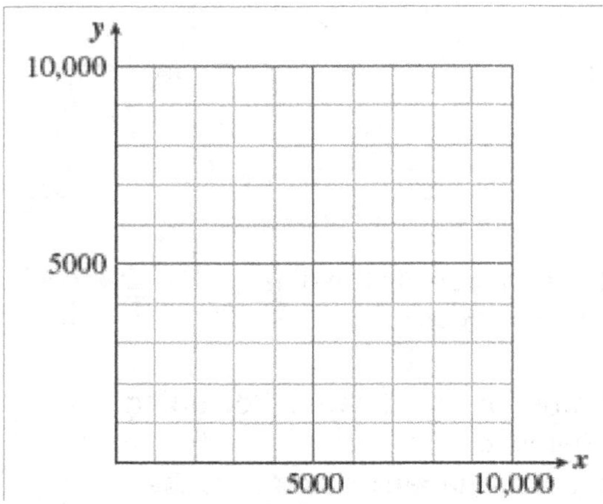

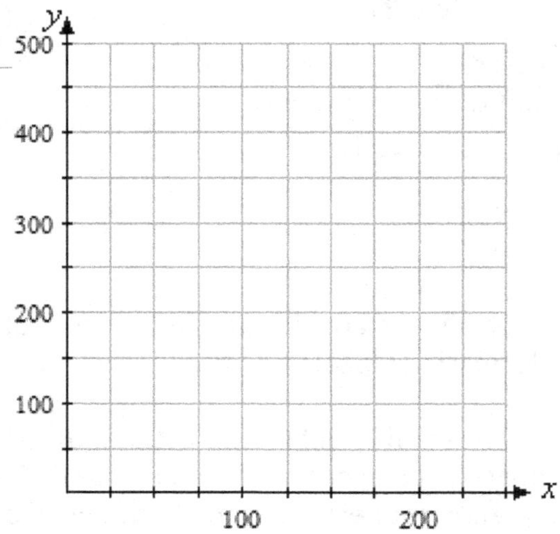

28. Delbert must increase his daily potassium intake by 1800 mg. He decides to eat a combination of figs and bananas, which are both low in sodium. There are 9 mg potassium per gram of fig, and 4 mg potassium per gram of banana.

a. How much potassium is in x grams of fig? How much potassium is in y grams of banana?

b. Write an equation in general form that relates the number of grams x of fig and the number of grams y of banana Delbert needs to get 1800 mg of potassium.

c. Find the intercepts and sketch the graph.

d. What do the intercepts tell us about Delbert's diet?

3.2 Using a Graphing Calculator

Solving for y in Terms of x

An equation of the form

$$Ax + By = C$$

where A, B, and C are constants, is a linear equation, as long as A and B are not both zero. We can use a calculator to graph a linear equation, but before we can enter the equation, we must solve for y in terms of x.

Example 1 Solve the equation $3x + 2y = 16$ for y in terms of x.

Solution We want to write the equation in the form $y =$ (expression in x). To do this, we must isolate the y-term.

$$3x + 2y = 16 \qquad \text{Subtract } 3x \text{ from both sides.}$$
$$2y = -3x + 16 \qquad \text{Divide both sides by 2.}$$
$$y = \frac{-3x + 16}{2} \qquad \text{Divide each term by 2.}$$
$$y = \frac{-3x}{2} + \frac{16}{2} \qquad \text{Simplify.}$$
$$y = -1.5x + 8$$

RQ1. In Example 1, how did we simplify the expression $\frac{-3x+16}{2}$?

The Standard Window

After we have solved the equation for y, we can use the graphing calculator. Here are the two steps for entering a graph in the standard window.

To graph an equation:
1. Press [Y=] and enter the equation(s) you wish to graph.
2. Select the standard graphing window by pressing [ZOOM] [6].

Example 2 Use a graphing calculator to graph the equation $y = -1.5x + 8$

Solution We press [Y=] and enter the right side of the equation, $-1.5\mathbf{X} + 8$, after $\mathbf{Y_1} =$. We use the [X,T,θ,n] key to enter **X**, so we type in

−1.5 [X,T,θ,n] [+] 8

Next we set a graphing window. Press the [ZOOM] key, and then the number [6] to select the standard graphing window. This window displays values from −10 to 10 on both axes. The resulting graph is shown at right.

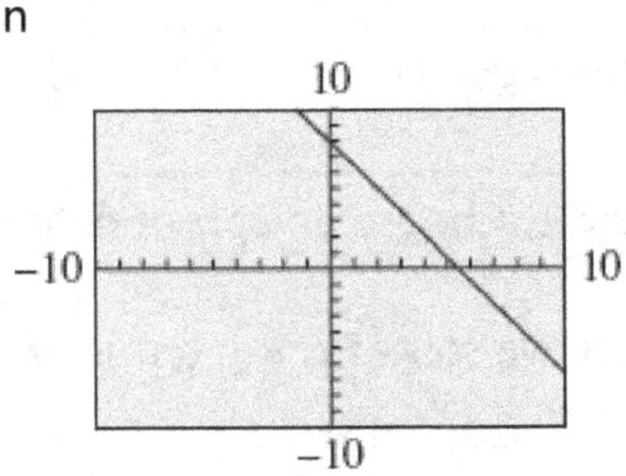

Caution! To enter a negative number, we use the negative key, [(−)]. Do not use the [−] key, which is for subtraction.

RQ2. How are the axes scaled in the standard window?

When you want to erase the graphs, press **Y=** and then press **CLEAR**. This deletes the expression following $\mathbf{Y_1}$. Use the down arrow key, ▽, and the **CLEAR** key to delete any other equations on the display.

Choosing a Graphing Window

Now consider the equation $6x - 5y = 90$. To solve for y, we first subtract $6x$ from both sides of the equation.

$$-5y = 90 - 6x \qquad \textbf{Divide both sides by } -5.$$
$$y = \frac{90}{-5} - \frac{6x}{-5} \qquad \textbf{Simplify.}$$
$$y = -18 + \frac{6}{5}x$$

If we graph this equation in the standard window, we see only a small piece of the graph in the lower corner of the screen, as shown at right. We should choose a window that shows more of the graph. How do we choose an appropriate window for a graph? The window should show the essential features of the graph and give the viewer a good idea of its overall shape. For the graph of a linear equation, the most informative points are the x- and y-intercepts. Both intercepts should be visible in the graphing window.

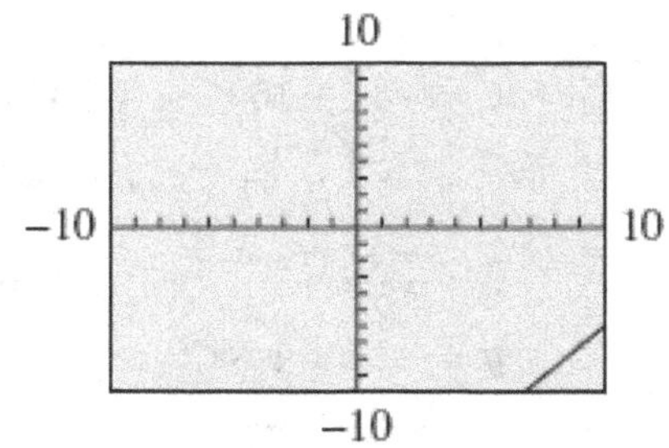

To graph a linear equation, we should choose scales for the axes so that both intercepts can be shown on the graph.

You can check that the intercepts of the graph of $6x - 5y = 90$ are $(15, 0)$ and $(0, -18)$. To show these intercepts, a good choice for the window might be

$$\text{Xmin} = -20 \qquad \text{Xmax} = 20$$
$$\text{Ymin} = -20 \qquad \text{Ymax} = 20$$

Press **WINDOW** and enter the values above for the edges of the window, using the ▽ key to move through the menu items. Then press **GRAPH**. The resulting graph is shown at right.

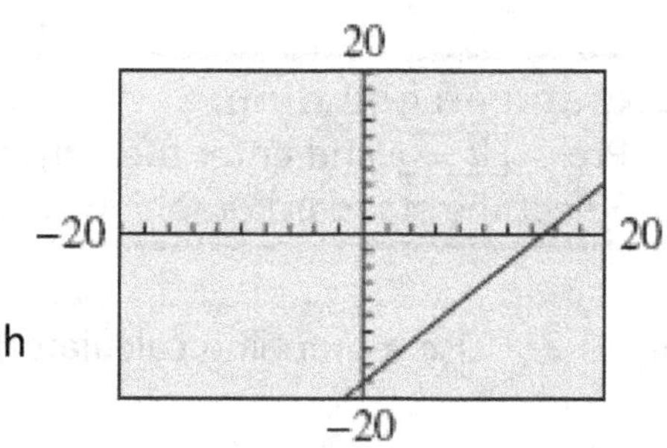

RQ3. Which points on a line should be visible in the viewing window?

■ **Example 3** Choose an appropriate window and graph the equation $10y - 15x = 6$

Solution First, we solve for y in terms of x. We add $15x$ to both sides to find

$$10y = 15x + 6 \qquad \textbf{Divide both sides by 10.}$$
$$y = 1.5x + 0.6$$

We press **Y=** and enter this equation.

To decide on a good window, we find the x- and y-intercepts of the graph. Setting x equal to zero, we find $y = 0.6$, and when $y = 0$ we find $x = -0.4$. The intercepts are thus

$(-0.4, 0)$ and $(0, 0.6)$. For this graph we choose a very small window so that we can distinguish the intercepts from the origin. We set x- and y-limits slightly larger than the intercept values. The figure shows the graph in the window

Xmin $= -1$ Xmax $= 1$
Ymin $= -1$ Ymax $= 1$

We have set Xscl and Yscl equal to 0.1, so that the tick marks occur at intervals of 0.1 unit.

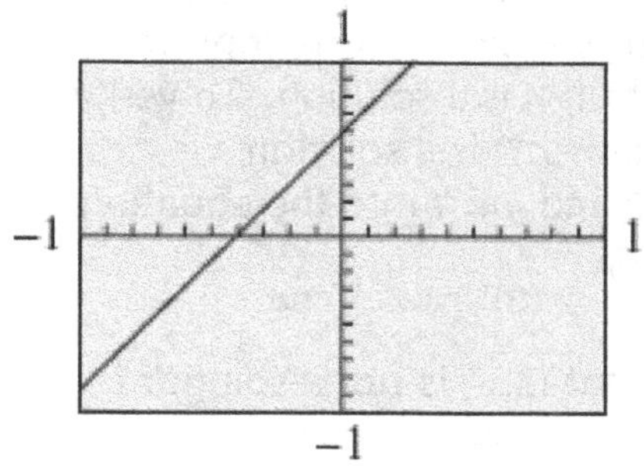

Finding Coordinates with a Graphing Calculator

We can use the **Trace** feature or the **value** feature of the calculator to find the coordinates of points on a graph.

Example 4 Graph the equation $y = -2.6x - 5.4$ in the window

Xmin $= -5$ Xmax $= 4.4$
Ymin $= -20$ Ymax $= 15$

Find the y-coordinate of the point with x-coordinate 2.

Solution We enter the equation and set the window, then press GRAPH. Next, we press TRACE, and a "bug" begins flashing on the display. The coordinates of the bug appear at the bottom of the display, as shown in the figure. We use the left and right arrows to move the bug along the graph. (You can check that the coordinates of the point $(2, -10.6)$ do satisfy the equation $y = -2.6x - 5.4$.)

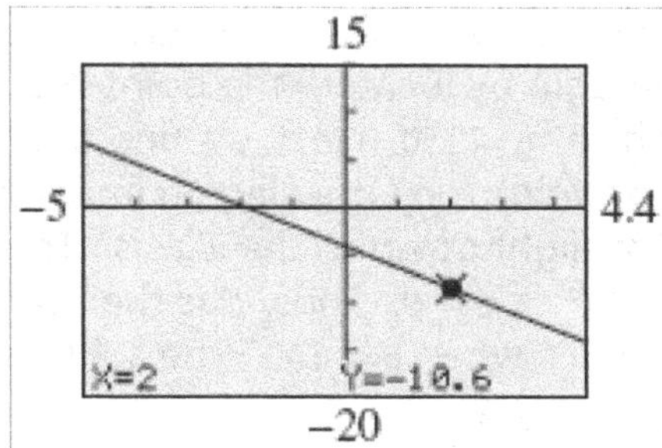

Look Closer: The points identified by the **Trace** bug depend on the window settings and on the type of calculator. If we want to find the y-coordinate for a particular x-value, we can use the **value** feature of the calculator. We press 2nd CALC, and choose the first menu item, **value.** Then we enter the x-coordinate of the desired point and press ENTER. Try entering 2 after **X** =, and you should find that $y = -10.6$, as before.

Graphical Solution of Equations

The graph of an equation in two variables is just a picture of its solutions. When we read the coordinates of a point on the graph, we are reading a pair of x- and y-values that make the equation true.

Example 5 Verify, algebraically and graphically, that the ordered pair $(2, 7)$ is a solution of the equation $y = 2x + 3$.

Solution In the figure, the point $(2,7)$ lies on the graph of $y=2x+3$. This confirms graphically that $(2,7)$ is a solution. To verify algebraically that $(2,7)$ is a solution, we substitute $x=2$ and $y=7$ into the equation:

$$7 \stackrel{?}{=} 2(2)+3 \quad \textbf{True}$$

Can you show that $(3,6)$ is **not** a solution?

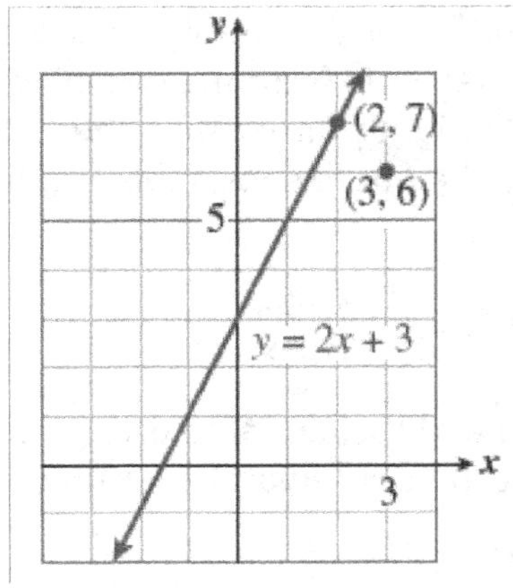

RQ4. What do the points on the graph of an equation represent?

Look Closer: In Example 5, we can also say that $x=2$ is a solution of the equation $2x+3=7$. Thus, we can use graphs to find solutions of equations in one variable.

Example 6 Use the graph of $y=285-15x$ to solve the equation $150=285-15x$.

Solution We compare the two equations in the problem, and observe that y has been replaced by 150. So we begin by locating the point P on the graph for which $y=150$. Next, we find the x-coordinate of point P by drawing an imaginary line from P straight down to the x-axis. The x-coordinate of P is $x=9$. Thus, P is the point $(9,150)$, and $x=9$ when $y=150$. The solution of the equation is $x=9$.

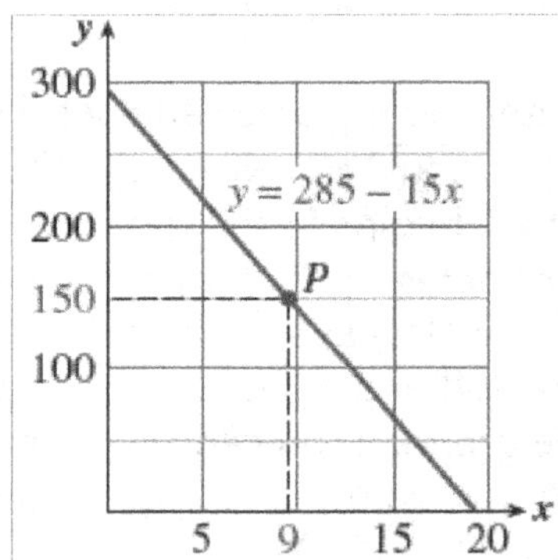

You can verify the solution to Example 6 algebraically by substituting $x=9$ into the equation:

Does $150 = 285 - 15(9)$?
$285 - 15(9) = 285 - 135 = 150$ **Yes.**

Look Closer: The relationship between an equation and its graph is an important one. For the previous example, make sure you understand that the following three statements are equivalent.

1. The point $(9,150)$ lies on the graph of $y=285-15x$.
2. The ordered pair $(9,150)$ is a solution of the equation $y=285-15x$.
3. $x=9$ is a solution of the equation $150=285-15x$.

RQ5. In Example 6, what point would you locate to solve the equation $200=285-15x$?

Graphical Solution of Inequalities

We can also use graphs to solve inequalities. Consider the inequality

$$285 - 15x \geq 150$$

To solve this inequality we must find all values of x that make the expression $285 - 15x$ greater than 150 or equal to 150. We could begin by trying some values of x. Here is a table obtained by evaluating $285 - 15x$.

x	0	2	4	6	8	10	12
$\mathbf{285 - 15x}$	285	255	225	195	165	135	115

From the table, we see that values of x less than or equal to 8 are solutions of the inequality, but we have not checked all possible x-values. We can get a more complete picture from a graph.

Example 7 The figure shows the graph of the equation $y = 285 - 15x$. Use the graph to solve the inequality $285 - 15x \geq 150$.

Solution We look for points on the graph with y-coordinates greater than or equal to 150. Some of these points are shown as dots. Which x-values produced these points with y-values greater than or equal to 150? We can read the x-coordinates by dropping straight down to the x-axis, as shown by the arrows. For example, the x-value corresponding to $285 - 15x = 150$ is $x = 9$. For larger values of $285 - 15x$ we must choose x-values less than 9. Thus, the solutions are all values of x less than or equal to 9, as shown on the x-axis in the figure.

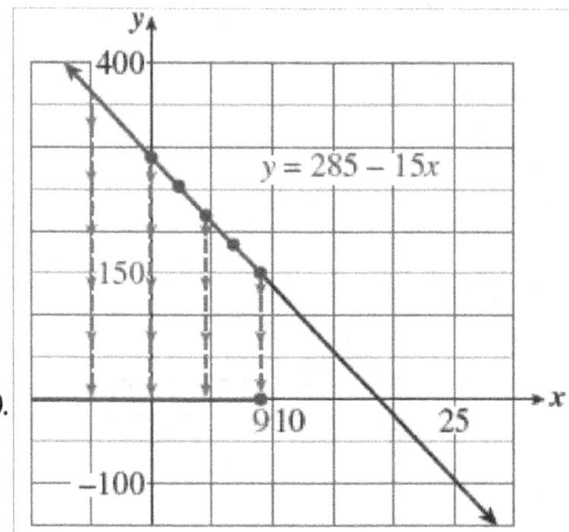

You can verify the solution in Example 7 by solving the inequality algebraically.

RQ6. What equation could you graph to solve $2x - 8 \geq 10$?

Using a Graphing Calculator to Solve Equations and Inequalities

With the aid of a graphing calculator, we can quickly find (at least approximate) solutions to equations and inequalities.

Example 8a. Use a graphing calculator to solve the equation $572 - 23x = 181$. Set the **WINDOW** to

Xmin $= -40$ Xmax $= 54$
Ymin $= -100$ Ymax $= 900$

b. Use the same graph to solve the inequality $572 - 23x \geq 181$

Solutions a. Because the graphing calculator does not display scales on the axes, it is often helpful to plot a horizontal line corresponding to the y-value of interest. After setting the **WINDOW**, we enter $\mathbf{Y_1} = 572 - 23\mathbf{X}$ and $\mathbf{Y_2} = 181$. We press **GRAPH** to see the graphs shown in the figure. Then we activate the **TRACE**, and use the arrow keys until the bug rests on the intersection of the two lines, that is, when the y-coordinate is 181. The corresponding value of x is $x = 17$. Thus, the solution to the equation $572 - 23x = 181$ is $x = 17$.

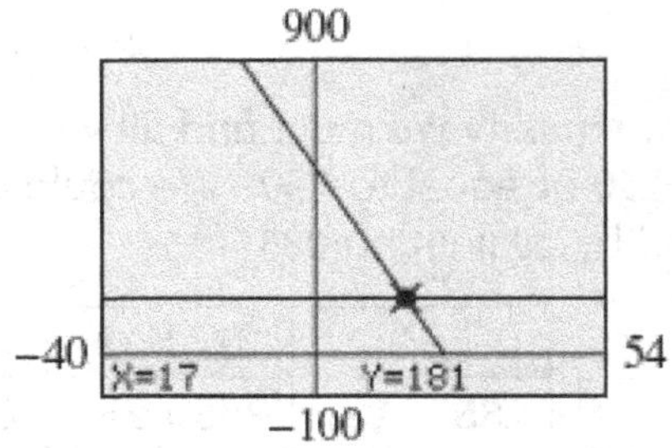

b. To solve the inequality

$$572 - 23x \geq 181$$

we want points with y-values greater than 181. Experiment by pressing the left and right arrow keys to move the bug to either side of the intersection point. Notice that points with y-coordinates greater than 181 have x-coordinates less than 17. Thus, the solution to the inequality is $x \leq 17$.

Skills Warm-Up

Solve the inequalities.

1. $18 - 5x > 4 - 3x$

2. $9 - \frac{2}{3}x \leq 18$

3. $-24 < -6 - 3x \leq 15$

4. $-9 < \frac{1 - 4x}{5} < -3$

Answers: **1.** $x < 7$ **2.** $x \geq -13.5$ **3.** $-7 \leq x < 6$ **4.** $4 < x < 11.5$

Homework 3.2

Skills Practice

■ For Problems 1-8,

a. Solve each equation for y in terms of x,

b. Graph the equation on your calculator in the specified window,

c. Make a pencil and paper sketch of the graph. Label the scales on your axes, and the coordinates of the intercepts.

1. $2x + y = 6$

Standard window

2. $3x - 4y = 1200$

Xmin $= -1000$ Ymin $= -1000$
Xmax $= 1000$ Ymax $= 1000$
Xscl $= 100$ Yscl $= 100$

3. $0.2x + 5y = 0.1$

Xmin $= -1$ Ymin $= -0.1$
Xmax $= 1$ Ymax $= 0.1$
Xscl $= 0.1$ Yscl $= 0.01$

4. $70x + 3y = y + 420$

Xmin $= 0$ Ymin $= 0$
Xmax $= 10$ Ymax $= 250$
Xscl $= 1$ Yscl $= 25$

■ For Problems 5-8,

a. Find the x- and y-intercepts.

b. Solve the equation for y.

c. Use the intercepts to select a suitable graphing window on your calculator. Both intercepts should be visible and distinguishable).

d. Graph the resulting equation using the window you specified.

5. $x + y = 100$

6. $25x - 36y = 1$

7. $\dfrac{y}{12} - \dfrac{x}{47} = 1$

8. $-2x = 3y + 84$

9. The figure shows a graph of $y = -2x + 6$.

a. Use the graph to find all values of x for which:

i) $y = 12$ ii) $y > 12$
iii) $y < 12$

b. Use the graph to solve:

i) $-2x + 6 = 12$ ii) $-2x + 6 > 12$
iii) $-2x + 6 < 12$

c. Explain why your answers to parts (a) and (b) are the same.

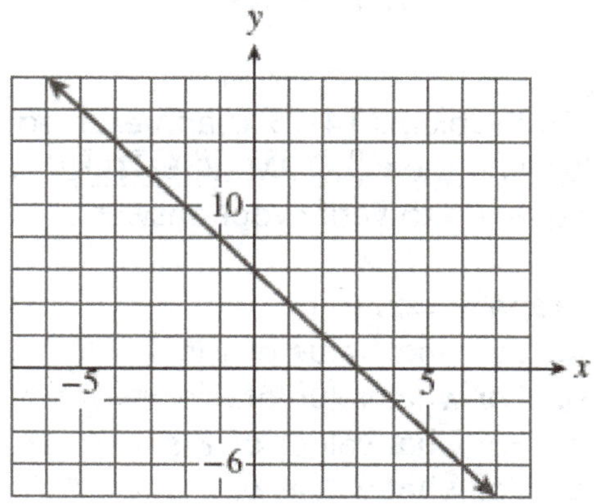

■ For Problems 10-12,

a. Use the graph to solve each of the following inequalities.

b. Indicate your solutions along the x-axis of the graph.

c. Then solve the inequality algebraically.

10. $-2x \geq 8$

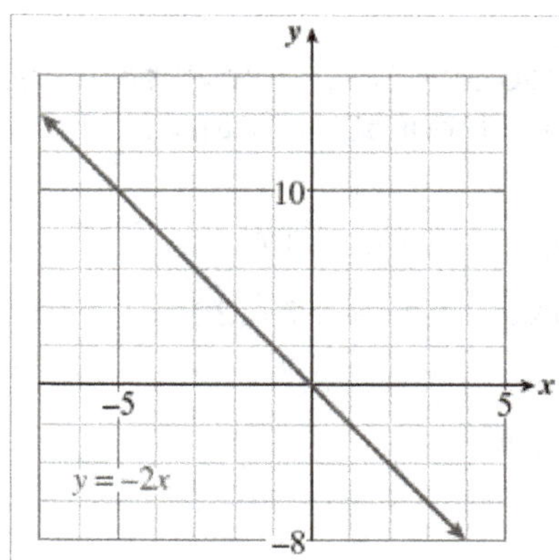

11. $3x + 5 \leq -4$

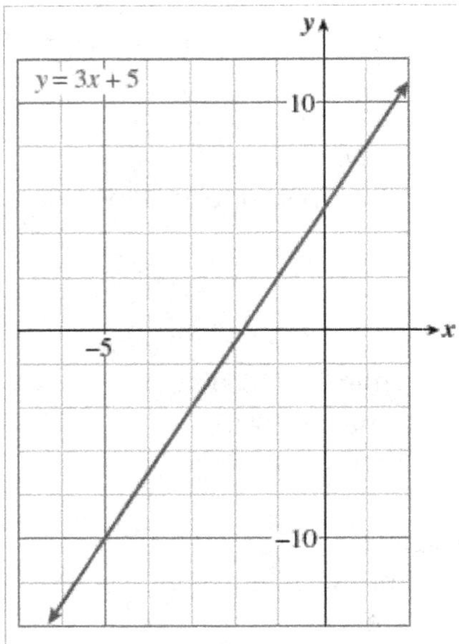

12. $7 - \frac{2x}{3} > 1$

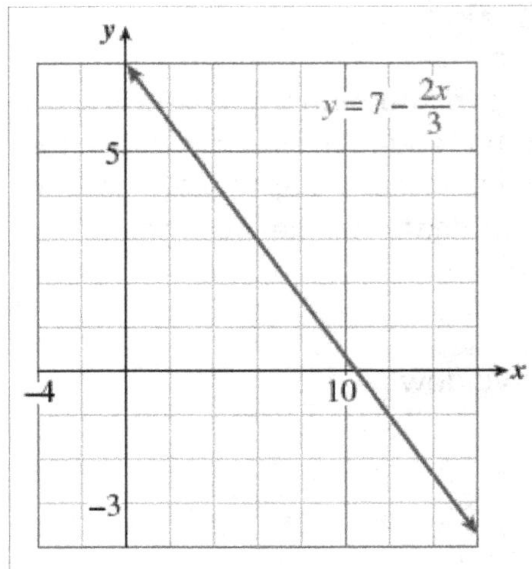

13. The figure shows the graph of

$$y = 1.4x - 0.64$$

Use the graph to solve each equation or inequality, then solve algebraically.

a. $1.4x - 0.64 = 0.2$

b. $-1.2 = 1.4x - 0.64$

c. $1.4x - 0.64 > 0.2$

d. $-1.2 > 1.4x - 0.64$

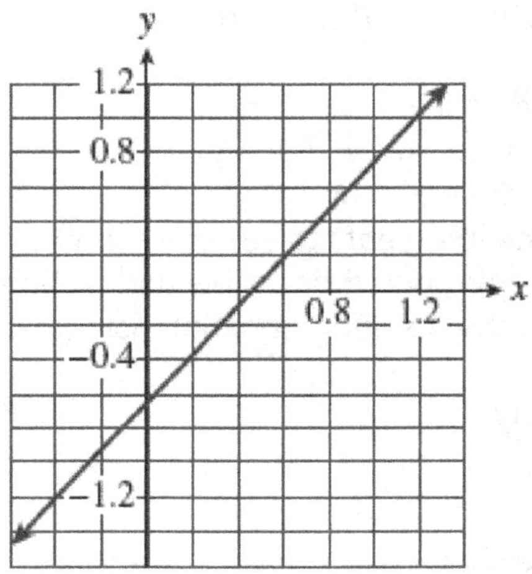

For Problems 14-15, graph each equation with the **ZInteger** setting. (Press ZOOM 6, and then ZOOM 8 ENTER.) Use the graph to answer each question. Use the equation to verify your answers.

14. Graph $y = 2x - 3$.

a. For what value of x is $y = 5$?
b. For what value of x is $y = -13$?
c. For what values of x is $y > -1$?
d. For what values of x is $y < 25$?

15. Graph $y = 6.5 - 1.8x$.

a. For what value of x is $y = -13.3$?
b. For what value of x is $y = 24.5$?
c. For what values of x is $y \leq 15.5$?
d. For what values of x is $y \geq -7.9$?

For Problems 16-17, graph each equation with the **ZInteger** setting. Use the graph to solve each equation or inequality. Check your solutions algebraically.

16. Graph $y = -0.4x + 3.7$

a. Solve $-0.4x + 3.7 = 2.1$

b. Solve $-0.4x + 3.7 > -5.1$

17. Graph $y = \frac{2}{3}x - 24$

a. Solve $\frac{2}{3}x - 24 = -10\frac{2}{3}$

b. Solve $\frac{2}{3}x - 24 \leq -19\frac{1}{3}$

Applications

18. Systolic blood pressure increases with increasing exercise intensity. Kieran's blood pressure (in mm Hg) is 120 at rest and rises by 6 for each minute he is on a treadmill with steadily increasing exercise intensity.

a. Fill in the table.

Time on treadmill (minutes)	0	1	2	5	10
Blood pressure (mm Hg)					

b. Find a formula that gives blood pressure, p, in terms of time, t.

c. Use your calculator to graph the equation for p in terms of t.

d. What is Kieran's blood pressure after 3.5 minutes?

e. Kieran's blood pressure should not exceed 165 mm Hg. When should the treadmill stop increasing his exercise intensity?

3.3 Ratio and Proportion

Ratios

A **ratio** is a type of quotient used to compare two numerical quantities.

For example, saying that the ratio of pupils to computers at a certain elementary school is 10 to 3 means that for every 10 pupils in the school there are 3 computers. We often display a ratio as a fraction because their properties are similar. For the example just mentioned we would write

$$\frac{\text{Number of pupils}}{\text{Number of computers}} = \frac{10}{3}$$

Caution! This ratio does *not* mean that there are exactly 10 pupils and 3 computers at the school. There could be 40 pupils and 12 computers or 200 pupils and 60 computers (or many other combinations) because

$$\frac{40}{12} = \frac{200}{60} = \frac{10}{3}$$

■

■ **Example 1** In a survey of 100 of his classmates, Greg found that 68 students supported a strict gun control law, and 32 students did not. What is the ratio of survey respondents who support gun control to those who oppose it?

Solution The ratio of those supporting gun control to those opposed is 68 to 32, or $\frac{68}{32}$. This fraction reduces to $\frac{17}{8}$, or 17 to 8. ■

Caution! When we write a ratio, it is important to notice which number is the numerator, and which is the denominator. When we write "the ratio of a to b" as a fraction, a is the numerator and b is the denominator. ■

In Example 1, we could also compute the ratio of opponents to supporters of the gun control law. In that case, the answer would be $\frac{32}{68}$ or $\frac{8}{17}$, the reciprocal of the answer to Example 1.

Ratio

A **ratio** is a type of quotient used to compare two numerical quantities. The **ratio of a to b** is written $\frac{a}{b}$.

Look Closer: A ratio can be expressed as a decimal instead of a common fraction. Because $\frac{17}{8}$ is equal to 2.125, we can write the ratio in Example 1 as 2.125 to 1. We usually express a ratio as a decimal fraction if the numbers being compared are decimal numbers. •

Example 2 A small circuit board measures 4.2 centimeters long by 2.4 centimeters wide. What is the ratio of its length to its width?

Solution The ratio of length to width is $\frac{4.2 \text{ cm}}{2.4 \text{ cm}}$, or 1.75, as you can verify with your calculator. (Can you express this ratio as a common fraction?) ■

RQ1. What is a ratio?
RQ2. When do we express a ratio as a decimal fraction?

Rates

A **rate** is a ratio that compares two quantities with different units.

You are familiar with many rates already. If you say that your average speed was 50 miles per hour, or that apples cost 89 cents per pound, you are using rates. Recall that *per* usually indicates division, so

50 miles per hour means $\frac{50 \text{ miles}}{1 \text{ hour}}$

89 cents per pound means $\frac{89 \text{ cents}}{1 \text{ pound}}$

Example 3 Sarah traveled 390 miles on 15 gallons of gas. Express her rate of fuel consumption as a ratio, and then as a rate.

Solution As a ratio, Sarah's rate of fuel consumption was $\frac{390 \text{ miles}}{15 \text{ gallons}}$. We simplify by dividing the denominator into the numerator, to get $\frac{26 \text{ miles}}{1 \text{ gallon}}$, or 26 miles per gallon. ■

Proportions

A **proportion** is a statement that two ratios are equal.

Thus, a proportion is an equation in which both sides are ratios. Here are some proportions:

$$\frac{6}{9} = \frac{8}{12} \quad \text{and} \quad \frac{3.2}{8} = \frac{1}{2.5}$$

A proportion may involve variables. How can we solve a proportion such as this one?

$$\frac{7}{5} = \frac{x}{6}$$

Because x is divided by 6, we could multiply both sides of the equation by 6 to clear the fraction. But we can clear both fractions at the same time if we multiply by the LCD (Lowest Common Denominator) of the two fractions. In this case, the LCD is $5 \cdot 6 = 30$. Multiplying both sides by 30 gives us

$$30\left(\frac{7}{5}\right) = \left(\frac{x}{6}\right)30$$
$$42 = 5x$$

Now we can divide both sides by 5 to get the answer,

$$x = \frac{42}{5} = 8.4$$

Look Ahead: There is a short-cut we can use that avoids calculating an LCD. Observe that we can arrive at the equation $42 = 5x$ directly from the original proportion by **cross-multiplying**:

$$\frac{7}{5} = \frac{x}{6}$$
$$42 = 5x$$

Then we complete the solution as before.

The cross-multiplying shortcut is a fundamental property of proportions.

Property of Proportions

If $\frac{a}{b} = \frac{c}{d}$, then $ad = bc$.

Example 4 Solve $\frac{2.4}{1.5} = \frac{8.4}{x}$

Solution We apply the property of proportions and cross-multiply to get

$$2.4x = 1.5(8.4) \qquad \text{Divide both sides by 2.4.}$$
$$x = \frac{1.5(8.4)}{2.4} \qquad \text{Simplify the right side.}$$
$$x = 5.25$$

We can check the solution by substituting $x = 5.25$ into the original proportion.

Caution! Cross-mutiplying works **only** for solving proportions! Do not try to use cross-multiplying on other types of equations, or for other operations on fractions. In particular, do not use cross-multiplying when multiplying fractions or adding fractions.

RQ3. What is a rate?
RQ4. What is a proportion?
RQ5. What is the short-cut for solving a proportion called?

30 cents per quart, and complete the following table.

	otal Price (cents)	$\frac{\text{Total Price}}{\text{Number of Quarts}}$
	80	$\frac{80}{1} = 80$
	160	$\frac{160}{2} =$
	240	
	320	

al price / r of quarts', or price per quart, is the same for
s with common sense: The price per quart of
ny quarts you buy.

l if their ratio is always the same.

pe juice costs 80 cents. How much grape

quarts that can be purchased for $10.00, or
variables *total price* and *number of quarts* are proportional, we
that their ratio is constant. Thus, the ratio

$$\frac{1000 \text{ cents}}{x \text{ quarts}} \quad \text{is equal to the ratio} \quad \frac{80 \text{ cents}}{1 \text{ quart}}$$

$$\frac{1000}{x} = \frac{80}{1}$$

We solve the proportion by cross-mutiplying to get

$$1000\,(1) = 80x \qquad \textbf{Divide both sides by 80.}$$

$$x = \frac{1000}{80} = 12.5$$

Thus, 12.5 quarts of grape juice can be purchased for $10.00. (Or, if you can only buy quart bottles, then you can buy 12 quarts and have 40 cents left over.) ■

Caution! When writing a proportion, we must be careful that *both* ratios have the *same units* in their numerators, and the same units in their denominators. In the example above, it would *not* be correct to equate $\frac{1000}{x}$ and $\frac{1}{80}$, because the ratios do not have the same units. ■

For the grape juice example, the ratio $\dfrac{\text{total price}}{\text{number of quarts}}$ was always equal to 80. The number 80 is called the **constant of proportionality**.

Look Closer: If we let P stand for the price of the grape juice and q stand for the number of quarts purchased, we have the equation

$$\frac{P}{q} = 80$$

or

$$P = 80q$$

The graph of this equation is a straight line that passes through the origin, as shown at right.

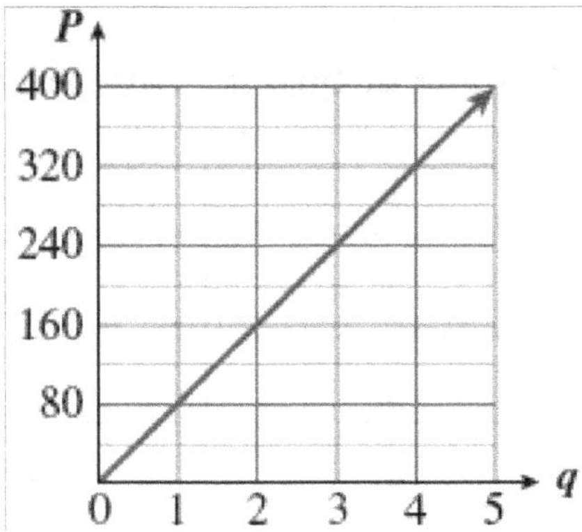

Proportional Variables

If two variables are proportional, they are related by the equation

$$y = kx$$

where k is the **constant of proportionality**.

RQ6. What are proportional variables?
RQ7. If x and y are proportional, what equation do they satisfy?

Skills Warm-Up

Convert each common fraction to a decimal, and each decimal to a common fraction.

1. $\frac{3}{8}$	**2.** $\frac{7}{5}$	**3.** $\frac{4}{3}$	**4.** $\frac{5}{12}$
5. 0.04	**6.** $0.\overline{6}$	**7.** 1.875	**8.** 2.2

Answers: **1.** 0.375 **2.** 1.4 **3.** $1.\overline{3}$ **4.** $0.41\overline{6}$ **5.** $\frac{1}{25}$ **6.** $\frac{2}{3}$ **7.** $\frac{15}{8}$ **8.** $\frac{11}{5}$

Homework 3.3

Skills Practice

■ For Problems 1-4, write ratios or rates.

1. In a survey of 300 employees at a large company, 125 used public transportation to commute to work. What is the ratio of employees who use public transportation to those who do not?

2. Erica tutored in the math lab for 6.5 hours last week and made $56.68. What is Erica's rate of pay in dollars per hour?

3. The instructions for mixing a rose fertilizer call for $\frac{3}{4}$ cup of potash and $1\frac{3}{8}$ cups of nitrogen. What is the ratio of nitrogen to potash?

4. An orange contains 0.14 milligram of thiamin and 0.6 milligram of niacin. What is the ratio of niacin to thiamin?

5. Which of the following expressions are proportions?

a. $\frac{3}{x} = \frac{8}{15}$ **b.** $\frac{v}{9} - \frac{12}{5}$ **c.** $\frac{w}{6} + \frac{2}{5} = \frac{3}{2}$ **d.** $\frac{2.6}{m} = \frac{m}{1.6}$

■ For Problems 6-9, solve the proportion.

6. $\frac{x}{16} = \frac{9}{24}$ **7.** $\frac{182}{65} = \frac{21}{w}$ **8.** $\frac{7}{b} = \frac{5}{9}$ **9.** $\frac{7/3}{3} = \frac{z}{9/2}$

■ For Problems 10-12, decide whether the two variables are proportional.

10.

Time	Distance
1	45
2	90
4	180
5	225

11.

Length	Area
3	9
4	16
8	64
10	100

12.

Rate	Time
20	40
40	20
50	16
80	10

■ For Problems 13-16, decide whether the two quantities are proportional.

13. A 9-inch pizza costs $6, and a 12-inch pizza costs $10.
14. A commuter train travels 10 miles in 20 minutes, and 15 miles in 30 minutes.
15. It takes $4\frac{1}{2}$ cups of flour to make two loaves of bread, and 18 cups to make eight loaves.
16. An 18-foot sailboat sleeps four, and a 32-foot sailboat sleeps six.

Applications

■ For Problems 17-22,

a. Write a proportion.
b. Solve the proportion and answer the question.

17. If 3 pounds of coffee makes 225 cups, how many pounds of coffee is needed to make 3000 cups of coffee?

18. Gunther's car uses 32 liters of gas to travel 184 kilometers. How many liters will he need to travel 575 kilometers?

19. On a map of Fairfield County, $\frac{3}{4}$ centimeter represents a distance of 10 kilometers. If Eastlake and Kenwood are 6 centimeters apart on the map, what is the actual distance between the two towns?

20. In a survey of 1200 voters, 863 favored the construction of a light rail system. If 8000 people vote, about how many will vote for the light rail system?

21. If 1 inch equals 2.54 centimeters, what is the length in inches of a wire 35 centimeters long?

22. A cinnamon bread recipe calls for $1\frac{1}{4}$ tablespoons of cinnamon and 5 cups of flour. How much cinnamon would be needed with 8 cups of flour?

For Problems 23-24,

a. Find the constant of proportionality and write an equation relating the variables.
b. Make a table of values and graph the equation.
c. Use your equation to answer the question.

23. Megan's wages, w, are proportional to the time, h, she works. Megan made \$60 for 8 hours work as a lab assistant. How long will it take her to make \$500?

h	w

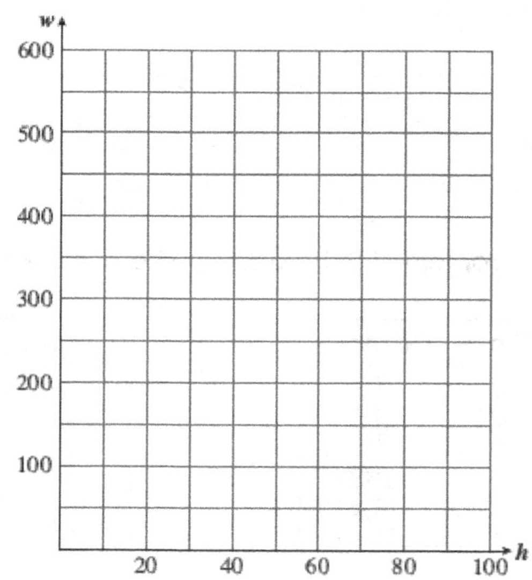

24. The amount of pure copper, c, produced by Copperfield Mine is proportional to the amount of ore extracted, g. A new lode produced 24 grams of copper from 800 grams of ore. How many grams of copper will one kilogram of ore yield?

g	c

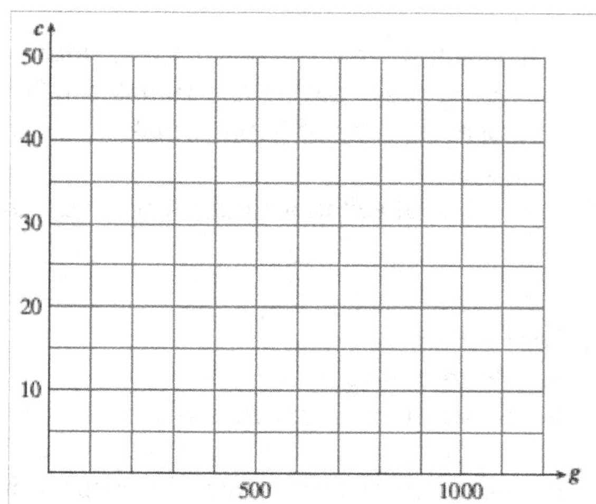

25. Trish's real-estate commission is proportional to the amount of the sale. She made \$3200 commission on a sale of \$80,000.

a. Write an equation relating Trish's sales and her commission.

b. What would her commission be on the sale of a \$200,000 property?

26. Everett can bicycle 16 miles in 2 hours and 24 miles in 3 hours. The distance d that Everett travels is proportional to the time t that he bicycles.

a. Write an equation for d in terms of t.

b. Graph your equation.

c. Complete the tables, and use them to answer the question below.

t	d
2	
3	
5	

t	d
4	
6	
10	

What happens to the distance Everett travels when he doubles the time he bicycles?

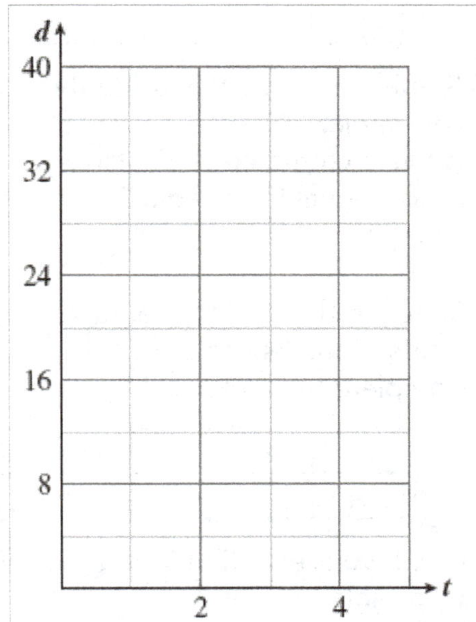

27. The sales tax on \$12 is 72 cents, and the sales tax on \$15 is 90 cents. The tax T is proportional to the price P.

a. Write an equation for T in terms of P.

b. Graph your equation.

c. Complete the tables, and use them to answer the question below.

P	T
12	
20	
30	

P	T
6	
10	
15	

What happens to the sales tax when the price is cut in half?

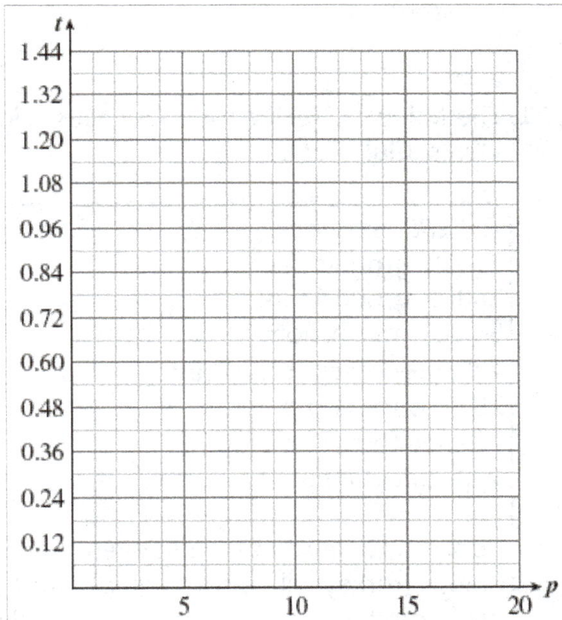

28. The cost of tuition at Walden College is given by the formula $T = 500 + 40u$, where u is the number of units you take.

a. Fill in the tables.

Units	Tuition
3	
5	
8	

Units	Tuition
6	
10	
16	

b. Does doubling the number of units you take double your tuition?

c. Is T proportional to u?

29. a. State a formula for the circumference of a circle in terms of its radius.

b. Complete the table.

r	2	5	7	10
C				

c. Graph your equation, using the values in your table.

d. Is the circumference of a circle proportional to its radius?

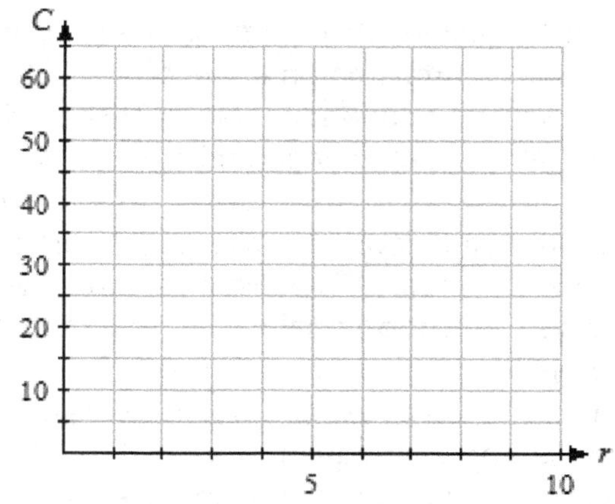

30. a. State a formula for the area of a square in terms of the length of its side.

b. Complete the table.

s	2	5	6	8
A				

c. Graph your equation, using the values in your table.

d. Is the area of a square proportional to the length of its side?

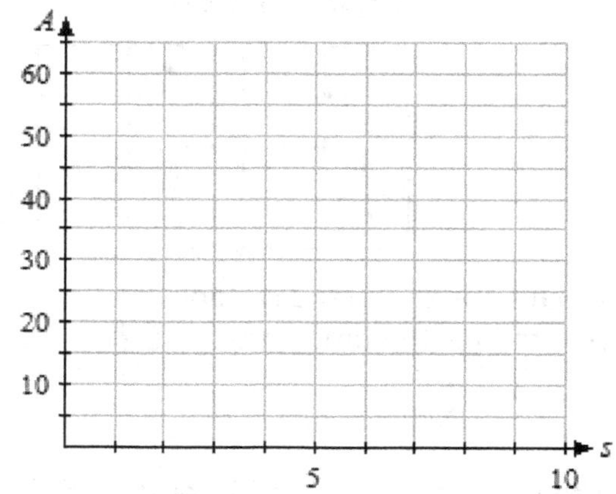

31. The following recipe for Cadet Mess Sloppy Joes comes from the West Point Officers Wives' Club Cookbook.

For 1000 servings:
310 lb. ground beef
100 lb. ground turkey
$5\frac{1}{2}$ cans mushrooms
4 lb. green peppers, sliced
18 cans tomato puree
12 cans tomato catsup
1 qt. cider vinegar
2 cans diced red peppers
20 lb. onions, diced
3 lb. chili powder
1 lb. garlic powder
$3\frac{1}{2}$ lb. sugar
1 qt. light flavor glow
28 oz. salt and pepper

Cook off turkey and beef in oven; place in pots. Braise off chopped peppers and diced onions (turkey will absorb liquid). Add stock to turkey to moisten well before adding other products. Stir together. Heat till piping hot. Serve on rolls.

a. What is the ratio of tomato puree to tomato catsup? How many servings of sloppy joes can you make if you have only 5 cans of tomato catsup? How many cans of tomato puree will you need?

b. How much chili powder (in ounces) will you need to serve sloppy joes to 50 people? How many pounds of ground beef?

c. The cookbook also lists the ingredients needed for 4500 servings. How many cans of mushrooms are required for 4500 servings? How many pounds of onions?

3.4 Slope

Rate of Change

A **rate of change** is a type of ratio that measures how one variable changes with respect to another.

Example 1 In order to fire a particular kind of pottery, the pieces must first be cured by raising the temperature slowly and evenly. Sonia checks the temperature in the drying oven at ten-minute intervals, and records the following data.

Time, x	0	10	20	30	40	50	60
Temperature, y	70	74	78	82	86	90	94

The heat in the oven should not increase any faster than 0.5 degree per minute. Is the temperature in the oven within the safe limits?

Solution A graph of the data is shown at right. Sonia calculates the rate at which the temperature is rising by finding the following ratio:

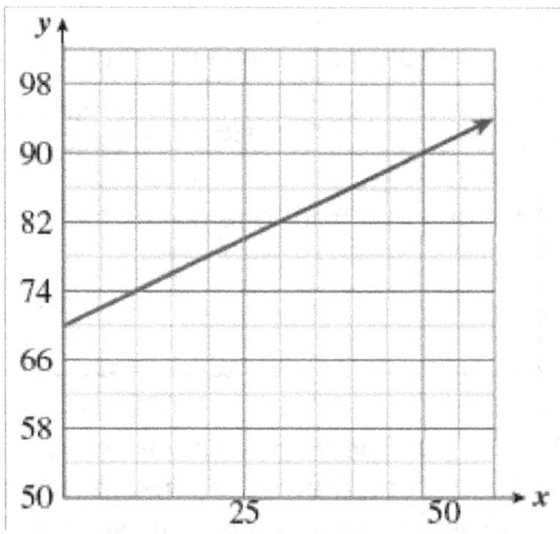

$$\frac{\text{change in temperature}}{\text{change in time}}$$

For example, over the first 10 minutes, the temperature rises from 70 degrees to 74 degrees, so

$$\frac{\text{change in temperature}}{\text{change in time}} = \frac{4 \text{ degrees}}{10 \text{ minutes}}$$

or 0.4 degree per minute. This is less than the maximum rate recommended for curing the pottery.

You can check that over each ten-minute interval the temperature again rises by four degrees, so it appears that the oven is heating up at an acceptable rate.

RQ1. What is a rate of change?
RQ2. What are the units of the rate of change in Example 1?

Slope

We introduce some new notation to use when calculating a rate of change.

The Greek letter Δ ("delta") is used in mathematics to indicate **change**.

In Example 1, we used the variable x to represent time and y to represent the temperature, so we denote the ratio $\frac{\text{change in temperature}}{\text{change in time}}$ by $\frac{\Delta y}{\Delta x}$. With this notation, we calculate the rate of change of temperature between the data points $(20, 78)$ and $(50, 90)$ as follows:

$$\frac{\Delta y}{\Delta x} = \frac{\text{12 degrees}}{\text{30 minutes}} = \text{0.4 degrees per minute}$$

We can illustrate the rate of change on a graph of the data, as shown at right. We move from the point $(20, 78)$ to the point $(50, 90)$ by moving horizontally a distance of $\Delta x = 30$ and then vertically a distance of $\Delta y = 12$.

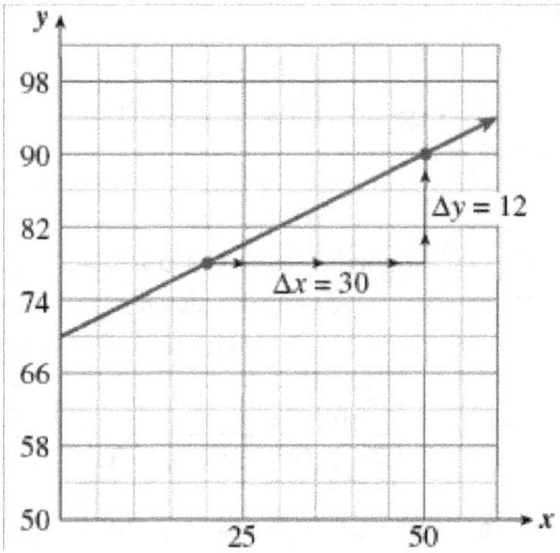

The rate of change of one variable with respect to another is so important in applications that the ratio $\frac{\Delta y}{\Delta x}$ is given a name; it is called **slope**, and is usually denoted by the letter m.

> **Slope**
>
> The **slope** of a line is defined by the ratio
>
> $$\textbf{slope} = \frac{\textbf{change in } y\textbf{-coordinate}}{\textbf{change in } x\textbf{-coordinate}}$$
>
> as we move from one point to another on the line. In symbols,
>
> $$m = \frac{\Delta y}{\Delta x}$$

Look Closer: The slope of a line measures how fast the y-coordinate changes as we increase the x-coordinate of points on the line. More specifically, when we move one unit in the x-direction, how many units should we move in the y-direction to get back tothe line?

Example 2 Use the points A and B to compute the slope of the line shown.

Solution The point A has coordinates $(1, 3)$, and B has coordinates $(5, 6)$. As we move along the line from $A(1, 3)$ to $B(5, 6)$, the y-coordinate changes by 3 units, and the x-coordinate changes by 4 units. The slope of the line is thus

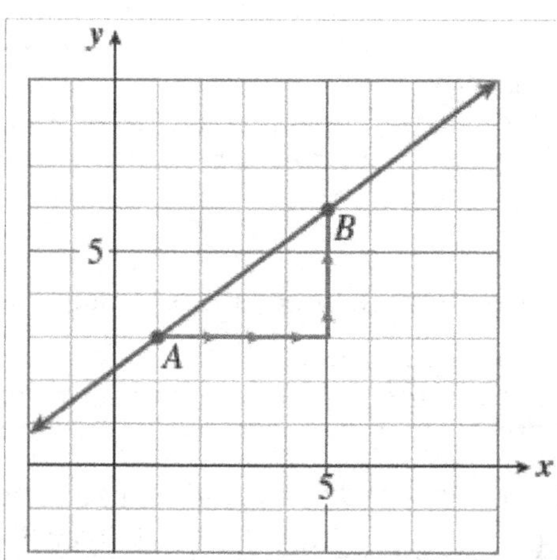

$$\frac{\Delta y}{\Delta x} = \frac{\text{change in } y\text{-coordinate}}{\text{change in } x\text{-coordinate}} = \frac{3}{4}$$

The slope tells us that if we start at any point on the line and move 1 unit in the x-direction, we must move $\frac{3}{4}$ unit in the y-direction to return to the line.

Caution! Note the difference between the statements $y = 3$ and $\Delta y = 3$; they are not the same. When we discuss a graph, $y = 3$ means that the y-coordinate of a particular point is 3, but $\Delta y = 3$ means that the y-coordinate changes by 3 units when we move from one point to another. ■

RQ3. What does Δ mean in mathematics?
RQ4. How do we indicate Δx on a graph?
RQ5. What is the name of the ratio $\dfrac{\Delta y}{\Delta x}$, and what letter is used to represent it?

Meaning of Slope

In Example 1, we graphed the temperature of a pottery oven over time. We calculated the slope of the graph as

$$\frac{\Delta y}{\Delta x} = 0.4 \text{ degrees per minute}$$

The slope gives us the rate of change of the temperature with respect to time: the temperature is increasing at a rate of 0.4 degrees per minute.

The **slope** of a line measures the **rate of change** of y with respect to x.

In different situations, this rate might be interpreted as a rate of growth or a speed. The slope of a graph can give us valuable information about the variables involved.

■ **Example 3** The graph shows the distance traveled by a driver for a cross-country trucking firm in terms of the number of hours she has been on the road.

a. Compute the slope of the graph.

b. What is the meaning of the slope for this problem?

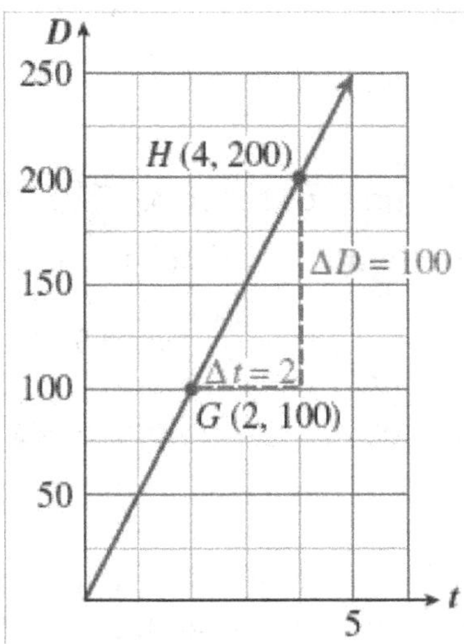

Solutions a. Choose any two points on the line, say $G(2, 100)$ and $H(4, 200)$ shown in the figure. As we move from G to H we find

$$m = \frac{\Delta D}{\Delta t} = \frac{100}{2} = 50$$

The slope of the line is 50.

b. The best way to understand the slope is to include units in the calculation.

$$\frac{\Delta D}{\Delta t} \text{ means } \frac{\text{change in distance}}{\text{change in time}} \quad \text{or} \quad \frac{\Delta D}{\Delta t} = \frac{100 \text{ miles}}{2 \text{ hours}} = 50 \text{ miles per hour}$$

The slope represents the trucker's average speed or velocity. ■

Caution! In Example 3, we refer to a point by a capital letter and the coordinates of the point, like this: $H(4, 200)$. This means that $t = 4$ and $D = 200$ at the point H. Do not confuse the coordinates (t, D) of a particular point with the values of Δt and ΔD obtained by moving from one point to a second point. ■

RQ6.	What does the slope of a line measure?
RQ7.	What does the slope of the line measure in Example 3?

Negative Slopes

The slope you calculated in Example 3 is positive because the graph is increasing. An increasing quantity has a positive rate of change, or slope, and a decreasing quantity has a negative rate of change. A negative slope might represent a rate of decrease or a rate of consumption.

Example 4 The value of new office equipment decreases, or depreciates, over time. The graph shows the value, V, in thousands of dollars, of a large photocopying machine t years after it was purchased.

a. Compute the slope of the line.

b. What does the slope tell you about the value of the machine?

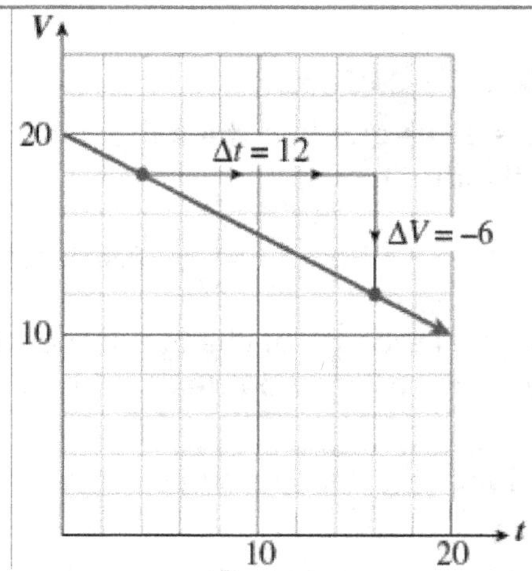

Solutions a. We choose two points on the graph, say, $(4, 18)$ and $(16, 12)$, and we imagine moving from the first point to the second point in two steps, first horizontally and then vertically, as shown in the figure. Each grid line on the horizontal axis counts for 2 units, and we move 6 spaces, so $\Delta t = 12$. To get to the second point, we must move vertically downward, so ΔV is negative. Each grid line crossing the vertical axis also counts for 2 units, so $\Delta V = -6$. Thus, the slope of the line is

$$\frac{\Delta V}{\Delta t} = \frac{-6}{12} = -0.5$$

b. Consider the units of the ratio $\frac{\Delta V}{\Delta t}$. The vertical axis is scaled in thousands, so the units of ΔV are thousands of dollars, and the units of Δt are years. Thus,

$$\frac{\Delta V}{\Delta t} = -0.5 \text{ thousand dollars per year}$$

or -500 dollars per year. The slope gives the rate of depreciation of the photocopying machine.

Look Closer: When we move downward on the graph in Example 4, ΔV is considered negative. Similarly, when we move to the left horizontally, Δt is considered negative. If we compute the slope in Example 3 by moving from the point $(16, 12)$ to the point $(4, 18)$ as shown at right, then we have

$$\frac{\Delta V}{\Delta t} = \frac{6}{-12} = -0.5,$$

the same answer as before. It doesn't matter which direction we move along a line to compute its slope; the answer will be the same.

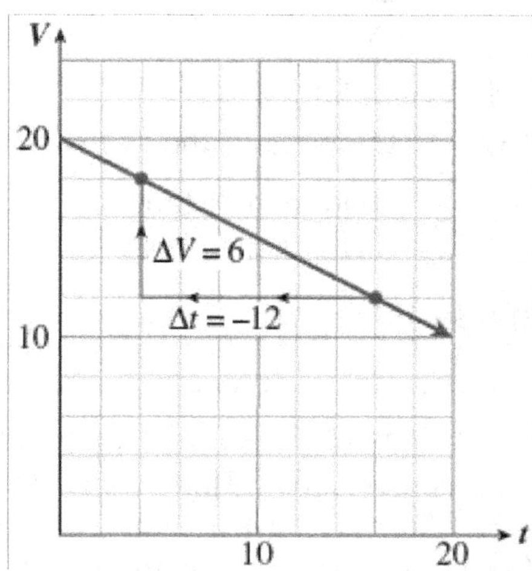

Lines Have Constant Slope

How do we know which two points to choose when we want to compute the slope of a line? It turns out that any two points on the line will do.

Example 5 Calculate the slope of the line $y = \frac{2}{3}x - 2$ shown in the figure:

a. By using the points $P(-3,-4)$ and $Q(3,0)$.
b. By using the points $R(6,2)$ and $S(0,-2)$.

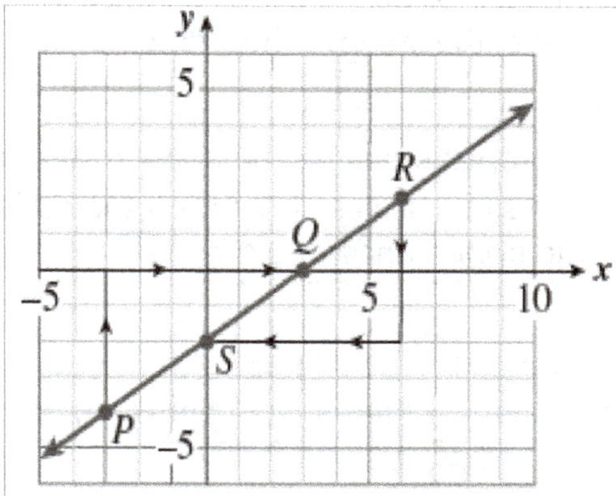

Solutions a. Starting at point $P(-3,-4)$, we move up 4 units in the y-direction and right 6 units in the x-direction to get to point $Q(3,0)$.

$$m = \frac{\Delta y}{\Delta x} = \frac{4}{6} = \frac{2}{3}$$

b. Starting at point $R(6,2)$, we move down 4 units in the y-direction and left 6 units in the x-direction to get to point $S(0,-2)$. Thus

$$m = \frac{\Delta y}{\Delta x} = \frac{-4}{-6} = \frac{2}{3}$$

We get the same value for the slope, $\frac{2}{3}$, by using either pair of points.

Example 4 illustrates an important property of lines: they have constant slope.

> No matter which two points on a line we use to calculate its slope, we will always get the same result.

We will see later that lines are the only graphs that have this property.

Geometrical Meaning of Slope

Suppose we graph two lines with positive slope on the same coordinate system. If we move along the lines from left to right, then the line with the larger slope will be steeper. This makes sense if we think of the slope as a rate of change: The line whose y-coordinate is increasing faster with respect to x is the steeper line.

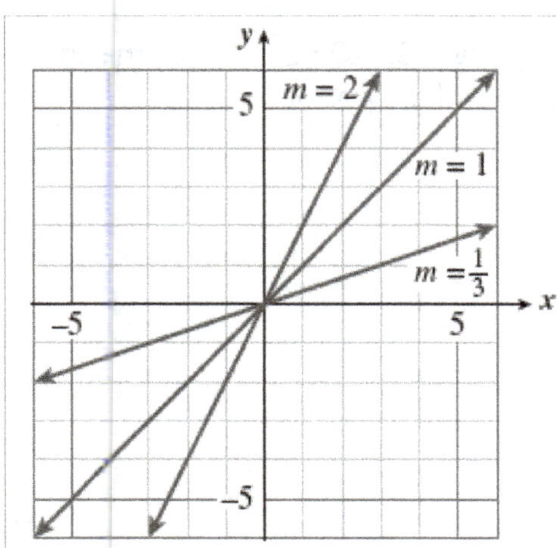

a.

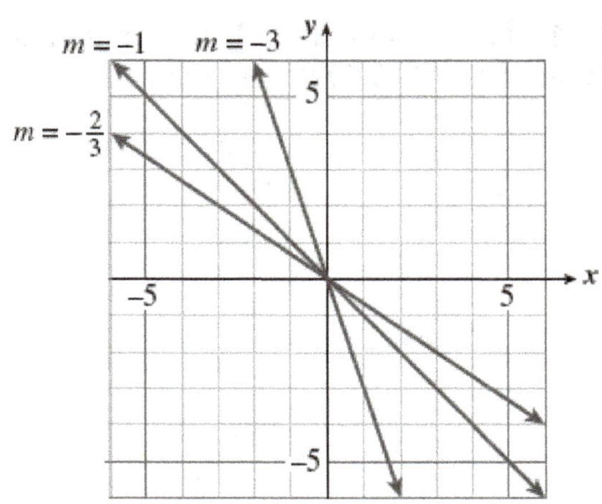

b.

Look Closer: You can verify the slope given for each line in Figure (a) by computing $\dfrac{\Delta y}{\Delta x}$. For each unit you increase in the x-direction, the steepest line increases 2 units in the y-direction, the middle line increases 1 unit in the y-direction, and the flattest line increases only $\frac{1}{3}$ unit.

Figure (b) shows several lines with negative slopes. These lines slant downwards or decrease as we move from left to right. The more negative the slope, the more sharply the line decreases. For both increasing and decreasing graphs, the larger the absolute value of the slope, the steeper the graph.

Caution! Slopes measure the relative steepness of two lines only if they are graphed on axes with the same scales. Changing the scale on either the x-axis or the y-axis can greatly alter the appearance of a graph.

RQ8. What sort of lines have negative slopes?
RQ9. One line is steeper than another if the absolute value of its slope is ______.

What about the slopes of horizontal and vertical lines? In the Homework Problems you will show that the slope of a horizontal line is zero, and the slope of a vertical line is undefined. We sometimes say that a vertical line has no slope, but a horizontal line does have a slope: Its slope is zero.

Skills Warm-Up

Write a rate for each of the following situations, including units.

1. Zack's average speed, if he drove 426 miles in 9 hours.
2. Zelda's average speed, if she ran 6.6 miles in 55 minutes.
3. The rate at which water flows through a pipe, if a 400-gallon storage tank fills in 20 minutes.
4. A baby whale's rate of growth, if it gains 3000 pounds in its first 40 days of life.
5. Earnest's rate of pay, if he earns $344 for a 40-hour week.
6. Meg's rate of pay, if she charges $90 to type a 40-page paper.

Answers: **1.** $47.\overline{3}$ miles per hour **2.** 0.12 miles per minute **3.** 20 gallons per minute **4.** 75 pounds per day **5.** 8.60 dollars per hour **6.** 2.25 dollars per page

Homework 3.4

Skills Practice

■ For Problems 1-2, find the slope of each line segment.

1a. 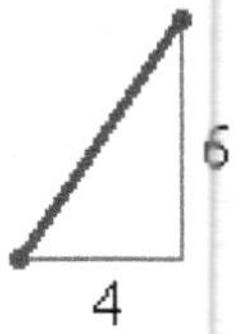

b. 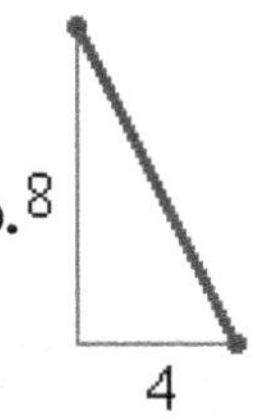

2a. 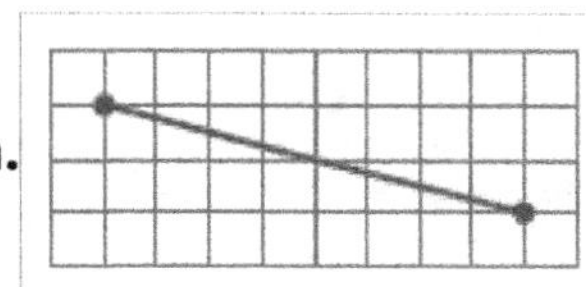

b. 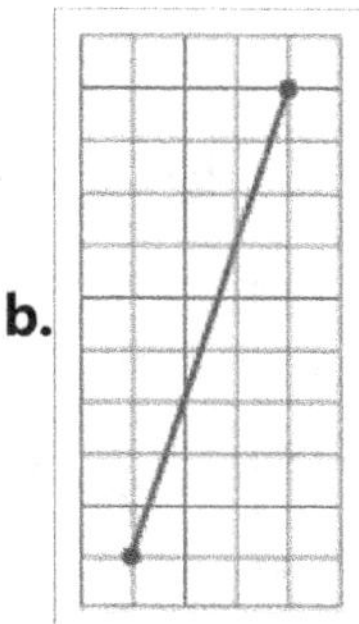

3. Choose two points from the table and compute the slope of the line.

x	0	2	6	8
y	−30	0	60	90

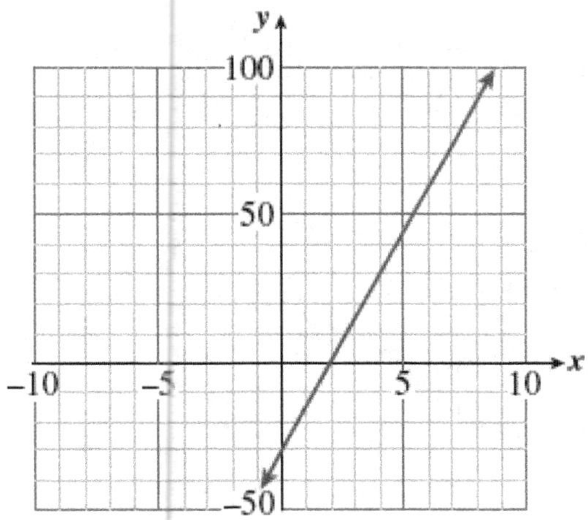

4. Graph the line and compute its slope: $y = -12x + 32$

x	−2	0	3	4
y				

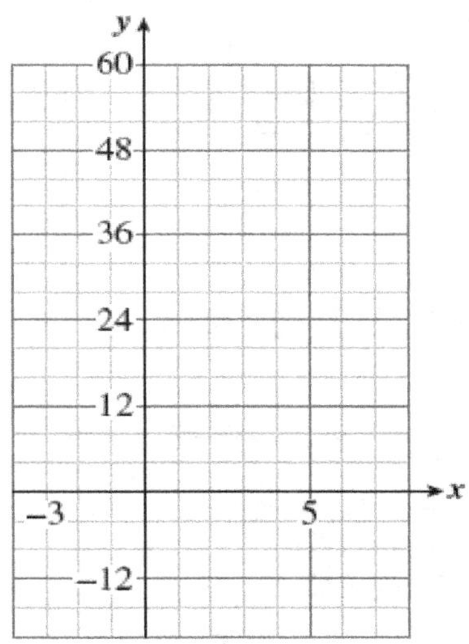

■ For Problems 5-6, find the slope of the line. Illustrate Δx and Δy on the graph.

5. $x + 2y = 6$

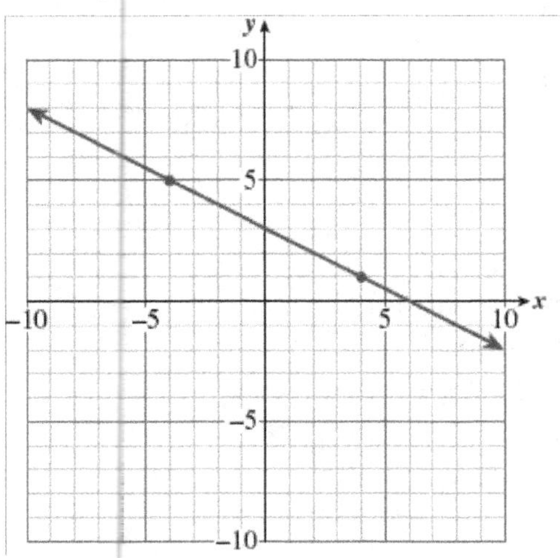

6. $3x - 2y = 0$

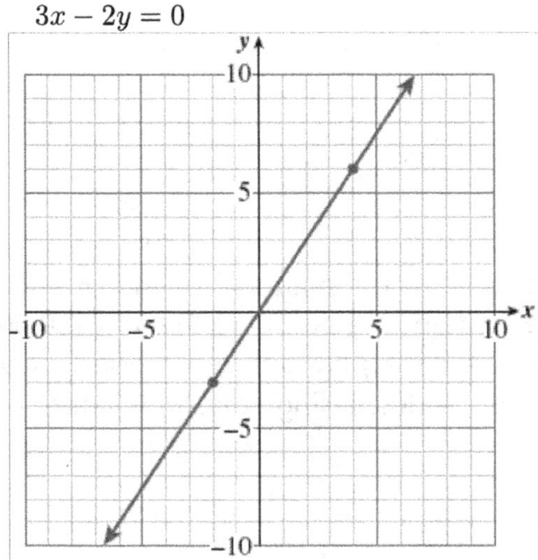

For Problems 7-10,

a. Find the intercepts of each line.
b. Graph the line on the grid provided. Use the intercept method.
c. Use the intercepts to calculate the slope of the line.
d. Calculate the slope again using the suggested points on the line.

7. $2x + 3y = 12$
$(-3, 6)$ and $(3, 2)$

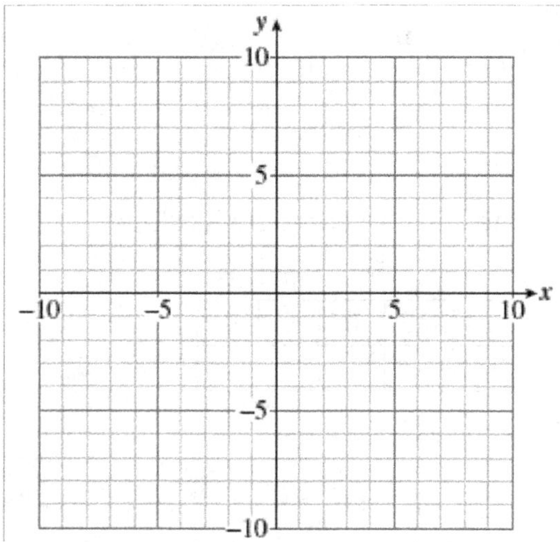

8. $5x - 2y = 10$
$(-2, -10)$ and $(4, 5)$

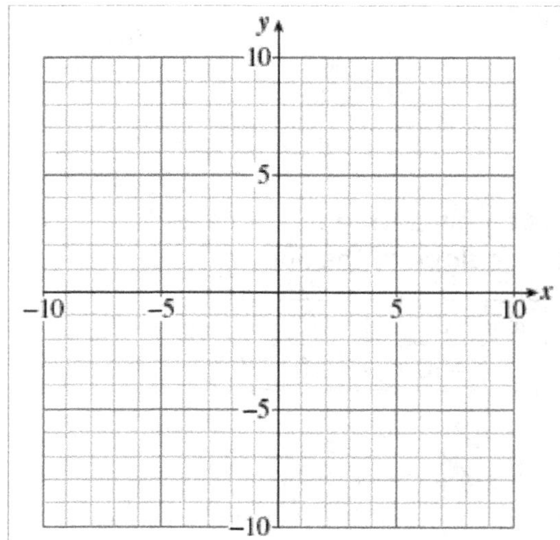

9. $x + y = 5$
$(-3, 8)$ and $(8, -3)$

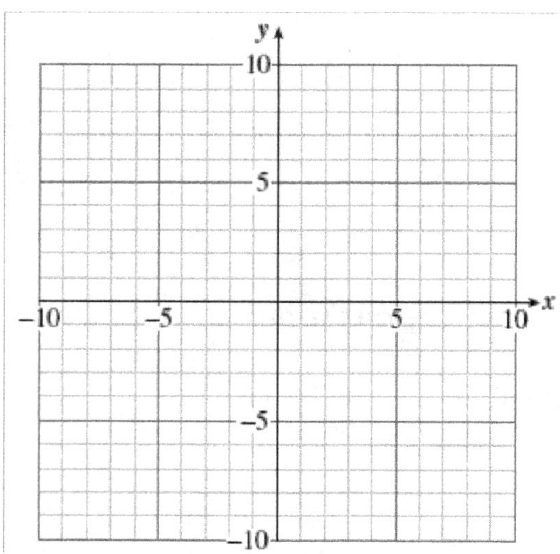

10. $x - 2y = 4$
$(6, 1)$ and $(-4, -4)$

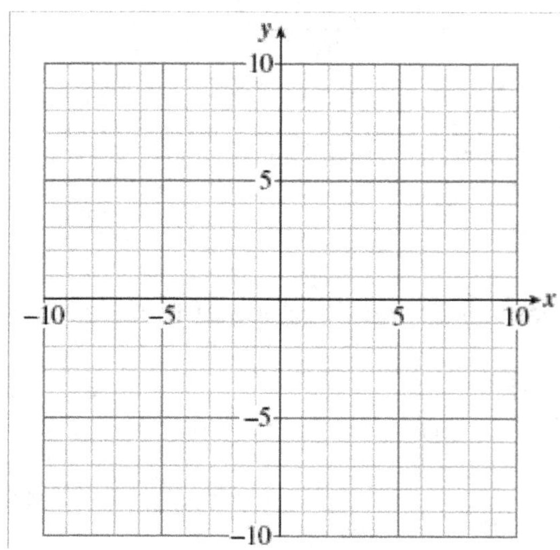

11. A line contains the points $(0, 0)$ and $(3, 2)$. What is its slope?
12. A line contains the points $(0, 0)$ and $(-30, 50)$. What is its slope?
13. Which line is steeper: one with slope $\frac{3}{5}$ or one with slope $\frac{5}{3}$?
14. Which line is decreasing: one with slope $\frac{1}{4}$ or one with slope -2?

15. The line shown in Figure (a) below has slope $\frac{5}{2}$. If $\Delta x = 7$, find Δy.

a.

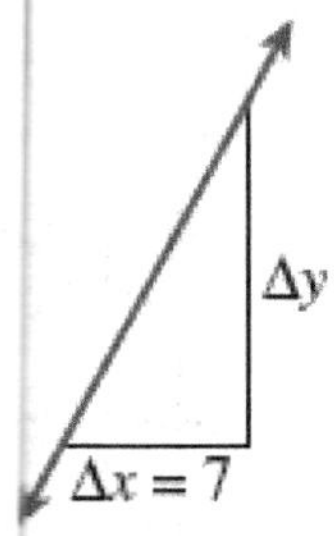

b.

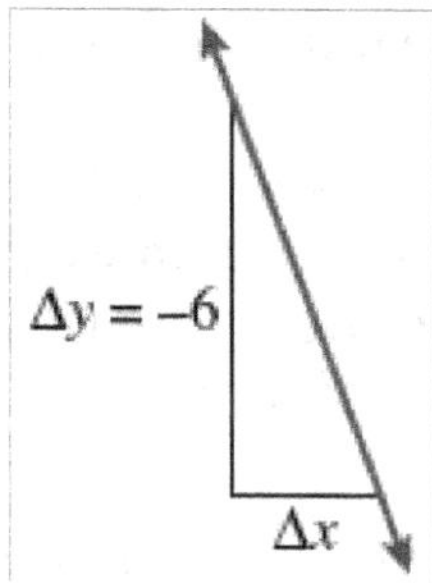

16. The line shown in Figure (b) above has slope -4. If $\Delta y = -6$, find Δx.

Applications

For Problems 17-18,

a. Compute the slope of the graph, including units.

b. Interpret the slope as a rate; what does it tell you about the problem?

17. Audrey can drive 150 miles on 6 gallons of gas, and 225 miles on 9 gallons of gas. Write an equation for the distance, d, that Audrey can drive on g gallons of gas.

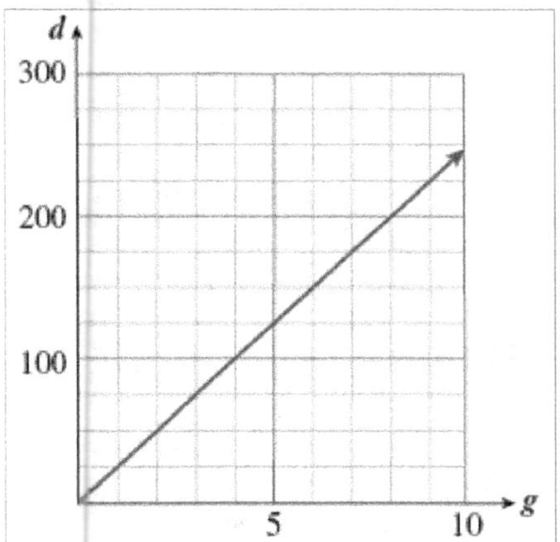

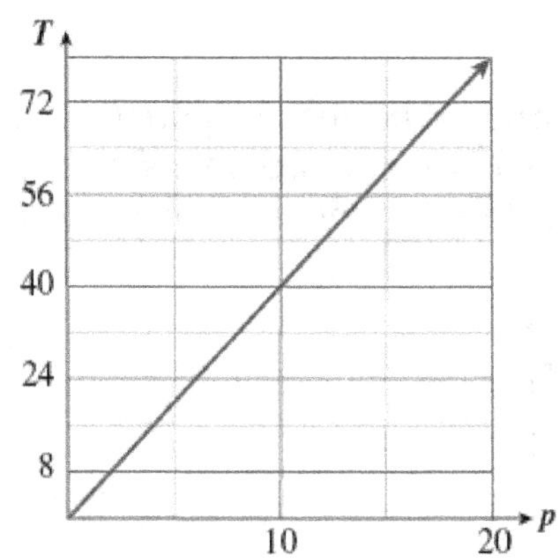

18. The sales tax on a \$15 purchase is 60 cents, and 80 cents on a \$20 purchase. Write an equation for the tax T on a purchase of p dollars.

19. Lynette is saving money for the down payment on a new car. The figure at right shows the amount A she has saved, in dollars, w weeks after the first of the year.

a. How much does Lynette save each week?

b. Give the coordinates (w, A) of any two points on the graph. Use those coordinates to compute the slope of the graph, $\frac{\Delta A}{\Delta w}$.

c. What are the units of the slope? What does the slope tell you about the problem?

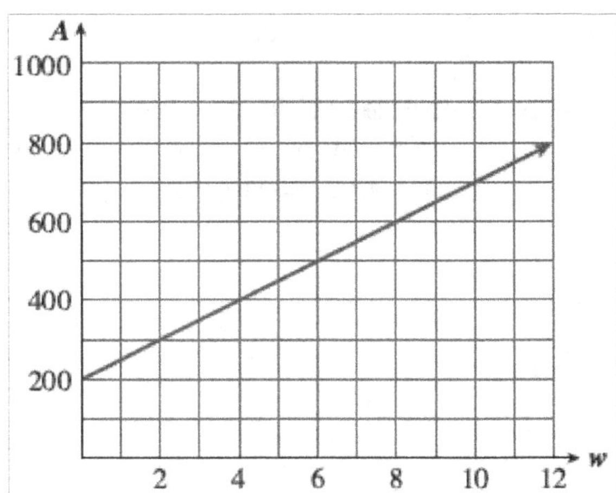

20. Jason is raising a rabbit for the county fair. The figure at right shows the rabbit's weight W when it was t weeks old.

a. How much did the rabbit's weight increase from the fourth week to the twelfth week? From the second week to the eighth week?

b. Compute the rabbit's rate of growth, including units.

c. Illustrate the rate of growth, $\dfrac{\Delta W}{\Delta t}$, on the graph.

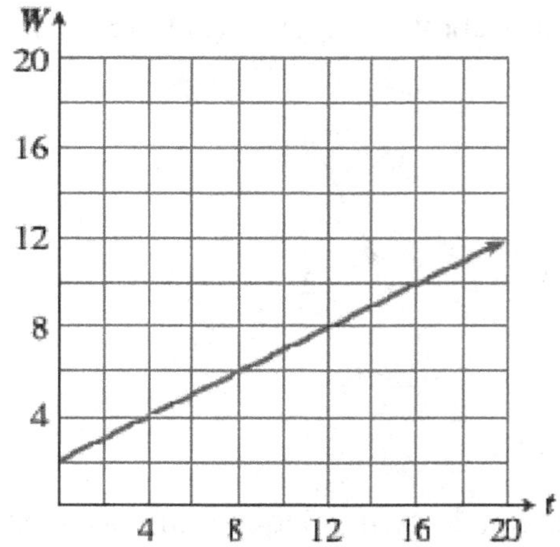

For Problems 21-22, draw and label a sketch for the situation. Use the definition of slope to answer the questions.

21. A sign on the highway says "6% grade, next 3 miles." This means that the slope of the road ahead is $\frac{6}{100}$. How much will you climb in elevation (in feet) over the next 3 miles?

22. A wheelchair ramp must have a slope of 0.125. If the ramp must reach a door whose base is 2 feet off the ground, how far from the building should the base of the ramp be placed?

23. a. Calculate the slope of the line in Figure (a) below.

b. Explain why $\Delta y = 0$ for any two points on the line.

c. Explain why the slope of any horizontal line is zero.

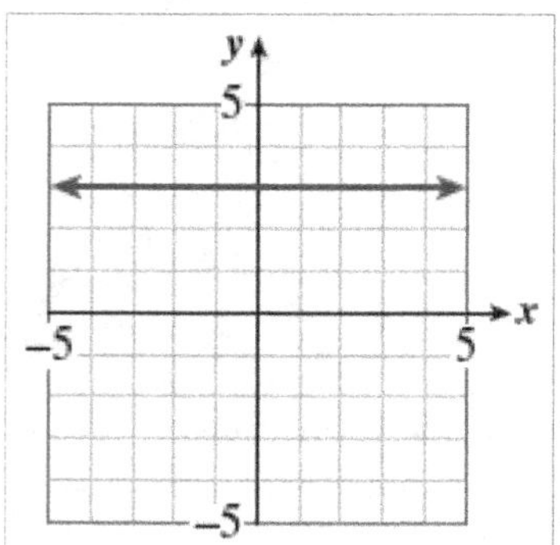

a.

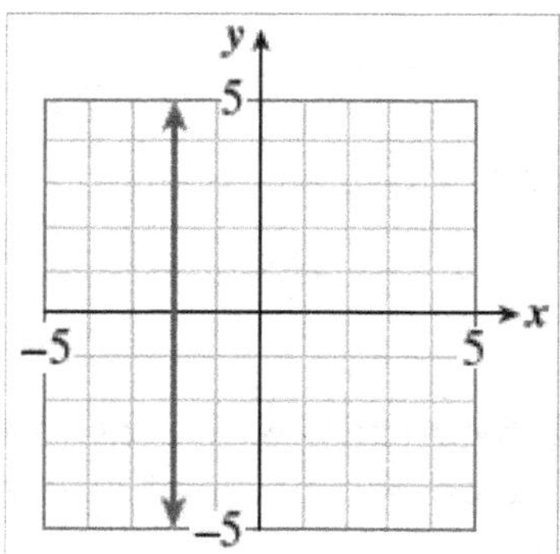

b.

24. a. Calculate the slope of the line in Figure (b) above.

b. Explain why $\Delta x = 0$ for any two points on the line.

c. Explain why the slope of any vertical line is undefined.

25. a. Which of the two graphs in the figure below appears steeper?

b. Compute the slopes of the two graphs. Which has the greater slope?

I

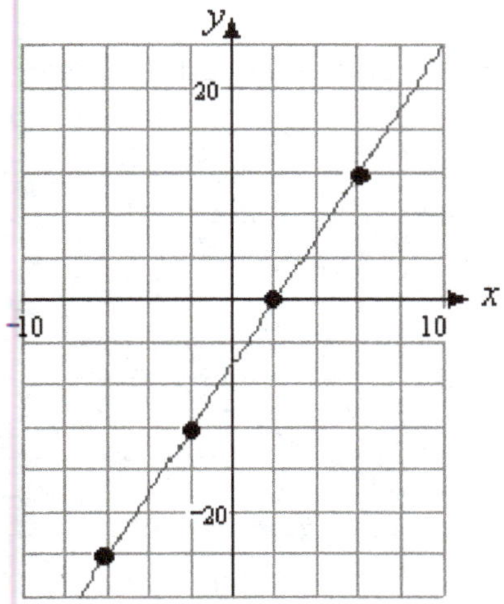

II

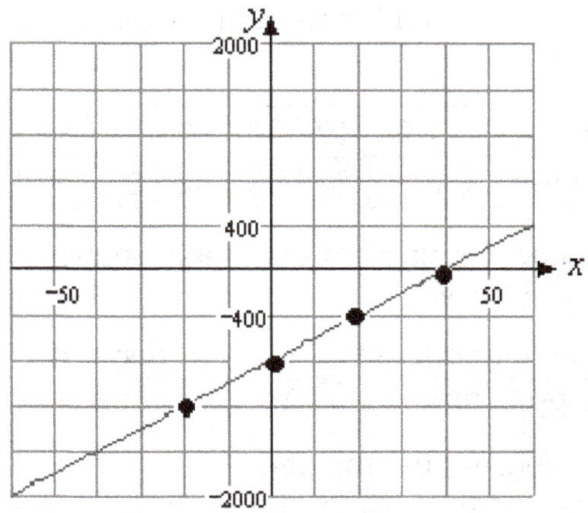

26. Kira buys granola in bulk at the health food store. There are two standard containers customers can use. The size of each container in ounces and its price in dollars are shown in the graph.

a. Read the coordinates of the two points shown on the graph.

b. Calculate the slope of the graph, including units. What does the slope tell us about the granola?

c. Extend the graph to include 25 ounces of granola. How much taller must you make the vertical axis?

d. Extend the graph to include $9.00 worth of granola. How far must you extend the horizontal axis?

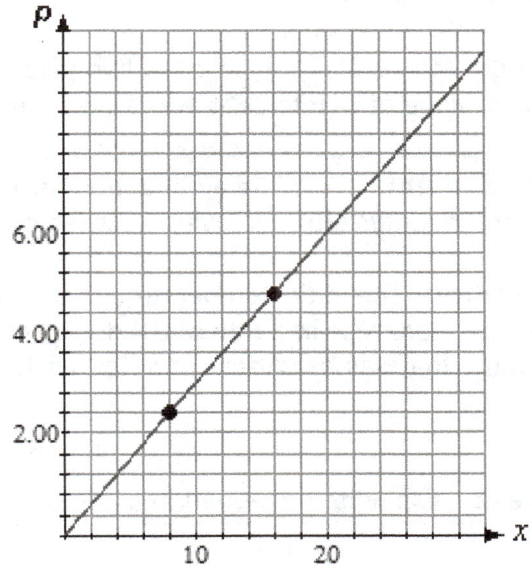

3.5 Slope-Intercept Form

Writing a Linear Equation

In Lesson 3.4, we plotted data for the temperature inside an oven used to cure pottery.

Time, x	0	10	20	30	40	50	60
Temperature, y	70	74	78	82	86	90	94

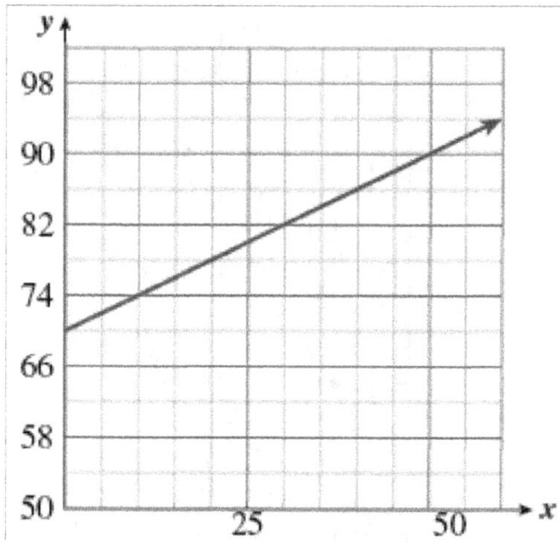

We computed the slope of the graph, shown in the figure. The slope is

$$\frac{\text{change in temperature}}{\text{change in time}} = \frac{\Delta y}{\Delta x} = \frac{4 \text{ degrees}}{10 \text{ minutes}}$$

or $m = 0.4$ degrees per minute.

You can see from either the table or the graph that the y-intercept is the point $(0, 70)$. This means that the initial temperature inside the drying oven was 70 degrees.

Look Ahead: If we know two pieces of information about a line: its slope and its initial value or y- intercept, we can write its equation.

Example 1 Write an equation for the temperature, H, inside the pottery drying oven t minutes after the oven is turned on.

Solution The initial temperature in the oven was 70 degrees, so $H = 70$ when $t = 0$. The temperature rose at a rate of 0.4 degrees per minute, so we add 0.4 degrees to H for each minute that passes. After t minutes, we have added $0.4t$ degrees, giving us a temperature of

$$H = 70 + 0.4t$$

We can also write the equation as $H = 0.4t + 70$.

RQ1. What does the constant term in an equation tell us about the graph?
RQ2. What does the coefficient of the input variable tell us?

From Example 1, we see that the coefficients of a linear equation tell us something about its graph.

1. The constant term tells us the vertical intercept of the graph.
2. The coefficient of the input variable tells us the slope of the graph.

Slope-Intercept Form

A linear equation written in the form

$$y = mx + b$$

is said to be in **slope-intercept** form. The coefficient m is the **slope** of the graph, and b is the **y-intercept**.

RQ3. What does the y-intercept tell us in Example 1?
RQ4. What is the slope-intercept form of an equation?

Slope-Intercept Method of Graphing

A linear equation has the form $Ax + By = C$, and its graph is a straight line. We have already studied two methods for graphing linear equations:

1. Make a table of values and plot points
2. Find and plot the intercepts (the intercept method)

There is a third graphing method that makes use of the slope of the line. We can use the slope-intercept form to sketch a graph quickly, without having to plot a lot of points.

Example 2 Graph the equation $y = \frac{3}{4}x - 2$

Solution The slope of the line is $\frac{3}{4}$ and its y-intercept is the point $(0, -2)$. We begin by plotting the y-intercept, as shown in the figure. Next, we use the slope to find another point on the line. The slope,

$$m = \frac{\Delta y}{\Delta x} = \frac{3}{4}$$

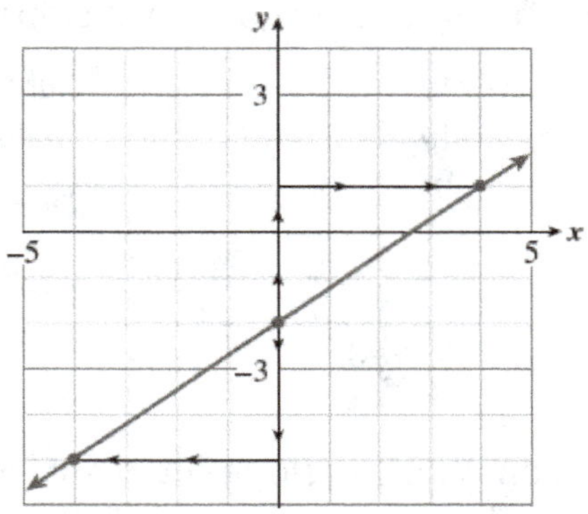

gives the ratio of the change in y-coordinate to the change in x-coordinate as we move from any point on the line to another. Thus, starting at the point $(0, -2)$, we move:

3 units up (the positive y-direction),
then **4 units right** (the positive x-direction)

to locate another point on the line. The coordinates of this new point are $(4, 1)$. Finally,we draw a line through the two points, as shown in the figure.

Look Closer: To improve the accuracy of the graph in Example 2, we can find a third point on the line by writing the slope in an equivalent form. We change the sign of both numerator and denominator of the slope to get

$$m = \frac{\Delta y}{\Delta x} = \frac{-3}{-4}$$

Starting again from the y-intercept $(0, -2)$, we now move 3 units down and 4 units left, and find the point $(-4, -5)$ on the graph.

RQ5. Name two methods of graphing that we have already studied.
RQ6. When using the slope-intercept method, what is the first point we plot?
RQ7. How do we use the slope to find a second point on the line?

The **slope-intercept method** can be used to graph any non-vertical line.

To Graph a Line Using the Slope-Intercept Method

1. Write the equation in the form $y = mx + b$.
2. Plot the y-intercept, $(0, b)$.
3. Write the slope as a fraction, $m = \dfrac{\Delta y}{\Delta x}$.
4. Use the slope to find a second point on the graph: Starting at the y-intercept, move Δy units in the y-direction, then Δx units in the x-direction.
5. Find a third point by moving $-\Delta y$ units in the y-direction, then $-\Delta x$ units in the x-direction, starting from the y-intercept.
6. Draw a line through the three plotted points.

Finding the Slope-Intercept Form

Not all linear equations appear in slope-intercept form. However, we can write the equation of any non-vertical line in slope-intercept form by solving the equation for y in terms of x.

Example 3 Find the slope and y-intercept of the graph of $3x - 4y = 8$

Solution To write the equation in slope-intercept form, we solve for y in terms of x.

$3x - 4y = 8$	**Subtract $3x$ from both sides.**
$-4y = -3x + 8$	**Divide both sides by -4.**
$\dfrac{-4y}{-4} = \dfrac{-3x+8}{-4}$	**Divide each term of the right side by -4.**
$y = \dfrac{-3x}{-4} + \dfrac{8}{-4}$	**Simplify each quotient.**
$y = \dfrac{3}{4}x - 2$	

The equation is now in slope-intercept form, with $m = \frac{3}{4}$ and $b = -2$. Thus, the slope of the graph is $\frac{3}{4}$ and the y-intercept is the point $(0, -2)$.

Caution! Do not confuse solving for y with finding the y-intercept. In Example 3, we do **not** set $x = 0$ before solving for y. When we find the y-intercept, we are looking for a specific point, namely, the point with x-coordinate zero, so we replace x by 0. When we "solve for y," we are writing the equation in another form, so both variables, x and y, still appear in the equation.

RQ8. How do we put an equation into slope-intercept form?

Skills Warm-Up

Solve for the indicated variable.

1.	$2q + p = 10$	for p	**2.**	$2l + 2w = 18$	for l
3.	$3a + 9 = -6b$	for a	**4.**	$2c = 2d + 22$	for d
5.	$5r - 4s = 24$	for s	**6.**	$2m = 11 - 3n$	for n

Answers: **1.** $10 - 2q$ **2.** $9 - w$ **3.** $-3 - 2b$ **4.** $c - 11$ **5.** $\dfrac{5r - 24}{4}$ **6.** $\dfrac{11 - 2m}{3}$

Homework 3.5

Skills Practice

■ For Problems 1-4,

a. Write the equation in slope-intercept form.

b. State the slope and y-intercept of the graph.

1. $y = 3x + 4$

2. $6x + 3y = 5$

3. $2x - 3y = 6$

4. $5x = 4y$

■ For Problems 5-8,

a. Find the slope and the y-intercept of the line.

b. Write an equation for the line.

5. 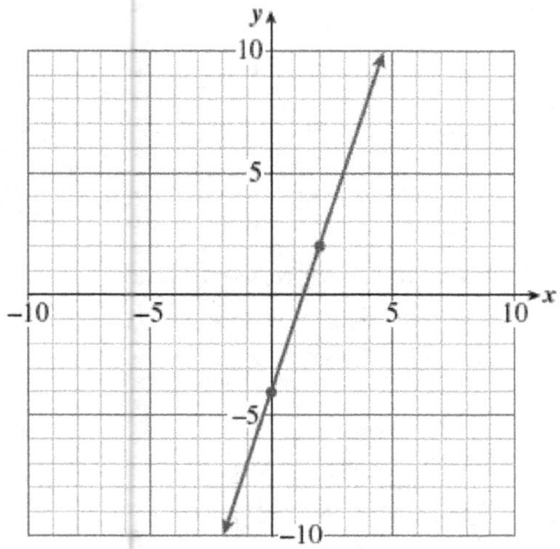

6. 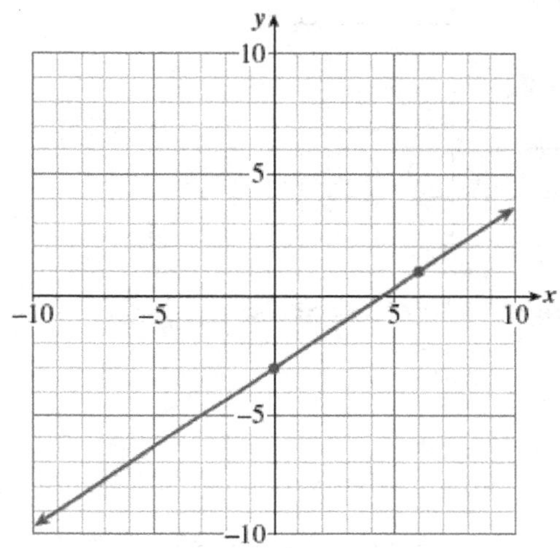

7. 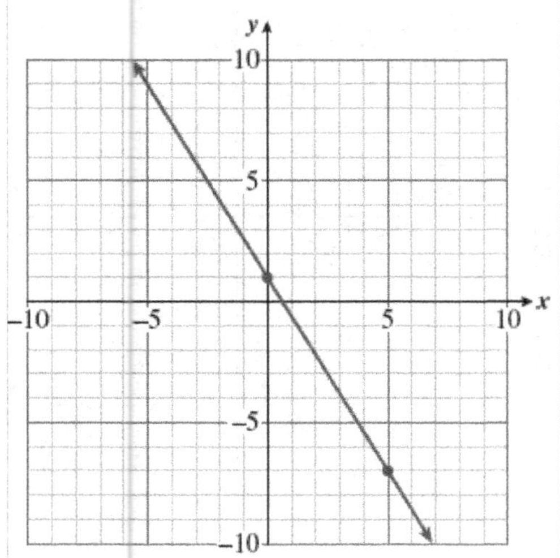

8. 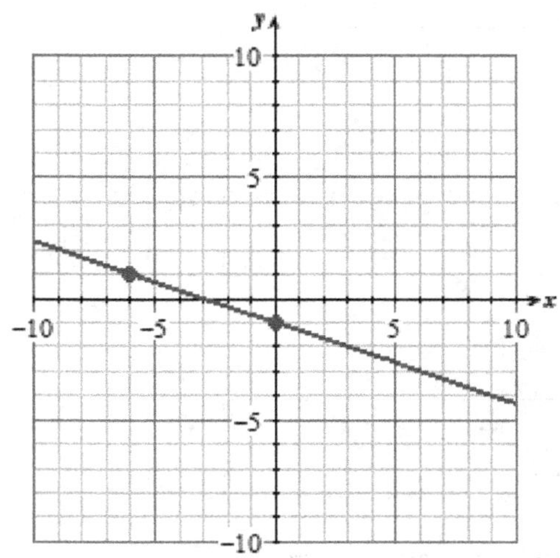

For Problems 9-10,

a. Fill in the y-values in the tables and graph the lines.
b. Choose two points on each line and compute its slope.
c. What is the y-intercept of each line?

9. I. $y = 2x - 6$

x	-1	0	1	2	3
y					

II. $y = 2x + 1$

x	-1	0	1	2	3
y					

III. $y = 2x + 3$

x	-1	0	1	2	3
y					

All three lines have the same ______.

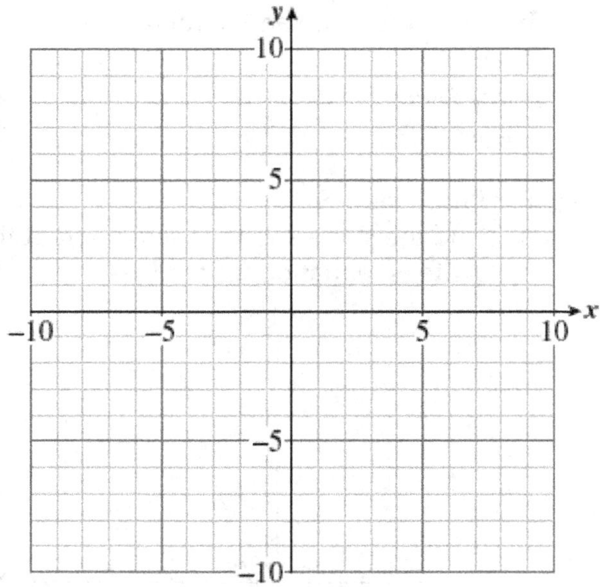

10. I. $y = \frac{1}{4}x + 2$

x	-4	-2	0	2	4
y					

II. $y = \frac{1}{2}x + 2$

x	-4	-2	0	2	4
y					

III. $y = x + 2$

x	-4	-2	0	2	4
y					

All three lines have the same ______.

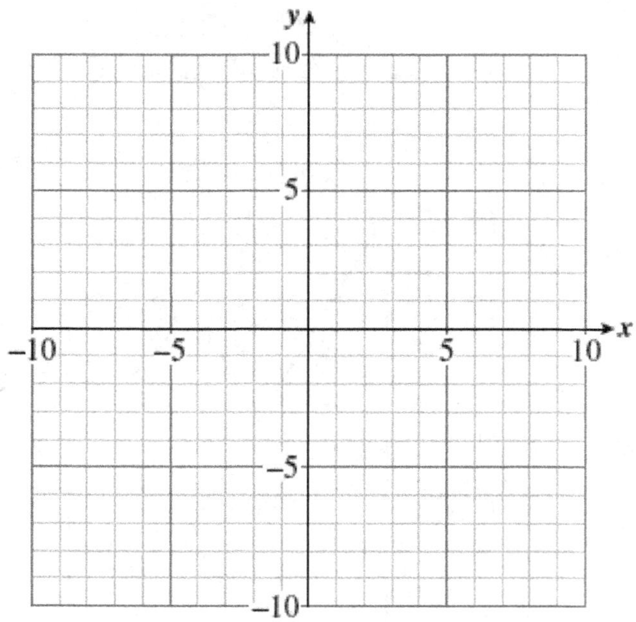

For Problems 11-12,

a. Find the intercepts of the graph and graph the line.
b. Compute the slope of the line.
c. Put the equation in slope-intercept form.

11. $3x + 4y = 12$

x	0	
y		0

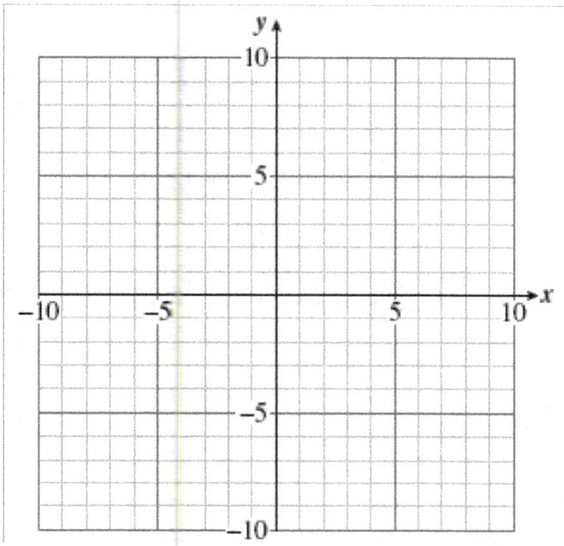

12. $y + 3x - 8 = 0$

x	0	
y		0

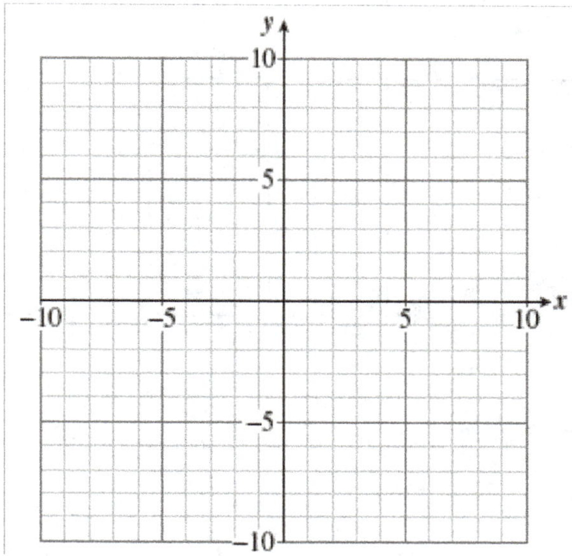

In Problems 13-14,

a. Put the equation in slope-intercept form.
b. What is the y-intercept of each line? What is its slope?
c. Use the slope to find two more points on the line.
d. Graph the line.

13. $3x - 5y = 0$

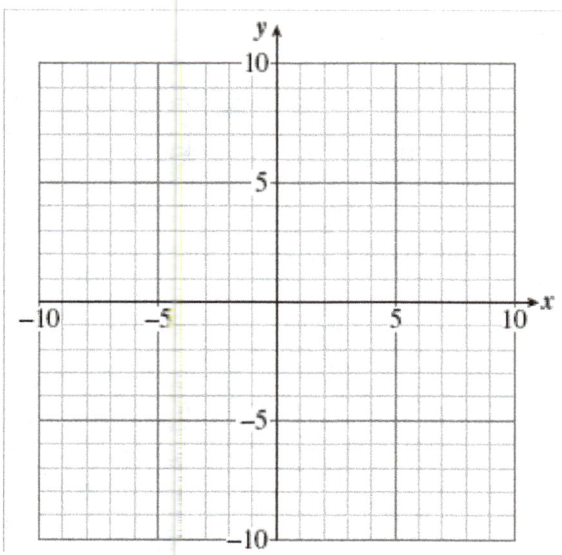

14. $5x + 4y = 0$

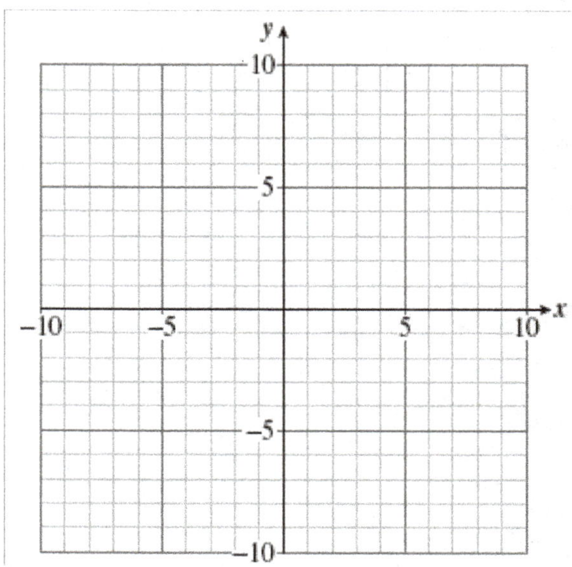

For Problems 15-18, graph the equation by using the slope-intercept method.

15. $y = 3 - x$

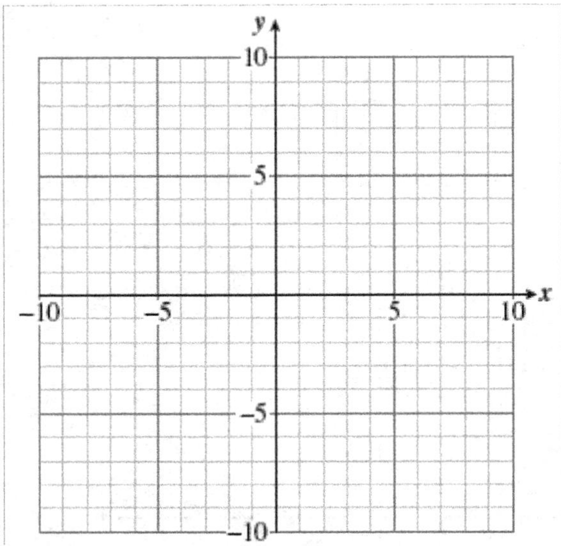

16. $y = 3x - 1$

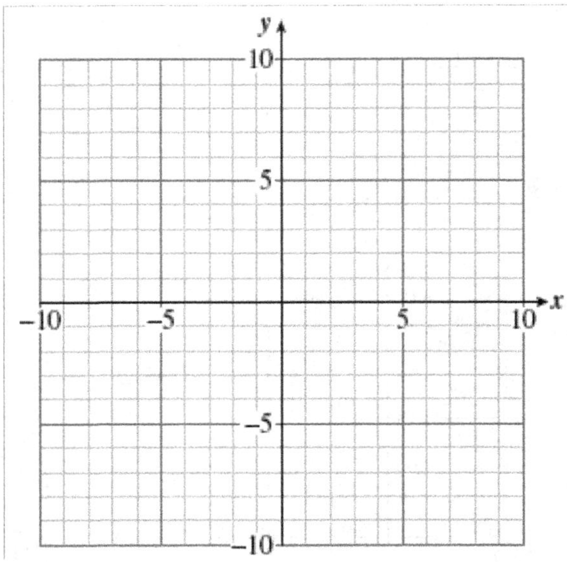

17. $y = \frac{3}{4}x + 2$

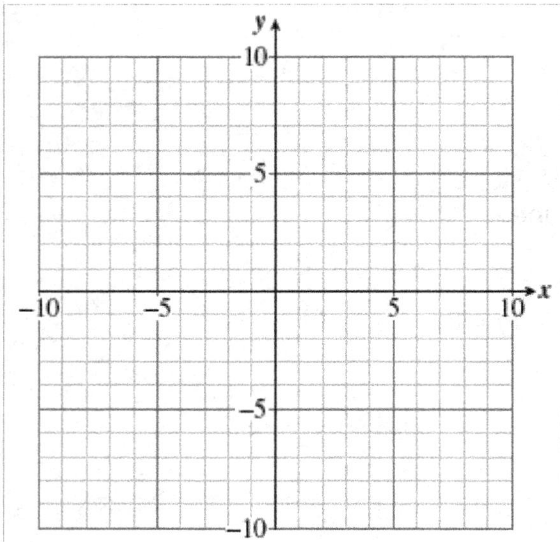

18. $y = -2 - \frac{4}{3}x$

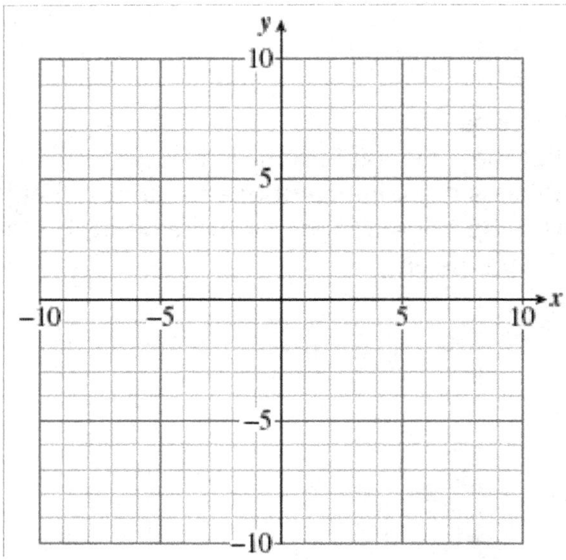

For problems 19-20, graph the line with the given slope and passing through the given point.

19. $m = \frac{3}{2},\ (-1, -2)$

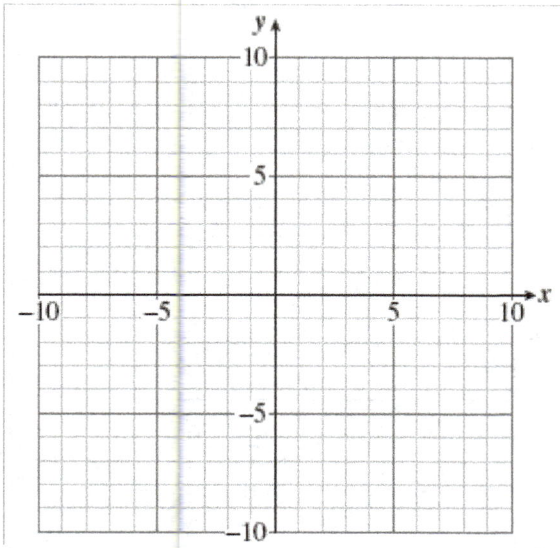

20. $m = -2,\ (-2, 6)$

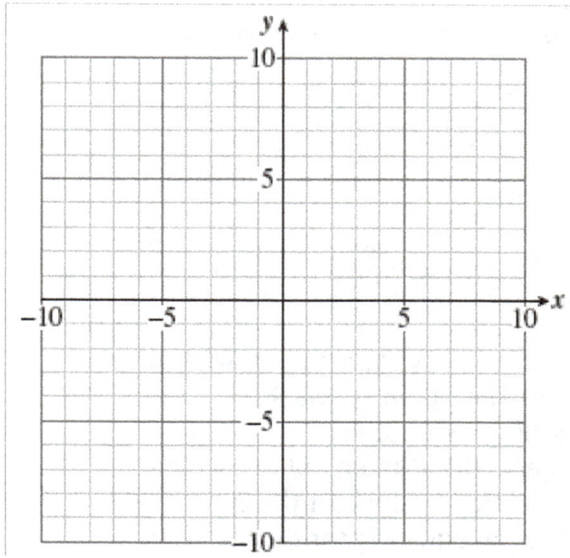

Applications

For Problems 21-23,

a. Graph the line by the slope-intercept method.

b. Explain what the slope and the vertical intercept tell us about the problem.

21. Serda's score on her driving test is computed by the equation $S = 120 - 4n$, where n is the number of wrong answers she gives.

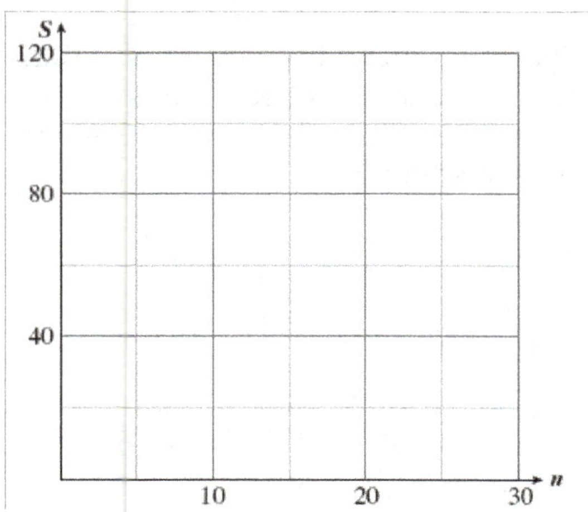

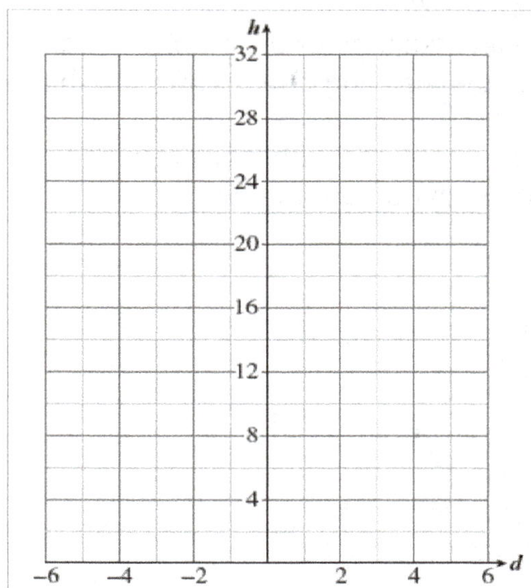

22. Greg is monitoring the growth of a new variety of string beans. The height of the vine each day is given in inches by $h = 18 + 3d$, where $d = 0$ represents today.

23. Cliff's score was negative at the end of the first round of College Quiz, but in Double Quiz his score improved according the equation $S = -400 + 20q$, where q is the number of questions he answered.

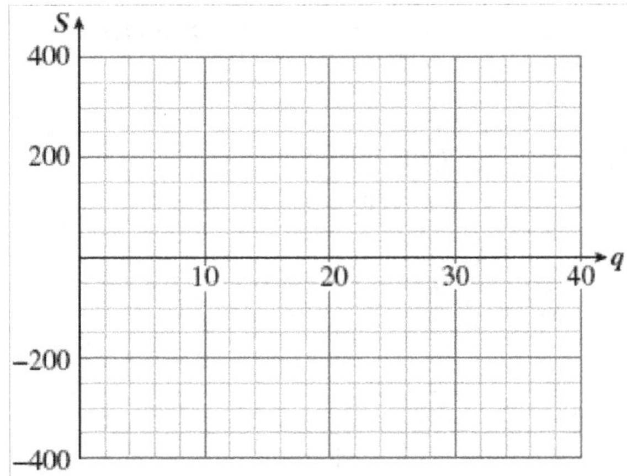

24. Robin opened a yogurt smoothie shop near campus. The graph shows Robin's profit P after selling s smoothies.
   a. What is the P-intercept of the line?
   b. Calculate the slope of the line.
   c. Write an equation for the line in slope-intercept form.
   d. What do the slope and the P-intercept tell us about the problem?

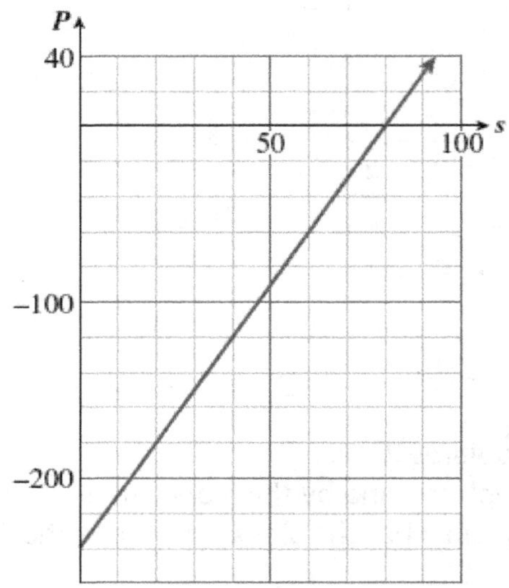

For Problems 25-27,
   a. Find an equation for the graph shown.
   b. State the meaning of the slope and the vertical intercept for the problem.

25. The figure shows the altitude a (in feet) of a skier t minutes after getting on a ski lift.

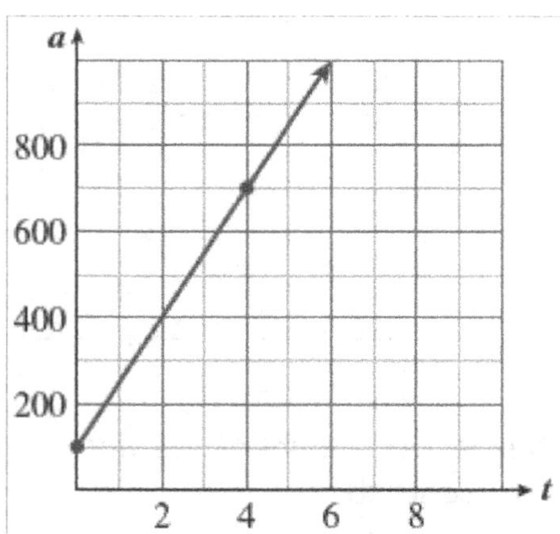

26. The figure shows the amount of garbage G (in tons) that has been deposited at a dump site t years after new regulations go into effect.

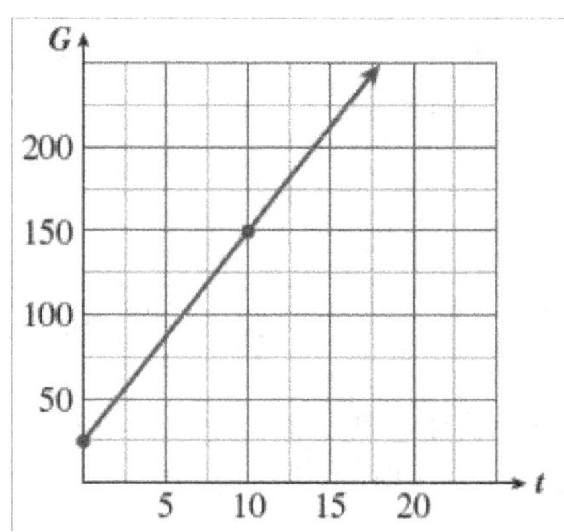

27. The figure shows the amount of money M (in dollars) in Tammy's bank account w weeks after she loses all sources of income.

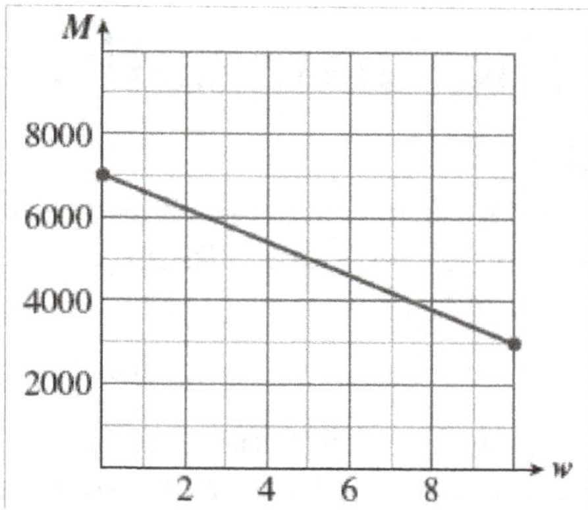

28. Match each graph with its equation.

a. $y = \frac{3}{4}x + 2$ b. $y = \frac{-3}{4}x + 2$

c. $y = \frac{3}{4}x - 2$ d. $y = \frac{-3}{4}x - 2$

I

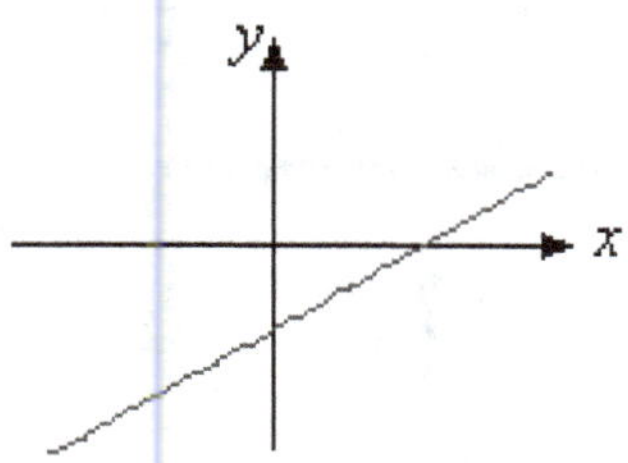

II

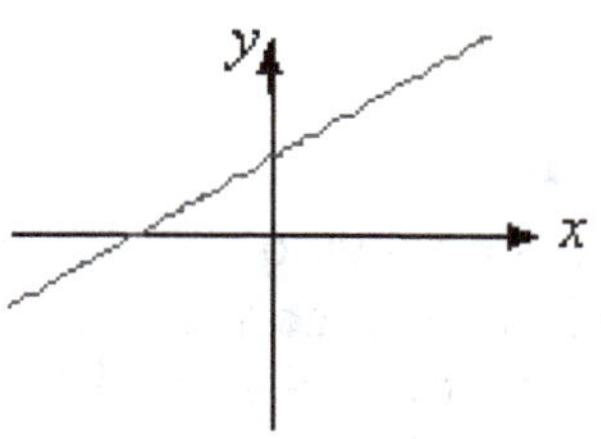

III

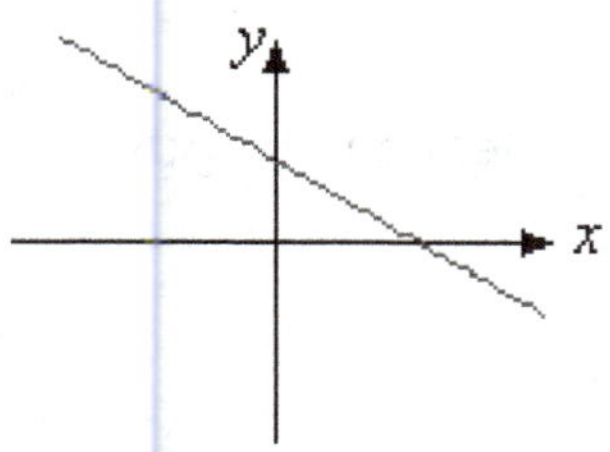

IV

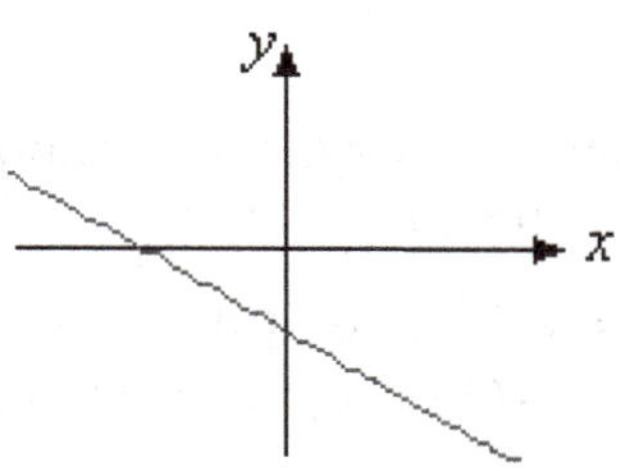

29. The formula $F = \frac{9}{5}C + 32$ converts the temperature in degrees Celsius to degrees Fahrenheit.

a. What is the Fahrenheit temperature when it is 10° Celsius?
b. What is the Celsius temperature when it is –4° Fahrenheit?
c. Graph the equation $y = \frac{9}{5}x + 32$ in the window

Xmin $= -40$ Ymin $= -50$
Xmax $= 40$ Ymax $= 100$

d. Find the slope and explain its meaning for this problem.
e. Find the C-intercept and explain its meaning for this problem. Find the F-intercept and explain its meaning for this problem.

3.6 The Distributive Law

Products

So far we have considered sums and differences of algebraic expressions. Now we take a look at simple products. To simplify a product such as $3(2a)$, we multiply the numerical factors to find

$$3(2a) = (3 \cdot 2)a = 6a$$

This calculation is an application of the associative law for multiplication. We can see why the product is reasonable by recalling that multiplication is repeated addition, so that

$$3(2a) = 2a + 2a + 2a = 6a$$

Three terms

Example 1a. $5(3x) = 15x$ **b.** $-2(-4h) = 8h$

RQ1. Explain the difference between the expressions $2b + 3b$ and $2(3b)$.

The Distributive Law

Another useful property for simplifying expressions is the distributive law. First consider a numerical example. Suppose we would like to find the area of the two rooms shown in the figure.

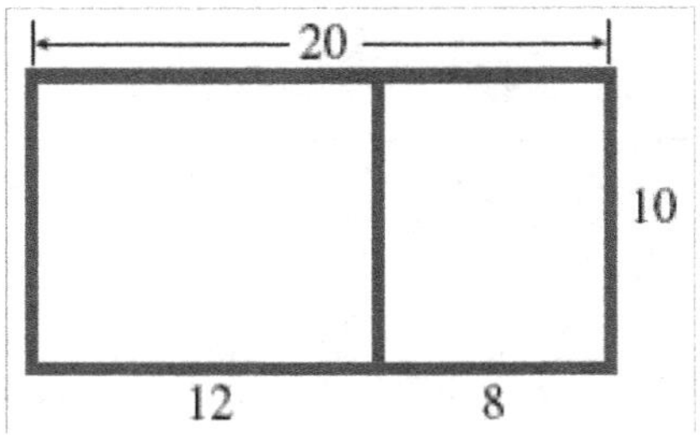

We can do this in two different ways:

Method 1: Think of the rooms as one large rectangle, with width 10 feet and length $12 + 8$ feet. The area is then given by

$$\text{Area} = \text{width} \times \text{length}$$
$$= \mathbf{10\,(12 + 8)} = 10\,(20) = 200 \text{ square feet}$$

Method 2: Find the area of each room separately and add the two areas together. This gives us

$$\mathbf{10(12) + 10(8)} = 120 + 80 = 200 \text{ square feet}$$

Of course, we get the same answer with either method. However, looking at the first step in each calculation, we see that

$$10(12 + 8) = 10(12) + 10(8)$$

These are two ways to compute the expression $10(12 + 8)$. The first way follows the order of operations, adding $12 + 8$ first. The second way distributes the multiplication to each term inside parentheses; that is, we multiply each term inside the parentheses by 10.

$$\mathbf{10}(12 + 8) = \mathbf{10}(12) + \mathbf{10}(8)$$

This equation is an example of the distributive law.

Distributive Law

If a, b, and c are any numbers, then

$$\boldsymbol{a(b + c) = ab + ac}$$

If the terms inside parentheses are not like terms, we have no choice but to use the distributive law to simplify the expression.

Example 2 Simplify $-2(3x-1)$

Solution We multiply each term inside parentheses by -2:

$$\begin{aligned}\mathbf{-2}(3x-1) &= \mathbf{-2}(3x)\mathbf{-2}(-1)\\ &= -6x+2\end{aligned}$$

Caution! Compare the three similar expressions:

$$-3x-5x, \qquad -3(-5x), \qquad -3(-5-x)$$

To simplify these expressions, we must first recognize the operations involved. The first is a sum of terms, the second is a product, and the third is an application of the distributive law. Thus,

$$-3x-5x=-8x, \qquad -3(-5x)=15x, \qquad -3(-5-x)=15+3x$$

RQ2. When do we need the distributive law to simplify an expression?

Solving Equations and Inequalities

If an equation or inequality contains parentheses, we use the distributive law to simplify each side before we begin to solve.

Example 3 Solve $25-6x=3x-2(4-x)$

Solution We apply the distributive law to the right side of the equation.

$25-6x=3x-2(4-x)$	**Apply the distributive law.**
$25-6x=\mathbf{3x}-8+\mathbf{2x}$	**Combine like terms.**
$25-6x=5x-8$	**Subtract $5x$ from both sides.**
$25-11x=-8$	**Subtract 25 from both sides.**
$-11x=-33$	**Divide both sides by -11.**
$x=3$	**The solution is 3.**

You should verify each step of the solution, and check the answer.

Caution! The expression on the right side of the equation in Example 3 is made up of two terms:

$$3x \quad \text{and} \quad -2(4-x)$$

Remember that we think of all terms as *added* together, so the minus sign in front of the 2 tells us that 2 is negative. When we apply the distributive law, we multiply each term inside parentheses by -2 to get $-8+2x$.

We know that a fraction bar often serves as a grouping device, like parentheses.

Example 4 Solve the proportion $\dfrac{7.6}{1.2}=\dfrac{x+3}{2x-1}$

Solution We apply the property of proportions and cross-multiply to get

$$7.6\,(2x-1)=1.2\,(x+3)$$ **Apply the Distributive Law.**

$$15.2\,x-7.6=1.2\,x+3.6$$ **Subtract $1.2x$ from both sides and add 7.6 to both sides.**

$$14x=11.2$$ **Divide both sides by 14.**

$$x=\frac{11.2}{14}=0.8$$

We can check the solution by substituting $x=0.8$ into the original proportion.

RQ3. In Example 4, why did we put parentheses around $2x-1$ and $x+3$?

Algebraic Expressions

We can often simplify algebraic expressions by combining like terms.

Example 5 Delbert and Francine are collecting aluminum cans to recycle. They will be paid x dollars for every pound of cans they collect. At the end of three weeks, Delbert collected 23 pounds of aluminum cans, and Francine collected 47 pounds.

a. Write algebraic expressions for the amount of money Delbert made, and the amount Francine made.

b. Write and simplify an expression for the total amount of money Delbert and Francine made from aluminum cans.

Solutions a. We multiply the number of pounds collected by the price per pound. Delbert made $23x$ dollars, and Francine made $47x$ dollars.

b. We add the amount of money Delbert made to the amount Francine made.

$$23x+47x=70x$$

RQ4. In Example 5, what does x stand for? What does $23x$ stand for?

The distributive law is also helpful for writing algebraic expressions.

Example 6 The length of a rectangle is 3 feet less than twice its width, w.

a. Write an expression for the length of the rectangle in terms of w.

b. Write and simplify an expression for the perimeter of the rectangle in terms of w.

c. Suppose the perimeter of the rectangle is 36 feet. Write and solve an equation to find the dimensions of the rectangle.

Solutions a. "Three feet less than twice the width" tells us to subtract 3 from twice the width. The length of the rectangle is $2w-3$.

b. The perimeter of a rectangle is given by the formula $P = 2l + 2w$. We substitute our expression for the length to get

$$\begin{aligned} P &= 2(2w - 3) + 2w && \text{Apply the distributive law.} \\ &= 4w - 6 + 2w && \text{Combine like terms.} \\ &= 6w - 6 \end{aligned}$$

c. We set our expression for the perimeter equal to 36.

$$\begin{aligned} 6w - 6 &= 36 && \text{Add 6 to both sides.} \\ 6w &= 42 && \text{Divide both sides by 7.} \\ w &= 7 \end{aligned}$$

The width of the rectangle is 7 feet, and its length is $2(7) - 3 = 11$ feet. ■

RQ5. In Example 6, why did we multiply $2w - 3$ by 2?

Skills Warm-Up

■ Use a formula to write an equation. Then solve the equation.

1. 32.5% of the class are engineering majors. If there are 91 engineering majors, how many students are in the class?

2. A rectangular cookie sheet is 40 cm long and has a perimeter of 116 cm. How wide is the cookie sheet?

■ Write an equation you could solve to answer the question.

3. Erika bought a 50-pound bag of dog food and after 28 days she had 29 pounds left. How much dog food did she use per day?

4. A hamburger contains 60 calories less than two bags of fries. A hamburger and one bag of fries contains 780 calories. How many calories are in a bag of fries?

Answers: **1.** $0.325x = 91;$ 280 **2.** $2w + 2(40) = 116;$ 18 cm **3.** $50 - 28d = 29;$ $\frac{3}{4}$ lb **4.** $x + (2x - 60) = 780;$ 280

Homework 3.6

Skills Practice

In Problems 1-4, use the distributive law to remove parentheses. For some of these exercises you will use the distributive law in the form

$$(b+c)a = ba + ca$$

1. $5(2y-3)$

2. $-2(4x+8)$

3. $-(5b-3)$

4. $(-6+2t)(-6)$

In problems 5-6, simplify each product. Which product in each pair requires the distributive law?

5. a. $8(4c)$
b. $8(4+c)$

6. a. $2(-8-t)$
b. $2(-8t)$

7. Which of the following is a correct application of the distributive law?
a. $5(3a) = 15a$
b. $5(3+a) = 15+a$
c. $5(3a) = 15(5a)$
d. $5(3+a) = 15+5a$

8. Simplify each expression if possible. Then evaluate for $x=3,\ y=9$.
a. $2(xy)$
b. $2(x+y)$
c. $2-xy$
d. $-2xy$

In Problems 9-12, simplify and combine like terms.

9. $-6(x+1)+2x$

10. $5-2(4x-9)+9x$

11. $-4(3+2z)+2z-3(2z+1)$

12. $3(-3a-4)-3a-4(3a-4)$

In Problems 13-20, solve each equation or inequality.

13. $6(3y-4) = -60$

14. $5w-64 = -2(3w-1)$

15. $-22c+5(3c+4) = 20+8c$

16. $6-4(2a+3) \geq -6a+2$

17. $4-3(2t-4) > -2(4-3t)$

18. $0.25(x+3)-0.45(x-3) = 0.30$

19. $\dfrac{a}{a+2} = \dfrac{2}{3}$

20. $\dfrac{0.3}{0.5} = \dfrac{b+2}{12-b}$

21. Choose a value for the variable and show that the following pairs of expressions are **not** equivalent.
a. $5(x+3)$ and $5x+3$
b. $-4(c-3)$ and $-4c-12$

22. a. Evaluate the expression $(2x+7)-4(4x-2)-(-2x+3)$ for $x=-3$.
b. Simplify the expression in part (a) by combining like terms.
c. Evaluate your answer to part (b) for $x=-3$. Check that you got the same answer for part (a).
d. Are the expressions in parts (a) and (b) equivalent? Why or why not?

Applications

23. The length of a rectangle is 8 feet shorter than twice its width. Write expressions in terms of w, the width of the rectangle.
- **a.** The length of the rectangle
- **b.** The perimeter of the rectangle
- **c.** The area of the rectangle

24. Risa bought a skirt and a sweater that together cost \$108. Then she bought a pair of shoes that cost twice as much as the sweater. Write and simplify expressions in terms of c, the cost of the skirt:
- **a.** The cost of the sweater
- **b.** The cost of the shoes
- **c.** The total amount Risa spent

25. Albert and Isaac left the same hotel at the same time, and each drove for 3 hours, but Albert drove 15 miles per hour faster than Isaac. Write and simplify expressions in terms of Isaac's speed, s.
- **a.** How far Isaac drove
- **b.** How far Albert drove
- **c.** If they drove in the same direction, how far apart they are now
- **d.** If they drove in opposite directions, how far apart they are now

26. A box of AlmondOats contains 15 ounces of cereal made of oat flakes and sliced almonds. Oat flakes cost 15 cents per ounce, and almonds cost 35 cents per ounce. Write and simplify expressions in terms of a, the number of ounces of almonds in the box.
- **a.** The number of ounces of oat flakes in the box
- **b.** The cost of the almonds
- **c.** The cost of the oat flakes
- **d.** The total cost of a box of AlmondOats

27. Trader Jim's Pomegranate Punch is 25% pomegranate juice, and Pomajoy is 60% pomegranate juice. Vera mixes some of each to make 8 quarts of juice drink. Write and simplify expressions in terms of x, the number of quarts of Pomajoy she uses.
- **a.** The number of quarts of Pomegranate Punch
- **b.** The number of quarts of pomegranate juice in the Pomajoy
- **c.** The number of quarts of pomegranate juice in the Pomegranate Punch
- **d.** The number of quarts of pomegranate juice in Vera's mixture

28. Premium ice cream is 12% butterfat, and chocolate syrup is 55% butterfat. A chocolate sundae, without whipped cream and a cherry, weighs 10 ounces. Write and simplify expressions in terms of x, the number of ounces of ice cream in the sundae.
- **a.** The number of ounces of chocolate syrup in the sundae
- **b.** The amount of fat in the ice cream
- **c.** The amount of fat in the chocolate syrup
- **d.** The total amount of fat in the sundae

In problems 29-30, find an algebraic expression for the area of the rectangle. Use the distributive law to write the expression in two ways.

29. 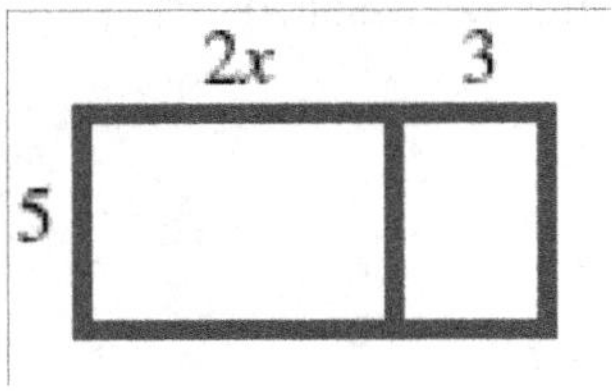

30. 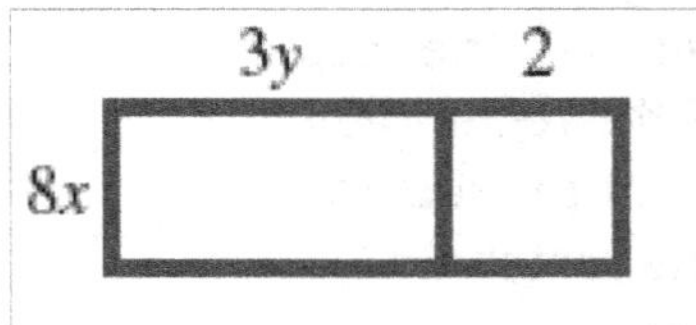

31. The length of a rectangular vegetable garden is 6 yards more than twice its width.
- **a.** Write an expression for the perimeter of the garden in terms of its width.
- **b.** Ann bought 42 yards of fence to enclose the garden. What are its dimensions?

32. Revenue from the state lottery is divided between education and administrative costs in the ratio of 4 to 3. If the lottery revenue this year is \$24,000,000, how much money will go to education?

33. An apple and a glass of milk together contain 260 calories.
- **a.** If an apple contains a calories, how many calories are in a glass of milk?
- **b.** Write an expression for the number of calories in two apples and three glasses of milk.
- **c.** If two apples and three glasses of milk contain 660 calories, find the number of calories in an apple.

34. Melody sold 47 tickets to a charity concert. Reserved seats cost \$10 and open seating was \$6 a ticket. Let x represent the number of reserved seats she sold. Write expressions in terms of x for:
- **a.** The number of open seating tickets Melody sold
- **b.** The amount of money Melody collected from reserved seating tickets
- **c.** The amount of money Melody collected from open seating tickets
- **d.** The total amount of money Melody collected

35. Melody from Problem 34 collected \$330 from the sale of concert tickets. Write and solve an equation to answer the question: How many reserved seats did she sell?

Chapter 3 Summary and Review

Section 3.1 Some Examples of Linear Models

- A **mathematical model** is a simplified representation of reality that helps us understand a process or phenomenon.
- We can represent the relationship by a table of values, by a graph, or by an algebraic equation.

Linear Model
A **linear model** describes a variable that increases or decreases at a constant rate. It has the form

$$y = \textbf{(starting value)} + \textbf{(rate)} \times t$$

Intercepts of a Graph
The points where a graph crosses the axes are called the **intercepts**.
1. To find the x-intercept, set $y = 0$ and solve for x.
2. To find the y-intercept, set $x = 0$ and solve for y.

To Graph a Linear Equation Using the Intercept Method:
1. Find the x- and y-intercepts of the graph.
2. Draw the line through the two intercepts.
3. Find a third point on the graph as a check.
(Choose any convenient value for x and solve for y.)

The **general form** for a linear equation is

$$Ax + By = C \qquad (A \text{ and } B \text{ cannot both be } 0)$$

Section 3.2 Using a Graphing Calculator

- Before we can enter the equation, we must solve for y in terms of x.
- To graph a linear equation, we should choose scales for the axes so that both intercepts can be shown on the graph.
- We can use the **Trace** feature or the **value** feature of the calculator to find the coordinates of points on a graph.
- When we read the coordinates of a point on the graph of an equation, we are reading a pair of x- and y-values that make the equation true.

Section 3.3 Ratio and Proportion

- A **ratio** is a type of quotient that allows us to compare two numerical quantities.

Ratio
The **ratio of** a **to** b is written $\frac{a}{b}$.

- A **rate** is a ratio that compares two quantities with different units.
- A **proportion** is a statement that two ratios are equal. In other words, a proportion is a type of equation in which both sides are ratios.

- We use cross-multiplying to solve proportions.

> **Property of Proportions**
>
> If $\dfrac{a}{b} = \dfrac{c}{d}$, then $ad = bc$.

- Two variables are said to be **proportional** if their ratios are always the same.

Section 3.4 Slope

- The slope of a line measures the rate of change of y with respect to x.

> **Slope**
>
> The slope of a line is defined by the ratio
>
> $$\textbf{slope} = \frac{\textbf{change in } y\textbf{-coordinate}}{\textbf{change in } x\textbf{-coordinate}}$$
>
> as we move from one point to another on the line. In symbols,
>
> $$m = \frac{\Delta y}{\Delta x}$$

- Lines have constant slope. No matter what two points on a line we use to calculate its slope, we always get the same result.
- The slope of a horizontal line is zero, and the slope of a vertical line is undefined.

Section 3.5 Slope-Intercept Form

> **Slope-Intercept Form**
>
> A linear equation written in the form
>
> $$y = mx + b$$
>
> is said to be in **slope-intercept** form. The coefficient m is the slope of the graph, and b is the y-intercept.

- The constant term in the slope-intercept form tells us the initial value of y, and the coefficient of x tells us the rate of change of y with respect to x.

> **To Graph a Line Using the Slope-Intercept Method**
>
> **1.** Write the equation in the form $y = mx + b$.
> **2.** Plot the y-intercept, $(0, b)$.
> **3.** Write the slope as a fraction, $m = \dfrac{\Delta y}{\Delta x}$.
> **4.** Use the slope to find a second point on the graph: Starting at the y-intercept, move Δy units in the y-direction, then Δx units in the x-direction.
> **5.** Find a third point by moving $-\Delta y$ units in the y-direction, then $-\Delta x$ units in the x-direction, starting from the y-intercept.
> **6.** Draw a line through the three plotted points.

Section 3.6 The Distributive Law

> **Distributive Law**
> If a, b, and c are any numbers, then
> $$a(b+c) = ab + ac$$

> **Steps for Solving Linear Equations**
> 1. Use the distributive law to remove any parentheses.
> 2. Combine like terms on each side of the equation.
> 3. By adding or subtracting the same quantity on both sides of the equation, get all the variable terms on one side and all the constant terms on the other.
> 4. Divide both sides by the coefficient of the variable to obtain an equation of the form $x = a$.

Review Questions

■ Use complete sentences to answer the questions in Problems 1-10.

1. Name three ways to represent a model.
2. State the general form for a linear model.
3. What is the difference between evaluating an expression and solving an equation?
4. Explain how to find the intercepts of a graph.
5. State the fundamental property of proportions.
6. How can we tell if two variables are proportional?
7. Give a formula for slope using the Δ notation.
8. State the slope-intercept formula, and explain the meaning of the coefficients.
9. Explain the slope-intercept method for graphing a line.
10. What is the easiest way to find the slope of a line from its equation?

Review Problems

■ For problems 1-6, write and graph a linear equation. Then answer the questions.

1. Last year Pinwheel Industries introduced a new model calculator. It cost \$2000 to develop the calculator and \$20 to manufacture each one.
   a. Complete the table of values showing the total cost C of producing n calculators.

n	100	500	800	1200	1500
C					

   b. Write an equation that expresses C in terms of n.
   c. Graph the equation.
   d. What is the cost of producing 1000 calculators? Illustrate this as a point on your graph.
   e. How many calculators can be produced for \$10,000? Illustrate this as a point on your graph.

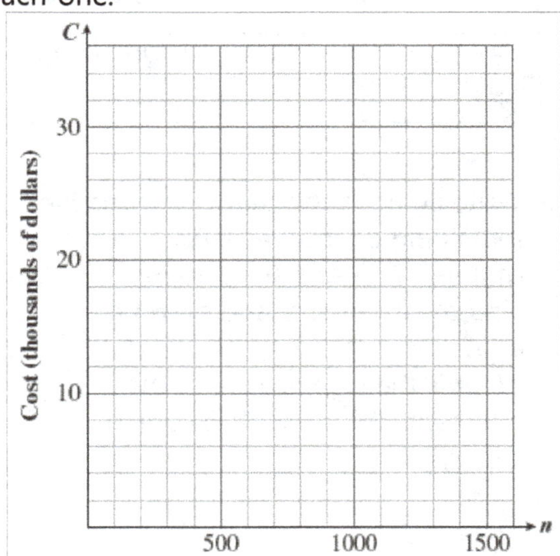

2. Megan weighed 5 pounds at birth and gained 18 ounces per month during her first year.
   a. Complete the table of values that shows Megan's weight w in terms of her age m in months.

m	2	4	6	9	12
w					

   b. Write an equation for w in terms of m.
   c. Graph the equation.
   d. How much did Megan weigh at 9 months? Illustrate this as a point on your graph.
   e. When did Megan weigh 9 pounds? Illustrate this as a point on your graph.

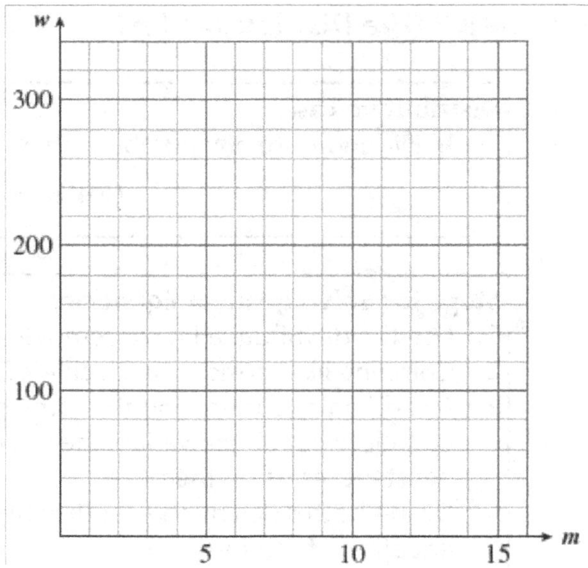

3. The world's oil reserves were 1660 billion barrels in 1976; total annual consumption is 20 billion barrels.
   a. Complete the table of values that shows the remaining oil reserves R in terms of time t (in years since 1976).

t	5	10	15	20	25
R					

   b. Write an equation that expresses R in terms of t.
   c. Find the intercepts and graph the equation.
   d. What is the significance of the intercepts to the world's oil supply?

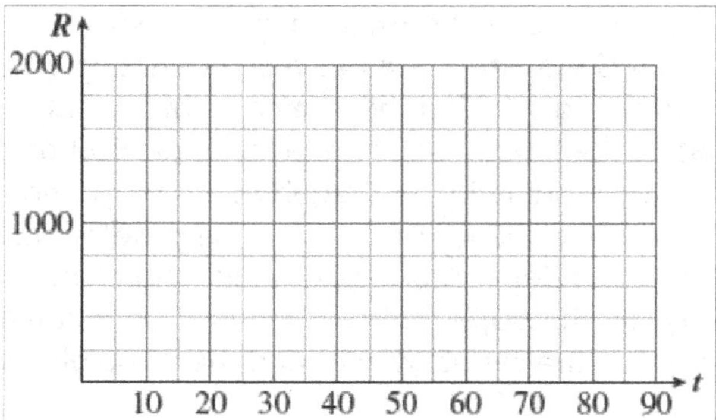

4. The world's copper reserves were 500 million tons in 1976; total annual consumption is 8 million tons.
   a. Complete the table of values that shows the remaining copper reserves R in terms of time t (in years since 1976).

t	5	10	15	20	25
R					

   b. Write an equation that expresses the remaining copper reserves R in terms of time t (in years since 1976).
   c. Find the intercepts and graph the equation by hand.
   d. What is the significance of the intercepts to the world's copper supply?

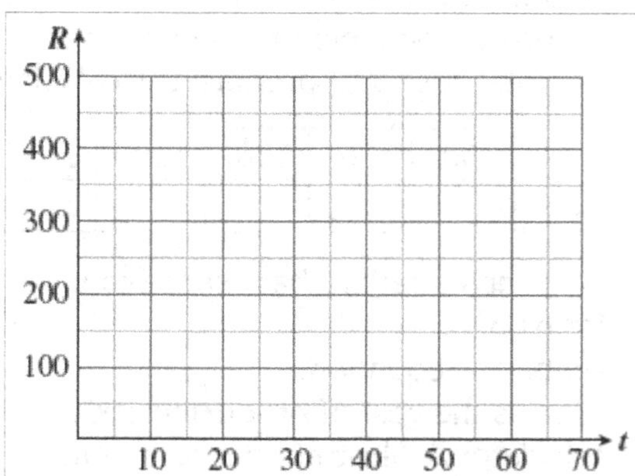

5. The owner of a movie theater needs to bring in $1000 at each screening in order to stay in business. He sells adult tickets at $5 apiece and children's tickets at $2 each.

a. Write an equation that relates the number of adult tickets, A, he must sell and the number of children's tickets, C.

b. Find the intercepts and graph the equation by hand.

c. If the owner sells 120 adult tickets, how many children's tickets must he sell?

d. What is the significance of the intercepts to the sale of tickets?

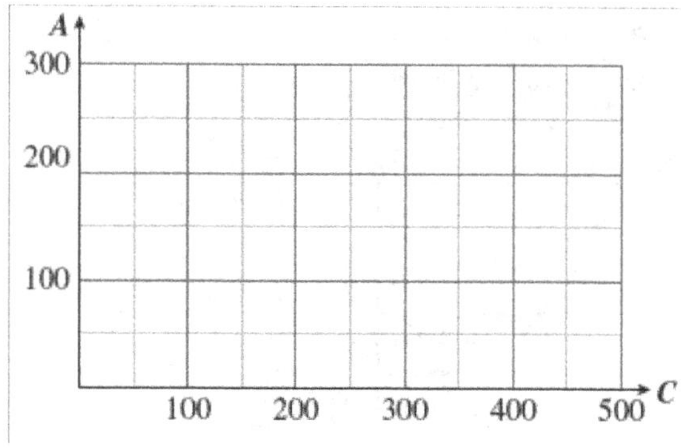

6. Alida plans to spend part of her vacation in Atlantic City and part in Saint-Tropez. She estimates that after airfare her vacation will cost $60 per day in Atlantic City and $100 per day in Saint-Tropez. She has $1200 to spend after airfare.

a. Write an equation that relates the number of days, C, Alida can spend in Atlantic City and the number of days, T, in Saint-Tropez.

b. Find the intercepts and graph the equation by hand.

c. If Alida spends 10 days in Atlantic City, how long can she spend in Saint-Tropez?

d. What is the significance of the intercepts to Alida's vacation?

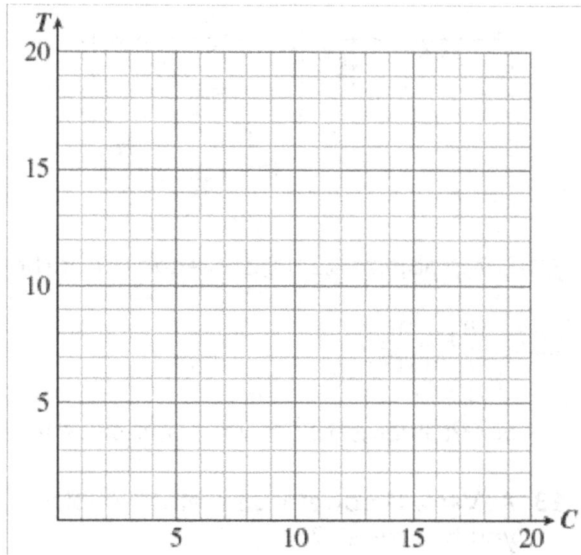

In part (b) of Problems 7-8, write and solve an equation to answer the question.

7. Monica has saved $7800 to live on while she attends college. She spends $600 a month.

a. Write an equation for the amount, S, in Monica's savings account after m months.

b. How long will it be before her savings are reduced to $2400?

c. Explain what the intercepts of your equation mean in terms of the problem.

8. Ronen is on a rock-climbing expedition. He is climbing out of a deep gorge at a rate of 4 feet per minute, and right now his elevation is -156 feet.

a. Write an equation for Ronen's elevation, h, after m minutes.

b. When will his elevation be -20 feet?

c. Explain what the intercepts of your equation mean in terms of the problem.

For problems 9-10, use the graph to solve each equation or inequality. Estimate your solutions if necessary.

9. **a.** $-3x + 9 = 24$
b. $-3x + 9 \geq 15$

10. **a.** $24x - 1800 = -1250$
b. $24x - 1800 < -2500$

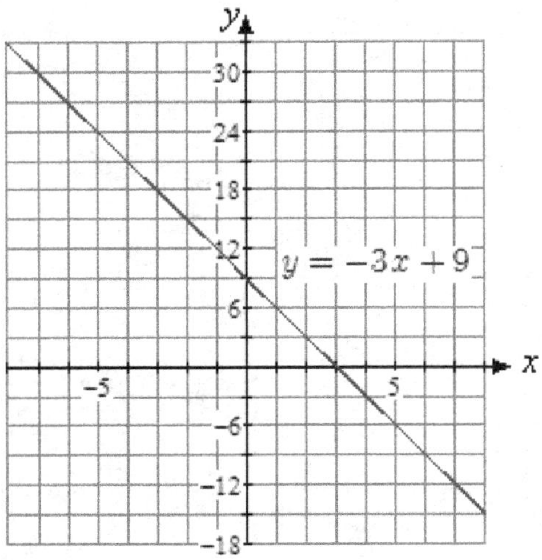

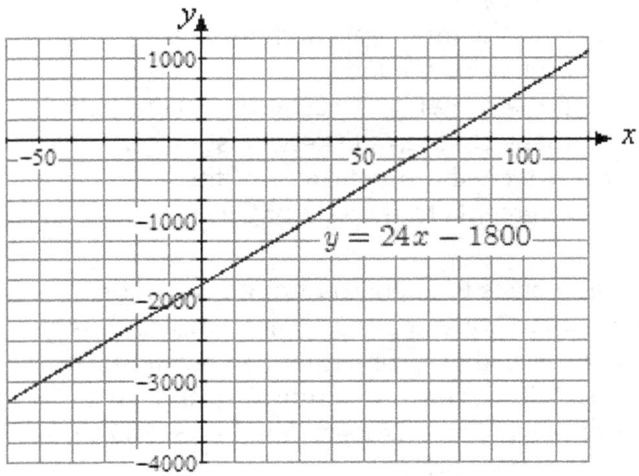

For Problems 11-12, solve the proportion.

11. $\dfrac{105}{y} = \dfrac{15}{17}$

12. $\dfrac{16}{q} = \dfrac{52}{86.125}$

For Problems 13-16, write and solve a proportion to answer the question.

13. At Van's Hardware, 36 metric frimbles cost \$4.86. How many metric frimbles can you buy for \$6.75?

14. Bob's weekly diet includes 70 grams of fat, 200 grams of protein, and 1800 grams of carbohydrates. Jenny's weekly diet includes 112 grams of fat. How many grams of protein and carbohydrates should she consume so that her diet has the same proportions as Bob's diet?

15. Simon joined a co-op that provides access to the Internet. His user bill, B, is proportional to the number of hours, h, that he is logged on to the net. Last month he was logged on for 28 hours, and his bill was \$21. This month Simon's bill was \$39. How many hours did he log on the Internet?

16. A small orchard that is 150 yards on each side contains 100 apple trees. The number of trees in an orchard is proportional to the area of the orchard. How many trees are in a larger orchard that is 600 yards on each side

For Problems 17-18, use the graph to find the slope of each line, and illustrate Δx and Δy on the graph.

17. $3x + 2y = -7$

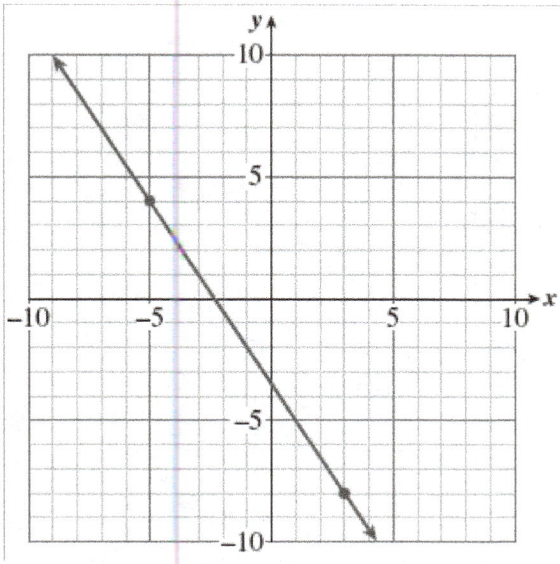

18. $3y = 5x$

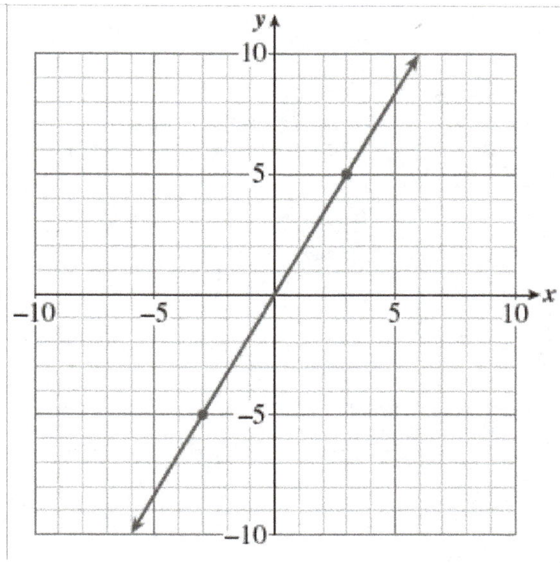

19. A spiked volleyball travels 6 feet in 0.04 seconds. A pitched baseball travels 66 feet in 0.48 seconds. Which ball travels faster?

20. Kendra needs $4\frac{1}{2}$ gallons of Luke's Brand primer to cover 1710 square feet of wall. She uses $5\frac{1}{3}$ gallons of Slattery's Brand primer for 2040 square feet of wall. Which brand covered more wall per gallon?

21. Which is steeper: Stone Canyon Drive, which rises 840 feet over a horizontal distance of 1500 feet, or Highway 33, which rises 1150 feet over a horizontal distance of 2000 feet?

22. The top of Romeo's ladder is on Juliet's window sill that is 11 feet above the ground, and the bottom of the ladder is 5 feet from the base of the wall. Is the incline of this ladder as steep as a firefighter's ladder that rises a height of 35 feet over a horizontal distance of 16 feet?

23. The planners at AquaWorld want the small water slide to have a slope of 25%. If the slide is 20 feet tall, how far should the end of the slide be from the base of the ladder?

24. In areas with heavy snowfall, the pitch (or slope) of the roof of an A-frame house should be at least 1.2. If a small ski chalet is 40 feet wide at its base, how tall is the center of the roof?

For Problems 25-28,

a. Find the intercepts of each line.
b. Use the intercepts to graph the line.
c. Use the intercepts to find the slope of the line.

25. $6x - 3y = 18$

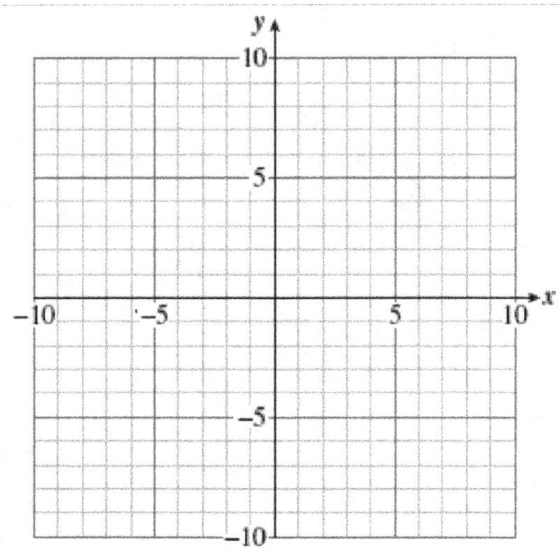

26. $4y + 9x = 36$

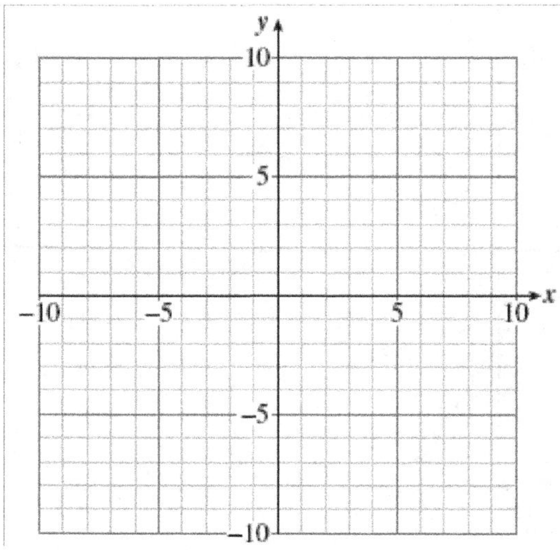

27. $\frac{x}{2} - \frac{y}{3} = 1$

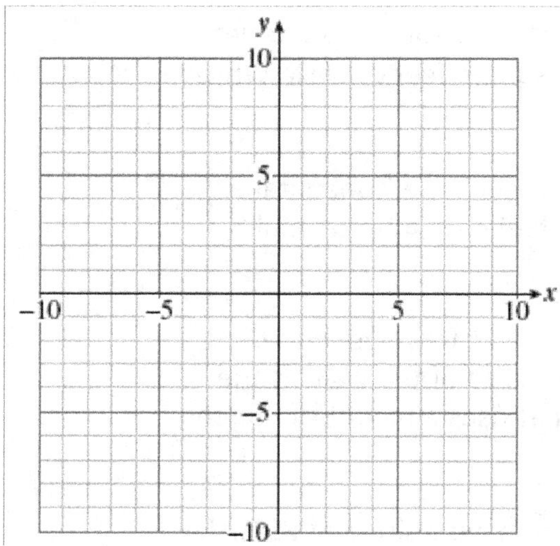

28. $y = 3x - 8$

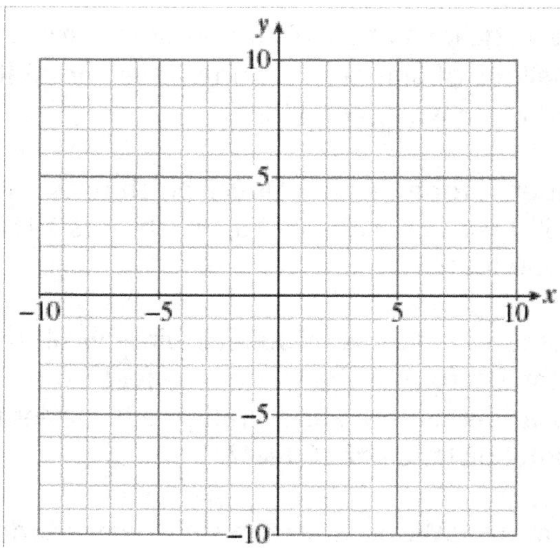

29. What is the slope of the line whose intercepts are $(-5, 0)$ and $(0, 3)$?

30. a. Find the x- and y-intercepts of the line $\frac{x}{4} - \frac{y}{6} = 1$.
b. What is the slope of the line in part (a)?

For Problems 31-34, find the slope and y-intercept of each line.

31. $2x - 4y = 5$

32. $\frac{1}{2}x + \frac{2}{3}y = \frac{5}{6}$

33. $8.4x + 2.1y = 6.3$

34. $y - 3 = 0$

For Problems 35-36,

a. Find the slope and y-intercept of the line.

b. Write an equation for the line.

35.

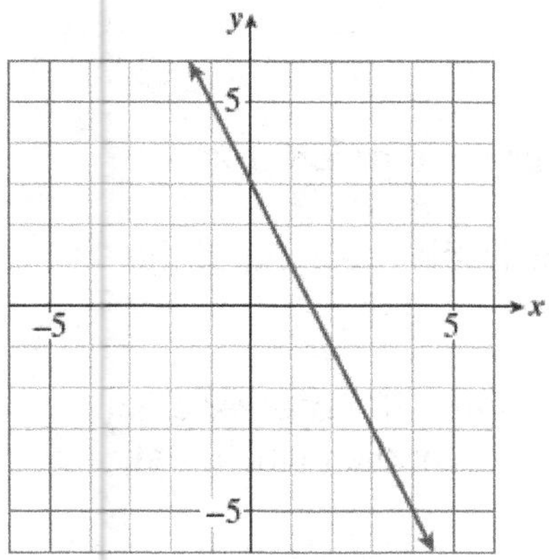

36.

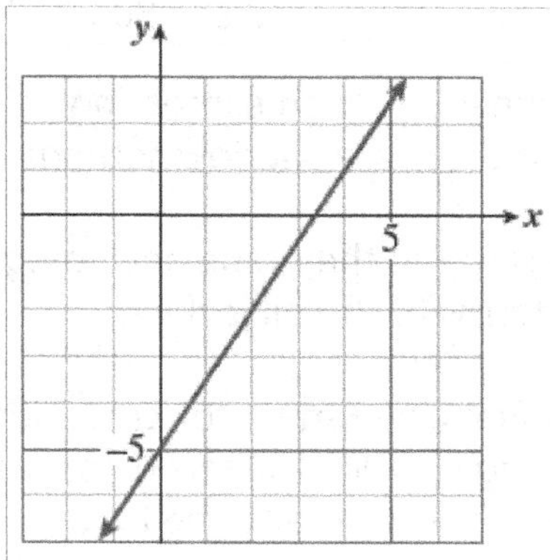

For Problems 37-40,

a. Find the slope and y-intercept of the line.

b. Graph the line using the slope-intercept method.

37. $2y + 5x = -10$

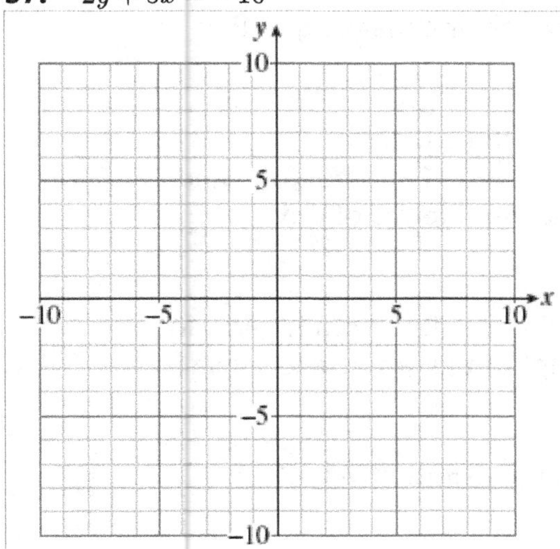

38. $y + 3x = 0$

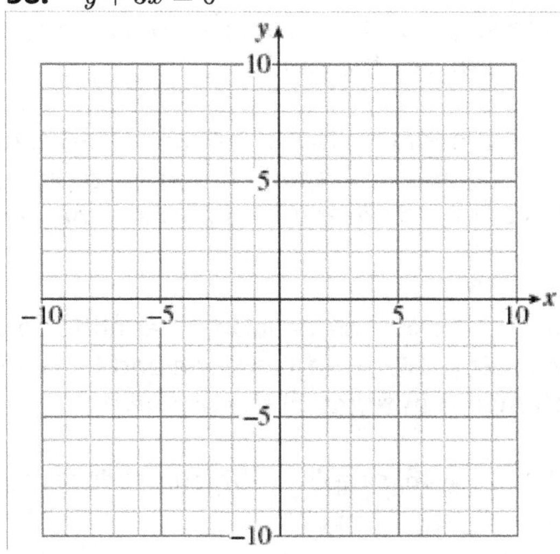

39. $3x = 4y - 8$

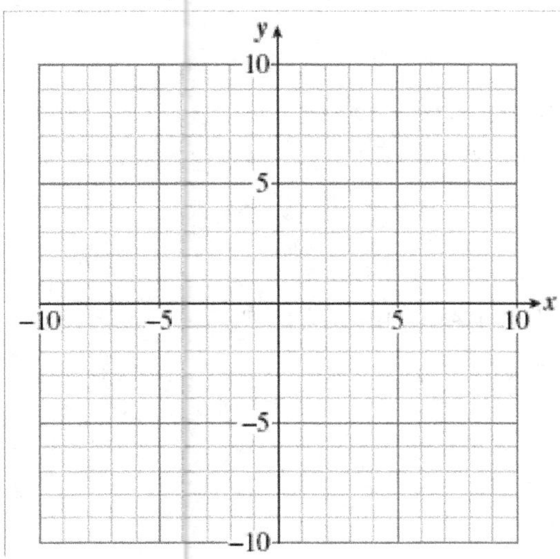

40. $\frac{x}{4} - \frac{y}{5} = 10$

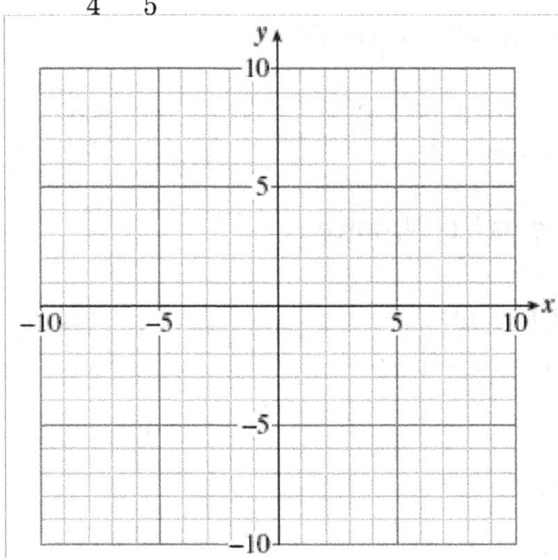

41. The table shows the amount of oil B (in thousands of barrels) left in a tanker t minutes after it hits an iceberg and springs a leak.

t	0	10	20	30
B	800	750	700	650

a. Write a linear equation expressing B in terms of t.
b. Choose appropriate window settings on your calculator and graph your equation.
c. Give the slope of the graph, including units, and explain the meaning of the slope in terms of the oil leak.

42. A traditional first experiment for chemistry students is to make 98 observations about a burning candle. Delbert records the height h of the candle in inches at various times t minutes after he lit it.

t	0	10	30	45
h	12	11.5	10.5	9.75

a. Write a linear equation expressing h in terms of t.
b. Choose appropriate window settings on your calculator and graph your equation.
c. Give the slope of the graph, including units, and explain the meaning of the slope in terms of the candle.

43. The price of a medium bowl of frozen yogurt is given by the equation $y = 1.35 + 0.85t$, where t is the number of toppings you select. Find the slope and the y-intercept of the graph, and explain what each means in terms of the problem.

44. Beryl is sailing in a hot air balloon. Her altitude at t minutes after 2 pm is given in feet by $h = 500 - 15t$. Find the slope and the y-intercept of the graph, and explain what each means in terms of the problem.

For Problems 45-46, simplify. Which product requires the distributive law?

45. a. $-5(-6m)$
b. $-5(6 - m)$

46. a. $9(-3 - w)$
b. $9(-3w)$

For Problems 47-50, simplify.

47. $(4m + 2n) - (2m - 5n)$

48. $(-5c - 6) + (-11c + 15)$

49. $-7w - 2(4w - 13)$

50. $4(3z - 10) + 5(-z - 6)$

For Problems 51-54, solve.

51. $5p + 10(17 - p) = 2p - 5$

52. $-3(k - 2) - 4(2k + 5) = 10 + 3k$

53. $4(3a - 7) < -18 + 2a$

54. $\dfrac{2x - 1}{x + 3} = \dfrac{3}{2}$

55. The length of a rectangle is 3 times its width.
  a. If the width of a rectangle is x, what is its length?
  b. Express the perimeter of the rectangle in terms of x.
  c. Suppose the perimeter of the rectangle is 48 centimeters. Find the dimensions of the rectangle.

56. In a city council election the winner received 132 votes more than her opponent.
  a. If the winner received y votes, how many did her opponent receive?
  b. Write an expression for the total number of votes cast for the two candidates.
  c. If 12,822 votes were cast, how many did each candidate receive?

57. Wheaton Elementary school plans to buy 30 computers. The computers with speakers cost $1200 each, and those without speakers cost $800 each. Let x represent the number of computers with speakers. Write expressions in terms of x for:
  a. The number of computers without speakers.
  b. The total cost of the computers with speakers.
  c. The total cost of the computers without speakers.
  d. The total cost of all 30 computers.
  e. If Wheaton Elementary has $28,800 to spend on the computers, how many of each kind can they buy?

58. a. Find the slope and y-intercept of the line $y = 4$.
  b. Explain why the equation of a vertical line cannot be written in slope-intercept form.

■ For Problems 59-66, use the most convenient method to graph the equation.

59.	$4x - 3y = 12$	**60.**	$\dfrac{x}{6} - \dfrac{y}{12} = 1$
61.	$50x = 40y - 20,000$	**62.**	$1.4x + 2.1y = 8.4$
63.	$3x - 4y = 0$	**64.**	$x = -4y$
65.	$4x = -12$	**66.**	$2y - x = 0$

67. The rate at which air temperature decreases with altitude is called the lapse rate. In the troposphere, the layer of atmosphere that extends from the earth's surface to a height of about 7 miles, the lapse rate is about 3.6°F for every 1000 feet.
  a. If the temperature on the ground is 62°F, write an equation for the temperature at an altitude of h feet.
  b. What is the temperature outside an aircraft flying at an altitude of 30,000 feet? How much colder is that than the ground temperature?
  c. What is the temperature at the top of the troposphere?

68. In his television program, *Notes from a Small Island*, aired in February of 1999, Bill Bryson discussed the future of the British aristocracy. Because not all families produce an heir, 4 or 5 noble lines die out each year. At this rate, Mr. Bryson says, and if no more peers are created, there will be no titled families left by the year 2175.
  a. Assuming that on average 4.5 titled families die out each year, write an equation for the number of noble houses left in year t, where $t = 0$ in 1999.
  b. Graph your equation.
  c. According to your graph, how many noble families existed in 1999? Which point on the graph corresponds to this information?

Chapter 4 Applications of Linear Equations

Lesson 4.1 Systems of Linear Equations

- Decide whether an ordered pair is a solution of a system
- Solve a system of equations by graphing
- Identify a system as consistent, inconsistent, or dependent
- Write a system of equations to solve an applied problem

Lesson 4.2 Algebraic Solution of Systems

- Solve systems by the substitution method
- Solve systems by the elimination method
- Identify a system as consistent, inconsistent, or dependent
- Write a system of equations to solve an applied problem

Lesson 4.3 Applied Problems

- Use a system to solve problems about interest
- Use a system to solve problems about mixtures
- Use a system to solve problems about motion

Lesson 4.4 Point-Slope Form

- Usethe two-point formula for slope
- Use the point-slope form to graph a line
- Use the point-slope form to find the equation of a line

Lesson 4.5 Using the Point-Slope Form

- Find the equation of a line through two points
- Use the point-slope form to find a linear model
- Find equations for parallel or perpendicular lines
- Find equations for horizontal or vertical lines

Lesson 4.6 Linear Regression

- Find a line of best fit
- Use interpolation and extrapolation
- Use a calculator to find the least squares regression line

4.1 Systems of Linear Equations

What is a System of Equations?

In many applications it is useful to write two or more linear equations.

Example 1 Delbert and Francine are buying appliances for their new home. They have a choice of two different refrigerators: a standard model that sells for \$1000, or an energy-efficient model at a price of \$1200. The standard model costs \$6 per month to run, and the energy-efficient model costs \$2 per month. Write linear equations for the total cost of each refrigerator after t months.

Solution Let S stand for the total cost of running the standard refrigerator for t months, so $S = mt + b$. The initial cost of the standard model is \$1000, so $b = \mathbf{1000}$. The cost increases at a rate of \$6 per month, so $m = \mathbf{6}$. Thus,

$$S = \mathbf{6}t + \mathbf{1000}$$

Let E stand for the total cost of running the energy-efficient refrigerator. For this model, the initial cost is $b = \mathbf{1200}$, and the total cost increases at a rate of $m = \mathbf{2}$ dollars per month. Thus,

$$E = \mathbf{2}t + \mathbf{1200}$$

Look Closer: When we consider two equations together, as in the example above, we often use the same variables for both equations, like this:

$$y = 6x + 1000$$
$$y = 2x + 1200$$

A pair of linear equations with the same variables is called a **system of linear equations**.

In Example 1, the standard refrigerator costs less initially, but costs more to run each month. Delbert and Francine want to know when the energy-efficient model will begin to pay for itself, or in other words, when the total cost of running the standard model will exceed the total cost of the energy-efficient model.

Example 2 When is the total cost of running the two refrigerators in Example 1 equal?

Solution We would like to know the value of t when $S = E$. We could use trial-and-error by evaluating both S and E for various values of t. The results of such a search are given in the table below.

t	10	20	30	40	50
S	1060	1120	1180	1240	1300
E	1220	1240	1260	1280	1300

We see that when $t = 50$, the total cost of both models is \$1300. Thus, at the end of 50 months, or 4 years 2 months, the cost of the standard model equals the cost of the energy-efficient model.

In Example 2 we found an ordered pair, $(50,\ 1300)$, that makes both equations true.

A **solution** to a system of equations is an ordered pair (x, y) that satisfies each equation in the system.

To check whether $(50, 1300)$ is really a solution of the system, we substitute the coordinates into each equation.

$S = 6t + 1000$	Does $\mathbf{1300} = 6(\mathbf{50}) + 1000$?	**True.**
$E = 2t + 1200$	Does $\mathbf{1300} = 2(\mathbf{50}) + 1200$?	**True.**

We have verified that the ordered pair $(50, 1300)$ is a solution to the system in Example 1.

RQ1. What is a solution to a system of equations?
RQ2. Decide whether the ordered pair $(2, 3)$ is a solution to the system

$$\begin{aligned} 3x - 4y &= -6 \\ x + 2y &= -4 \end{aligned}$$

Solving a System of Equations by Graphing

The solution of a system satisfies both equations in the system, so the point that represents the solution must lie on both graphs. It is the intersection point of the two lines described by the system. Thus, we can solve a system of equations by graphing the equations and looking for the point (or points) where the graphs intersect.

Example 3 Solve the system in Example 1 by graphing.

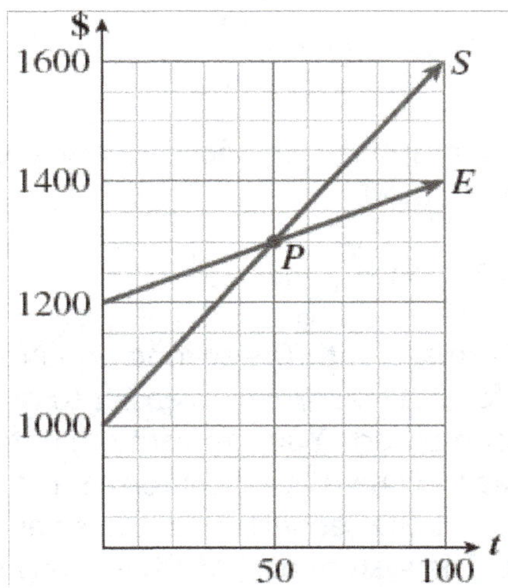

Solution The graph of the two equations is shown at right. The graph of S has a smaller initial value than E, but it increases more rapidly because its slope is greater. The two graphs intersect at the point labeled P. At this point the values of S and E are the same, approximately 1300. The t-coordinate of point P is 50, so the solution of the system is $(50,\ 1300)$. The total cost of each model is about \$1300 over the first 50 months of operation.

After 50 months, the total cost of the energy-efficient model will be less than the cost of the standard model.

RQ3. How do we find the solution to a system of equations by graphing?

Inconsistent and Dependent Systems

Not every system of equations has a solution.

Example 4 Delbert and Francine are also shopping for homeowners insurance. HomeLife offers a policy for semiannual premiums of \$408, or they can join a co-op for

$50 and pay just $68 per month for comparable coverage. Which option is less expensive in the long run?

Solution We'll write a system of equations and graph them, as we did in Example 3. If t stands for the number of months, then the total cost of the co-op insurance is

$$C = 50 + 68t$$

and the HomeLife policy costs

$$H = \frac{408}{6}t = 68t$$

(because a semiannual premium covers 6 months). The solution of the system represents the time when the costs of the two policies are equal. As you can see, the two lines are parallel, so the graphs do not intersect. This system has no solution. There is no point where the two policies cost the same amount; the HomeLife policy is always less expensive.

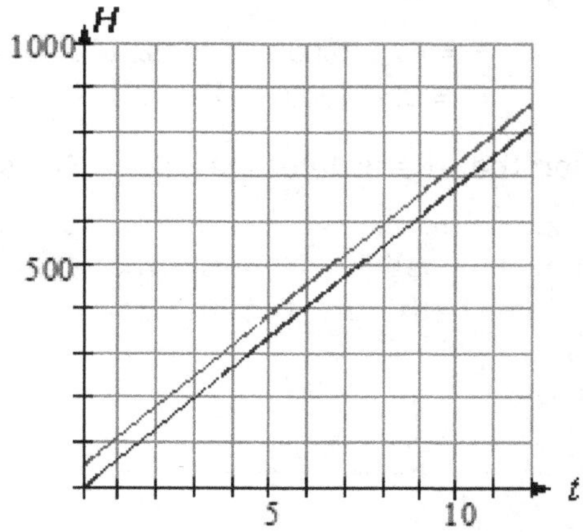

In Example 4, the two graphs never intersect, so the system has no solution. Such a system is called **inconsistent**. There is a third possibility: Both equations may have the same graph.

Example 5 Solve the system

$$y = 2x + 2$$
$$6x - 3y = -6$$

Solution We graph the first equation by the slope-intercept method:

$$b = 2 \quad \text{and} \quad m = \frac{\Delta y}{\Delta x} = \frac{2}{1}$$

Its graph is shown at right. The second equation, which has the form $Ax + By = C$, is easier to graph by the intercept method. You can verify that its intercepts are $(0, 2)$ and $(-1, 0)$. The graph of the second line is identical to the first graph. Every solution to the first equation is also a solution to the second equation, so every point on the line is a solution of the system as well. The system has infinitely many solutions.

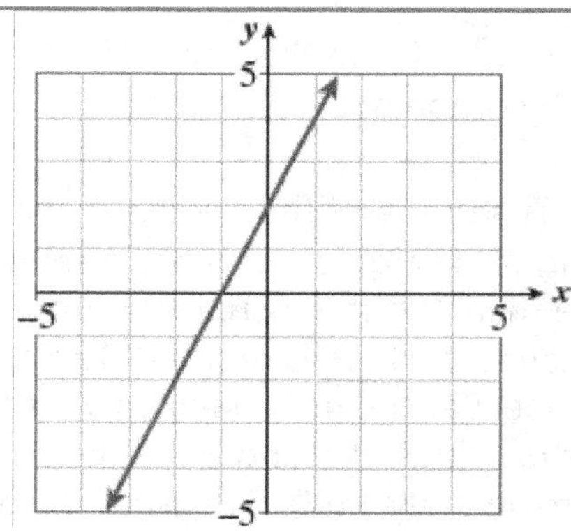

We can verify that the two equations in Example 5 are actually equivalent by solving the second equation for y in terms of x.

$$6x - 3y = -6 \qquad \textbf{Add } 3y \textbf{ and 6 to both sides.}$$
$$6x + 6 = 3y \qquad \textbf{Divide both sides by 3.}$$
$$2x + 2 = y$$

The second equation is the same as the first equation. A system in which the two equations are equivalent is called **dependent**. A dependent system has infinitely many solutions: Every point on the graph is a solution to the system.

We have now seen three types of systems. These are illustrated in the figure on the next page. Most of the systems we will study have exactly one solution. Such systems are called **consistent and independent**.

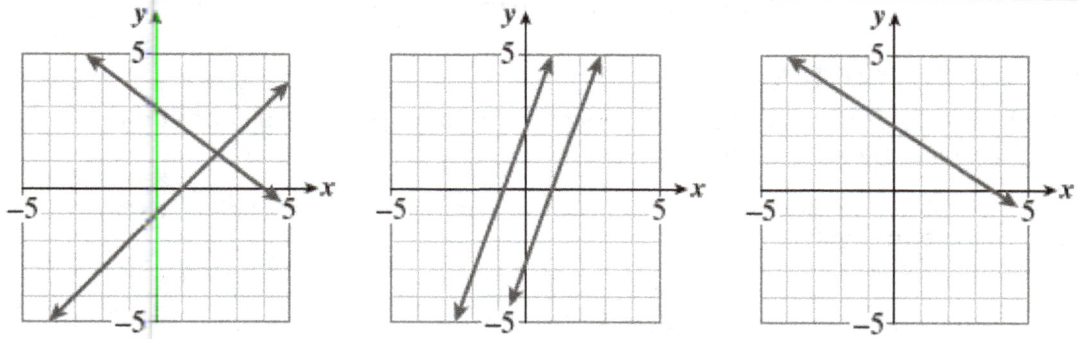

Consistent and independent; one solution

Inconsistent; no solution

Dependent; infinitely many solutions

1. **Consistent and independent system.** The graphs of the two lines intersect in exactly one point. The system has exactly one solution.
2. **Inconsistent system**. The graphs of the equations are parallel lines and hence do not intersect. An inconsistent system has no solutions.
3. **Dependent system**. All the solutions of one equation are also solutions to the second equation, and hence are solutions of the system. The graphs of the two equations are the same line. A dependent system has infinitely many solutions.

RQ4. What does the graph of an inconsistent system look like? How many solutions does it have?
RQ5. What does the graph of a dependent system look like? How many solutions does it have?

Using a Graphing Calculator

We use a calculator to graph the equations in a system quickly and easily.

■ **Example 6** Use your calculator to solve the system by graphing.

$$y = 1.7x + 0.4$$
$$y = 4.1x + 5.2$$

Solution Set the graphing window to

Xmin $= -9.4$ Xmax $= 9.4$
Ymin $= -10$ Ymax $= 10$

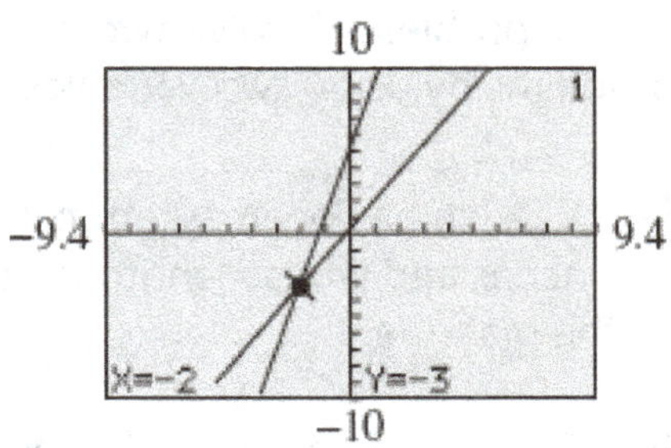

and enter the two equations We see that the two lines intersect in the third quadrant, and we use the TRACE key to find the coordinates of the intersection point. You can check that the point $(-2, -3)$ lies on both graphs. (Use the up and down arrow keys to move the bug from one graph to the other.) The solution to the system is $x = -2,\ y = -3$. ■

The values we obtain from a calculator may be only approximations, so it is a good idea to check the solution algebraically. In Example 6 we find that both equations are true when we substitute $x = -2$ and $y = -3$.

$$-3 = 1.7\,(-2) + 0.4 \qquad \textbf{True}$$
$$-3 = 4.1\,(-2) + 5.2 \qquad \textbf{True}$$

The Intersect Feature

Because the **Trace** feature does not show every point on a graph, we may not find the exact solution to a system by tracing the graphs.

■ **Example 7** Solve the system

$$3x - 2.8y = 21.06$$
$$2x + 1.2y = 5.3$$

Solution We graph the system in the standard window by solving each equation for y. Enter

$$\mathbf{Y_1} = (21.06 - 3\mathbf{X})/-2.8$$
$$\mathbf{Y_2} = (5.3 - 2\mathbf{X})/1.2$$

and then press ZOOM 6. If we trace along the graph to the intersection point, we will not find the same coordinates on both lines. The intersection point is not displayed in this window. Instead, we can use the **intersect** feature of the calculator.

Using the arrow keys, we position the **Trace** bug as close to the intersection point as we can. We then press 2nd CALC to see the **Calculate** menu. Press 5 for **intersect**, and respond to the calculator's questions, **First curve?, Second curve?**, and **Guess?** by pressing ENTER. The calculator will then display the intersection point, $x = 4.36$, $y = -2.85$. ■

We substitute these values into the original system to check that they satisfy both equations.

$$3(\mathbf{4.36}) - 2.8(\mathbf{-2.85}) = 21.06 \qquad \textbf{True}$$
$$2(\mathbf{4.36}) + 1.2(\mathbf{-2.85}) = 5.3 \qquad \textbf{True}$$

Thus, $(4.36, -2.85)$ is the exact solution to the system. How does the calculator find the exact coordinates of the intersection point? In the next section we'll learn how to find the solution of a system using algebra.

Applications

Many practical problems involve two or more unknown quantities. Often it is easier to solve these problems by using two variables and writing a system of equations.

■ **Example 8** Allen has been asked to design a rectangular Plexiglas plate whose perimeter is 28 inches and whose length is three times its width. What should the dimensions of the plate be?

Solution **Step 1** The dimensions of a rectangle are its length and width, so we let x represent the width of the plate and y represent its length.

Step 2 We must write two equations about the length and width of the plate. We know a formula for the perimeter of a rectangle, $P = 2l + 2w$, so the first equation is

$$2x + 2y = 28$$

We also know that the length is three times the width, or $y = 3x$. These two equations make a system.

$$2x + 2y = 28$$
$$y = 3x$$

Step 3 We use the intercept method to graph the first equation, $2x + 2y = 28$:

x	y
0	14
14	0

When $x = 0$, $y = 14$, and when $y = 0$, $x = 14$.

We use the slope-intercept method to graph $y = 3x$:

$$b = 0 \qquad m = \frac{\Delta y}{\Delta x} = \frac{3}{1}$$

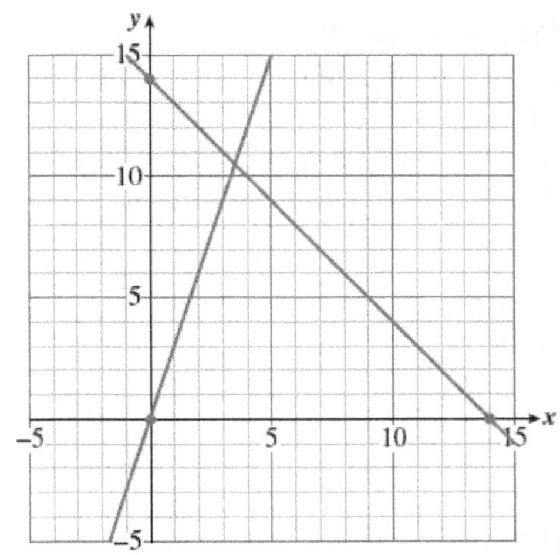

Their graphs are shown at right. The graphs intersect at approximately $(3.5, 10.5)$. You can verify that $x = 3.5$ and $y = 10.5$ are a solution to the system by checking that these values make *both* equations true.
Step 4 The width of the plate should be 3.5 inches and its length should be 10.5 inches.

RQ6. When might you use a system of equations to solve an applied problem?

An Application from Economics

The owner of a retail business must try to balance the consumer demand for his product with the supply that he can obtain from manufacturers. Supply and demand both vary with the price of the product: Consumers usually buy fewer items if the price increases, but manufacturers will be willing to supply more units of the product if its price increases.

The **demand equation** gives the number of units of the product that consumers will buy in terms of the price per unit. The **supply equation** gives the number of units that the producer will supply in terms of the price per unit. The price at which the supply and demand are equal is called the **equilibrium price.** This is the price at which the consumer and the producer agree to do business.

Example 9 A woolen mill can produce $400x$ yards of suit fabric if they can charge x dollars per yard. Their clients in the garment industry will buy $6000 - 100x$ yards of suit fabric at a price of x dollars per yard. Find the equilibrium price and the amount of fabric that will change hands at that price.

Solution **Step 1** price per yard: x
number of yards: y

Step 2 The supply equation tells us how many yards of fabric the mill will produce for a price of x dollars per yard.

$$y = 400x$$

The demand equation tells us how many yards of fabric the garment industry will buy at a price of x dollars per yard.

$$y = 6000 - 100x$$

Step 3 We graph the two equations on the same set of axes, as shown above. We set the **Window** values to

Xmin $= 0$ Xmax $= 94$
Ymin $= 0$ Ymax $= 6200$

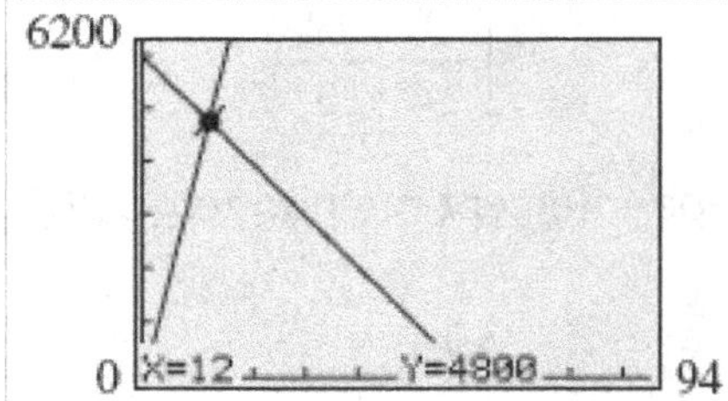

and use the **Trace** to locate the solution. (Or use the **intersect** command.) The graphs intersect at the point $(12, 4800)$.

Step 4 The equilibrium price is \$12 per yard, and the mill will sell 4800 yards of fabric at that price.

RQ7. The equilibrium price is the intersection of the graphs of what two equations?

Skills Warm-Up

Write algebraic expressions to answer the questions.

1. How much interest is earned after 2 years on $d + 50$ dollars at 6% annual interest rate?
2. How much manganese is there in $8 - z$ grams of an alloy that is 35% manganese?
3. How far will a boat travel in 2 hours at a speed of $r + 3$ miles per hour?
4. How far will a train moving at 40 miles per hour travel in $3 + x$ hours?
5. A small plane has a top airspeed of v miles per hour. How far can the plane travel in 5 hours against a headwind of 15 miles per hour?
6. A fishing boat has a top speed in still water of 26 miles per hour. How far can the boat travel downstream in 3 hours if the speed of the current is w miles per hour?

Answers: **1.** $0.12(d + 50)$ **2.** $0.35(8 - z)$ **3.** $2(r + 3)$ **4.** $40(3 + x)$ **5.** $5(v - 15)$ **6.** $3(26 + w)$

Homework 4.1

Skills Practice

For Problems 1-2, decide whether the given ordered pair is a solution of the system.

1. $x + 2y = -8$, $2x - y = 4$ $\quad (4, -2)$

2. $x = 5y + 13$, $2x + 3y = 0$ $\quad (3, -2)$

For Problems 3-4, solve each system of equations using the graphs given. Verify algebraically that your solution satisfies both equations.

3. $2.3x - 3.7y = 6.9$
$1.1x + 3.7y = 3.3$

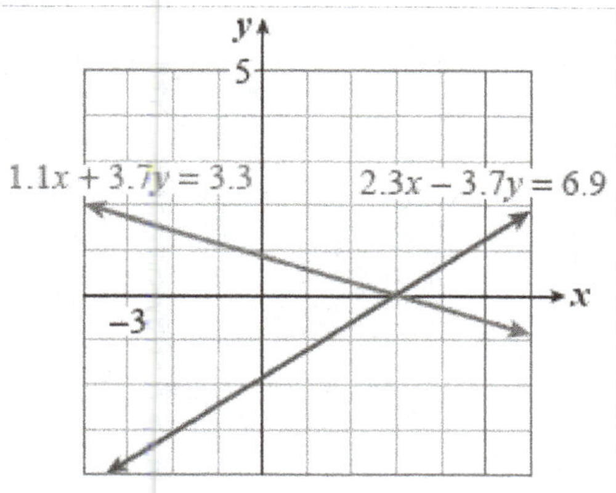

4. $35s - 17t = 560$
$24s + 15t = 2250$

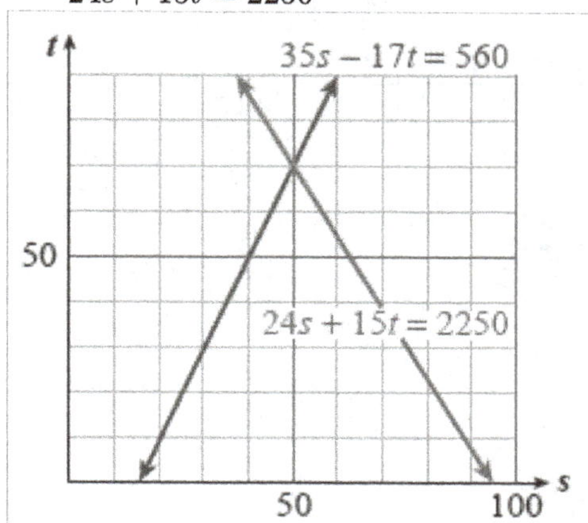

For problems 5-6, solve the system by graphing.

5. $y = -3x$
$y = 3x - 6$

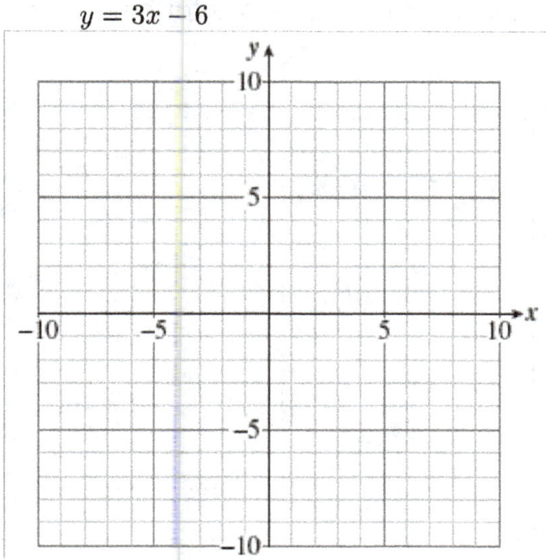

6. $y = \frac{3}{4}x - 2$
$2y - 5x = 10$

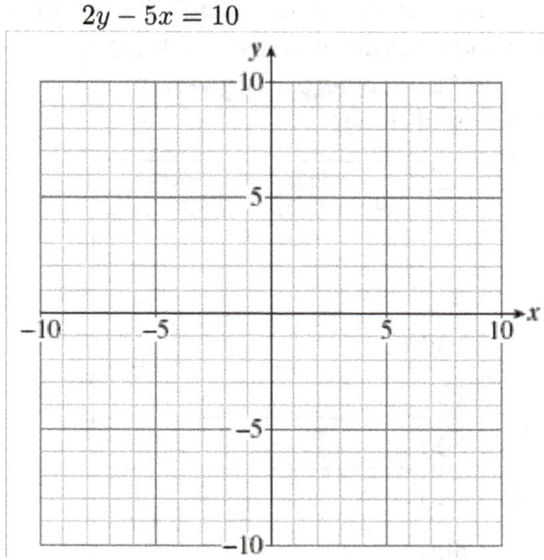

Solve the systems in Problems 7-8 by graphing. Use the "friendly" window

Xmin $= -9.4$ Xmax $= 9.4$
Ymin $= -10$ Ymax $= 10$

Verify algebraically that your solution satisfies both equations.

7. $y = 2.6x + 8.2$
$y = 1.8 - 0.6x$

8. $y = 7.2 - 2.1x$
$-2.8x + 3.7y = 5.5$

For problems 9-10, graph each system by hand, using either the intercept method or the slope-intercept method. Identify each system as dependent, inconsistent, or consistent and independent.

9. $x + y = 4$
$2x + 2y = 12$

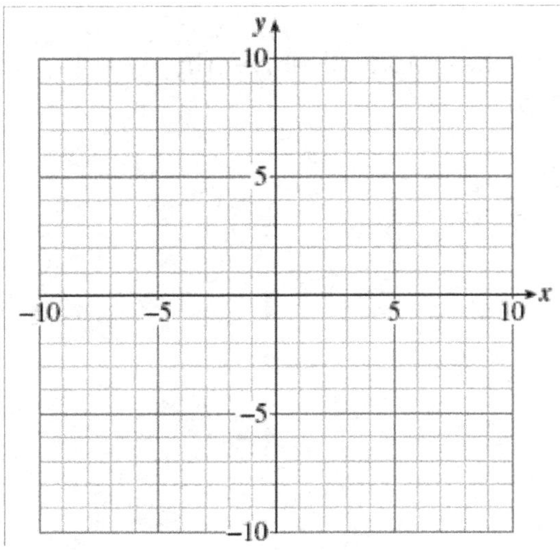

10. $2y - 12 = -6x$
$12x + 4y = 24$

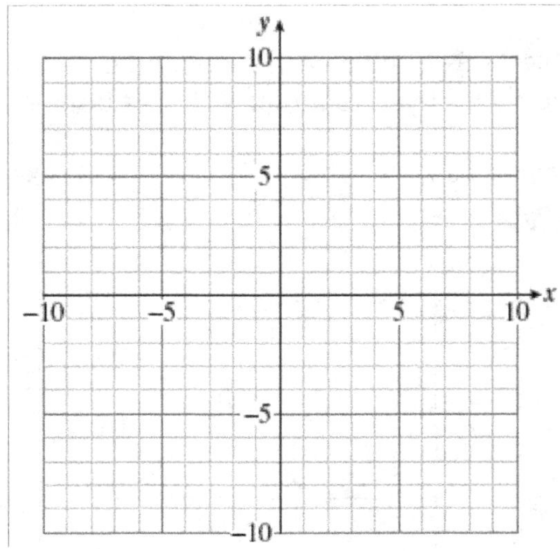

Applications

11. Delbert has accepted a sales job and is offered a choice of two salary plans. Under Plan A he receives $20,000 a year plus a 3% commission on his sales. Plan B offers a $15,000 annual salary plus a 5% commission.

a. Let x stand for the amount of Delbert's sales in one year. Write equations for his total annual earnings under each plan.

b. Fill in the table, where x is given in thousands of dollars.

x	Earnings Under Plan A	Earnings Under Plan B
0		
50		
100		
150		
200		
250		
300		
350		
400		

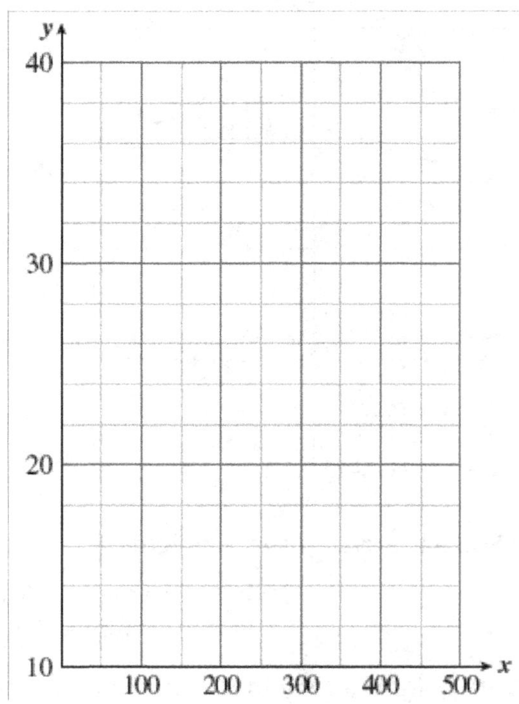

c. Graph both of your equations on the grid. (Both axes are scaled in thousands of dollars.)

d. For what sales amount do the two plans result in equal earnings for Delbert?

12. The Aquarius jewelry company determines that each production run to manufacture a pendant involves an initial setup cost of $200 and $4 for each pendant produced. The pendants sell for $12 each.

a. Express the cost C of production in terms of the number x of pendants produced.

b. Express the revenue R in terms of the number x of pendants sold.

c. Complete the table of values for the cost and revenue.

x	5	10	15	20	25	30
C						
R						

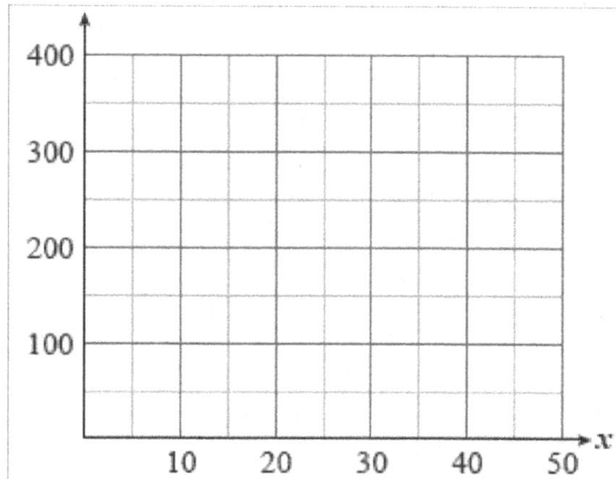

d. Graph the revenue and cost on the same set of axes.

e. How many pendants must be sold for the Aquarius company to break even on a particular production run?

13. Yasuo can afford to produce $50x$ bushels of wheat if he can sell them at x cents per bushel, and the market will buy $2100 - 20x$ bushels at x cents per bushel.

a. What is the supply equation?

b. What is the demand equation?

c. Graph both equations in the window

$$\text{Xmin} = 0 \quad \text{Xmax} = 120$$
$$\text{Ymin} = 0 \quad \text{Ymax} = 2500$$

d. Find the equilibrium price and the number of bushels Yasuo can sell at that price.

■ For Problems 14-15, write a system of equations in two variables to model the problems, and solve the system graphically. Verify your solution algebraically.

14. A bouquet of 4 roses and 8 carnations costs $14. A bouquet of 6 roses and 9 carnations costs $18. How much do one rose and one carnation cost?

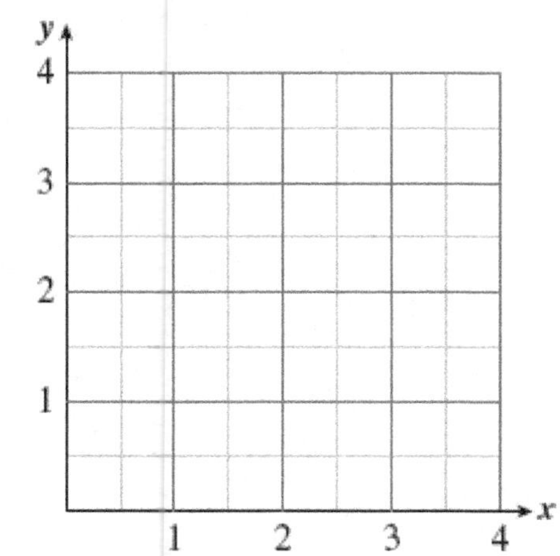

Grid for #14

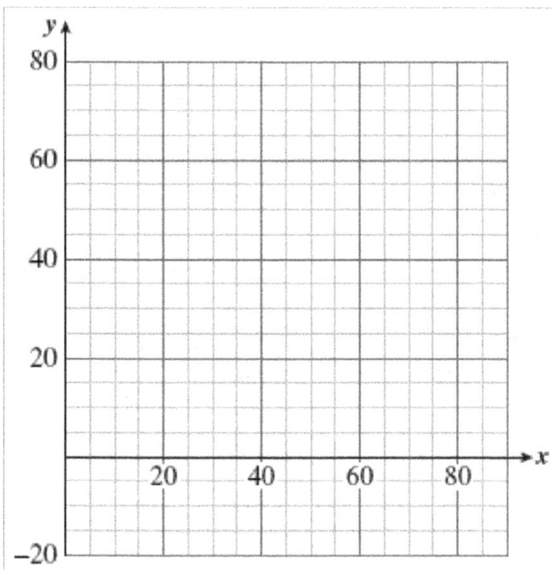

Grid for #15

15. The vertex angle of an isosceles triangle is 15° less than each base angle. Find the measure of each angle of the triangle. (**Hint:** For your second equation, recall that the sum of the angles in any triangle is 180°.)

16. Without graphing, explain how you can tell that the system has no solution.

$$3x = y + 3$$
$$6x - 2y = 12$$

4.2 Algebraic Solution of Systems

In the previous Lesson, we compared the costs of operating two different refrigerators, a standard model and an energy-efficient model. We wrote a system of equations for this problem,

$$y = 6x + 1000$$
$$y = 2x + 1200$$

where x is the number of months the refrigerator has been running. We looked for a point on the graphs where the two y-values, which represent the costs, are equal. This point is the intersection point of the graphs, as shown in the figure. The x-coordinate of the intersection point shows that the costs are equal after 50 months.

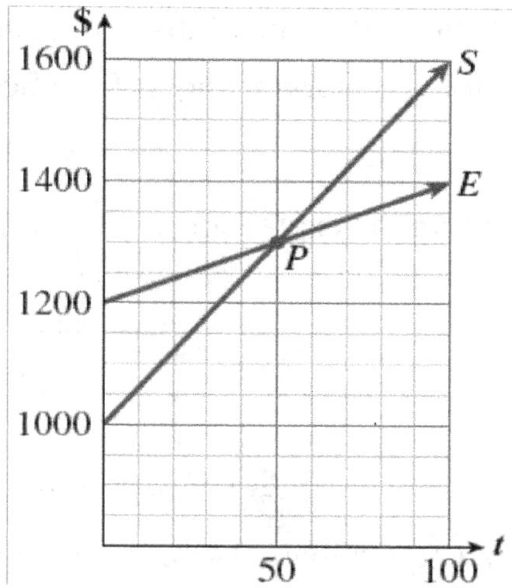

We can also solve the system using algebra.

Example 1 Solve the system algebraically.

$$y = 6x + 1000$$
$$y = 2x + 1200$$

Solution We are looking for the point where the two y-values are equal. Therefore, we set the two expressions for y equal, which gives us an equation in x to solve:

$6x + 1000 = 2x + 1200$	**Subtract $2x$ from both sides.**
$4x + 1000 = 1200$	**Subtract 1000 from both sides.**
$4x = 200$	**Divide both sides by 4.**
$x = 50$	

We find that $x = 50$, the same answer we got using graphing. Substituting $x = 50$ into either equation gives $y = 1300$. The solution is $(50, 1300)$.

Substitution Method

In Example 1, we solved the system by equating the two expressions for y. We can think of this procedure as substituting the expression for y from one equation into the other equation, like this:

$$y = \underbrace{6x + 1000}$$ Substitute into second equation.

$$y = 2x + 1200$$

to get

$$6x + 1000 = 2x + 1200$$

Look Closer: The method works well in this example because both equations in the system were already in the form $y = mx + b$. Sometimes we must solve one of the equations for y in terms of x before we can substitute.

Example 2 Use substitution to solve the system

$$y + 3x = 1$$
$$2y + 5x = 5$$

Solution We first solve one of the equations for y in terms of x. It is easier to solve the first equation for y, by subtracting $3x$ from both sides to get:

$$y = 1 - 3x$$

We now substitute the expression $1 - 3x$ for y in the *other* equation:

$$y = \underbrace{1 - 3x}$$

$$2y + 5x = 5$$

$$2(1 - 3x) + 5x = 5$$

This gives us an equation in one variable, which we can solve as usual. We begin by simplifying the left side.

$2(1 - 3x) + 5x = 5$	**Apply the distributive law.**
$2 - 6x + 5x = 5$	**Combine like terms.**
$2 - x = 5$	**Subtract 2 from both sides.**
$-x = 3$	**Divide both sides by -1.**
$x = -3$	

This is the x-coordinate of the solution point. To find the y-coordinate, we substitute $x = -3$ into either of the original equations. It is easiest to use the equation we solved for y, namely $y = 1 - 3x$:

$$y = 1 - 3(-3) = 10 \qquad \textbf{Substitute } -3 \textbf{ for } x.$$

The solution to the system is $x = -3$, $y = 10$, or the point $(-3, 10)$. You can check that these values satisfy both of the original equations in the system.

RQ1. What is the first step in the substitution method?
RQ2. How can you check your answer for the solution of a system?

Here is a summary of the substitution method.

To Solve a System by Substitution:

1. Choose one of the variables in one of the equations. (It is best to choose a variable whose coefficient is 1 or -1.) Solve the equation for that variable.
2. Substitute the result of Step 1 into the *other* equation. This gives an equation in one variable.
3. Solve the equation obtained in Step 2. This gives the solution value for one of the variables.
4. Substitute this value into the result of Step 1 to find the solution value of the other variable.

Elimination Method

A second algebraic method for solving systems is called **elimination**. As with the substitution method, we try to obtain an equation in a single variable, but we do it by eliminating one of the variables in the system. We first put both equations into the general linear form, $Ax + By = C$.

Example 3 Solve the system

$$\begin{aligned} 5x &= 2y + 21 \\ 2y &= 19 - 3x \end{aligned}$$

Solution First, we rewrite each equation in the form $Ax + By = C$.

$$\begin{array}{rl} 5x = & 2y + 21 \\ -2y = & -2y \\ \hline 5x - 2y = & 21 \end{array} \quad \textbf{Subtract } 2y. \qquad \begin{array}{rl} 2y = & 19 - 3x \\ +3x = & +3x \\ \hline 3x + 2y = & 19 \end{array} \quad \textbf{Add } 3x.$$

We add the equations together by adding the left side of the first equation to the left side of the second equation, and then adding the two right sides together, as follows:

$$\begin{array}{rl} 5x - 2y &= 21 \\ 3x + 2y &= 19 \\ \hline 8x \qquad &= 40 \end{array}$$

Note that the y-terms canceled, or were eliminated. We are left with an equation in x that is easy to solve:

$$\begin{aligned} 8x &= 40 && \textbf{Divide both sides by 8.} \\ x &= 5 \end{aligned}$$

We are not finished yet, because we must still find the value of y. We can substitute our value for x into either of the original equations, and solve for y. We'll use the second equation, $3x + 2y = 19$:

$$\begin{aligned} 3(5) + 2y &= 19 && \textbf{Subtract 15 from both sides.} \\ 2y &= 4 && \textbf{Divide by 2.} \\ y &= 2 \end{aligned}$$

Thus, the solution is the point $(5, 2)$. ■

Look Ahead: You may have noticed that this method worked because the coefficients of y in the two equations were opposites, 2 and -2. This caused the y-terms to cancel out when we added the two equations together. What if the coefficients of neither x nor y are opposites? Then we must multiply one or both of the equations in the system by a suitable constant. ●

Example 4 Solve the system

$$\begin{aligned} 4x + 3y &= 7 \\ 3x + y &= -1 \end{aligned}$$

Solution We can choose to eliminate either the x-terms or the y-terms in a system. For this example, it will be faster to eliminate the y-terms. If we multiply each term of the second equation by -3, then the coefficients of y will be opposites:

$$-\mathbf{3}(3x + y = -1) \quad \rightarrow \quad -9x - 3y = 3$$

Because we are applying the multiplication property of equality, we must multiply *each* term by -3, not just the y-term. We can then replace the second equation by its new version to obtain a new system, and we add the equations together:

$$\begin{aligned} 4x + 3y &= 7 \\ \underline{-9x - 3y} &\underline{= 3} \\ -5x \qquad &= 10 \end{aligned}$$

The y-terms were eliminated, and we solve the resulting equation for x to get $x = -2$. Finally, we substitute $x = -2$ into either of the equations to find $y = 5$. The solution is $(-2, 5)$. ■

Caution! In Example 4, we multiplied both sides of the equation $3x + y = -1$ by -3. (This is an application of the Multiplication Property of Equality.) We must be careful to multiply *every term on both sides of the equation* by the same constant. Otherwise, we won't have an equivalent equation -- its solutions will not be the same. ■

When we add a multiple of one equation to the other we are making a **linear combination** of the equations.

The method of elimination is also called the method of linear combinations. Sometimes it is necessary to multiply *both* equations by suitable constants in order to eliminate one of the variables.

■ **Example 5** Use linear combinations to solve the system

$$\begin{aligned} 5x - 2y &= 22 \\ 2x - 5y &= 13 \end{aligned}$$

Solution This time we choose to eliminate the x-terms. We must arrange things so that the coefficients of the x-terms are opposites, so we look for the smallest integer that both 2 and 5 divide into evenly. (This number is called the *lowest common multiple*, or LCM, of 2 and 5.) The LCM of 2 and 5 is 10. We want one of the coefficients of x to be 10, and the other to be -10. To achieve this, we multiply the first equation by 2 and the second equation by -5.

$$\begin{aligned} \mathbf{2}\,(5x - 2y = 22) &\quad\rightarrow\quad 10x - 4y = 44 \\ \mathbf{-5}\,(2x - 5y = 13) &\quad\rightarrow\quad -10x + 25y = -65 \end{aligned}$$

Adding these new equations eliminates the x-term and yields an equation in y.

$$\begin{aligned} 10x - \;\; 4y &= 44 \\ \underline{-10x + 25y} &\underline{= -65} \\ 21y &= -21 \end{aligned}$$

We solve for y to find $y = -1$. Finally, we substitute $y = -1$ into the first equation and solve for x.

$$\begin{aligned} 5x - 2(-1) &= 22 \\ 5x + 2 &= 22 \\ x &= 4 \end{aligned}$$

The solution to the system is $(4, -1)$. ■

RQ3. What is a linear combination of expressions?
RQ4. What is the first step in the elimination method?

Here are the steps for solving a system by elimination.

To Solve a System by Elimination:

1. Write each equation in the form $Ax + By = C$.
2. Decide which variable to eliminate. Multiply each equation by an appropriate constant so that the coefficients of that variable are opposites.
3. Add the equation from Step 2 and solve for the remaining variable.
4. Substitute the value found in Step 3 into one of the original equations and solve for the other variable.

Look Ahead: How do you know which method to use to solve a system, substitution or elimination? Both methods will work on any linear system. However, substitution will be easier if one of the variables in one of the equations has a coefficient of 1 or -1. Otherwise, the elimination method is usually more efficient. ●

RQ5. How do we eliminate a variable from a system?
RQ6. When is the substitution method easier than elimination?

Inconsistent and Dependent Systems

Recall that a system of two parallel lines has no solution and is called inconsistent. If the two equations in a system have the same graph, then every point on the graph is a solution and the system is called dependent. The elimination method will reveal whether the system falls into one of these two cases.

Example 6 Solve each system by elimination.

a. $3x - y = 2$
$-6x + 2y = 3$

b. $x - 2y = 3$
$2x - 4y = 6$

Solutions a. To eliminate the y-terms, we multiply the first equation by 2 and add:

$$\begin{array}{r} 6x - 2y = 4 \\ -6x + 2y = 3 \\ \hline 0x + 0y = 7 \end{array}$$

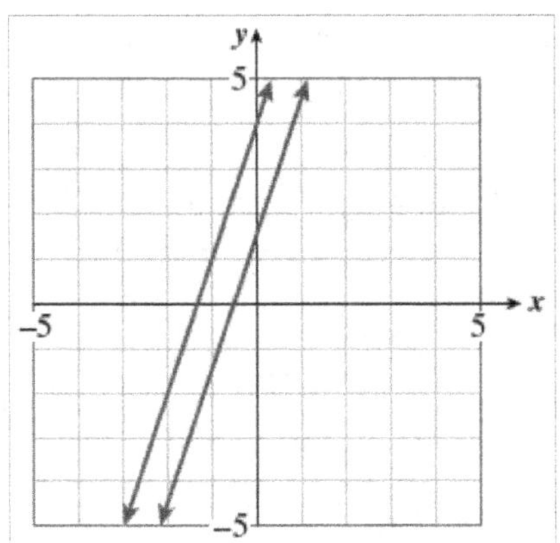

Both variables are eliminated, and we are left with the false statement $0 = 7$. There are no values of x or y that will make this equation true, so the system has no solutions. The graph shows that the system is inconsistent.

b. To eliminate the x-terms we multiply the first equation by -2 and add:

$$\begin{array}{r} -2x + 4y = -6 \\ 2x - 4y = 6 \\ \hline 0x + 0y = 0 \end{array}$$

We are left with the true but unhelpful equation $0 = 0$. The two equations are in fact equivalent (one is a constant multiple of the other), so the system is dependent. The graph of both equations is shown in the figure. ■

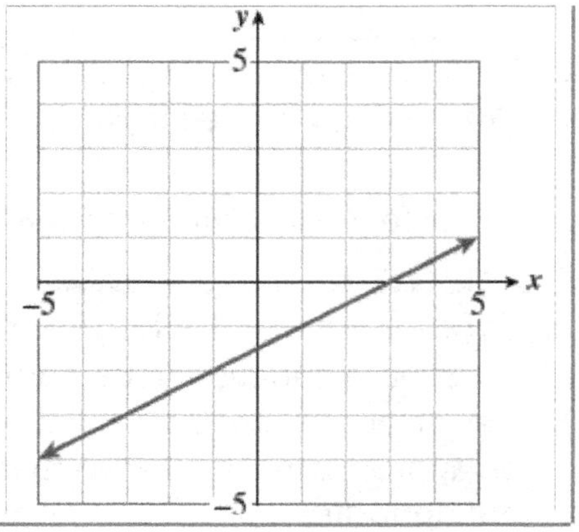

Example 5 illustrates a rule for identifying inconsistent and dependent systems.

When Using Elimination to Solve a System:

1. If combining the two equations results in an equation of the form

$$\mathbf{0x + 0y = k \quad (k \neq 0)}$$

then the system is inconsistent.

2. If combining the two equations results in an equation of the form

$$\mathbf{0x + 0y = 0}$$

then the system is dependent.

RQ7. How can you tell if a system is inconsistent?
RQ8. How can you tell if a system is dependent?

Skills Warm-Up

■ Solve for the indicated variable.

1. $2q + p = 10$	for p	**2.** $2l + 2w = 18$	for l
3. $3a + 9 = -6b$	for a	**4.** $2c = 2d + 22$	for d
5. $5r - 4s = 24$	for s	**6.** $2m = 11 - 3n$	for n

Answers: **1.** $10 - 2q$ **2.** $9 - w$ **3.** $-3 - 2b$ **4.** $c - 11$ **5.** $\dfrac{5r - 24}{4}$ **6.** $\dfrac{11 - 2m}{3}$

Homework 4.2

Skills Practice

In Problems 1-6, solve the system by the substitution method.

1. $y = 2x$
$3x + y = 10$

2. $2a + 3b = 4$
$b = 2a + 4$

3. $2x - 3y = 4$
$x + 1 = 3y$

4. $2a + b = 16$
$4a - 4 = -3b$

5. $8r + s = 4$
$2r + 7s = -8$

6. $36a = 6 - 3b$
$2b = 3a - 5$

In Problems 7-12, solve the system by elimination.

7. $x + y = 5$
$x - y = 1$

8. $2a + b = 3$
$a + b = 2$

9. $3x + 2y = 7$
$x + y = 3$

10. $3a = 1 + 5b$
$2b = 14 + 6a$

11. $2x + 3y = -1$
$3x + 5y = -2$

12. $5z = 1 - 3w$
$5w = 2 - 7z$

In Problems 13-15, decide whether the system is inconsistent, dependent, or consistent.

13. $3x - 6y = 6$
$x - 2y = 3$

14. $8a = 6 + 12b$
$4 = 6a - 9b$

15. $3p = 1 + 7q$
$21q = 9p - 3$

In Problems 16-17, which method, elimination or substitution, would be easier to apply to each system?

16. $2u + 3v = -8$
$2u - 3v = -6$

17. $17a - b = 0$
$11a + 3b = 4$

Applications

For Problems 18-22,

a. Choose variables for the two unknown quantities in the problem.
b. Write a system of equations to model the problem.
c. Solve the system and answer the question.

18. A hamburger and a chocolate shake together contain 1030 calories. Two shakes and three hamburgers contain 2710 calories. How many calories are there in one hamburger and in one chocolate shake?

19. Darryl is in charge of buying furniture for a new restaurant. Chairs cost $175 and tables cost $250. Darryl's budget allows $11,400 for tables and chairs, and he plans to buy four chairs for each table. How many of each can he buy?

20. The perimeter of a rectangle is 42 meters, and its width is 13 meters less than its length. Find the dimensions of the rectangle.

21. Three pounds of bacon and 2 pounds of coffee costs \$17.80. Two pounds of bacon and 5 pounds of coffee costs \$32.40. How much do 1 pound of bacon and 1 pound of coffee cost?

22. Sanaz can afford to produce $35x$ pairs of sunglasses if she can sell them at x dollars per pair, and the market will buy $1700 - 15x$ at x dollars a pair.

a. Write the supply and demand equations for the sunglasses.

b. Find the equilibrium price and the number of sunglasses Sanaz will produce and sell at that price.

4.3 Applications of Systems

In this section we explore some applications of systems of equations. We begin by reviewing some familiar formulas.

Interest

Recall the formula for calculating interest,

$$I = Prt$$

I is the interest you will earn after t years if you invest a principal of P dollars in an account that earns simple interest at an annual rate r. If we make two or more investments at the same time, then the total interest we earn is the sum of the interests earned on each investment separately.

Caution! Notice the difference between interest, I, and interest rate, r. The **interest rate** is a *percentage*, such as 5%. The **interest** is the *amount* of money you earn, usually in dollars. ■

Warm-Up 1 Harvey deposited \$1200 in two accounts. He put \$700 in a savings account that pays 6% annual interest rate, and the rest in his credit union, which pays 7% annual interest rate. How much will Harvey's investments earn in two years?

Step 1 Calculate the interest earned by each account.

Savings: $I = Prt =$

Credit union: $I = Prt =$

Step 2 Add the earnings from the two investments.

$$\begin{pmatrix}\text{Total}\\ \text{interest}\end{pmatrix} = \begin{pmatrix}\text{Interest from}\\ \text{stocks}\end{pmatrix} + \begin{pmatrix}\text{Interest from}\\ \text{bonds}\end{pmatrix}$$

Here is a similar example with variables.

Warm-Up 2 You have \$5000 to invest for one year. You want to put part of the money into bonds that pay 7% interest rate and the rest of the money into stocks, which involve some risk but will pay 12% if the investment is successful.

a. Fill in the table.

Amount invested in stocks	Amount invested in bonds	Interest from stocks	Interest from bonds	Total interest
\$500				
\$1000				
\$3200				
\$4000				

Now suppose you invest x dollars in the stocks and y dollars in the bonds.

b. Sum of amounts invested: $x + y =$

c. Interest earned on the stocks: $I = Prt =$

Interest earned on the bonds: $I = Prt =$

d. Finally, suppose that you earned a total of \$345 in interest from your two investments.

Write an equation about this, using your expressions from part (c).

$$\begin{pmatrix}\text{Total}\\ \text{interest}\end{pmatrix} = \begin{pmatrix}\text{Interest from}\\ \text{stocks}\end{pmatrix} + \begin{pmatrix}\text{Interest from}\\ \text{bonds}\end{pmatrix}$$

RQ1. What formula do we need for interest problems?
RQ2. What is the difference between principal and interest?

Example 1 Mort invested money in two accounts, a savings plan that pays 8% interest and a mutual fund that pays 7% interest. He put twice as much money in the savings plan as in the mutual fund. At the end of the year Mort's total interest income was \$345. How much did he invest in each account?

Solution The unknown quantities are the amounts Mort invested in his two accounts.

Amount invested in savings: x
Amount invested in mutual fund: y

Do not confuse the amount Mort invested in each account (the principal) with the amount he earned in interest! A table is a good way to keep these amounts straight.

	Principal	Interest Rate	Interest
Savings plan	x	0.08	$0.08x$
Mutual fund	y	0.07	$0.07y$

We must write two equations for the problem, one about the principal and one about the interest. From the statement "He put twice as much money in the savings plan as in the mutual fund," we write the equation

$$\begin{pmatrix}\text{Amount invested}\\ \text{in savings plan}\end{pmatrix} = 2 \cdot \begin{pmatrix}\text{Amount invested}\\ \text{in mutual fund}\end{pmatrix}$$

Equation about principal: $x = 2y$

From the statement "Mort's total interest income was \$345," we write the equation

$$\begin{pmatrix}\text{Interest from}\\ \text{savings plan}\end{pmatrix} + \begin{pmatrix}\text{Interest from}\\ \text{mutual fund}\end{pmatrix} = \text{Total interest}$$

Equation about interest: $0.08x + 0.07y = 345$

This gives us a system of equations.

$$\begin{aligned} x &= 2y \\ 0.08x + 0.07y &= 345 \end{aligned}$$

Because the first equation is already solved for x in terms of y, we'll use the substitution method to solve the system. We substitute $2y$ for x in the second equation and solve
for y.

$$\begin{aligned} 0.08(2y) + 0.07y &= 345 \\ 0.16y + 0.07y &= 345 && \text{Combine like terms.} \\ 0.23y &= 345 && \text{Divide both sides by 0.23.} \\ y &= 1500 \end{aligned}$$

Finally, we substitute $y = 1500$ into the first equation to find

$$x = 2y = 2(1500) = 3000$$

Thus, Mort invested \$3000 in the savings plan and \$1500 in the mutual fund. ■

RQ3. What were the two equations in Example 1 about?
RQ4. Which method did we use to solve the system in Example 1?

Mixtures

A pharmacist has on hand 20 ounces of a certain drug at 40% strength, but she needs a small quantity of the drug at 75% strength for a prescription. She decides to add a pure form of the drug to the 40% solution. How much should she add to make a mixture of 75% strength?

To solve such problems, we review some properties of percent. We will need the percent formula,

$$P = rW$$

P stands for the part obtained when we take r percent of a whole amount, W.

Warm-Up 3 You have two jars of marbles. The first contains 40 marbles, of which 10 are red, and the second contains 60 marbles, of which 30 are red.

a. What percent of the marbles in the first jar are red? $r = \dfrac{P}{W} =$

What percent of the marbles in the second jar are red? $r = \dfrac{P}{W} =$

You pour both jars of marbles into a larger jar and mix them together.

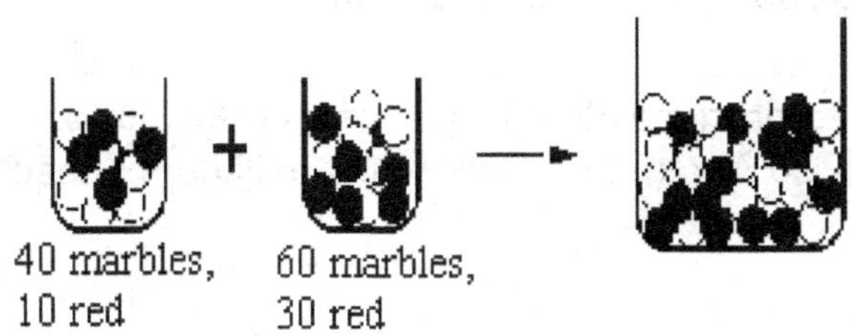

b. How many marbles total are in the larger jar?

How many red marbles are in the larger jar?

What percent of marbles in the larger jar are red? $r = \dfrac{P}{W} =$

c. Can we add the percents for the first two jars to get the percent red marbles in the mixture?

Caution! In Warm-Up 3, you should find that adding the percent of red marbles in the first two jars does **not** give the percent of red marbles in the mixture. That is,

$$255\% + 50\% \neq 40\%$$ ■

In general, **we cannot add percents** unless they are percents of the same whole amount.

RQ5. What formula do we need for mixture problems?
RQ6. When can we add percents?

Warm-Up 4 In a local city council election, your favored candidate, Justine Honest, ran in a small district with two precincts. Ms. Honest won 30% of the 500 votes cast in Precinct 1 and 70% of the 300 votes cast in Precinct 2. Did Candidate Honest win a majority (more than 50%) of the votes in her district?

Step 1 Fill in the first two rows of the table, using information from the problem and the formula $P = rW$.

	Total votes (W)	Percent for Honest (r)	Votes for Honest (P)
Precinct 1			
Precinct 2			
Entire district			

Step 2 Add down to complete the first and third columns of the table.

Step 3 Fill in the last entry in the table to answer the question in the problem. Use the formula $P = rW$ again.

So far we have considered mixture problems involving discrete objects, such as marbles or votes. The same methods apply to mixtures of liquids.

Example 2 A chemist wants to produce 45 milliliters of a 40% solution of carbolic acid by mixing a 20% solution with a 50% solution. How many milliliters of each should he use?

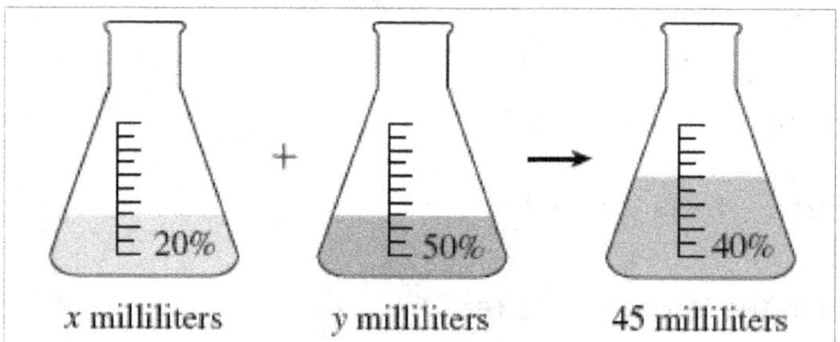

Solution We let x represent the number of milliliters of the 20% solution he needs and y the number of milliliters of the 50% solution. We use a table to organize the information. The first two columns contain the variables and information given in the problem: the number of milliliters of each solution and its strength as a percent.

	Number of Milliliters (W)	Percent Acid (r)	Amount of Acid (P)
20% Solution	x	0.20	
50% Solution	y	0.50	
Mixture	45	0.40	

We fill in the last column of the table by using the formula $P = rW$. The entries in this last column give the amount of the important ingredient (in this case, milliliters of acid) in each component solution and in the mixture.

	Number of Milliliters (W)	Percent Acid (r)	Amount of Acid (P)
20% Solution	x	0.20	$\mathbf{0.20x}$
50% Solution	y	0.50	$\mathbf{0.50x}$
Mixture	45	0.40	$\mathbf{0.40(45)}$

Now we can write two equations about the mixture problem. The first equation is about the total number of milliliters mixed together. The chemist must mix x milliliters of one solution with y milliliters of the other solution and end up with 45 milliliters of the mixture, so

Total amount of mixture: $x + y = 45$

The second equation uses the fact that the acid in the mixture can only come from the acid in each of the two original solutions. We used the last column of the table to calculate how much acid was in each component, and we add these quantities to get the amount of acid in the mixture.

Amount of acid: $0.20x + 0.50y = 0.40(45)$

These two equation make up a system:

$$\begin{aligned} x + y &= 45 \\ 0.20x + 0.50y &= 18 \end{aligned}$$

To simplify the system, we first multiply the second equation by 100 to clear the decimals.

$$\begin{aligned} x + y &= 45 \\ 20x + 50y &= 1800 \end{aligned}$$

We solve the system by elimination. We multiply the first equation by −20, and add the equations together.

$$\begin{aligned} -20x - 20y &= -900 \\ 20x + 50y &= 1800 \\ \hline 30y &= 900 \end{aligned}$$

Solving for y, we find $y = 30$. Substitute $y = 30$ into the first equation to find

$$x + 30 = 45$$

or $x = 15$. The chemist needs 15 milliliters of the 20% solution and 30 milliliters of the 50% solution for the mixture. ■

Look Closer: Notice that, in Example 2, once we have completed the table, it is easy to write a system of equations; we simply add down the first and third columns of the table. (Remember that we cannot add down the middle column, because percents don't add!) •

RQ7. What were the two equation in Example 2 about?
RQ8. Which method did we use to solve the system in Example 2?

Mixture problems are easy to solve if you complete a table first. The rows of the table represent the two components and the final mixture, and the columns are used to calculate the amount of the important ingredient in each. Here is a sample table that you can customize for the specifics of a particular problem.

	Total Amount (W)	Percent of Important Ingredient (r)	Amount of Important Ingredient (P)
First Component			
Second Component			
Mixture			

Motion

We can also use systems of equations to solve problems involving motion at a constant speed. We'll need the formula $\boldsymbol{D = RT}$.

Example 3 Geologists can calculate the distance from their seismograph to the epicenter of an earthquake by timing the arrival of the P and S waves. They know that P waves travel at about 5.4 miles per second and S waves travel at 3.0 miles per second. If the P waves arrived 3 minutes before the S waves, how far away is the epicenter of the quake?

Solution Let x represent the distance from the seismograph to the epicenter. The time it took for the waves to arrive is also unknown, so we'll let y be the travel time for the P waves, in seconds. The travel time for the S waves is then $y + 180$ seconds.

We organize all this information in a table.

	Rate	Time	Distance
P waves	5.4	y	x
S waves	3.0	$y + 180$	x

We can now write two equations about the problem, one for the P waves and one for the S waves, using the formula $RT = D.$

$$\begin{aligned} 5.4\,y &= x \\ 3.0\,(y + 180) &= x \end{aligned}$$

We solve the system by substitution. Substitute $\mathbf{5.4}\boldsymbol{y}$ for x in the second equation, and then solve for y.

$$\begin{aligned} 3.0\,(y + 180) &= \mathbf{5.4}\boldsymbol{y} && \textbf{Apply the distributive law.} \\ 3y + 540 &= 5.4y && \textbf{Subtract } 3y \textbf{ from both sides.} \\ 540 &= 2.4y && \textbf{Divide both sides by 2.4.} \\ 225 &= y \end{aligned}$$

Thus, it took the P waves 225 seconds to arrive at the seismograph. To solve for the distance x, substitute $y = \mathbf{225}$ into the first equation to find

$$x = 5.4\,(\mathbf{225}) = 1215$$

The epicenter is located 1215 miles from the seismograph. ■

RQ9. What formula do we need for motion problems?
RQ10. What did x and y represent in Example 3?

Warm-up Answers: **1.** \$154 **2b.** \$5000 **c.** $0.07x$, $0.12y$ **d.** $0.07x + 0.12y$
3a. 25%, 50% **b.** 100, 40, 40% **c.** No **4.** No, 45%

Skills Warm-Up

For part (b) of each problem, write an algebraic expression in two variables.

1. **a.** How much interest will you earn in 1 year if you invest \$2400 in a T-bill that pays $7\frac{1}{2}$% interest and \$800 in a savings account that pays 4.8% interest?
b. How much will you earn if you invest x dollars in the T-bill and y dollars in the savings account?

2. **a.** How far will you travel if you jog for 40 minutes at 9 miles per hour, and then jog for 30 minutes at 5 miles per hour?
b. How far will you travel if you jog for 40 minutes at x miles per hour, and the jog for 30 minutes at y miles per hour?

3. **a.** How long will it take you to drive 180 miles on the highway at an average speed of 60 miles per hour, and then 30 miles on a gravel road at an average speed of 40 miles per hour?
b. How long will it take you if you drive on the highway at x miles per hour and on the gravel road at y miles per hour?

4. **a.** How much nitrogen is in a mixture of 10 pounds of fertilizer that is 6% nitrogen and 4 pounds of fertilizer that is 60% nitrogen?
b. How much nitrogen is in a mixture of x pounds of the first fertilizer and y pounds of the second fertilizer?

Answers: **1a.** \$218.40 **b.** $0.075x + 0.048y$ **2a.** 8.5 miles **b.** $\frac{2}{3}x + \frac{1}{2}y$
3a. 3.75 hours **b.** $\dfrac{180}{x} + \dfrac{30}{y}$ **4a.** 3 lbs. **b.** $0.06x + 0.60y$

Homework 4.3

Applications

Use systems to solve Problems 1-3 about interest.

1. Goodlife Insurance Company has $150,000 in fees from its clients to invest. They deposited part of the money into bonds that pay 6.5% annual interest and the rest into a mutual fund that pays 11.8% annual interest.
   a. Assign variables to the amount of money Goodlife deposited into each account, and write an equation about the sum of the deposits.
   b. Write expressions for the interest earned on each account after 1 year.
   c. Goodlife earned $12,930 interest in 1 year. Write an equation about that.
   d. Solve your system of equations to find out how much Goodlife invested in each account.

2. Mario borrowed $30,000 from two banks to open a print shop. The first loan charges 12% annual interest, and the second charges 15% interest. Mario's annual interest payment on both loans together is $3750. How much did he borrow at each rate?
   a. Assign variables to the unknown quantities and make a table like the one in Example 1.
   b. Write two equations for the problem, one about the principals and one about the interests.
   c. Solve your system and answer the question in the problem.

3. Stefan borrowed twice as much on his 7% car loan as on his 4% student loan. The annual interest on the car loan is $500 more than the interest on the student loan. How much did Stefan borrow on each loan?

Solve Problems 4-5 about mixtures. (No variables are needed!)

4. The chemistry department has 80 students, of whom 35% are women. The physics department has 60 students, of whom 15% are women.
   a. How many chemistry students are women? How many physics students are women?
   b. How many students are there in both departments? How many of them are women?
   c. What percent of the students in chemistry and physics are women?

	Number of Students (W)	Percent Women (r)	Number of Women (P)
Chemistry			
Physics			
Total			

5. Pipette, a French chemistry student, has 30 milliliters of a 50% solution of acid. She wants to reduce the strength by adding 12 milliliters of a 15% solution of the same acid.
   a. How much acid is in the 30 milliliters of 50% solution? How much acid is in the 12 milliliters of 15% solution?
   b. How much acid is in the mixture? How many milliliters of the mixture are there?
   c. What percent of the mixture is acid?
   d. Fill in the table with your answers to parts (a)-(c).

	Number of Milliliters (W)	Percent Acid (r)	Amount of Acid (P)
50% Solution			
15% Solution			
Mixture			

Use a system of equations to solve Problems 6-7 about mixtures.

6. A pet store owner wants to mix a 12% saltwater solution and a 30% saltwater solution to obtain 45 liters of a 24% solution. How many liters of each ingredient does he need?
   a. Choose variables and make a table for the problem.
   b. Use your table to write two equations about the mixture.
   c. Solve your system and answer the question in the problem.

7. A newspaper poll of 400 people stated that 58% were in favor of a recycling program. It also said that 50% of the men and 70% of the women polled favored the program. How many women were polled?

Solve Problems 8-9 about motion.

8. Delbert and Francine leave Cedar Rapids at the same time and drive in opposite directions for 6 hours.
   a. Choose variables for Delbert and Francine's speeds and fill in the table.

	Rate	Time	Distance
Delbert			
Francine			

   b. Make a sketch showing Cedar Rapids, Delbert, and Francine. Label your sketch with the distance that each traveled.
   c. Francine drove 5 miles per hour slower than Delbert. After 6 hours, they are 570 miles apart. Write two equations about the problem.
   d. Solve your system to find Delbert's and Francine's speeds.

9. Bonnie left Dallas and drove north at 40 miles per hour. Three hours later Clyde headed north form Dallas on the same road at 70 miles per hour until he caught up with Bonnie.

a. Make a sketch showing Dallas, Bonnie, and Clyde.

b. Complete the table.

	Rate	Time	Distance
Bonnie		t	d
Clyde			

c. Use your table to write two equations about the problem.

d. Solve your system. How long did Bonnie drive before Clyde caught up? How far had she driven?

■ Use a system of equations to solve Problems 10-11.

10. A yacht leaves San Diego and heads south, traveling at 25 miles per hour. Six hours later a Coast Guard cutter leaves San Diego traveling at 40 miles per hour and pursues the yacht. How long will it take the cutter to catch the yacht? How far will they have traveled?

11. Byron and Ada conduct sight-seeing tours of New England by bicycle. Byron leads the tour group at an average speed of 10 miles per hour, while Ada goes ahead at a speed of 12 miles per hour to prepare lunch. Ada arrives at the lunch stop 40 minutes ($\frac{2}{3}$ of an hour) before the tour group. How far did the tour bicycle that morning?

■ Solve by writing a single equation in two variables.

12. In Julio's history class, the final grade is computed by adding 60% of the test average to 40% of the term paper grade. Julio's test average is 72.

a. Write an equation for Julio's final grade, g, in terms of his term paper grade, t.

b. What will Julio's final grade be if he earns a grade of 65 on the term paper? What if he earns a grade of 80?

c. Use the graph at right to estimate what grade Julio must make on the term paper in order to earn afinal grade of 80 in the class.

d. Use your equation from part (a) to verify your answer algebraically.

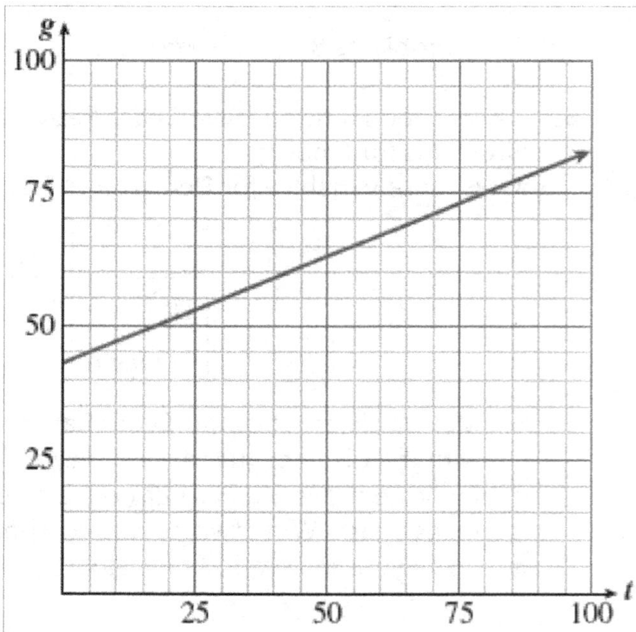

4.4 Point-Slope Form

Distance Between Points

It is easy to compute the distance between two points that lie on the same horizontal or vertical line. We subtract the smaller coordinate from the larger one.

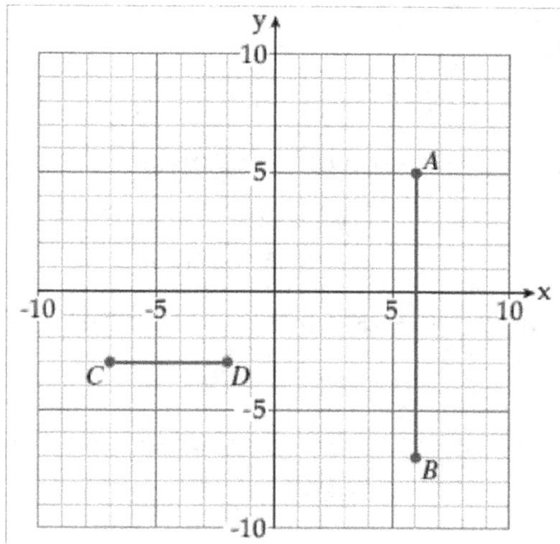

■ **Example 1** The distance between the points $A(6,5)$ and $B(6,-7)$ in the figure is

$$\Delta y = 5 - (-7) = 12$$

and the distance between points $C(-7,-3)$ and $D(-2,-3)$ is

$$\Delta x = -2 - (-7) = 5$$ ■

Note that the distance between two points is always a positive number.

Look Closer: When we compute slope, the direction in which we move on the line makes a difference. We will call this the **directed distance**, and it can be either positive or negative. If we move from C to D, we have moved in the positive x-direction, so the directed distance is positive:

$$\Delta x = -2 - (-7) = 5$$

If we move from D to C, we have moved in the negative x-direction, so the directed distance is negative:

$$\Delta x = -7 - (-2) = -5$$ ●

To find the **directed distance** between two points on a number line, we subtract the initial coordinate from the final coordinate.

RQ1. How do we find the distance between two points that lie on the same vertical line?
RQ2. A directed distance can be either ________ or ___________.

Look Ahead: So far we have computed the slope of a line by finding Δy and Δx on the graph. Now we use directed distance develop a formula for slope. ●

■ **Example 2** The figure on the next page shows a line segment joining P and R. Compute the slope of the line segment in two ways:

a. Find Δy and Δx using the graph.

b. Find Δy and Δx using coordinates.

Solutions **a.** As we move from P to Q, we move up 2 squares on the graph, so $\Delta y = 2$. As we move from Q to R, we move 8 squares to the right, so $\Delta x = 8$. Thus, the slope of the line is

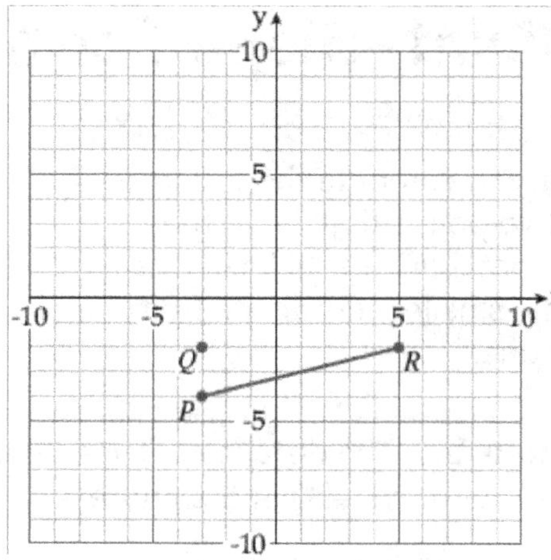

$$m = \frac{\Delta y}{\Delta x} = \frac{2}{8} = \frac{1}{4}$$

b. First we write down the coordinates of P and Q:

$$P(-3, \mathbf{-4}) \quad \text{and} \quad Q(-3, \mathbf{-2})$$

and compute the directed distance from P to Q:

$$\Delta y = -2 - (-4) = 2$$
final – initial

Then we write down the coordinates of Q and R:

$$Q(\mathbf{-3}, -2) \quad \text{and} \quad R(\mathbf{5}, -2)$$

and compute the directed distance from Q to R:

$$\Delta x = 5 - (-3) = 8$$
final – initial

We get the same value for the slope as in part (a), $m = \dfrac{\Delta y}{\Delta x} = \dfrac{2}{8} = \dfrac{1}{4}$.

Subscript Notation

When computing the slope of a line joining two points, we must be careful to subtract their x-coordinates and their y-coordinates in the same order, final minus initial. To distinguish between the coordinates of different points, we use a new notation called **subscripts**.

We denote the coordinates of a particular point by (x_1, y_1) and the coordinates of a second point by (x_2, y_2).

Example 3 Compute the slope between points D and F in the figure.

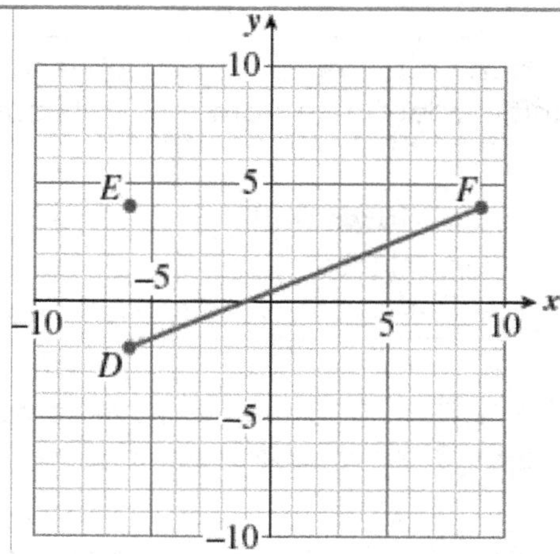

Solution We'll call D the first point and F the second point. Then their coordinates are

$$D: \; x_1 = -6 \quad \text{and} \quad y_1 = -2$$
$$F: \; x_2 = 9 \quad \text{and} \quad y_2 = 4$$

Now we can describe the formula for slope in an organized way. We compute Δy and Δx as directed distances. First, we observe that the coordinates of point E are $(-6, 4)$, or (x_1, y_2). From the figure, we can see that

$$\Delta y = y_2 - y_1 = 4 - (-2) = 6$$

and

$$\Delta x = x_2 - x_1 = 9 - (-6) = 15$$

Thus, the slope of the line segment joining D and F is

$$m = \frac{y_2 - y_1}{x_2 - x_1} = \frac{6}{15} = \frac{2}{5}$$

■

Caution! The subscript 1 on x_1, for instance, has nothing to do with the value of the coordinate; it merely identifies this coordinate as the x-coordinate of the first point. ■

RQ3. What do the subscripts in (x_1, y_1) mean?
RQ4. Write a formula for computing Δy.

A New Formula for Slope

The method described in Example 3 gives us a new formula for computing slope.

Two-Point Formula for Slope

The slope of the line joining points $P_1(x_1, y_1)$ and $P_2(x_2, y_2)$ is

$$\boldsymbol{m = \frac{y_2 - y_1}{x_2 - x_1}} \quad \text{if} \quad x_2 \neq x_1$$

We don't need a graph in order to use this formula, just the coordinates of two points on the line.

■ **Example 4** Compute the slope of the line joining the points $(-6, 2)$ and $(3, -1)$.

Solution It doesn't matter which point is P_1 and which is P_2, so we choose P_1 to be $(-6, 2)$. Then $(x_1, y_1) = (-6, 2)$ and $(x_2, y_2) = (3, -1)$. Thus,

$$\begin{aligned} m &= \frac{y_2 - y_1}{x_2 - x_1} \\ &= \frac{-1 - 2}{3 - (-6)} = \frac{-3}{9} = \frac{-1}{3} \end{aligned}$$

■

Caution! In Example 6, we can reverse the order of **both** subtractions to find

$$\begin{aligned} m &= \frac{y_1 - y_2}{x_1 - x_2} \\ &= \frac{2 - (-1)}{-6 - 3} = \frac{3}{-9} = \frac{-1}{3} \end{aligned}$$

the same answer as before. The order of the points does not matter, but we must be consistent and use the same order when computing Δy and Δx. ■

RQ5. What is wrong with this formula for slope: $m = \frac{y_2 - y_1}{x_1 - x_2}$

Finding an Equation for a Line

The slope of a line is the same, no matter which points we use to compute it. Therefore, if we know the slope m of a line and any one point (x_1, y_1) on the line, then all other points on the line must satisfy:

$$\frac{y - y_1}{x - x_1} = m$$

This is, in fact, an equation for the line.

Example 5a. Graph the line that passes through the point $(1, 3)$ and has slope -2.
b. Find an equation for the line that passes through the point $(1, 3)$ and has slope -2.

Solutions a. We plot the point $(1, 3)$, then use the slope, $m = \frac{\Delta y}{\Delta x} = \frac{-2}{1}$, to find another point on the line. From the point $(1, 3)$, we move 2 units down and 1 unit to the right, arriving at $(2, 1)$. We draw the line through these two points.
b. We use the formula

$$\frac{\Delta y}{\Delta x} = \frac{y - y_1}{x - x_1}$$

with $\frac{\Delta y}{\Delta x} = \frac{-2}{1}$ and $(x_1, y_1) = (1, 3)$ to get

$$\frac{-2}{1} = \frac{y - 3}{x - 1}$$

To simplify the equation, we cross-multiply.

$$1(y - 3) = -2(x - 1)$$
$$y - 3 = -2x + 2$$
$$y = -2x + 5$$

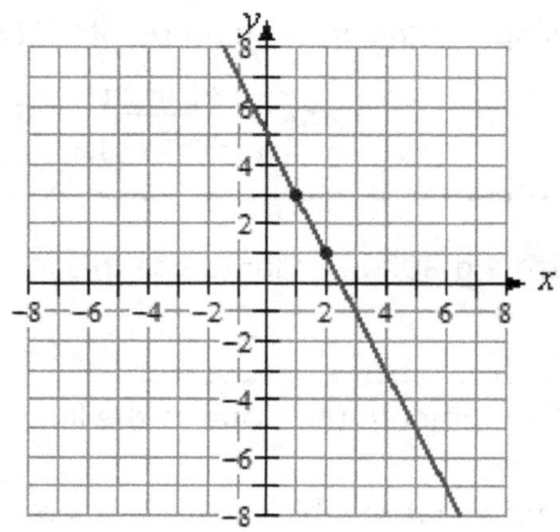

In Example 5 we used the slope formula in a new way: to find the equation of a line. We can simplify the formula by clearing the denominator:

$$(x - x_1)\frac{y - y_1}{x - x_1} = m(x - x_1)$$

to get

$$y - y_1 = m(x - x_1)$$

We call this the **point-slope formula** for linear equations.

Point-Slope Formula

To find an equation for the line of slope m passing through the point (x_1, y_1), use the **point-slope formula**

$$\boldsymbol{\frac{y - y_1}{x - x_1} = m}$$

or

$$\boldsymbol{y - y_1 = m(x - x_1)}$$

Look Closer: What is the difference between the slope formula and the point-slope formula?

the **slope formula** $m = \dfrac{y_2 - y_1}{x_2 - x_1}$

the **point-slope formula** $m = \dfrac{y - y_1}{x - x_1}$

They are really the same formula, but they are used for different purposes:

1. We use the slope formula to calculate the slope of a line when we know two points on the line. That is, we know (x_1, y_1) and (x_2, y_2), and we are looking for m.
2. We use the point-slope formula to find the equation of a line. That is, we know (x_1, y_1) and m, and we are looking for $y = mx + b$.

Caution! If we happen to know the y-intercept of a line, we can write its equation using the slope-intercept formula, $y = mx + b$. But if we don't know the y-intercept and instead know a different point, it is easier to use the point-slope formula.

RQ6. Give two versions of the point-slope formula.
RQ7. What is the point-slope formula used for?

Skills Warm-Up

Use cross-multiplying to solve each proportion for y in terms of x.

1. $\dfrac{y}{x} = \dfrac{-5}{2}$ **2.** $\dfrac{y-3}{4} = \dfrac{x}{2}$ **3.** $\dfrac{x+1}{5} = \dfrac{y-1}{3}$

4. $-2 = \dfrac{y+6}{x}$ **5.** $\dfrac{y+2}{x-5} = \dfrac{3}{4}$ **6.** $\dfrac{-1}{3} = \dfrac{4-y}{1-x}$

Answers: **1.** $y = \dfrac{-5}{2}x$ **2.** $y = 2x + 3$ **3.** $y = \dfrac{3}{5}x + \dfrac{8}{5}$ **4.** $y = -2x - 6$
5. $y = \dfrac{3}{4}x - \dfrac{23}{4}$ **6.** $y = \dfrac{-1}{3}x + \dfrac{13}{3}$

Homework 4.4

Skills Practice

For Problems 1-3, simplify.

1. $\dfrac{10-2}{2-8}$

2. $\dfrac{-5-(-5)}{2-9}$

3. $\dfrac{3}{2}(4-7)+\dfrac{1}{2}$

For Problems 4-7, compute the slope of the line joining the given points.

4. $(5,2)$, $(8,7)$

5. $(3,-2)$, $(0,1)$

6. $(6,-2)$, $(-3,-3)$

7. $(3,\ -5)$, $(8,\ -5)$

For Problems 8-10,

a. Find the x- and y-intercepts of each line.

b. Use the intercepts to compute the slope of the line.

8. $\dfrac{x}{5}+\dfrac{y}{7}=1$

9. $\dfrac{-x}{2.4}+\dfrac{y}{1.6}=1$

10. $3x+\dfrac{2}{7}y=1$

For Problems 11-16,

a. Use the point-slope method to graph the line with the given slope and passing through the given point.

b. Find an equation for the line. Write your equation in slope-intercept form.

c. Find the x-intercept of the line.

11. $m=-2,\ \ (-3,4)$

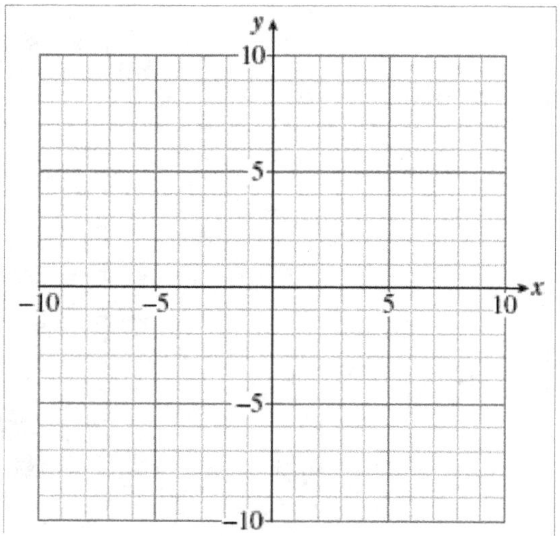

12. $m=\dfrac{1}{2},\ \ (4,-3)$

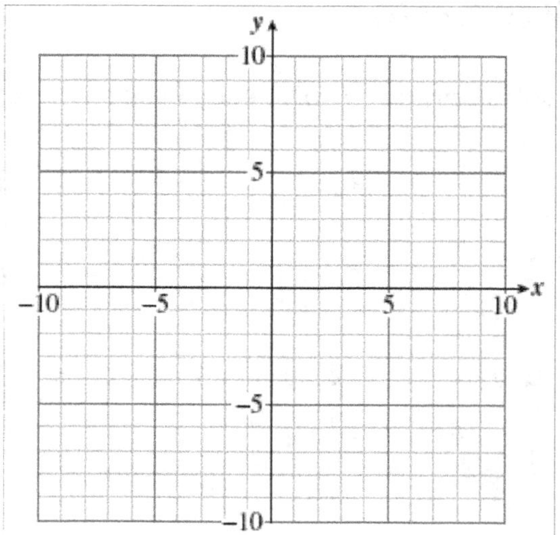

13. $m = \frac{-2}{3}$; $(-6, 2)$

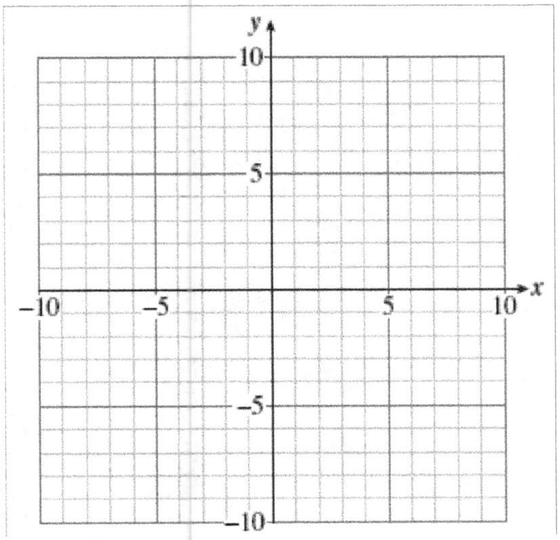

14. $m = 0$, $(-3, 5)$

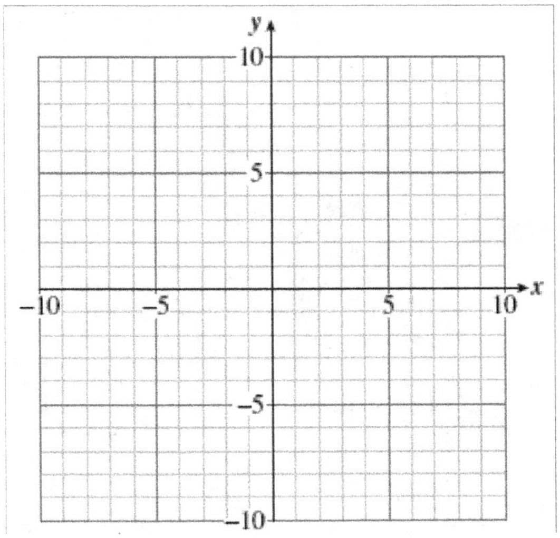

15. $m = \frac{-5}{2}$, $(6, -10)$

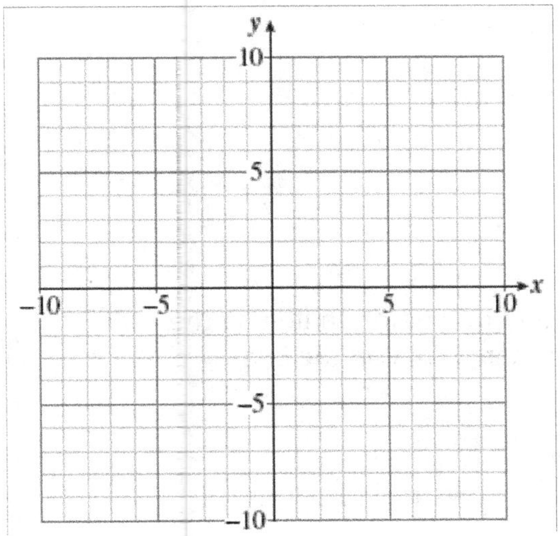

16. $m = 3$, $(1, -1)$

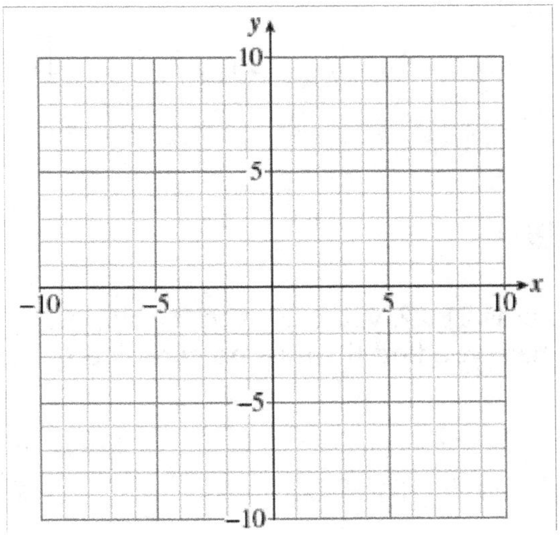

For Problems 17-19, without doing any calculations, give the slope of each line and the coordinates of one point on the line.

17. $y = \frac{3}{5}x - 7$

18. $y - 2 = 3(x + 5)$

19. $\frac{y+8}{x-1} = \frac{-7}{4}$

20. Match each equation with its graph.

a. $y - 1 = 2(x + 3)$ **b.** $y + 1 = 2(x - 3)$

c. $y + 1 = 2(x + 3)$ **d.** $y - 1 = 2(x - 3)$

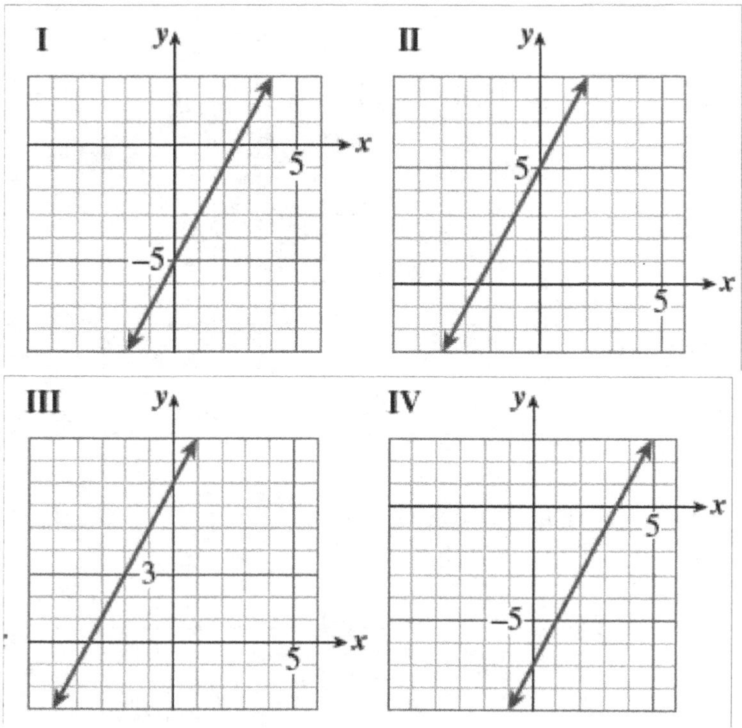

Applications

For Problems 21-22, you are given the slope of a line and one point on it. Use the grids to help you find the missing coordinates of the other points on the line.

21. $m = 2$, $(1, -1)$

a. $(0, ?)$ **b.** $(?, -7)$

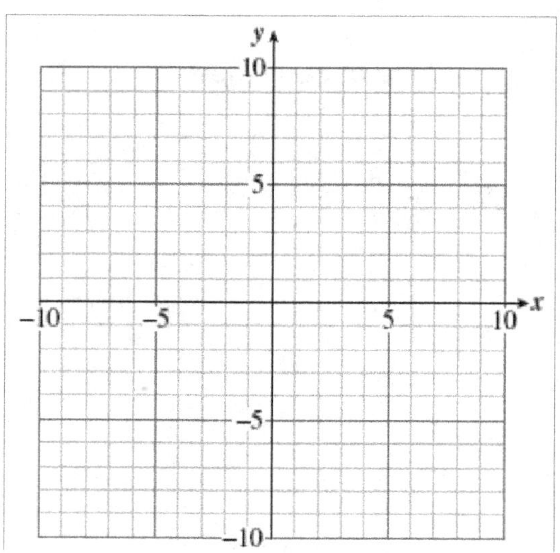

22. $m = \frac{-1}{2}$, $(2, -6)$

a. $(6, ?)$ **b.** $(?, -3)$

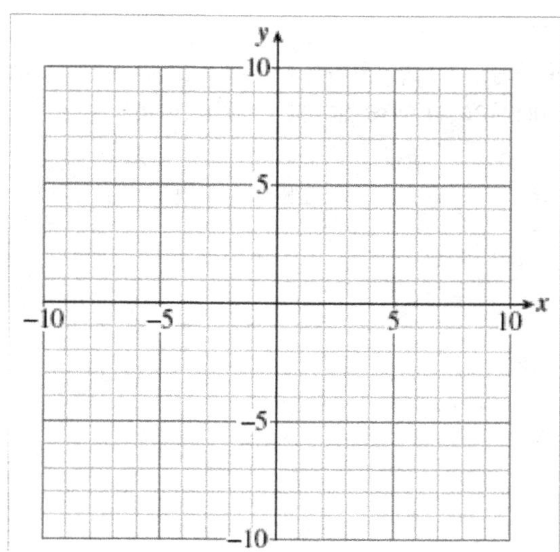

For Problems 23-26, calculate the slope of the line described, including units, and explain its meaning for the problem.

23. Rani and Larry start a college fund for their son, Colby. When Colby is five years old, the fund has $11,000. When he is 10 years old, the fund has grown to $17,000. The graph of the amount, A, in the fund when Colby is n years old is a straight line.

24. It cost Melanie $6.88 to copy and bind her 26-page paper at CopyQuik. It cost Nelson $8.80 to copy and bind his 50-page portfolio. The graph of the cost, C, of producing a document of length p pages CopyQuik is a straight line.

25. Dean is cycling down Mt. Whitney. After five minutes, his elevation is 9200 feet, and after 21 minutes, his elevation is 1200 feet. The graph of Dean's elevation, h, after cycling for t minutes is approximately a straight line.

26. At 2 pm the temperature was 16°, and by midnight it had dropped to $-14°$. The graph of temperature, T, at h hours after noon is approximately a straight line.

27. The figure shows the wholesale cost, C, in dollars, of p pounds of dry-roasted peanuts.

a. What does the slope of this graph tell you about dry-roasted peanuts?

b. Compute the slope, including units.

c. If the Lone Star Barbecue increases its weekly order of peanuts from 100 to 120 pounds, how much will the cost increase?

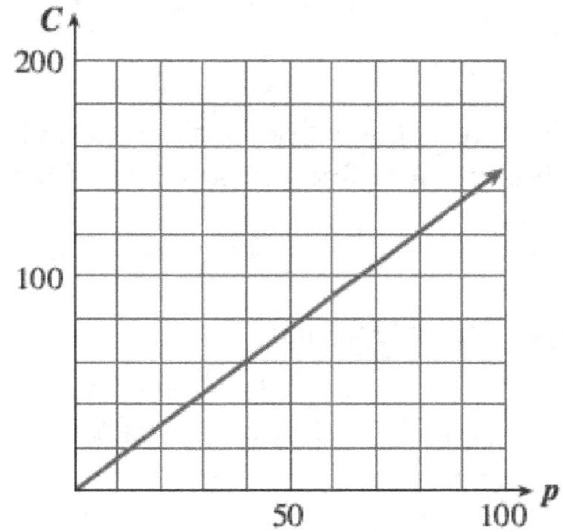

28. Roy is traveling home by train. The figure shows his distance d from home in miles h hours after the train started.

a. Compute the slope of the graph, including units.

b. What does the slope tell you about Roy's journey?

c. Find the vertical intercept of the graph. What is its meaning for the problem?

d. Compute the horizontal intercept of the graph. What is its meaning for the problem?

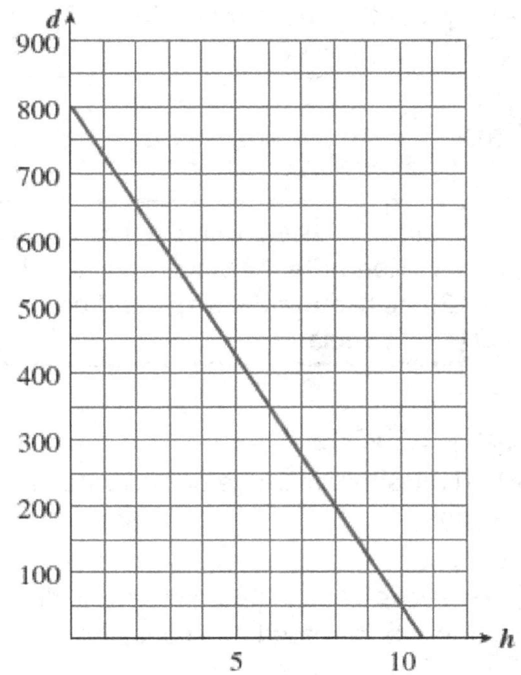

4.5 Using the Point-Slope Formula

The Line Through Two Points

How many lines pass through two given points? There is only one. We can use the point-slope formula to can find its equation.

Example 1 Find an equation for the line that passes through $(2, -1)$ and $(-1, 3)$.

Solution We solve this problem in two steps: First, we find the slope of the line, and then we use the point-slope formula.

Step 1 Let $(x_1, y_1) = (2, -1)$ and $(x_2, y_2) = (-1, 3)$. Using the slope formula, we find that

$$m = \frac{y_2 - y_1}{x_2 - x_1} = \frac{3 - (-1)}{-1 - 2} = \frac{4}{-3} = \frac{-4}{3}$$

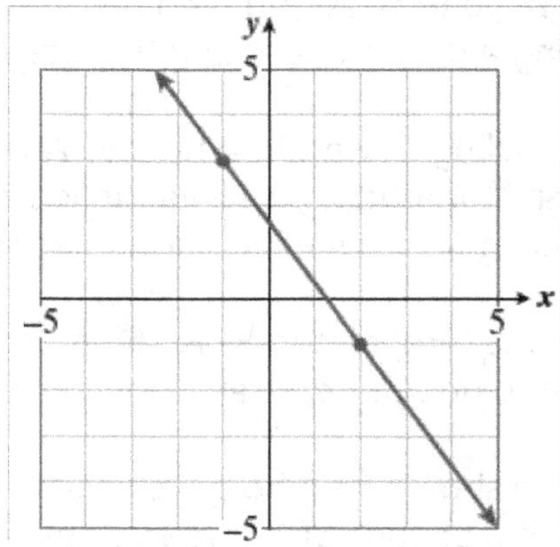

Step 2 We apply the point-slope formula with $m = \frac{-4}{3}$ and $(x_1, y_1) = (2, -1)$. (We can use either point to find the equation of the line.) Then

$$\frac{y - y_1}{x - x_1} = m \qquad \text{becomes} \qquad \frac{y - (-1)}{x - 2} = \frac{-4}{3}$$

We cross-multiply to find

$$\begin{aligned} 3(y + 1) &= -4(x - 2) && \textbf{Apply the distributive law.} \\ 3y + 3 &= -4x + 8 && \textbf{Solve for } y. \\ 3y &= -4x + 5 \\ y &= \frac{-4}{3}x + \frac{5}{3} \end{aligned}$$

The graph of the line is shown in the figure above. ■

RQ1. How many different lines pass through two given points?
RQ2. What formula do we use to find the equation of a line through two points?

To Fit a Line through Two Points:
1. Compute the slope between the two points.
2. Substitute the slope and either point into the point-slope formula.

Applications

Variables that increase or decrease at a constant rate can be described by linear equations.

Example 2 In 1993, Americans drank 188.6 million cases of wine. Wine consumption increased at a constant rate over the next decade, and in 2003 we drank 258.3 million cases of wine. (Source: LA Times, Adams Beverage Group)

a. Find a formula for wine consumption, W, in millions of cases, t years after 1990.
b. State the slope as a rate of change. What does the slope tell us about this problem?

Solutions a. We have two data points of the form (t, W), namely $(t_1, W_1) = (3,\ 188.6)$ and $(t_2, W_2) = (13,\ 258.3)$. We use the point slope formula to fit a line through these two points. First we compute the slope.

$$\frac{\Delta W}{\Delta t} = \frac{W_2 - W_1}{t_2 - t_1} = \frac{258.3 - 188.6}{13 - 3} = 6.97$$

Next, we substitute the slope $m = 6.97$ and either of the two data points into the point-slope formula.

$$\begin{aligned} W &= W_1 + m(t - t_1) \\ W &= 188.6 + 6.97(t - 3) \\ W &= 167.69 + 6.97t \end{aligned}$$

Thus, $W = 167.69 + 6.97t$.
b. The slope gives us the rate of change of W with respect to t. The units of the variables can help us interpret the slope in context.

$$\frac{\Delta W}{\Delta t} = \frac{258.3 - 188.6 \text{ millions of cases}}{13 - 3 \text{ years}} = 6.97 \text{ millions of cases/year}$$

Over the ten years between 1993 and 2003, wine consumption in the US increased at a rate of 6.97 million cases per year.

RQ3. What are the two steps to fit a line through two points?

Parallel and Perpendicular Lines

Lines that lie in the same plane but never intersect are called **parallel lines**. It is easy to understand that parallel lines have the same slope.

Exercise Calculate the slopes of the two parallel lines in the figure. (We denote the slope of the first line by m_1 and the slope of the second line by m_2.)

$$m_1 = \frac{\Delta y}{\Delta x} =$$

$$m_2 = \frac{\Delta y}{\Delta x} =$$

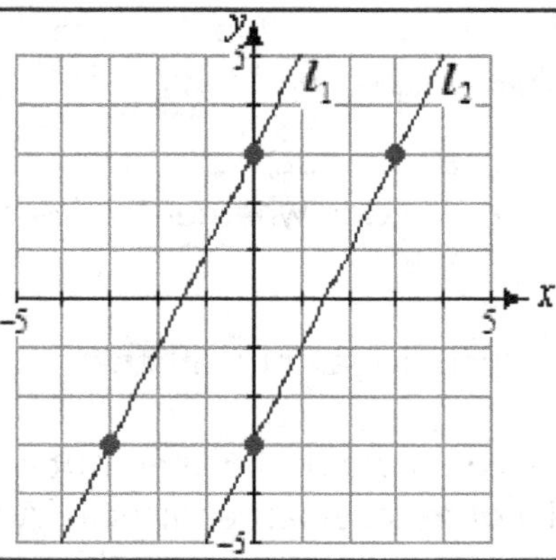

Lines that intersect at right angles are called **perpendicular lines**. It is a little harder to see the relationship between the slopes of perpendicular lines.

Exercise Calculate the slopes of the two perpendicular lines in the figure.

$$m_1 = \frac{\Delta y}{\Delta x} =$$

$$m_2 = \frac{\Delta y}{\Delta x} =$$

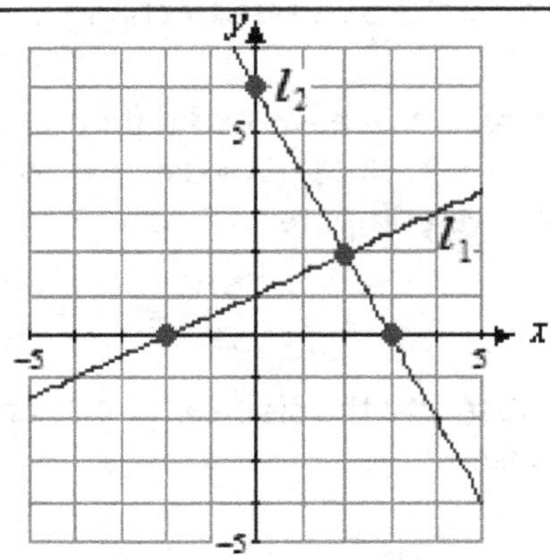

In the Exercise above, note that the product of m_1 and m_2 is -1, that is,

$$m_1 m_2 = \frac{1}{2}(-2) = -1$$

This relationship holds for any pair of perpendicular lines.

Parallel and Perpendicular Lines

1. Two lines are **parallel** if their slopes are equal, that is, if
$$\boldsymbol{m_1 = m_2}$$
or if both lines are vertical.
2. Two lines are **perpendicular** if the product of their slopes is -1, that is, if
$$\boldsymbol{m_1 m_2 = -1}$$
or if one of the lines is horizontal and one is vertical.

RQ4. What do we call two lines that lie in the same plane but never intersect?
RQ5. What are perpendicular lines?

Example 3 Decide whether the lines

$$2x + 3y = 6 \qquad \text{and} \qquad 3x - 2y = 6$$

are parallel, perpendicular, or neither.

Solution We could graph the lines, but we can't be sure from a graph if the lines are exactly parallel or exactly perpendicular. A more accurate way to answer the question is to find the slope of each line. To do this we write each equation in slope-intercept form, that is, we solve for y.

$$\begin{aligned} 2x + 3y &= 6 \\ 3y &= -2x + 6 \\ y &= \frac{-2}{3}x + 2 \end{aligned} \qquad\qquad \begin{aligned} 3x - 2y &= 6 \\ -2y &= -3x + 6 \\ y &= \frac{3}{2}x - 3 \end{aligned}$$

The slope of the first line is $m_1 = \frac{-2}{3}$, and the slope of the second line is $m_2 = \frac{3}{2}$. The slopes are not equal, so the lines are not parallel. However, the product of the slopes is

$$m_1m_2 = \left(\frac{-2}{3}\right)\left(\frac{2}{3}\right) = -1$$

so the lines are perpendicular. ■

Another way to state the condition for perpendicular lines is

$$m_2 = \frac{-1}{m_1}$$

Because of this relationship, we often say that the slope of one perpendicular line is the **negative reciprocal** of the other.

■ **Example 4** Find an equation for the line that passes through the point $(1, 4)$ and is perpendicular to the line $4x - 2y = 6$.

Solution We first find the slope of the desired line, then use the point-slope formula to write its equation. The line we want is perpendicular to the given line, so its slope is the negative reciprocal of $m_1 = 2$, the slope of the given line. Thus

$$m_2 = \frac{-1}{m_1} = \frac{-1}{2}$$

Now use the point-slope formula with $m = \frac{-1}{2}$ and $(x_1, y_1) = (1, 4)$.

$$\begin{aligned} y - y_1 &= m(x - x_1) \\ y - 4 &= \frac{-1}{2}(x - 1) && \textbf{Apply distributive law.} \\ y - 4 &= \frac{-1}{2}x + \frac{1}{2} && \textbf{Add 4 to both sides.} \\ y &= \frac{-1}{2}x + \frac{9}{2} && \frac{1}{2} + 4 = \frac{1}{2} + \frac{8}{2} = \frac{9}{2} \end{aligned}$$

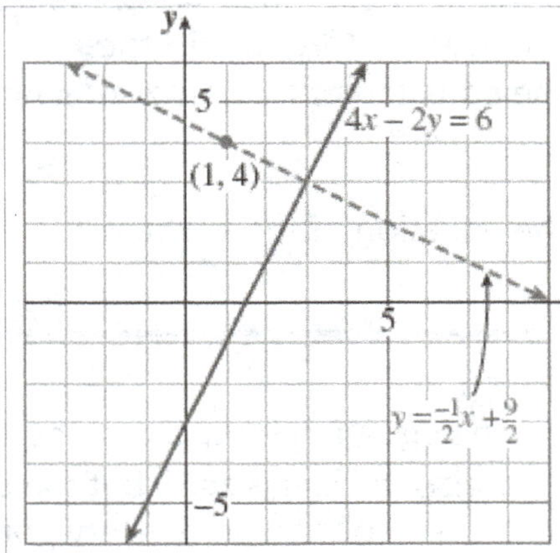

The given line and the perpendicular line are shown in the figure. ■

RQ6. What is true about the slopes of parallel lines?
RQ7. What do we call the slopes of perpendicular lines?

Equations for Horizontal and Vertical Lines

In Lesson 3.4 we learned that the slope of a horizontal line is zero. What does this tell us about the equation of a horizontal line that passes through a particular point?

■ **Example 5** Find the equation of the horizontal line that passes through $(5, 3)$.

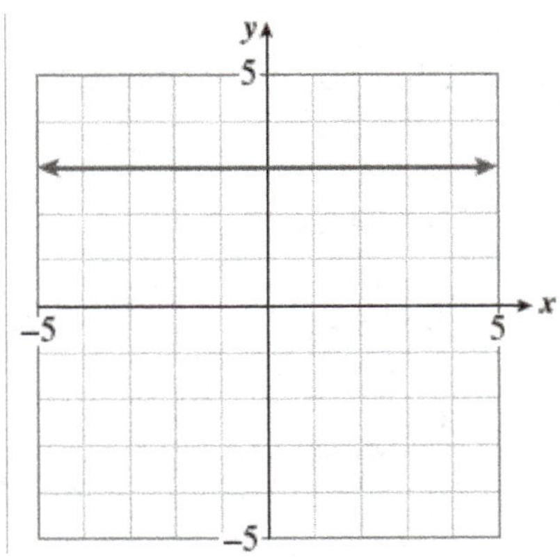

Solution Looking at the graph of the line shown at right, we see that the y-coordinate of every point on the line is 3. In particular, the y-intercept of the line is the point $(0, 3)$, so $b = 3$. As we noted above, the slope of the line is $m = 0$. Substituting these values into the slope-intercept form gives us the equation $y = 0 \cdot x + 3$, or just $y = 3$. The fact that x does not appear in the equation means that $y = 3$ for every point on the line, no matter what the value of x is. ■

Look Closer: What about the equation of a vertical line? The slope of a vertical line is undefined; a vertical line does not have a slope. We cannot use the slope-intercept form to write the equation of a vertical line. However, we can use what we learned about horizontal lines in Example 2.

Look at the graph of the vertical line at right. Every point on the line has x-coordinate -2, no matter what the y-coordinate is. An equation for this line is $x = -2$. The value of x does not depend upon y; it is constant, so y does not appear in the equation. ●

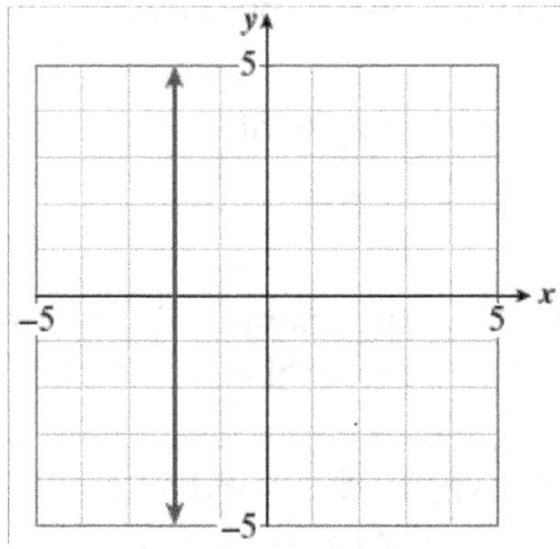

The y-intercept of the horizontal line $y = 3$ is $(0, 3)$; it has no x-intercept. The x-intercept of the vertical line $x = -2$ is $(-2, 0)$; it has no y-intercept.

Horizontal and Vertical Lines

1. The equation of the **horizontal line** passing through $(0, b)$ is

$$y = b$$

2. The equation of the **vertical line** passing through $(a, 0)$ is

$$x = a$$

Caution! The equation for a line is not the same thing as the slope of the line! The slope of every horizontal line is zero, but the equation of a horizontal line has the form $y = b$, where b is the y-coordinate of every point on the line. Similarly, the equation of a vertical line has the form $x = a$, but the slope of a vertical line is undefined ■

RQ8. Give an example of an equation of a vertical line.
RQ9. Give an example of an equation of a horizontal line.
RQ10. Why can't we use the point-slope formula to find the equaiton of a vertical line?

Skills Warm-Up

For Problems 1-8, find the negative reciprocal.

1. $\frac{3}{4}$ **2.** $\frac{-1}{3}$ **3.** -6 **4.** $\frac{11}{8}$

5. $1\frac{2}{3}$ **6.** $-2\frac{1}{2}$ **7.** -3.2 **8.** 0.625

Answers: **1.** $\frac{-4}{3}$ **2.** 3 **3.** $\frac{1}{6}$ **4.** $\frac{-8}{11}$ **5.** $\frac{-3}{5}$ **6.** $\frac{2}{5}$ **7.** 0.3125 **8.** -1.6

Homework 4.5

Skills Practice

In Problems 1-3, decide whether the lines are parallel perpendicular, or neither.

1. $3x - 4y = 2$
$8y - 6x = 6$

2. $5x + 3y = 1$
$3y - 5x = 5$

3. $2x = 4 - 5y$
$2y = 4x - 5$

4. **a.** Use your calculator to graph the equations $y = 3x + 8$ and $y = 3.1x + 6$ together in the standard window. Do you think the lines are parallel?
b. Find the slope of each line in part (a). Are the lines parallel?
c. Find the y-value for each equation when $x = 20$. What do your answers tell you about the two lines?

5. The slopes of several lines are given below. Which of the lines are parallel to the graph of $y = 0.75x + 2$, and which are perpendicular to it?

a. $m = \frac{3}{4}$ **b.** $m = \frac{8}{6}$ **c.** $m = \frac{-20}{15}$ **d.** $m = \frac{-39}{52}$

e. $m = \frac{4}{3}$ **f.** $m = \frac{-16}{12}$ **g.** $m = \frac{36}{48}$ **h.** $m = \frac{9}{12}$

In Problems 6-8,
a. Sketch a rough graph of each equation, and label its intercept.
b. State the slope of each line.

6. $y = -3$ **7.** $2x = 8$ **8.** $x = 0$

In Problems 9-13, give the equation of the line described.

9. Horizontal, passes through $(6, -5)$
10. m is undefined, passes through $(2, 1)$
11. Parallel to $x = 4$, passes through $(-8, 3)$
12. The x-axis
13. Perpendicular to the y-axis, passes through $(4, -12)$

In Problems 14-16, find an equation for the line passing through the given points. Write your answer in slope-intercept form.

14. $(-2, 4), (1, 7)$ **15.** $(3, 5), (-3, -5)$ **16.** $(6, 4), (-2, 5)$

In Problems 17-20, find the equation of the line shown in the graph.

17. 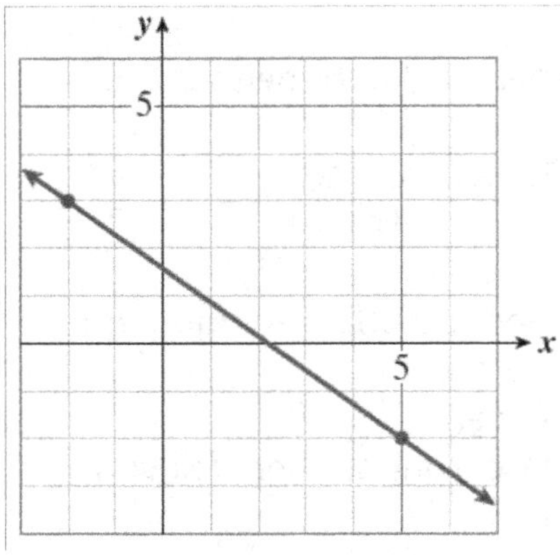

18. 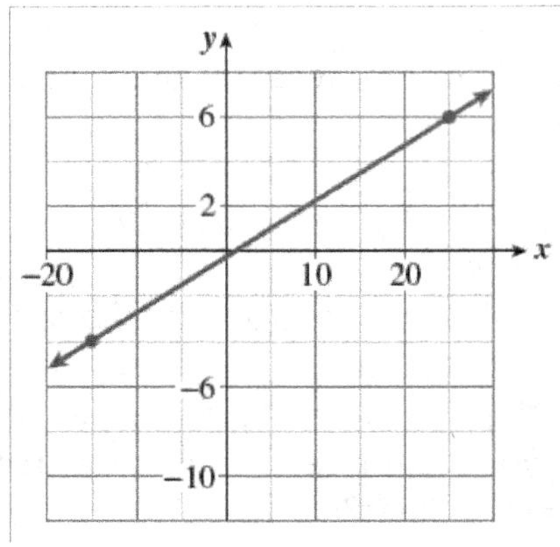

19. 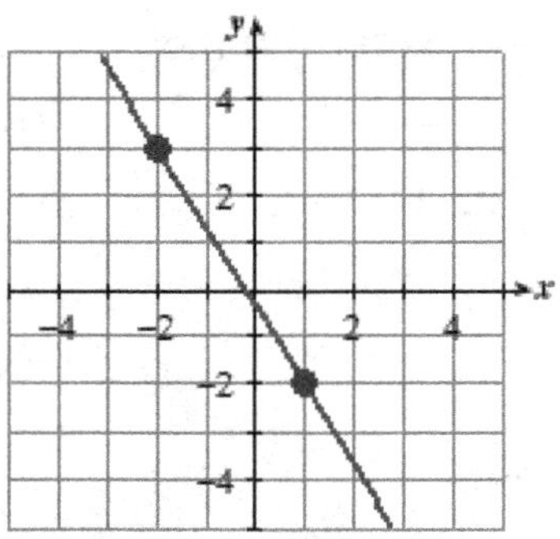

20. 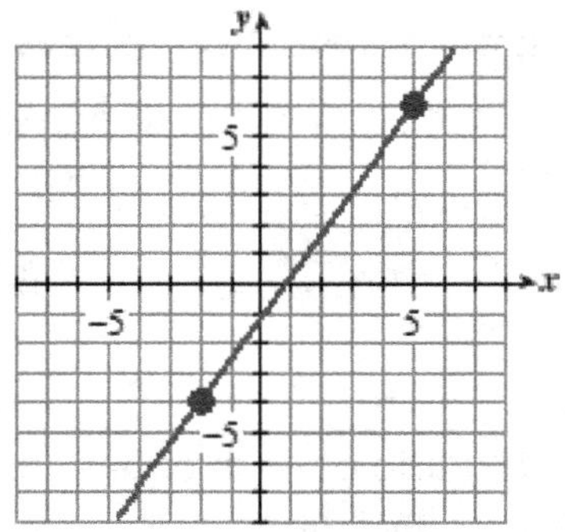

21. **a.** Put the equation $2y - 3x = 5$ into slope-intercept form, and graph the equation.

b. What is the slope of any line that is parallel to $2y - 3x = 5$?

c. On your graph for part (a), sketch by hand a line that is parallel to $2y - 3x = 5$ and passes through the point $(-3, 2)$.

d. Use the point-slope formula to write an equation for the line that is parallel to the graph of $2y - 3x = 5$ and passes through the point $(-3, 2)$.

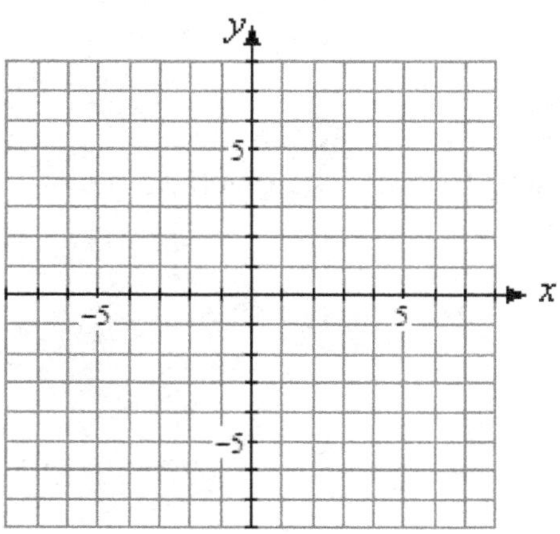

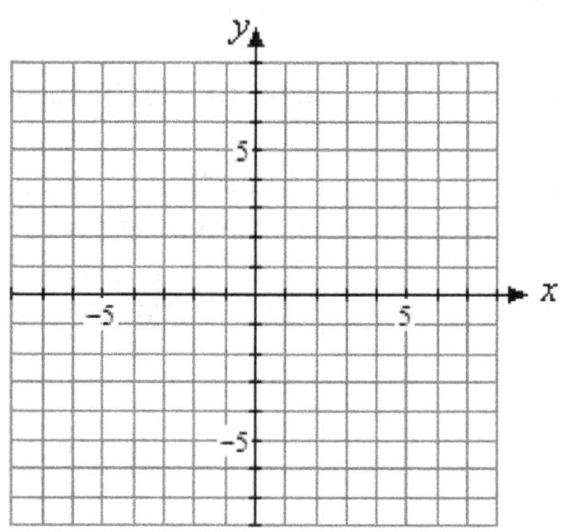

22. **a.** Put the equation $x - 2y = 5$ into slope-intercept form, and graph the equation.

b. What is the slope of any line that is perpendicular to $x - 2y = 5$?

c. On your graph for part (a), sketch by hand a line that is perpendicular to $x - 2y = 5$ and passes through the point $(4, -3)$.

d. Use the point-slope formula to write an equation for the line that is perpendicular to the graph of $x - 2y = 5$ and passes through the point $(4, -3)$.

23. a. What is the slope of the line $y = 2x + 1$?
b. What is the slope of a line parallel to $y = 2x + 1$?
c. A line is parallel to $y = 2x + 1$ and passes through $(-2, 1)$. What is its equation?

24. a. What is the slope of the line $y = -3x - 2$?
b. What is the slope of a line perpendicular to $y = -3x - 2$?
c. A line is perpendicular to $y = -3x - 2$ and passes through $(1, 3)$. What is its equation?

Applications

In Problems 25-28, we'll find a linear model from two data points.

Step 1 Make a table showing the coordinates of two data points for the model.
Step 2 Find a linear equation in slope-intercept form relating the variables.
Step 3 Use your equation to answer the questions.

25. Francine is driving into the mountains and stopping periodically to record the temperature, T, at various altitudes, h. The temperature at an altitude of 3200 feet is 77°, and the temperature at 8000 feet is 65°.
a. What is the slope, including units? What does it tell you about the problem?
b. What will the temperature be at 10,000 feet?
c. What was the temperature at sea level?

26. Envirotech is marketing a new line of microwave clothes dryers. It cost them $45,000 to produce the first 100 dryers. When they had produced 180 dryers, their total cost was up to $61,000. They would like to know the total cost, C, of producing x dryers.
a. What is the slope, including units? What does it tell you about the problem?
b. Envirotech has budgeted $100,000 for microwave clothes dryers this year. How many can they produce?
c. How much did Envirotech invest in development before they made the first dryer?

27. Flying lessons cost $645 for an 8-hour course and $1425 for a 20-hour course. Both prices include a fixed insurance fee. Express the cost, C, of flying lessons in terms of the length, h, of the course in hours.
a. What is the slope, including units? What does it tell you about the problem?
b. How much does a 10-hour course cost?
c. How much is the fixed insurance fee?

28. On an international flight a passenger may check two bags each weighing 70 kilograms, or 154 pounds, and one carryon bag weighing 50 kilograms, or 110 pounds. Express the weight, p, of a bag in pounds in terms of its weight, k, in kilograms.
a. What is the slope, including units? What does it tell you about the problem?
b. What does a 50-pound bag weigh in kilograms?
c. Why is the constant term in your equation equal to zero?

4.6 Linear Regression

Scatterplots

Empirical data points in a linear relation may not lie exactly on a line. There are many factors that can affect experimental data, including measurement error, the influence of environmental conditions, and the presence of related variable quantities.

Example 1 A consumer group wants to test the gas mileage of a new model SUV. They test-drive six vehicles under similar conditions and record the distance each drove on various amounts of gasoline.

Gasoline used (gal)	9.6	11.3	8.8	5.2	10.3	6.7
Miles driven	155.8	183.6	139.6	80.4	167.1	99.7

a. Are the data linear?
b. Draw a line that fits the data.
c. What does the slope of the line tell us about the data?

Solutions a. No, the data are not strictly linear. If we compute the slopes between successive data points, the values are not constant. We can see from an accurate plot of the data, shown at right, that the points lie close to, but not precisely on, a straight line.

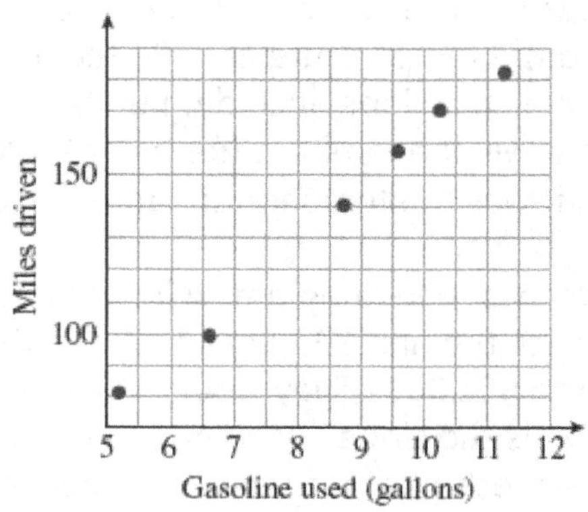

b. We'd like to draw a line that comes as close as possible to all the data points, even though it may not pass precisely through any of them. In particular, we try to adjust the line so that we have the same number of data points above the line and below the line. One possible solution is shown at right.

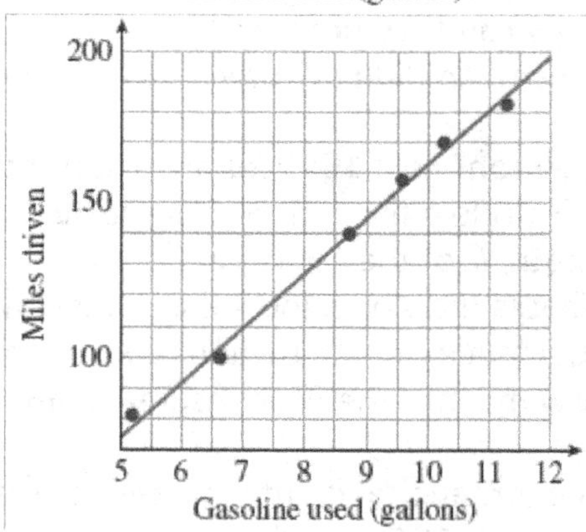

c. To compute the slope of the line of best fit, we first choose two points on the line. Our line appears to pass through one of the data points, $(8.8, 139.6)$. Look for a second point on the line whose coordinates are easy to read, perhaps $(6.5, 100)$. The slope is

$$m = \frac{139.6 - 100}{8.8 - 6.5} = 17.2 \text{ miles per gallon.}$$

According to our data, the SUV gets about 17.2 miles to the gallon.

RQ1. How should you choose a line of best fit?

Look Closer: The graph in Example 1 is called a **scatterplot**. The points on a scatterplot may or may not show some sort of pattern. Consider the three plots shown below. In figure (a) the data points resemble a cloud of gnats; there is no apparent pattern to their locations. In figure (b) the data follow a generally decreasing trend, but certainly do not all lie on the same line. The points in figure c are even more organized; they seem to lie very close to an imaginary line.

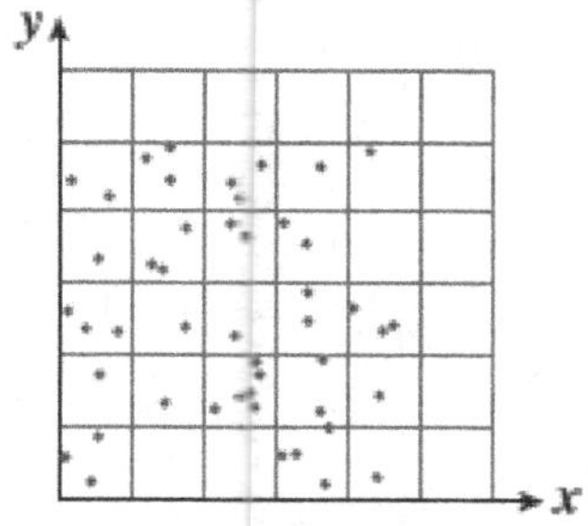

a.

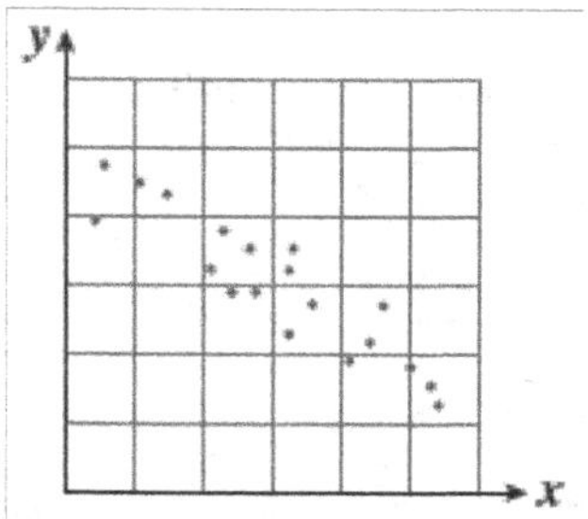

b.

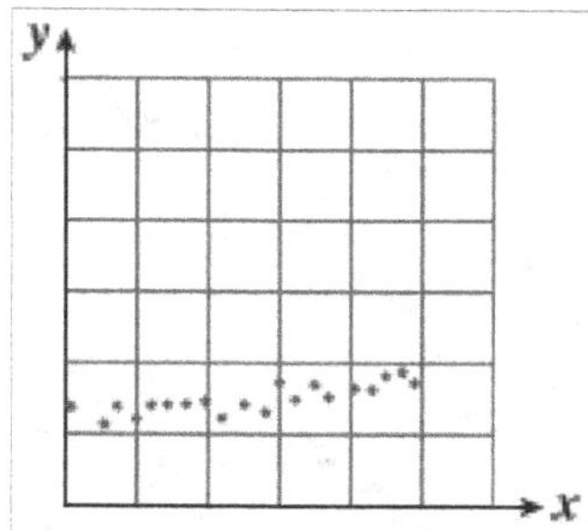

c.

If the data in a scatterplot are roughly linear, we can estimate the location of an imaginary **line of best fit** that passes as close as possible to the data points. We can then use this line to make predictions about the data.

Linear Regression

The process of predicting an output value based on a straight line that fits the data is called **linear regression**, and the line itself is called the **regression line**. Here is an example of finding and using a regression line.

One measure of a person's physical fitness is the body mass index, or BMI. Your body mass index is the ratio of your weight in kilograms to the square of your height in centimeters. Thus, thinner people have lower BMI scores, and fatter people have higher scores. The Center for Disease Control considers a BMI between 18.5 and 24.9 to be healthy. The points on the scatterplot below show the BMI of Miss America from 1921 to 1991. From the data in the scatterplot, can we see a trend in Americans' ideal of female beauty?

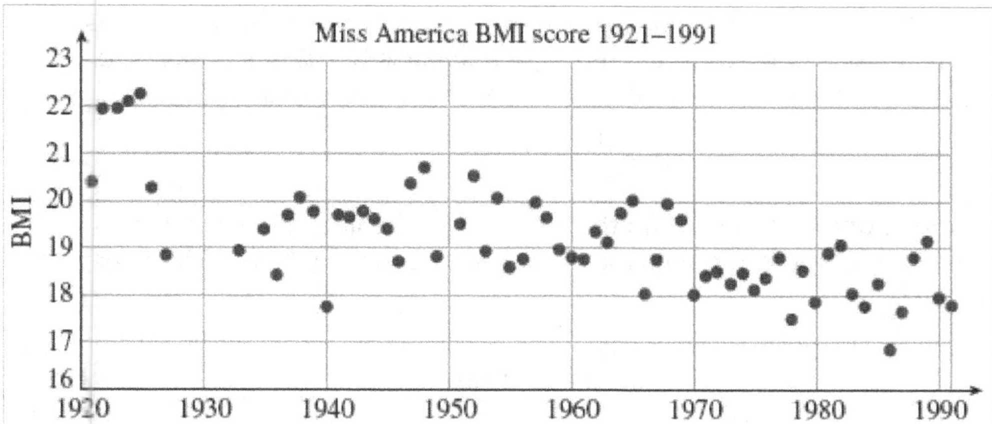

Example 2a. Estimate a line of best fit for the Miss America scatterplot. (Source: http://www.pbs.org)

b. Use your line to estimate the BMI of Miss America 1980.

Solutions a. We draw a line that fits the data points as best we can, as shown below. (Note that we have set $t = 0$ in 1920 on this graph.) We try to end up with roughly equal numbers of data points above and below our line.

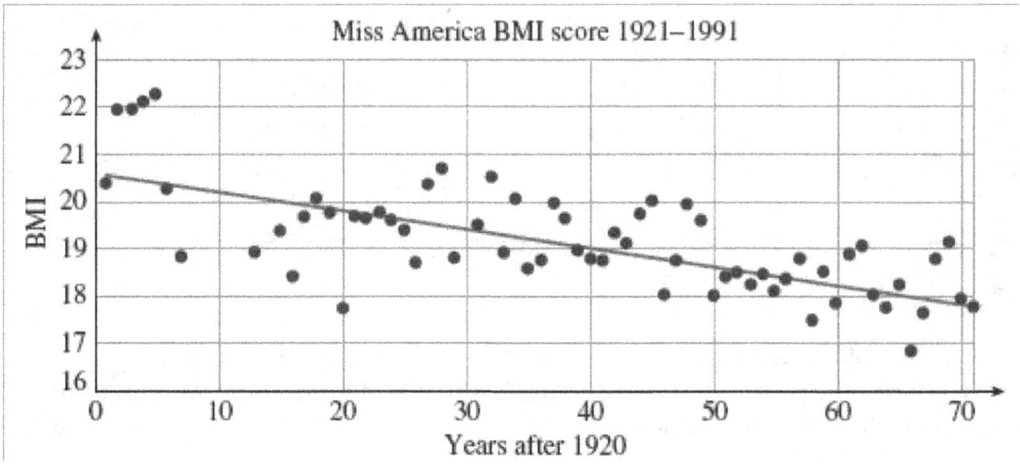

b. We see that when $t = 60$ on this line, the y-value is approximately 18.3. We therefore estimate that Miss America 1980 had a BMI of 18.3. (Her actual BMI was 17.85.) ■

In Example 2, we used the graph of the regression line to estimate on output value. However, the equation of the regression line is usually used (instead of a graph) to predict values.

■ **Example 3a.** Find the equation of the regression line in Example 2.
b. Use the regression equation to predict the BMI of Miss America 1980.

Solutions a. We first calculate the slope by choosing two points on the regression line we drew in Example 2. The points we choose are not necessarily any of the original data points; instead they should be points on the regression line itself. The line appears to pass through the points $(17, 20)$ and $(67, 18)$. The slope of the line is then

$$m = \frac{18 - 20}{67 - 17} \approx -0.04$$

Now we use the point-slope formula to find the equation of the line. We substitute $m = -0.04$ and use either of the two points for (x_1, y_1); we'll choose $(17, 20)$. The equation of the regression line is

$$y = y_1 + m(x - x_1)$$
$$y = 20 - 0.04\,(x - 17) \qquad \textbf{Simplify.}$$
$$y = 20.68 - 0.04t$$

b. We'll use the regression equation to make our prediction. For Miss America 1980, $t = 60$ and

$$y = 20.68 - 0.04(60) = 18.28$$

This value agrees well with the estimate we made in Example 2. ■

RQ2. Should we choose two of the data points to find the equation for a line of best fit? Why or why not?

Linear Interpolation and Extrapolation

Using a regression line to estimate values between known data points is called **interpolation**. Making predictions beyond the range of known data is called **extrapolation.**

Example 4a. Use linear interpolation to estimate the BMI of Miss America 1960.
b. Use linear extrapolation to predict the BMI of Miss America 2001.

Solutions a. For 1960, we substitute $t = 40$ into the regression equation we found in Example 3.

$$y = 20.68 - 0.04(40) = 19.08$$

We estimate that Miss America 1960 had a BMI of 19.08. (Her BMI was actually 18.79.)
b. For 2001, we substitute $t = 81$ into the regression equation.

$$y = 20.68 - 0.04(81) = 17.44$$

Our model predicts that Miss America 2001 had a BMI of 17.44. In fact, her BMI was 20.25. By the late 1990's, public concern over the self-image of young women had led to a reversal of the trend towards ever-thinner role models.

Caution! Example 4b illustrates an important fact about extrapolation: If we try to extrapolate too far, we may get unreasonable results. For example, if we use our model to predict the BMI of Miss America 2520 (when $t = 600$), we get

$$y = 20.68 - 0.04(600) = -3.32$$

Even if the Miss America pageant is still operating in 600 years, the winner cannot have a negative BMI. Our linear model provides a fair approximation for 1920-1990, but if we try to extrapolate too far beyond the known data, the model may no longer apply.

RQ3. Explain the difference between interpolation and extrapolation.
RQ4. In general, should you have more confidence in figures obtained by interpolation or by extrapolation? Why?

Using a Calculator for Linear Regression

Estimating a line of best fit is a subjective process. Rather than base their estimates on such a line, statisticians often use the **least squares regression line**. This regression line minimizes the sum of the squares of all the vertical distances between the data points and the corresponding points on the line, as shown at right. Many calculators are programmed to find the least squares regression line, using an algorithm that depends only on the data, and not on the appearance of the graph.

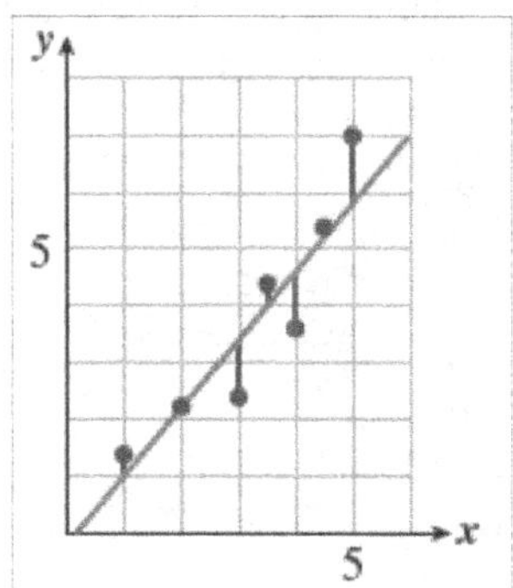

Look Ahead: You can use a graphing calculator to make a scatterplot, find a regression line, and graph the regression line with the data points. On the TI-84 calculator we use the statistics mode, which you can access by pressing STAT. You will see the display shown below. Choose 1 to **Edit** (enter or alter) data.

Now follow the instructions in Example 5 for using your calculator's statistics features.

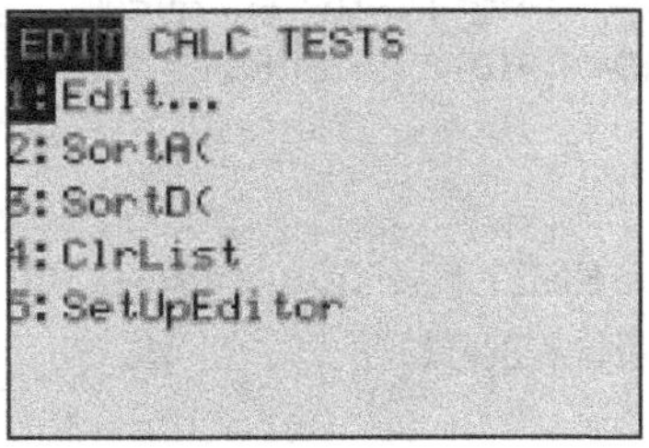

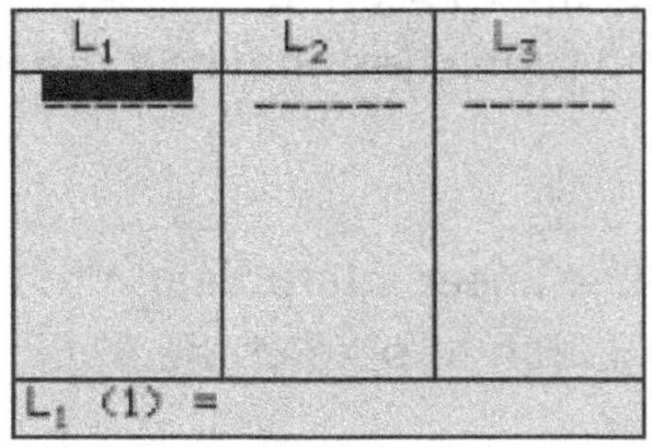

Example 5a. Find the equation of the least squares regression line for the following data.

$$(10,12),\ (11,14),\ (12,14),\ (12,16),\ (14,20)$$

b. Plot the data points and the least squares regression line on the same axes.

Solutions a. Step 1: We must first enter the data. Press [STAT] [ENTER] to select **Edit**. If there are data in column $\mathbf{L_1}$ or $\mathbf{L_2}$, clear them out: Use the [△] key to select $\mathbf{L_1}$, press [CLEAR] [ENTER], then do the same for $\mathbf{L_2}$. Enter the x-coordinates of the data points in the $\mathbf{L_1}$ column and the y-coordinates in the $\mathbf{L_2}$ column, as shown at left below.

Step 2: Now we are ready to find the regression equation for our data. Press [STAT] [▷] [4] to select linear regression, or **LinReg** $(ax+b)$, then press [ENTER]. The calculator will display the equation $y = ax + b$ and values for a and b, as shown below right. You should find that your regression line is approximately $y = 1.95x - 7.86$.

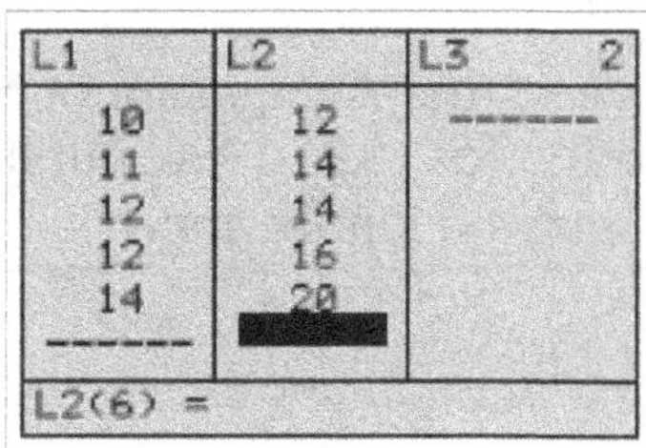

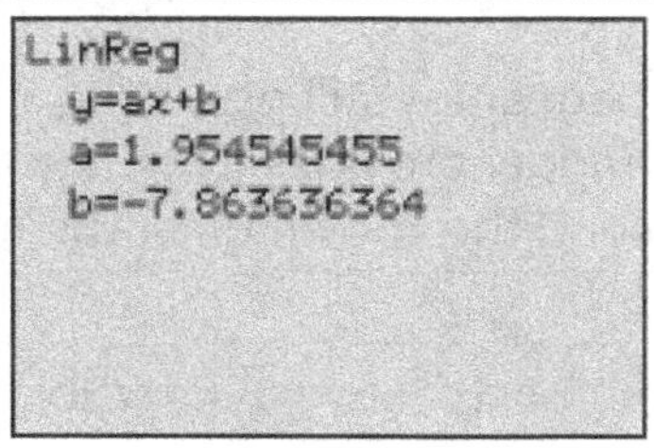

b. First, we first clear out any old definitions in the [Y=] list. Position the cursor after $\mathbf{Y_1}$= and copy in the regression equation as follows: Press [VARS] 5 [▷] [▷] [ENTER]. To draw a scatterplot, press [2nd] [Y=] [1] and set the **Plot1** menu as shown below left. Finally, press [ZOOM] [9] to see the scatterplot of the data and the regression line. The graph is shown below right.

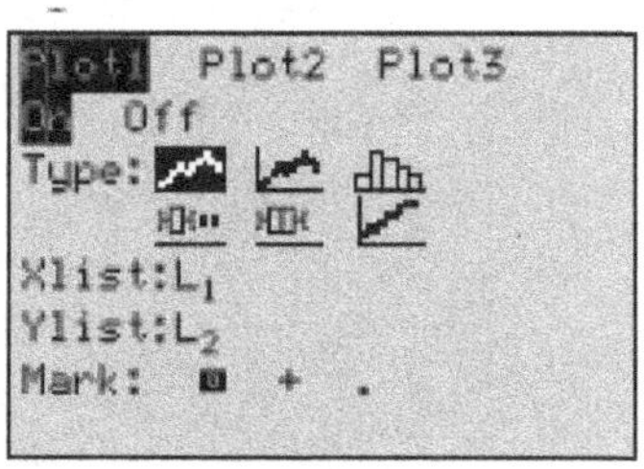

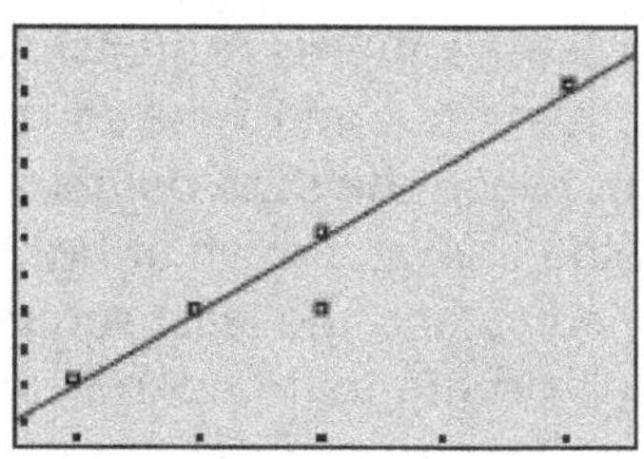

Caution! When you are through with the scatterplot, press [Y=] [△] [ENTER] to turn off the **Stat Plot**. If you neglect to do this, the calculator will continue to show the scatterplot even after you ask it to plot a new equation.

RQ5. What method does your calculator use to find a regression line?

Skills Warm-Up

1. Find the equation of the regression line in Example 1.
2. Explain the meaning of the y-intercept for this equation. Does it make sense in this situation?

Find the equation of the line through the two points.

3. $(18, 8.36)$, $(25, 10.25)$
4. $(2.8, 379.2)$, $(6, 264)$

Answers: **1.** $y = 17.2x - 11.9$ **2.** A car drove −11.9 miles on 0 gallons of gasoline. No. **3.** $y = 0.27x + 3.5$ **4.** $y = 480 - 36x$

Homework 4.6

Skills Practice

For Problems 1-2, the regression line can be improved by adjusting either m or b. Draw a line that fits the data points more closely.

1. 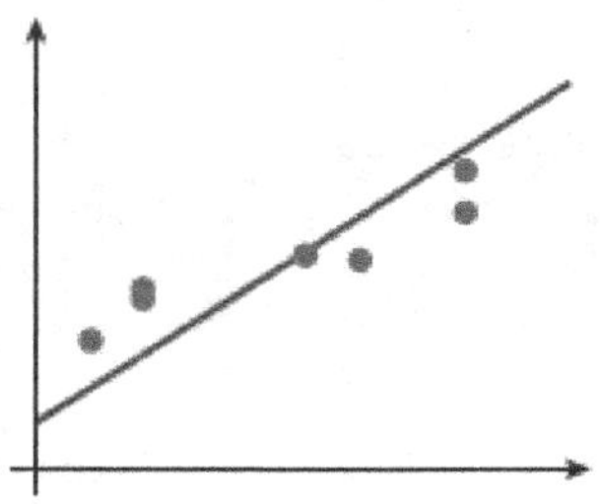

2. 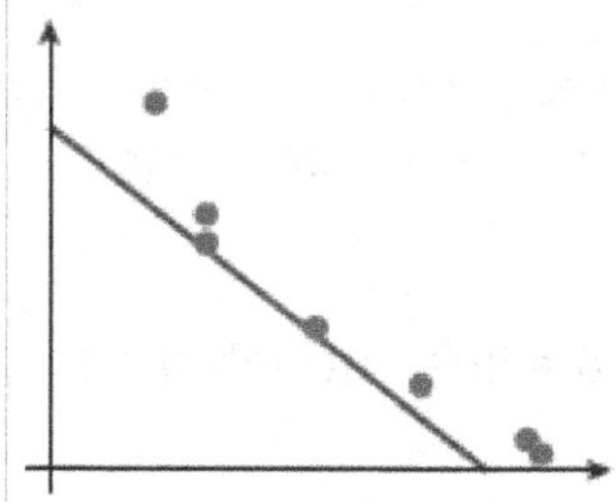

3. The scatterplot shows the ages of ten army drill sergeants and the time it took each to run 100 meters, in seconds.
   a. What was the hundred-meter time for the 25-year-old drill sergeant?
   b. How old was the drill sergeant whose hundred-meter time was 12.6 seconds?
   c. Use a straightedge to draw a line of best fit through the data points.
   d. Use your line of best fit to predict the hundred-meter time of a 28-year-old drill sergeant.
   e. Choose two points on your regression line and find its equation.
   f. Use the equation to predict the hundred-meter time of a 40-year-old drill sergeant and a 12 year-old drill sergeant. Are these predictions reasonable?

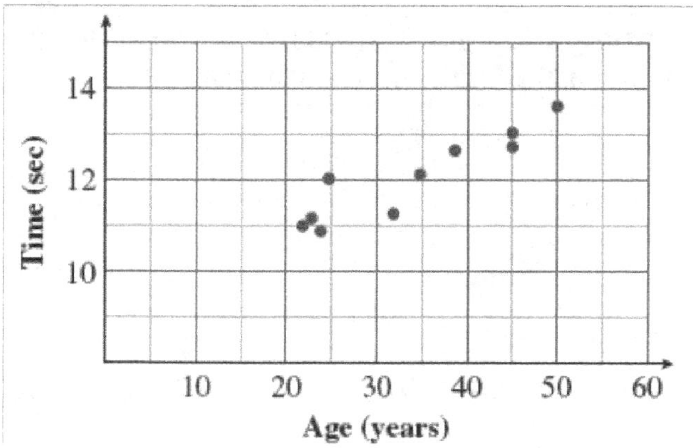

4. With Americans' increased use of faxes, pagers, and cell phones, new area codes are being created at a steady rate. The table shows the number of area codes in the US each year. (USA Today, NeuStar, Inc.)

Year	1997	1998	1999	2000	2001	2002	2003
Number of Area Codes	151	186	204	226	239	262	274

   a. Let t represent the number of years after 1995, and plot the data. Draw a line of best fit for the data points.
   b. Find an equation for your regression line.
   c. How many area codes do you predict for 2010?

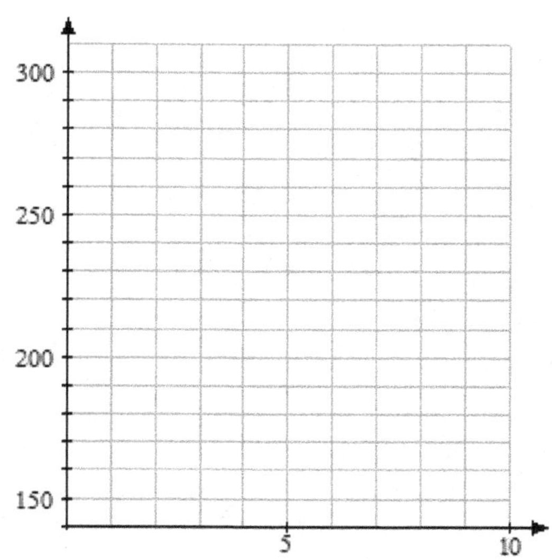

5. Teenage birth rates in the US declined from 1991 to 2000. The table shows the number of births per 1000 women in selected years. (Source: US National Health Statistics)

Year	1991	1993	1995	1996	1997	1998
Births	62.1	59.6	56.8	54.4	52.3	51.1

a. Let t represent the number of years after 1990, and plot the data. Draw a line of best fit for the data points.
b. Find an equation for your regression line.
c. Estimate the teen birth rate in 1994.
d. Predict the teen birth rate in 2010.

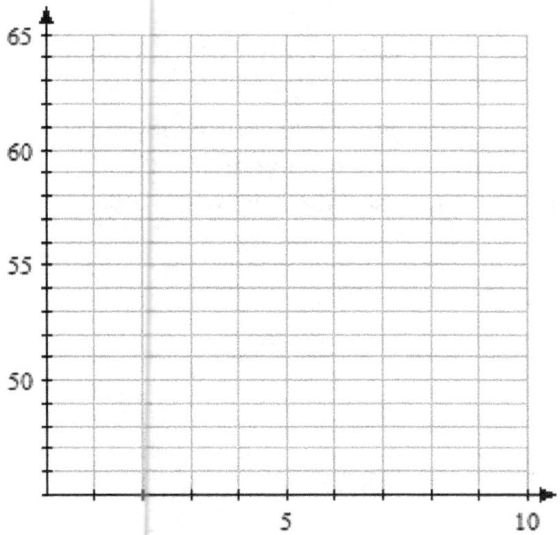

Grid for #5

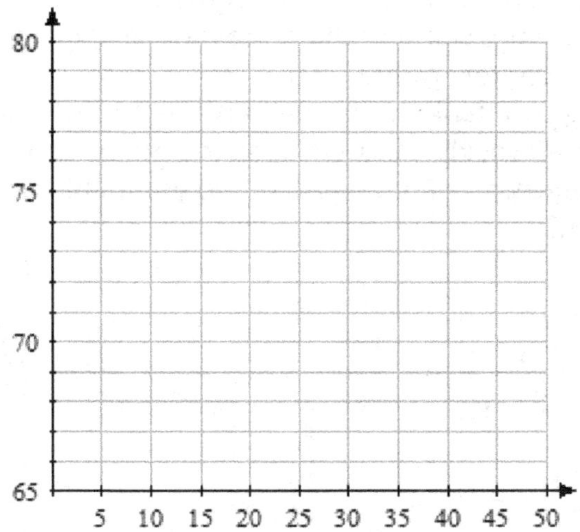

Grid for #6

6. Life expectancy in the US has been rising since the nineteenth century. The table shows the US life expectancy in selected years. (Source: http://www.infoplease.com)

Year	1950	1960	1970	1980	1990	2000
Life Expectancy at Birth	68.2	69.7	70.8	73.7	75.4	77

a. Let t represent the number of years after 1950, and plot the data. Draw a line of best fit for the data points.
b. Find an equation for your regression line.
c. Estimate the life expectancy of someone born in 1987.
d. Predict the life expectancy of someone born in 2010.

7. "The earnings gap between high-school and college graduates continues to widen, the Census Bureau says. On average, college graduates now earn just over $51,000 a year, almost twice as much as high-school graduates. And those with no high-school diploma have actually seen their earnings drop in recent years." The table on the next page shows the unemployment rate and the median weekly earnings for employees with different levels of education. (Source: Morning Edition, March 28, 2005)

	Years of Education	Unemployment Rate	Weekly Earnings ($)
Some high-school, no diploma	10	8.8	396
High school graduate	12	5.5	554
Some college, no degree	13	5.2	622
Associate degree	14	4.0	672
Bachelor's degree	16	3.3	900
Master's degree	18	2.9	1064
Professional degree	20	1.7	1307

a. Plot years of education on the horizontal axis and weekly earnings on the vertical axis.
b. Find an equation for the regression line.
c. State the slope of the regression line, including units, and explain what it means in the context of the data.
d. Do you think this model is useful for extrapolation or interpolation? For example, what weekly earnings does the model predict for someone with 15 years of education? For 25 years? Do you think these predictions are valid? Why or why not?

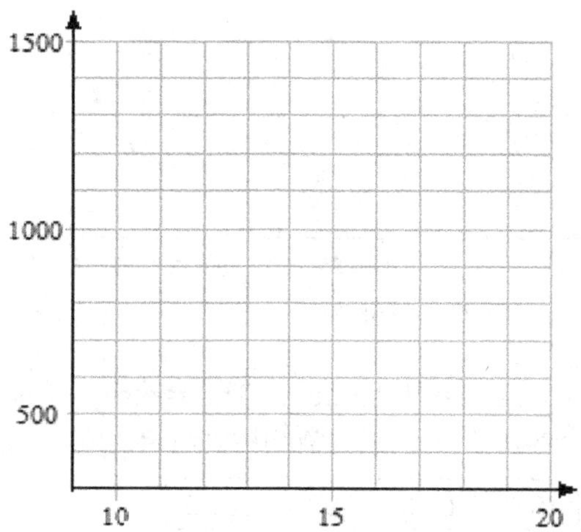

Grid for #7

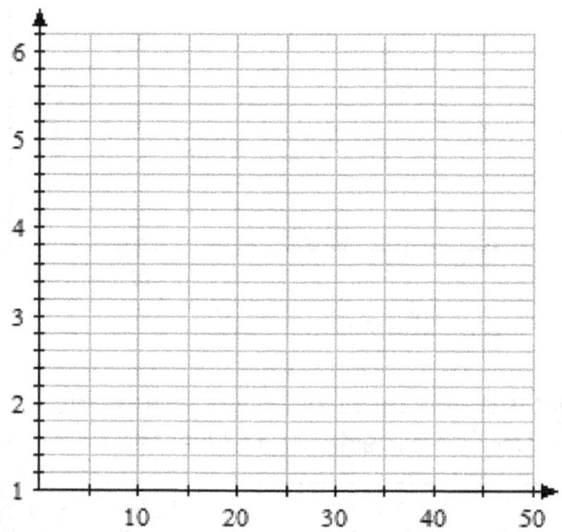

Grid for #8

8. The table shows the amount of carbon released into the atmosphere annually from burning fossil fuels, in billions of tons, at 5-year intervals from 1950 to 1995. (Source: www.worldwatch.org)

Year	50	55	60	65	70	75	80	85	90	95
Carbon emissions	1.6	2.0	2.5	3.1	4.0	4.5	5.2	5.3	5.9	6.2

a. Let t represent the number of years after 1950, and plot the data. Draw a line of best fit for the data points.
b. Find an equation for your regression line.
c. Estimate the amount of carbon released in 1992.

■ For Problems 9-10,

a. Use linear interpolation to give approximate answers.
b. What is the meaning of the slope in the context of the problem?

9. The temperature in Encino dropped from 81° at 1 a.m. to 73° at 5 a.m. Estimate the temperature at 4 a.m.

10. A car starts from a standstill and accelerates to a speed of 60 miles per hour in 6 seconds. Estimate the car's speed 2 seconds after it began to accelerate.

■ For Problems 11-12, use linear interpolation or extrapolation to answer the questions.

11. The temperature of an automobile engine is 9° Celsius when the engine is started and 51° seven minutes later. Use a linear model to predict the engine temperature for both two minutes and two hours after it started. Are your predictions reasonable?

12. Ben weighed 8 pounds at birth and 20 pounds at age 1 year. How much will he weigh at age 10 if his weight increases at a constant rate?

Applications

13. Bracken, a type of fern, is one of the most successful plants in the world, growing on every continent except Antarctica. New plants, which are genetic clones of the original, spring from a network of underground stems, or rhizomes, to form a large circular colony. The graph shows the diameters of various colonies plotted against their age. (Source: Chapman et al., 1992)

a. Calculate the rate of growth of the diameter of a bracken colony, in meters per year.
b. Find an equation for the line of best fit. (What should the vertical intercept of the line be?)
c. In Finland, bracken colonies over 450 meters in diameter have been found. How old are these colonies?

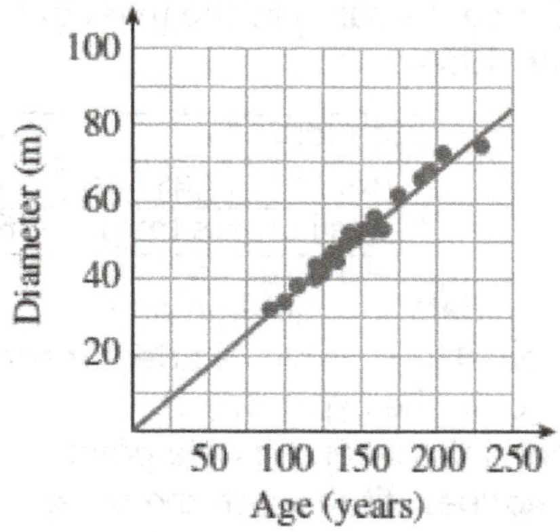

14. One of the factors that determines the strength of a muscle is its cross-sectional area. The data below show the cross-sectional area of the arm flexor muscle for several men and women, and their strength, measured by the maximum force they exerted against a resistance. (Source: Davis, Kimmet, Autry, 1986)

Women	Area (sq cm)	11.5	10.8	11.7	12.0	12.5	12.7	14.4	14.4	15.7
	Strength (kg)	11.3	13.2	13.2	14.5	15.6	14.8	15.6	16.1	18.4
Men	Area (sq cm)	13.5	13.8	15.4	15.4	17.7	18.6	20.8	–	–
	Strength (kg)	15.0	17.3	19.0	19.8	20.6	20.8	26.3	–	–

a. Plot the data for both men and women on the same graph, using different symbols for the data points for men and the data points for women.
b. Are the data for both men and women described reasonably well by the same regression line? Draw a line of best fit through the data.
c. Find the equation of your line of best fit, or use a calculator to find the regression line for the data.
d. What does the slope mean in this context?

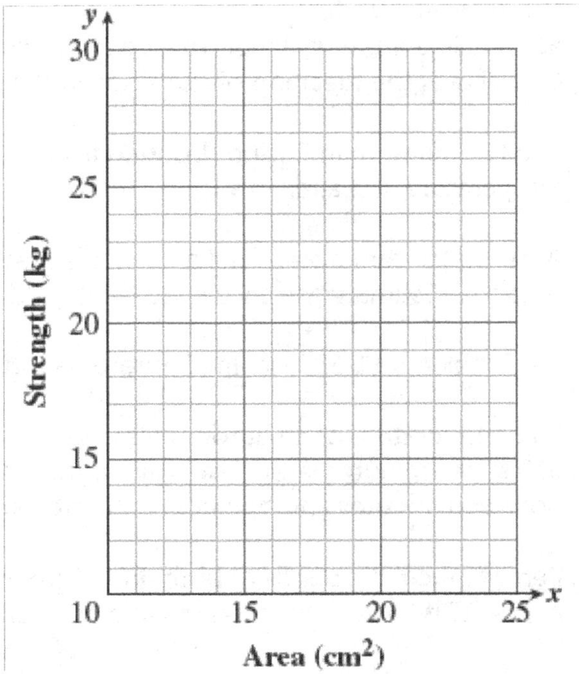

15. Six students are trying to identify an unknown chemical compound by heating the substance and measuring the density of the gas that evaporates. (Density = mass/volume.) The students record the mass lost by the solid substance and the volume of the gas that evaporated from it. They know that the mass lost by the solid must be the same as the mass of the gas that evaporated. (Source: Hunt and Sykes, 1984)

Student	A	B	C	D	E	F
Volume of gas (cm^3)	48	60	24	81	76	54
Loss in mass (mg)	64	81	32	107	88	72

a. Plot the data with volume on the horizontal axis. Which student made an error in the experiment?
b. Ignoring the incorrect data point, draw a line of best fit through the other points.
c. Find an equation of the form $y = kx$ for the data. Why should you expect the regression line to pass through the origin?
d. Use your equation to calculate the mass of 1000 cm^3 (one liter) of the gas.
e. Here are the densities of some gases at room temperature.

Hydrogen	8 mg/liter
Nitrogen	1160 mg/liter
Oxygen	1330 mg/liter
Carbon dioxide	1830 mg/liter

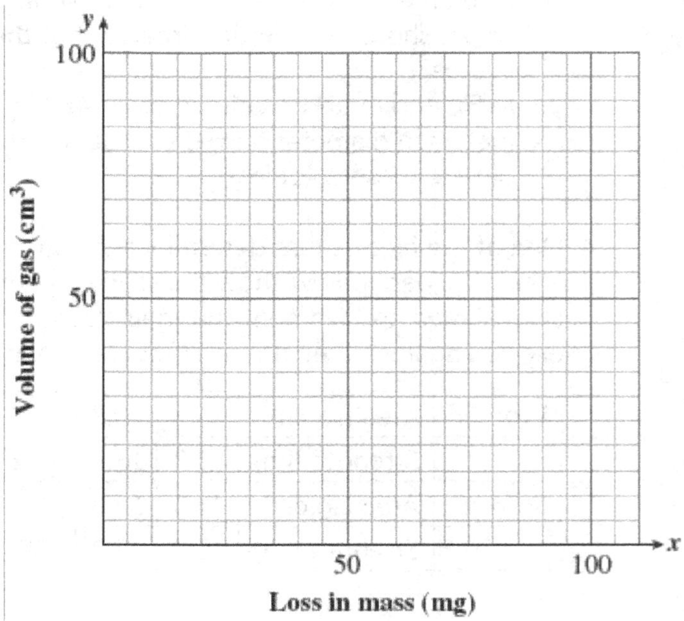

Which of these might have been the gas that evaporated from the unknown substance? **(Hint:** Use your answer to part (d) to calculate the density of the gas. 1 cm^3 = 1 millimeter.)

16. Birds' nests are always in danger from predators. If there are other nests close by, the chances of predators finding the nest increase. The table shows the probability of a nest being found by predators and the distance to the nearest neighboring nest. (Source: Perrins, 1979)

Distance to nearest neighbor (meters)	20	40	60	80	100
Probability of predators (%)	47	34	32	17	1.5

a. Plot the data and the least squares regression line.

b. Use the regression line to estimate the probability of predators finding a nest if its nearest neighbor is 50 meters away.

c. If the probability of predators finding a nest is 10%, how far away is its nearest neighbor?

d. What is the probability of predators finding a nest if its nearest neighbor is 120 meters away? Is your answer reasonable?

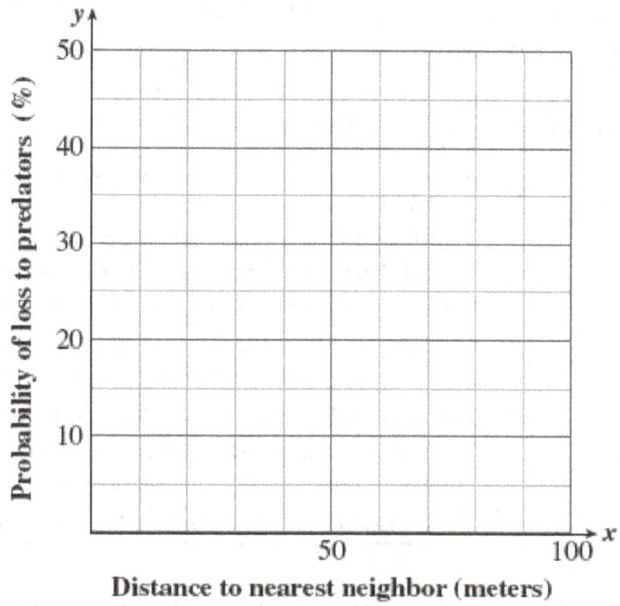

17. A swimmer generates power to maintain a constant speed in the water. If she must swim against an opposing force, the power increases. The table shows the power expended by a swimmer while working against different amounts of force. (A positive force opposes the swimmer, and a negative force helps her.) (Source: di Prampero et al, 1974; Alexander, 1992)

Force (newtons)	−3.5	0	0	6	8	10	17	17
Metabolic power (watts)	100	190	230	320	380	450	560	600

a. Plot the data on the grid, or use the **StatPlot** feature on your calculator. Use your calculator to find the least squares regression line. Graph the regression line on top of the data.

b. Use your regression line to estimate the power needed for the swimmer to overcome an opposing force of 15 newtons.

c. Use your regression line to estimate the power generated by the swimmer when there is no force either hindering or helping her.

d. Estimate the force needed to tow the swimmer at 0.4 meters per second while she rests. (If she is resting she is not generating any power).

e. The swimmer's mechanical power (or rate of work) is computed by multiplying her speed times the force needed to tow her at rest. Use your answer to part (d) to calculate the mechanical power she generates by swimming at 0.4 meters per second.

f. The ratio of mechanical power to metabolic power is a measure of the swimmer's efficiency. Compute the efficiency of the swimmer when there is no external force opposing or helping her.

Chapter 4 Summary and Review

Section 4.1 Systems of Linear Equations

- A pair of linear equations in two variables

$$a_1x + b_1y = c_1$$
$$a_2x + b_2y = c_2$$

considered together is called a **system of linear equations**, or a **linear system**.

- A **solution** to a system of linear equations is an ordered pair (x, y) that satisfies each equation in the system. Because the solution is a point that lies on both graphs, it is the intersection point of the two lines described by the system.

> **Linear Systems**
> There are three types of linear systems:
> 1. **Consistent and independent system.** The graphs of the two lines intersect in exactly one point. The system has exactly one solution.
> 2. **Inconsistent system**. The graphs of the equations are parallel lines and hence do not intersect. An inconsistent system has no solutions.
> 3. **Dependent system**. All the solutions of one equation are also solutions to the second equation, and hence are solutions of the system. The graphs of the two equations are the same line. A dependent system has infinitely many solutions.

- In economics, the **demand equation** gives the number of units of the product that consumers will buy in terms of the price per unit. The **supply equation** gives the number of units that the producer will supply in terms of the price per unit. The price at which the supply and demand are equal is called the **equilibrium price**.

Section 4.2 Algebraic Solution of Systems

> **To Solve a System by Substitution:**
> 1. Choose one of the variables in one of the equations. (It is best to choose a variable whose coefficient is 1 or -1.) Solve the equation for that variable.
> 2. Substitute the result of Step 1 into the *other* equation. This gives an equation in one variable.
> 3. Solve the equation obtained in Step 2. This gives the solution value for one of the variables.
> 4. Substitute this value into the result of Step 1 to find the solution value of the other variable.

> **To Solve a System by Elimination:**
> 1. Write each equation in the form $Ax + By = C$.
> 2. Decide which variable to eliminate. Multiply each equation by an appropriate constant so that the coefficients of that variable are opposites.
> 3. Add the equation from Step 2 and solve for the remaining variable.
> 4. Substitute the value found in Step 3 into one of the original equations and solve for the other variable.

When Using Elimination to Solve a System:

1. If combining the two equations results in an equation of the form
$$0x + 0y = k \quad (k \neq 0)$$
then the system is inconsistent.
2. If combining the two equations results in an equation of the form
$$0x + 0y = 0$$
then the system is dependent.

Section 4.3 Applications of Systems

- We can use tables to organize the information in applications involving interest, mixtures, or motion.
- The formula for calculating interest is

$$I = Prt$$

I is the interest you will earn after t years if you invest a principal of P dollars in an account that earns simple interest at an annual rate r.

- The formula for calculating percents is

$$P = rW$$

P stands for the part obtained when we take r percent of a whole amount, W.

- In general, **we cannot add percents** unless they are percents of the same whole amount.
- To solve problems involving motion at a constant speed we use the formula

$$D = RT$$

D stands for the distance you travel at speed R in time T.

Section 4.4 Point-Slope Form

Two-Point Formula for Slope

The slope of the line joining points $P_1(x_1, y_1)$ and $P_2(x_2, y_2)$ is

$$m = \frac{y_2 - y_1}{x_2 - x_1} \quad \text{if} \quad x_2 \neq x_1$$

Point-slope Formula

The equation for the line that passes through the point (x_1, y_1) and has slope m is

$$y = y_1 + m(x - x_1)$$

or

$$y - y_1 = m(x - x_1) \quad \text{or} \quad \frac{y - y_1}{x - x_1} = m$$

Section 4.5 Using the Point-Slope Form

Parallel and Perpendicular Lines

1. Two lines are **parallel** if their slopes are equal, that is, if

$$m_1 = m_2$$

or if both lines are vertical.

2. Two lines are **perpendicular** if the product of their slopes is -1, that is, if

$$m_1 m_2 = -1$$

or if one of the lines is horizontal and one is vertical.

To Fit a Line through Two Points :

1. Compute the slope between the two points.
2. Substitute the slope and either point into the point-slope formula.

Horizontal and Vertical Lines

1. The equation of the **horizontal line** passing through $(0, b)$ is

$$y = b$$

2. The equation of the **vertical line** passing through $(a, 0)$ is

$$x = a$$

Section 4.6 Linear Regression

- Empirical data points in a linear relation may not lie exactly on a line, but instead form a **scatterplot**.
- If the data in a scatterplot are roughly linear, we can estimate the location of a **line of best fit** that passes as close as possible to the data points.
- A straight line that fits the data is called a **regression line.**
- Using a regression line to estimate values between known data points is called **interpolation**. Making predictions beyond the range of known data is called **extrapolation.**

Review Questions

■ Use complete sentences to answer the questions in Problems 1-15.

1. State the two-point formula for slope.
2. How can you find the equation of a line when you know the slope of the line and one point on the line?
3. **a.** What is true about the slopes of parallel lines?
   b. What is true about the slopes of perpendicular lines?
4. **a.** What does the equation of a vertical line look like? What is the slope of a vertical line?
   b. What does the equation of a horizontal line look like? What is the slope of a horizontal line?
5. How can you find the solution to a linear system by graphing?
6. Name two algebraic methods for solving linear systems.

7. Suppose you are using the elimination method to solve a system. How can you tell if the system is dependent or inconsistent?
8. How would you label the columns when making a table for a problem about interest?
9. How would you label the columns when making a table for a problem about a mixture?
10. How would you label the columns when making a table for a problem about motion?
11. Under what conditions can we add percents?
12. What is the intersection of the demand equation and the supply equation called?
13. How can you estimate a line of best fit for a scatterplot?
14. What is the least squares regression line?
15. What is the difference between interpolation and extrapolation?

Review Problems

■ For Problems 1-2, decide whether the given point is a solution of the system.

1. $x + y = 8$ $(-2, 10)$
$x - y = 2$

2. $8x + 3y = 21$ $(-3, 1)$
$5x = y - 16$

■ For Problems 3-4, solve the system by graphing.

3. $x + y = 5$
$2x - y = 4$

4. $x - y = 7$
$y = \frac{-2}{3}x - 2$

■ For Problems 5-6, solve the system by graphing. Use the **ZDecimal** window.

5. $y = -2.9x - 0.9$
$y = 1.4 - 0.6x$

6. $y = 0.6x - 1.94$
$y = -1.1x + 1.29$

■ For Problems 7-8, solve by substitution.

7. $y = 2x + 1$
$2x + 3y = -21$

8. $x + 4y = 1$
$2x + 3y = -3$

■ For Problems 9-10, solve by elimination.

9. $2x + 7y = -19$
$5x - 3y = 14$

10. $4x + 3y = -19$
$5x + 15 = -2y$

■ For Problems 11-14, solve using substitution or elimination.

11. $x + 5y = 18$
$x - y = -3$

12. $x + 5y = 11$
$2x + 3y = 8$

13. $\frac{2}{3}x - 3y = 8$
$x + \frac{3}{4}y = 12$

14. $3x = 5y - 6$
$3y = 10 - 11x$

For Problems 15-18, decide whether each system is inconsistent, dependent, or consistent and independent.

15. $2x - 3y = 4$
$x + 2y = 7$

16. $2x - 3y = 4$
$6x - 9y = 4$

17. $2x - 3y = 4$
$6x - 9y = 12$

18. $x - y = 6$
$x + y = 6$

For Problems 19-30, solve by writing and solving a system of equations.

19. A health food store wants to produce 30 pounds of granola worth 80 cents per pound. They plan to mix cereal worth 65 cents per pound with dried fruit worth 90 cents per pound. How much of each should they use?

20. The perimeter of a rectangle is 50 yards and its length is 9 yards greater than its width. Find the dimensions of the rectangle.

21. Last year Veronica made $93 in interest from her savings accounts, one of which paid 6% interest, and the other paid 9%. This year her interest rates dropped to 4% and 8%, respectively, and she made $76 interest. How much does Veronica have invested in each account?

22. Marvin invested $300 more at 6% than he invested at 8%. His total annual income from his two investments is $242. How much did he invest at each rate?

23. How many pounds of an alloy containing 60% copper must be melted with an alloy containing 20% copper to obtain 8 pounds of an alloy containing 30% copper?

24. Jerry Glove came to bat only 20 times in the first half of the season and got hits 15% of the time. During the second half of the season, Jerry came to bat 140 times and got hits 35% of the time.

a. How many hits did Jerry get in the first half of the season? How many hits did he get in the second half?
b. How many times did Jerry come to bat all season? How many hits did he get?
c. What percent of Jerry's at-bats resulted in hits?
d. Who had a better batting average in the first half of the season, Joe Cleat from Problem 8 in Homework 4.3, or Jerry? Who had a better batting average in the second half? Who had the better batting average overall?

25. Alida and Steve are moving to San Diego. Alida is driving their car, and Steve is driving a rental truck. They start together, but Alida drives twice as fast as Steve. After 3 hours they are 93 miles apart. How fast is each traveling?

26. Jake rides for the Pony Express, covering his route in 6 hours and returning home, 8 miles per hour slower, in 9 hours. How far does Jake ride?

27. A math contest exam has 40 questions. A contestant scores 5 points for each correct answer, but loses 2 points for each wrong answer. Lupe answered all the questions and her score was 102. How many questions did she answer correctly?

28. A game show contestant wins $25 for each correct answer he gives but loses $10 for each incorrect response. Roger answered 24 questions and won $355. How many answers did he get right?

29. Barbara wants to earn $500 a year by investing $5000 in two accounts, a savings plan that pays 8% annual interest and a high-risk option that pays 13.5% interest. How much should she invest in each account?

30. An investment broker promises his client a 12% return on her funds. If the broker invests $3000 in bonds paying 8% interest, how much must he invest in stocks paying 15% interest to keep his promise?

■ For Problems 31-34, find the slope of the line segment joining the points.

31. $(-1, 4)$, $(3, -2)$

32. $(5, 0)$, $(2, -6)$

33. $(6.2, 1.4)$, $(-2.1, 4.8)$

34. $(0, -6.4)$, $(-5.6, 3.2)$

■ Which of the tables in Problems 35-36 represent variables that are related by a linear equation?

35. a.

r	E
1	5
2	$\frac{5}{2}$
3	$\frac{5}{3}$
4	$\frac{5}{4}$
5	1

b.

s	t
10	6.2
20	9.7
30	12.6
40	15.8
50	19.0

36. a.

w	A
2	−13
4	−23
6	−33
8	−43
10	−53

b.

x	C
0	0
2	5
4	10
8	20
16	40

■ For Problems 37-38, the table gives values for a linear equation in two variables. Fill in the missing values.

37.

d	V
−5	−4.8
−2	−3
	−1.2
6	1.8
10	

38.

q	S
−8	−8
−4	56
3	
	200
9	264

39. a. Graph the line with slope $\frac{-5}{3}$ that passes through the point $(-2, 1)$.
b. Find an equation in point-slope form for the line in part (a)

40. a. Graph the line of slope $\frac{6}{5}$ that passes through the point $(-3, -4)$.
b. Find an equation in point-slope form for the line in part (a).

41. An interior decorator bases her fee on the cost of a remodeling job. The table below shows her fee, F, for jobs of various costs, C, both given in dollars.

C	5000	10,000	20,000	50,000
F	1000	1500	2500	5500

a. Write a linear equation for F in terms of C.
b. Choose appropriate window settings on your calculator and graph your equation.
c. Give the slope of the graph, and explain the meaning of the slope in terms of the decorator's fee.

42. Auto registration fees in Connie's home state depend on the value of the automobile. The table below shows the registration fee R for a car whose value is V, both given in dollars.

V	5000	10,000	15,000	20,000
R	135	235	335	435

a. Write a linear equation for R in terms of V.
b. Choose appropriate window settings on your calculator and graph your equation.
c. Give the slope of the graph, and explain the meaning of the slope in terms of the registration fee.

■ For Problems 43-44, find an equation for the line passing through the two given points.

43. $(3, -5),\ (-2, 4)$

44. $(0, 8),\ (4, -2)$

■ For Problems 45-46,
a. Make a table of values showing two data points.
b. Find a linear equation relating the variables.
c. State the slope of the line, including units, and explain its meaning in the context of the problem.

45. The population of Maple Rapids was 4800 in 1982 and had grown to 6780 by 1997. Assume that the population increases at a constant rate. Express the population P of Maple Rapids in terms of the number of years t since 1982.

46. Cicely's odometer read 112 miles when she filled up her 14-gallon gas tank and 308 when the gas gauge read half full. Express her odometer reading m in terms of the amount of gas g she used.

■ For Problems 47-48, decide whether the lines are parallel, perpendicular, or neither.

47. $y = \frac{1}{2}x + 3;\quad x - 2y = 8$

48. $4x - y = 6;\quad x + 4y = -2$

49. Which pair of lines are parallel, which are perpendicular, and which are neither?

a. $5x + 4y = 3$
$3y - 5x = 4$

b. $5x + 3y = 1$
$5y - 3x = 6$

c. $4y - 5x = 1$
$2y - \frac{5}{2}x = 2$

50. a. Write an equation for any line that is parallel to $2y = 5x - 3$.
b. Write an equation for any line that is perpendicular to $2y = 5x - 3$.

51. Write an equation for the line that is parallel to the graph of $2x + 3y = 6$ and passes through the point $(1, 4)$.

52. Write an equation for the line that is perpendicular to the graph of $2x + 3y = 6$ and passes through the point $(1, 4)$.

■ For Problems 53-60, find an equation for the line described. Write your answers in slope-intercept form if possible.

53. With x-intercept $(5, 0)$ and y-intercept $(0, -1)$
54. With slope $-\frac{1}{4}$ and passing through the point $(3, -5)$
55. Parallel to the y-axis and passing through the point $(-4, 2)$
56. Horizontal and passing through the point $(-3, 8)$
57. Passing through the points $(-3, 1)$ and $(-1, -1)$
58. Passing through the points $(1, -2)$ and $(-7, 3)$.
59. Perpendicular to the line $2x + 3y = 1$ and passes through $(-4, 1)$
60. Perpendicular to the line $y = \frac{1}{3}x - 2$

61. In 1986 the space shuttle Challenger exploded because of O-ring failure on a morning when the temperature was about 30° F. Previously there had been 1 incident of O-ring failure when the temperature was 70° F and 3 incidents when the temperature was 54° F. Use linear extrapolation to estimate the number of incidents of O-ring failure you would expect when the temperature is 30°F.

62. Thelma typed a 19-page technical report in 40 minutes. She required only 18 minutes for an 8-page technical report. Use linear interpolation to estimate how long Thelma would require to type a 12-page technical report.

63. The scatterplot shows weights (in pounds) and heights (in inches) for a team of distance runners.

a. Use a straightedge to draw a line that fits the data.

b. Use your line to predict the weight of a 65-inch-tall runner and the weight of a 71-inch-tall runner.

c. Use your answers from part (b) to approximate the equation of a regression line.

d. Use your answer to part (c) to predict the weight of a runner who is 68 inches tall.

e. The points on the scatterplot are $(65.5, 130)$, $(66.5, 133)$, $(67, 135)$, $(69, 140)$, and $(70, 143)$. Use your calculator to find the least squares regression line.

f. Use the regression line to predict the weight of a runner who is 68 inches tall.

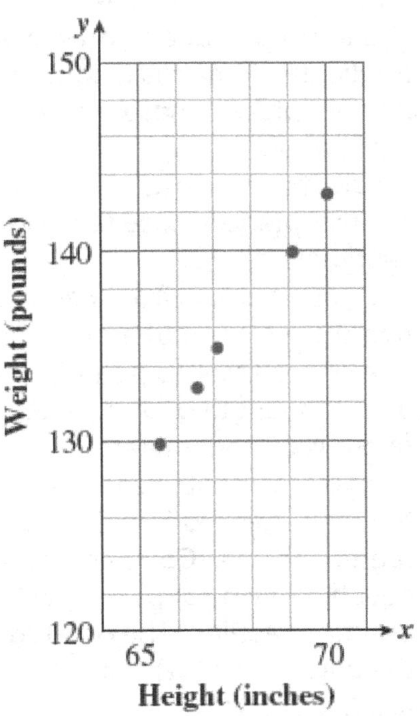

64. The scatterplot shows best times for various women running 400 meters and 100 meters.

a. Use a straightedge to draw a line that fits the data.

b. Use your line to predict the 400-meter time of a woman who runs the 100-meter dash in 11.2 seconds, and the 400-meter time of a woman who runs the 100-meter dash in 13.2 seconds.

c. Use your answers from part (b) to approximate the equation of a regression line.

d. Use your answer to part (c) to predict the 400-meter time of a woman who runs the 100-meter dash in 12.1 seconds.

e. The points on the scatterplot are $(11.1, 52.4)$, $(11.5, 54.7)$, $(11.9, 57.4)$, $(12.2, 57.9)$, $(12.7, 61.3)$, and $(13.0, 63.0)$. Use your calculator to find the least squares regression line.

f. Use the regression line to predict the 400-meter time of a woman who runs the 100-meter dash in 12.1 seconds.

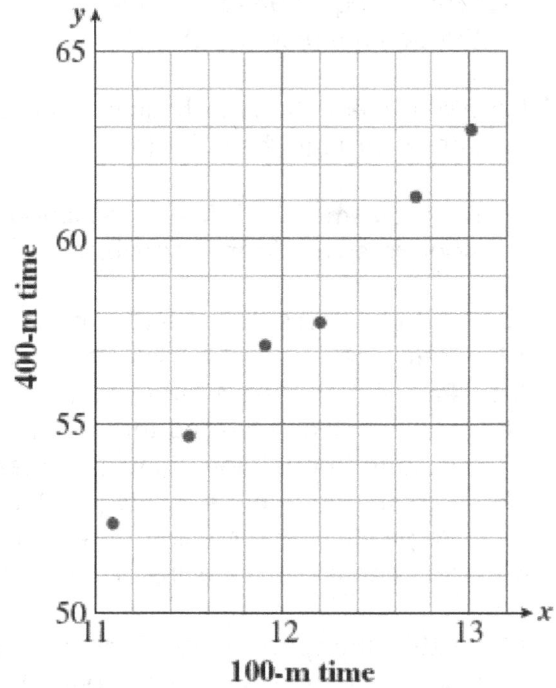

65. Archaeopteryx is an extinct creature with characteristics of both birds and reptiles. Only six fossil specimens are known, and only five of those include both a femur (leg bone) and a humerus (forearm bone) The scatterplot shows the lengths of femur and humerus for the five Archaeopteryx specimens.

a. Use a straightedge to draw a line that fits the data.

b. Predict the humerus length of an Archaeopteryx whose femur is 40 centimeters.

c. Predict the humerus length of an Archaeopteryx whose femur is 75 centimeters.

d. Use your answers from parts (a) and (b) to find the equation of a regression line.

e. Use your answer to part (d) to predict the humerus length of an Archaeopteryx whose femur is 60 centimeters

f. Use your calculator and the given points on the scatterplot to find the least squares regression line. Compare the score this equation gives for part (d) with what you predicted earlier. The ordered pairs defining the data are $(38, 41)$, $(56, 63)$, $(59, 70)$, $(64, 72)$, $(74, 84)$.

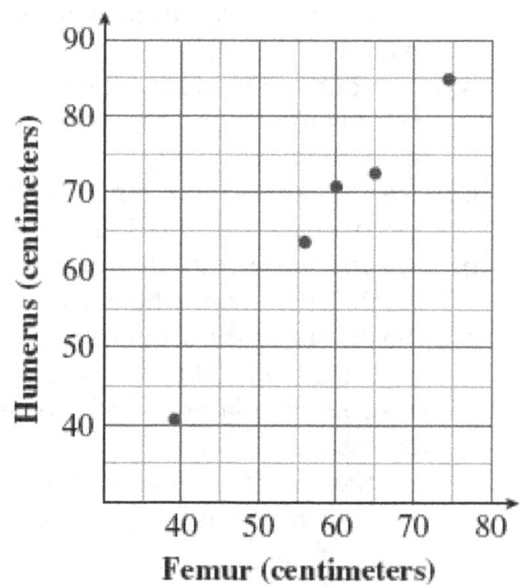

66. The scatterplot shows the boiling temperature of various substances on the horizontal axis, and their heats of vaporization on the vertical axis. (The heat of vaporization is the energy needed to change the substance from liquid to gas at its boiling point.)

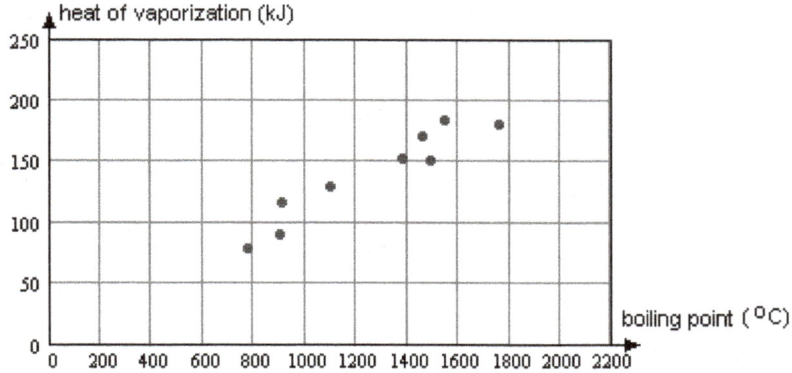

a. Use a straightedge to estimate a line of best fit for the scatterplot.

b. Use your line to predict the heat of vaporization of silver, whose boiling temperature is 2160°C.

c. Find the equation of the regression line.

d. Use the regression line to predict the heat of vaporization of potassium bromide, whose boiling temperature is 1435°C.

Chapter 5 Quadratic Equations

Lesson 1 Exponents and Formulas

- Compute powers
- Simplify expressions involving exponents
- Evaluate expressions involving exponents
- Combine like terms
- Use formulas that involve exponents
- Solve a formula for one variable in terms of the others

Lesson 2 Square Roots

- Compute square roots and cube roots
- Use radical notation
- Distinguish between rational and irrational numbers
- Distinguish between exact values and approximations
- Simplify expressions involving radicals

Lesson 3 Quadratic Equations

- Use formulas to compute volume and surface area
- Use the Pythagorean theorem
- Solve an equation in x^2
- Solve a formula for one variable in terms of the others

Lesson 4 Products of Binomials

- Apply the distributive law to simplify products
- Simplify products involving variables
- Compute the product of two binomials
- Compute the square of a binomial

Lesson 5 Solving Quadratic Equations

- Apply the Zero-Factor principle
- Factor out a common factor
- Solve a quadratic equation by factoring
- Solve a quadratic equation by extracting roots

Lesson 6 Graphing Parabolas

- Identify features of a parabola
- Graph parabolas of the form $y = ax^2$, $y = x^2 + c$ and $y = ax^2 + bx$
- Find the vertex of a parabola

5.1 Exponents and Formulas

What is an Exponent?

The area of a rectangle is given by the formula $A = lw$. The variable l stands for the length of the rectangle, and w stands for for its width. The area tells us how many square tiles, one unit on a side, will fit inside the rectangle, as shown below.

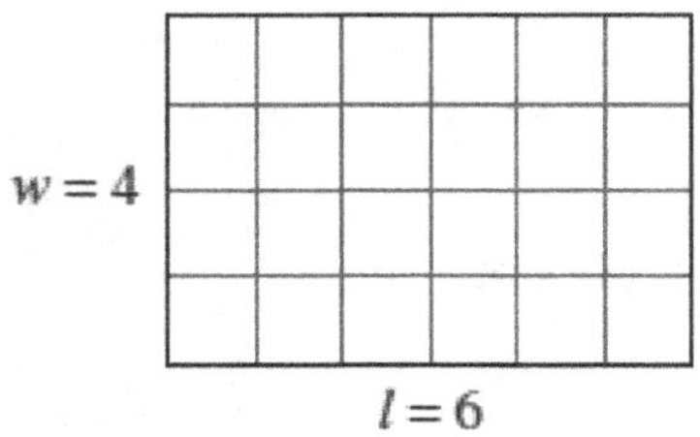

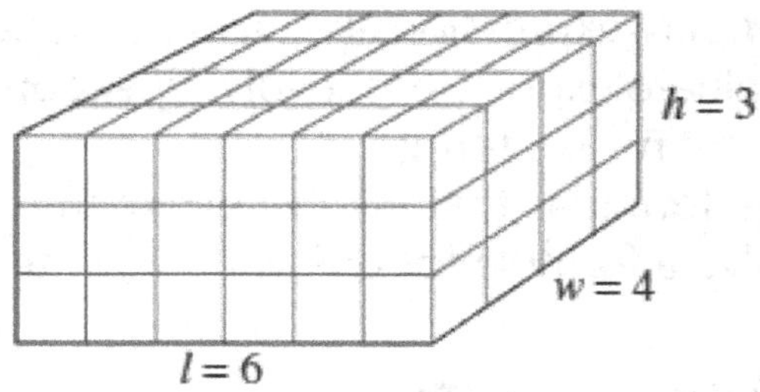

Similarly, the volume of a box (measured in cubic units) is given by the formula $V = lwh$, where l, w, and h stand for the length, width, and height of the box. The volume tells us how many blocks, one unit on a side, will fit inside the box. The volume of the box shown above is

$$V = lwh = 6 \cdot 4 \cdot 3 = 72 \text{ cubic inches}$$

A **square** is a rectangle whose length and width are equal. If we use s, for side, to stand for both the length and width of the square, its area is given by

$$A = s \cdot s$$

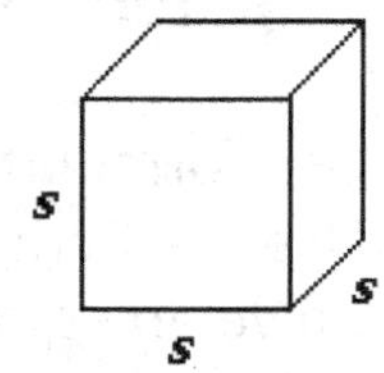

A **cube** is a box whose length, width, and height are all equal, so its volume is given by

$$V = s \cdot s \cdot s$$

RQ1. What is the formula for the volume of a box?

Finding the area of a square or the volume of a cube involves repeated multiplication by the same number. We indicate repeated multiplication with a symbol called an **exponent**.

An **exponent** is a number that appears above and to the right of a particular factor. It tells us how many times that factor occurs in the expression. The factor to which the exponent applies is called the **base**, and the product is called a **power** of the base.

For example,

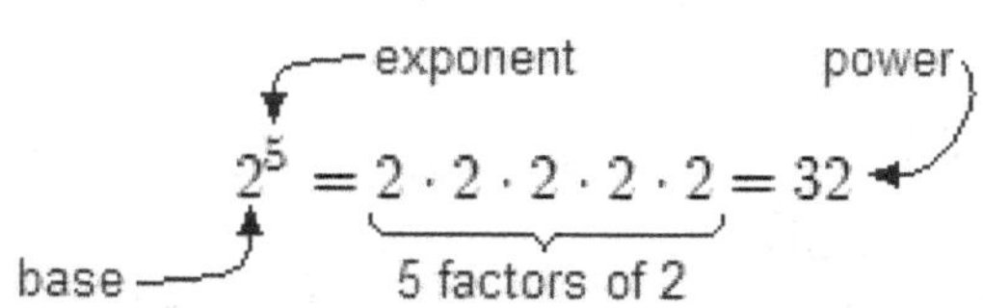

We read the expression 2^5 as *the fifth power of 2*, or as *2 raised to the fifth power*, or simply as *2 to the fifth*.

Exponents

An **exponent** indicates repeated multiplication.

$$a^n = a \cdot a \cdot a \cdot \cdots \cdot a \qquad (n \text{ factors of } a)$$

where n is a positive integer.

RQ2. What does an exponent tell us?

Example 1 Compute the following powers.

a. $6^2 = 6 \cdot 6 = 36$

b. $3^4 = 3 \cdot 3 \cdot 3 \cdot 3 = 81$

c. $\left(\frac{2}{3}\right)^3 = \left(\frac{2}{3}\right)\left(\frac{2}{3}\right)\left(\frac{2}{3}\right) = \frac{8}{27}$

d. $1.4^4 = 1.4(1.4)(1.4)(1.4) = 3.8416$

Caution! Note that 3^4 does not mean 3 times 4 or 12; we already have a way to write $3(4)$. Remember that an exponent or power indicates repeated multiplication of the base.

Look Closer: Because the exponents 2 and 3 are used frequently, they have special names.

5^2 means $5 \cdot 5$, and is read **5 squared** or the square of 5
5^3 means $5 \cdot 5 \cdot 5$, and is read **5 cubed** or the cube of 5

These names come from the formulas for the area, A, of a square with side s and the volume, V, of a cube with edge s:

$$A = s^2 \quad \text{and} \quad V = s^3$$

Powers of Negative Numbers

To show that a negative number is raised to a power, we enclose the negative number in parentheses.

For example, to indicate the square of –5, we write

$$(-5)^2 = (-5)(-5) = 25 \qquad \textbf{Exponent applies to } (-5).$$

If the negative number is *not* enclosed in parentheses, then the exponent applies only to the positive number, and the negative sign tells us that the power is negative. For example,

$$-5^2 = -5 \cdot 5 = -25 \qquad \textbf{Exponent applies only to } 5.$$

Look Closer: Note how the placement of parentheses changes the meaning of the expressions in Example 2.

Example 2 Compute each power.

a. -4^2 **b.** $(-4)^2$ **c.** $-(4)^2$ **d.** (-4^2)

Solutions a. Only 4 is squared: $-4^2 = -4 \cdot 4 = -16$

b. The negative number is squared: $(-4)^2 = (-4)(-4) = 16$

c. Only 4 is squared: $-(4)^2 = -(4)(4) = -16$

d. Only 4 is squared, and the entire expression appears within parentheses:

$$(-4^2) = (-4 \cdot 4) = -16$$

RQ3. How do we indicate that a negative number should be raised to a power?

Using a Calculator

Scientific calculators usually have a key labeled [x^y] or [y^x], called the power key, for computing powers. To compute 7^4 using the power key, we enter

7 [y^x] 4 [=]

Graphing calculators have a caret key, [^], for entering powers. On a graphing calculator, we enter 7^4 as

7 [^] 4 [ENTER]

Also, many calculators have a key labeled [x^2] for computing squares of numbers.

Example 3 Use a calculator to compute the powers.

a. $(1.2)^3$ **b.** $(-12)^4$

Solutions a. We enter the following key strokes:

1.2 [y^x] 3 [=] or 1.2 [^] 3 [ENTER]

to find that $(1.2)^3 = 1.728$.

b. If your calculator has a [+/−] key, you can enter

12 [+/−] [y^x] 4 [=]

to find that $(-12)^4 = 20,736$. On a graphing calculator, we enter

[(] [−] 12 [)] [^] 4

Caution! When using a calculator to compute a power of a negative number, we must remember to enclose the number in parentheses. For example, which sequence will calculate the square of −3?

[−] 3 [x^2] = −9 **Exponent applies only to 3.**

[(] [−] 3 [)] [x^2] = 9 **Exponent applies to (−3).**

The first sequence tells the calculator to square 3, then make the result negative. The second sequence tells us to square square −3 (multiply −3 by itself).

Powers of Variables

We can also use exponents with variables. We must be careful to distinguish between a ***product*** of a variable and a ***power*** of a variable.

> An exponent on a variable indicates repeated multiplication, while a coefficient in front of a variable indicates repeated addition.

Example 4 Compare the expressions x^4 and $4x$.

Solution These two expressions are not the same!

$$x^4 = x \cdot x \cdot x \cdot x$$

but

$$4x = x + x + x + x$$

The expressions x^4 and $4x$ are not equivalent; they are not equal for all values of x. (Can you think of one value of x for which they are equal?) ■

> **RQ4.** What is the difference between an exponent and a coefficient?

Order of Operations

How do exponents fit into the order of operations?

Example 5 Compare the expressions $2x^3$ and $(2x)^3$.

Solution In the first expression, only the x is cubed. Thus,

$2x^3$	means	$2xxx$	**Exponent applies to the base, x, only.**
$(2x)^3$	means	$(2x)(2x)(2x)$	**Exponent applies to the base, $2x$.**

The two expressions are not equivalent, as you can see by evaluating each for, say, $x = 5$:

$$2x^3 = 2(5^3) = 2(125) = 250$$
$$(2x)^3 = (2 \cdot 5)^3 = 10^3 = 1000$$ ■

> An exponent applies only to its base, and not to any other factors in the product. If we want an exponent to apply to more than one factor, we must enclose those factors in parentheses.

Look Closer: Think about the operations in Example 5.

- To evaluate $2x^3$, we compute the power x^3 first, and then the product, $2 \cdot x^3$.
- To evaluate $(2x)^3$, we compute the product $2 \cdot x$ inside parentheses first, and then compute the power. •

When simplifying an expression, we perform powers before multiplications, but after operations within parentheses. Thus, we include exponents in the order of operations as follows.

Order of Operations

1. Perform any operations inside parentheses, or above or below a fraction bar.
2. Compute all indicated powers.
3. Perform all multiplications and divisions in the order in which they occur from left to right.
4. Perform additions and subtractions in order from left to right.

Example 6 Simplify $-2(8-3\cdot4)^2$

Solution We simplify the expression within parentheses first.

$$\begin{aligned} -2(8-\mathbf{3\cdot4})^2 &= -2(8-\mathbf{12})^2 && \textbf{Multiply first, then subtract.}\\ &= -2(\mathbf{-4})^2 && \textbf{Compute the power, then multiply.}\\ &= -2(16) = -32 \end{aligned}$$

■

RQ5. In the order of operations, when do we evaluate powers?

It is especially important to follow the order of operations when evaluating an expression.

When we substitute a negative number for the variable, we enclose the negative number in parentheses.

Example 7 Evaluate each expression for $x=-6$.

a. x^2 **b.** $-x^2$ **c.** $2-x^2$ **d.** $(2x)^2$

Solutions a. We replace x by (-6), and then square. This means that -6 gets squared, not just 6. Thus,

$$x^2 = (-6)^2 = (-6)(-6) = 36$$

The parentheses are essential for this calculation. It would be incorrect to write $x^2 = -6^2 = -36$.

b. In this expression, only x is squared. The negative sign is applied to the result.

$$-x^2 = -(-6)^2 = -36$$

c. We replace x by (-6) to get

$$\begin{aligned} 2-x^2 &= 2-(-6)^2 && \textbf{Compute the power first.}\\ &= 2-36 = -34 \end{aligned}$$

d. When we replace x by (-6), for clarity we also change the existing parentheses to brackets.

$$\begin{aligned} (2x)^2 &= [2(-6)]^2 && \textbf{Multiply inside brackets first.}\\ &= [-12]^2 = 144 \end{aligned}$$

■

Like Terms

In Chapter 2 we learned how to add or subtract like terms. For example,

$$8x - 3x = 5x$$

but $\quad 8x - 3y \quad$ cannot be simplified

We can also combine like powers of the same variable. For instance,

$$8x^2 - 3x^2 = 5x^2$$

Look Closer: Notice that when we add like terms, we do not alter the exponent; only the coefficient of the power changes. Can we add different powers of the same variable? The answer to this question is No. For example,

$8x^2 - 3x \quad$ cannot be simplified

For most values of x, the numbers x and x^2 are different. Thus, $8x^2$ and $3x$ are not like terms, and they cannot be combined. •

Example 8 Combine like terms where possible.

a. $-6a^3 + 10a^3$ **b.** $5w^2 + 3w^3$

Solutions a. The exponents on the two terms are the same, so they can be combined. We add the coefficients, $-6 + 10 = 4$, and leave the powers unchanged:

$$-6a^3 + 10a^3 = 4a^3$$

b. The exponents on the terms are different, so they are not like terms. They cannot be combined.

Caution! When adding like terms, we do not add the exponents. For example,

$4x^2 + 3x^2 = 7x^4 \quad$ is incorrect.

RQ6. How do we combine like terms?

Formulas

Many useful formulas involve exponents. The figure below shows several solid objects, along with formulas for their volumes and surface areas.

- **Volume** is the amount of space contained within a three-dimensional object. It is measured in cubic units, such as cubic feet or cubic centimeters.
- **Surface area** is the sum of the areas of all the faces or surfaces that contain a solid. It is measured in square units.

Sphere	Cylinder	Cone	Pyramid

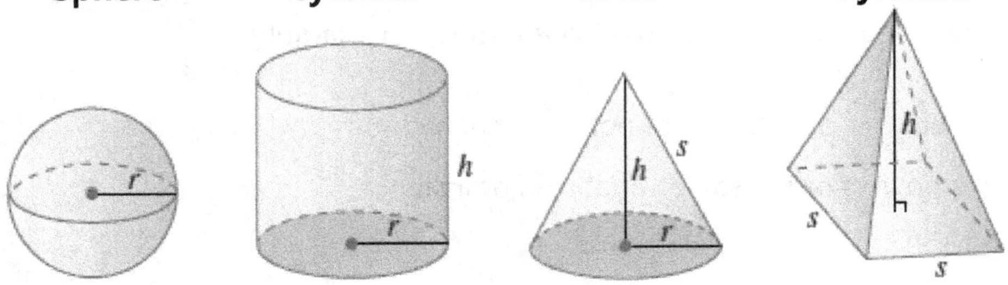

Sphere	Cylinder	Cone	Pyramid
$V = \frac{4}{3}\pi r^3$	$V = \pi r^2 h$	$V = \frac{1}{3}\pi r^2 h$	$V = \frac{1}{3}s^2 h$
$S = 4\pi r^2$	$S = 2\pi r^2 + 2\pi rh$	$S = \pi r^2 + \pi rs$	

Example 9 At the Red Deer Pub and Microbrewery there is a spherical copper tank in which beer is brewed. If the tank is 4 feet in diameter, how much beer does it hold?

Solution The formula for the volume of a sphere is $V = \frac{4}{3}\pi r^3$, where r is the radius of the sphere. The tank has diameter 4 feet, so its radius is 2 feet. We substitute $r = 2$ into the formula, and simplify, following the order of operations. (If your calculator does not have a key for π, you can use the approximation $\pi \approx 3.14$.)

$$
\begin{aligned}
V &= \frac{4}{3}\pi(2)^3 && \textbf{Compute the power first.}\\
&= \frac{4\pi(8)}{3} && \textbf{Multiply by } 4\pi\textbf{, divide by 3.}\\
&= 33.51...
\end{aligned}
$$

The tank holds approximately 33.5 cubic feet of beer, or about 251 gallons. ■

RQ7. What are the units of volume?
RQ8. What are the units of surface area?

Solving for One Variable

In the formula in Example 10, the variable P occurs more than once. We substitute the value for P each time it occurs.

Example 10 If a principal of P dollars is invested at interest rate r, the formula $A = P + Prt$ gives the amount of money in an account after t years. (The "amount," A, is the original principal plus the interest earned.) How long will it take an investment of \$5000 to grow to \$7000 in an account that pays $6\frac{1}{2}$% interest?

Solution We substitute 7000 for A, 5000 for P, and 0.065 for r in the formula, and then solve the resulting equation for t.

$$
\begin{aligned}
7000 &= 5000 + 5000\,(0.065)\,t && \textbf{Simplify the right side.}\\
2000 &= 5000 + 325\,t && \textbf{Subtract 5000 from both sides.}\\
2000 &= 325t && \textbf{Divide both sides by 325.}\\
6.15 &\approx t
\end{aligned}
$$

However, the answer to the question in the problem is not 6.15 years! Because interest is paid only once a year, the investor must wait until the end of the seventh year before the account contains at least \$7000. ■

Caution! To simplify the right side of the equation

$$7000 = 5000 + 5000(0.65)t$$

it is *not* correct to add the two 5000's and write $7000 = 10,000(0.65)t$. Remember that multiplication comes before addition in the order of operations.

If we plan to use a formula more than once with different values for the variables, it may be faster in the long run to solve for one of the variables in terms of the others. We have already done something like this when put a linear equation into slope-intercept form: in that situation we solved for y in terms of x.

Example 11 Solve the formula $A = P + Prt$ for t.

Solution For this problem, we treat t as the unknown and treat all the other variables as if they were constants. We begin by isolating the term containing t on one side of the equation.

$$A = P + Prt \qquad \textbf{Subtract } P \textbf{ from both sides.}$$

$$A - \boldsymbol{P} = P + Prt - \boldsymbol{P}$$

$$A - P = Prt \qquad \textbf{Divide both sides by } Pr.$$

$$\frac{A-P}{\boldsymbol{Pr}} = \frac{Prt}{\boldsymbol{Pr}}$$

$$\frac{A-P}{Pr} = t$$

We now have a new formula for t in terms of the other variables.

Caution! In the expression $P + Prt$ in Example 11, the terms P and Prt are not like terms. They cannot be combined.

Skills Warm-Up

Follow the order of operations to simplify each expression.

1. $-2[-3(-5) - 8(4)]$

2. $[-8 + 6(-4)(-3)][5 - (-2)]$

3. $\frac{4}{3}(-6)(6-9)(6-9)$

4. $\frac{3}{8}(-7-5)(-4-4)$

5. $-2.4(-3) + (8-4.5)(-7.2)$

6. $-9.6 - 3.2(-8-2.4)(-3)$

Answers: **1.** 34 **2.** 448 **3.** –72 **4.** –36 **5.** –18 **6.** –109.44

Homework 5.1

Skills Practice

Compute the powers in Problems 1-2.

1a. 4^3 **b.** 5^3 **c.** 5^4 **2a.** $\left(\frac{2}{3}\right)^4$ **b.** $\left(\frac{4}{5}\right)^3$ **c.** $\left(\frac{11}{9}\right)^2$

3. Use a calculator to compute the powers. Round your answers to the nearest hundredth.
a. $(3.1)^3$ **b.** $(2.6)^4$ **c.** $(0.8)^4$

For Problems 4-5, simplify.

4a. -5^2 **b.** -5^3 **c.** $(-5)^2$ **d.** $(-5)^3$
5a. $-(-2)^2$ **b.** $-(-2)^3$ **c.** $-2^3 - 2^2$ **d.** $-(2^3 - 2)^2$

6. Evaluate for $x = -2$.
a. $5x^3$ **b.** $5x^2$ **c.** $5 - x^2$ **d.** $5 - x^3$

7. Evaluate for $a = -3,\ b = -4$.
a. ab^3 **b.** $a - b^3$ **c.** $(a - b^2)^2$ **d.** $ab(a^2 - b^2)$

Simplify each pair of expressions in Problems 8-11.

8. a. $x + x + x$
b. $x \cdot x \cdot x$

9. a. $2a \cdot a$
b. $2a + a$

10. a. $-q - q - q$
b. $-q(-q)(-q)$

11. a. $-3m - m$
b. $(-3m)(m)$

In Problems 12-15, one of the two statements is true for all values of x, and the other is not. By trying some values of x, decide which statement is true.

12. a. $x + x = 2x$
b. $x + x = x^2$

x	$x+x$	$2x$	x^2
3			
5			
−4			
−1			

13. a. $x \cdot x = 2x$
b. $x \cdot x = x^2$

x	$x \cdot x$	$2x$	x^2
4			
6			
−3			
−1			

14. a. $x^2 + x^2 = x^4$
b. $x^2 + x^2 = 2x^2$

x	x^2+x^2	x^4	$2x^2$
2			
3			
−2			
−1			

15. a. $x + x^2 = x^3$
b. $x \cdot x^2 = x^3$

x	$x+x^2$	$x \cdot x^2$	x^3
1			
4			
−3			
−1			

■ Using what you learned in Problems 12-15, simplify the expressions in Problems 16-21 if possible, or state "cannot be simplified."

16. $5a^2 - 7a^2$ **17.** $3t - 2t^2$ **18.** $-m^2 - m^2$
19. $3k(k)$ **20.** $3k + 4k^2$ **21.** $3k^2 + 4k^2$

■ In Problems 22-24, simplify by combining like terms.

22. $6b^3 - 2b^3 - (-8b^3)$ **23.** $(2y^3 - 4y^2 - y) + (6y^2 + 2y + 1)$
24. $(5x^3 + 3x^2 - 4x + 8) - (2x^3 - 4x - 3)$

■ In Problems 25-28, explain why the calculation is **incorrect**, and give the correct answer.

25. $6w^3 + 8w^3 \rightarrow 14w^6$ **26.** $6 + 3x^2 \rightarrow 9x^2$
27. $4t^2 + 7 - (3t^2 - 5) \rightarrow t^2 + 2$ **28.** $5b^2 - 3b \rightarrow 2b$

■ For Problems 29-32, simplify mentally, without using pencil, paper, or calculator!

29. a. Multiply 3.5 by 100.
b. Multiply 3.5 by 1000.
c. Multiply 3.5 by 10,000.

30. a. 0.074×10^2 **b.** 0.074×10^3 **c.** 0.074×10^4
31. a. 24×10^2 **b.** 8.91×10^5 **c.** 0.003×10^4
32. a. $3 \cdot 10 + 9$ **b.** $2 \cdot 10^2 + 3 \cdot 10 + 4$
c. $3 \cdot 10^3 + 4 \cdot 10^2 + 5 \cdot 10 + 6$

Applications

■ For Problems 33-36, translate into an algebraic expression, then simplify. Round to two decimal places if necessary.

33. The square of the sum of 3 and 4
34. 5 more than x to the third power
35. 25% of the cube of h
36. The square of a added to the square of b

■ Problems 37-40 involve an area, a circumference, or a volume. Decide which measure is appropriate, and then solve the problem.

37. A circular pizza pan has a diameter of 14 inches. How much pizza dough is needed to cover it?

38. To find the diameter of a large tree in his yard, Delbert wraps a string around the trunk and then measures its length. If the string is 72.25 inches long, what is the diameter of the tree?

39. The first solo transatlantic balloon crossing was completed in 1984 in a helium-filled balloon called Rosie O'Grady. The diameter of the balloon was 58.7 feet. Assuming that the balloon was approximately spherical, calculate its volume.

40. To calculate the area of a large circular fish pond, Graham measures the distance around the edge of the pond as 75.4 feet. Calculate the area of the pond.

■ In Problems 41-42, use the formulas for volume and surface area.

41. The dome on the new planetarium is a hemisphere (half a sphere) of radius 40 feet.
- **a.** Draw a sketch of the dome, and label its radius.
- **b.** How much space is enclosed within the dome?
- **c.** What is the surface area of the dome?

42.
- **a.** How much sheet steel is needed to make a cylindrical can with radius 5 centimeters and height 16 centimeters? (**Hint:** Should you calculate the volume or the surface area of the can?)
- **b.** How long is a cylindrical section of concrete pipe if its diameter is 3 feet and its surface area is approximately 848.25 square feet? Note: a pipe does not have ends.
- **c.** Draw a sketch of the pipe in part (b), and label its dimensions.

■ For Problems 43-48, solve each formula for the variable indicated.

43.	$v = lwh$	for w	**44.**	$E = \dfrac{mv^2}{2}$	for m
45.	$A = \dfrac{h}{2}(b + c)$	for h	**46.**	$F = \dfrac{9}{5}C + 32$	for C
47.	$A = \pi rh + 2\pi r^2$	for h	**48.**	$\dfrac{x}{a} + \dfrac{y}{b} = 1$	for x

49. Myra sells mugs at the sidewalk fair every week. Her revenue from selling x mugs is $12x - 0.3x^2$ dollars, and the cost of producing x mugs is $50 + 3x$ dollars.
- **a.** Write an expression for the profit Myra earns from selling x mugs.
- **b.** Find Myra's profit from the sale of 10 mugs, from 15 mugs, and from 20 mugs.

50. The owner of the Koffee Shop pours the remainder of her old house blend, which is 30% Colombian beans, into a 50-pound bin, and fills it up with her new house blend, which is 25% Colombian beans. Let h stand for the number of pounds of the old house blend. Write algebraic expressions to answer the questions below.
- **a.** How many pounds of Colombian beans are in the old house blend?
- **b.** How many pounds of new blend does she pour into the bin?
- **c.** How many pounds of Colombian beans are in this amount of new house blend?
- **d.** How many pounds of Colombian beans are in the 50-pound bin?

For Problems 51-52, recall that two algebraic expressions are called **equivalent** if they are equal for every value of their variables.

51. a. Explain why the expressions $3z^2$ and $(3z)^2$ are not equivalent.
b. Explain why the expressions $(3z)^2$ and $9z^2$ are equivalent.

52. a. Find two values of x for which $2x = x^2$.
b. Find four values of x for which $2x \neq x^2$.
c. Is $2x$ equivalent to x^2?

5.2 Square Roots and Cube Roots

Suppose we would like to draw a square whose area is 25 square inches. How long should each side of the square be? Because the formula for the area of a square is $A = s^2$, the side s should satisfy the equation $s^2 = 25$. We want a number whose square is 25. As you can probably guess, the length of the square should be 5 inches, because $5^2 = 25$. We say that 5 is a **square root** of 25.

> **Square Root**
> The number s is called a **square root** of a number b if $s^2 = b$.

> Finding a square root of a number is the opposite of squaring a number.

Example 1 Find a square root of 25, and a square root of 144.

Solution 5 is a square root of 25 because $5^2 = 25$
12 is a square root of 144 because $12^2 = 144$ ■

Look Closer: 5 is not the only square root of 25, because $(-5)^2 = 25$ as well. Thus, 25 has two square roots, 5 and −5.

> Every positive number has two square roots, one positive and one negative.

> **RQ1.** What is the square root of a number n?

Radicals

The positive square root of a number is called the **principal square root**. The symbol $\sqrt{}$ denotes the positive or principal square root. Thus, we may write

$$\sqrt{25} = 5 \quad \text{and} \quad \sqrt{144} = 12$$

The symbol $\sqrt{}$ is called a **radical** sign, and the number inside is called the **radicand**. Square roots are often called **radicals**.

What about the other square root, the negative one? If we want to indicate the negative square root of a number, we place a negative sign outside the radical sign, like this:

$$-\sqrt{16} = -4 \quad \text{and} \quad -\sqrt{49} = -7$$

If we want to refer to both square roots, we use the symbol $\pm$, read plus or minus. For example,

$$\pm\sqrt{36} = \pm 6, \quad \text{which means} \quad 6 \text{ or } -6$$

Note that zero has only one square root: $\sqrt{0} = 0$.

Example 2 Find each square root.

a. $-\sqrt{81} = -9$ **b.** $\pm\sqrt{\dfrac{64}{121}} = \pm\dfrac{8}{11}$ ■

RQ2. What is the positive square root of a number called?
RQ3. What is a radicand?

Every positive number has two square roots, and zero has exactly one square root. What about the square root of a negative number? For example, can we find $\sqrt{-4}$? The answer is No, because the square of any number is positive (or zero). Try this yourself: The only reasonable candidates for $\sqrt{-4}$ are 2 and −2, but

$$2^2 = \_\_\_\_\_\_\_\_\_\_$$
$$(-2)^2 = \_\_\_\_\_\_\_\_\_\_$$

We cannot find the square root of a negative number. We say that the square root of a negative number is **undefined**.

RQ4. How do we find the square root of a negative number?

Rational and Irrational Numbers

A **rational number** is one that can be expressed as a quotient (or ratio) of two integers, where the denominator is not zero.

The term "rational" has nothing to do with being reasonable or logical; it comes from the word ratio. Thus, any fraction such as

$$\frac{2}{3}, \quad \frac{-4}{7}, \quad \text{or} \quad \frac{15}{6}$$

is a rational number. Integers are also rational numbers, because any integer can be written as a fraction with a denominator of 1. (For example, $6 = \frac{6}{1}$.) All of the numbers we have encountered before this chapter are rational numbers.

RQ5. What is a rational number?

Every fraction can be written in decimal form.

Decimal Form of a Rational Number
The decimal representation of a rational number has one of two forms.
1. The decimal representation terminates, or ends.
2. The decimal representation repeats a pattern.

Example 3 Write the decimal form for each rational number.

a. $\frac{3}{4}$ **b.** $\frac{4}{11}$

Solutions a. We can use a calculator or long division to divide the numerator by the denominator to find $\frac{3}{4} = 0.75$.
b. We divide 4 by 11 to find $\frac{4}{11} = 0.363636... = 0.\overline{36}$. The line over the digits 36 is called a repeater bar, and it indicates that those digits are repeated forever.

RQ6. How can you recognize a rational number in its decimal form?

What about the decimal form for $\sqrt{5}$? If you use a calculator with an eight-digit display, you will find

$$\sqrt{5} \approx 2.236068$$

However, this number is only an approximation, and not the exact value of $\sqrt{5}$. (Try squaring 2.236068 and you will see that

$$2.236068^2 = 5.000000100624$$

which is not exactly 5, although it is close.) In fact, no matter how many digits your calculator or computer can display, you can never find an exact decimal equivalent for $\sqrt{5}$. $\sqrt{5}$ is an example of an irrational number.

An **irrational number** is one that cannot be expressed as a quotient of two integers.

Look Closer: There is no terminating decimal fraction that gives the exact value of $\sqrt{5}$. The decimal representation of an irrational number never ends, and does not repeat any pattern! The best we can do is round off the decimal form and give an approximate value. Nonetheless, an irrational number still has a precise location on the number line, just as a rational number does. The figure below shows the locations of several rational and irrational numbers on a number line.

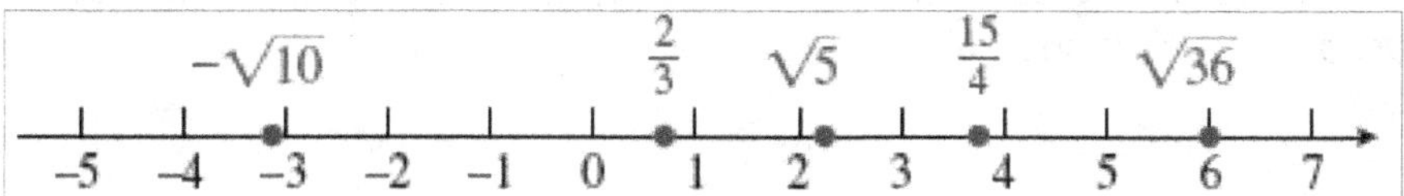

Each point on a number line corresponds either to a rational number or an irrational number, and these numbers fill up the number line completely. The rational and irrational numbers together make up the **real numbers**, and the number line is sometimes called the **real line**.

RQ7. What are the rational and irrational numbers together called?

It is important that you understand the distinction between an exact value and an approximation.

■ **Example 4a.** We cannot write down an exact decimal equivalent for an irrational number.

$\sqrt{5}$ indicates the exact value of the square root of 5
2.236068 is an approximation to the square root of 5

b. We often use a decimal approximation for a rational number.

$\frac{2}{3}$ indicates the exact value of 2 divided by 3

0.666667 is an approximation for $\frac{2}{3}$ ■

Of course, even though many radicals are irrational numbers, some radicals, such as $\sqrt{16} = 4$ and $\sqrt{\frac{9}{25}} = \frac{3}{5}$, represent integers or fractions. Integers such as 9 and 25, whose square roots are whole numbers, are called **perfect squares**.

Order of Operations

When we evaluate algebraic expressions that involve radicals, we must follow the order of operations as usual. Square roots occupy the same position as exponents in the order of operations: They are computed after parentheses but before multiplication.

Example 5 Find a decimal approximation to three decimal places for $8 - 2\sqrt{7}$

Solution You may be able to enter this expression into your calculator just as it is written. If not, you must enter the operations in the proper order. The expression has two terms, 8 and $-2\sqrt{7}$, and the second term is the product of $\sqrt{7}$ with -2. We should not begin by subtracting 2 from 8, because multiplication precedes subtraction. First, we find an approximation for $\sqrt{7}$:

$$\sqrt{7} \approx 2.6457513$$

Do not round off your approximations at any intermediate steps in the problem or you will lose accuracy at each step! You should be able to work directly with the value on your calculator's display.

Next, we multiply our approximation by -2 to find

$$-2\sqrt{7} \approx -5.2905026$$

Finally,we add 8 to get

$$8 - 2\sqrt{7} \approx 2.7084974$$

Rounding to three decimal places gives 2.708.

RQ8. When do we evaluate roots in the order of operations?

Caution! In Example 5, it is not true that $8 - 2\sqrt{7}$ is equal to $6\sqrt{7}$. The order of operations tells us that we must perform the multiplication $2\sqrt{7}$ first, then subtract the result from 8. You can verify that $6\sqrt{7} \approx 15.874$, which is not the same answer we got in Example 5.

We can now update the order of operations by modifying Steps 1 and 2 to include radicals.

Order of Operations

1. Perform any operations inside parentheses, under a radical, or above or below a fraction bar.
2. Compute all indicated powers and roots.
3. Perform all multiplications and divisions in the order in which they occur from left to right.
4. Perform additions and subtractions in order from left to right.

Cube Roots

Imagine a cube whose volume is 64 cubic inches. What is the length, c, of one side of this cube? Because the volume of a cube is given by the formula $V = c^3$, we must find a number that satisfies

$$c^3 = 64$$

We are looking for a number c whose cube is 64. With a little trial and error we can soon discover that $c = 4$. The number c is called the **cube root** of 64, and is denoted by $\sqrt[3]{64}$.

> **Cube Root**
> The number c is called a **cube root** of a number b if $c^3 = b$.

Example 6a. $\sqrt[3]{-64} = -4$ because $(-4)^3 = -64$

b. $\sqrt[3]{9}$ is an irrational number approximately equal to 2.08, because $2.08^3 = 8.998912$

Recall that every positive number has two square roots, and that negative numbers do not have square roots. The situation is different with cube roots.

> Every number has exactly one cube root. The cube root of a positive number is positive, and the cube root of a negative number is negative.

Just as with square roots, some cube roots are irrational numbers and some are not. Cube roots are treated the same as square roots in the order of operations.

> **RQ9.** What is the cube root of a number n?
> **RQ10.** How many cube roots does a negative number have?

Skills Warm-Up

Solve each equation for x.

1. $3x + 5 = 17$
2. $3x + k = 17$
3. $\dfrac{x}{4} - 9 = -4$
4. $\dfrac{x}{4} - m = -4$
5. $2x - 3 = 4x + 7$
6. $2x - c = 4x + 7$
7. $6x + 1 = 3(2 - x)$
8. $bx + 1 = 3(2 - x)$

Answers: **1.** 4 **2.** $\dfrac{17-k}{3}$ **3.** 20 **4.** $4(-4+m)$ **5.** –5 **6.** $\dfrac{-c-7}{2}$ **7.** $\dfrac{5}{9}$ **8.** $\dfrac{5}{b+3}$

Homework 5.2

Skills Practice

For Problems 1-3, simplify. Do not use a calculator!

1a. $4 - 2\sqrt{64}$ **b.** $\dfrac{4-\sqrt{64}}{2}$

2a. $\sqrt{9-4(-18)}$ **b.** $\sqrt{\dfrac{4(50)-56}{16}}$

3a. $5\sqrt[3]{8} - \dfrac{\sqrt[3]{64}}{8}$ **b.** $\dfrac{3+\sqrt[3]{-729}}{6-\sqrt[3]{-27}}$

For Problems 4-9, give a decimal approximation rounded to thousandths.

4. $5\sqrt{3}$ **5.** $\dfrac{-2}{3}\sqrt{21}$ **6.** $-3+2\sqrt{6}$

7. $2+6\sqrt[3]{-25}$ **8.** $\dfrac{8-2\sqrt{2}}{4}$ **9.** $3\sqrt{3}-3\sqrt{5}$

In Problems 10-12, choose the best approximation. Do not use a calculator!

10. $\sqrt{72}$ **a.** 7 **b.** 8 **c.** 36 **d.** 64

11. $\sqrt{13}$ **a.** 6 **b.** 6.5 **c.** 3.5 **d.** 4

12. $\sqrt{134}$ **a.** 11 **b.** 11.5 **c.** 12 **d.** 15

13. Each number below is approximately the square root of a whole number. Find the whole number.
a. 9.220 **b.** 12.961 **c.** 63.891

14. a. Evaluate $\sqrt{x^2}$ for $x = 3, 5, 8,$ and $12.$
b. Simplify $\sqrt{a^2}$ if $a \geq 0.$

For Problems 15-18, use the definitions of square root and cube root to simplify each expression. Do not use a calculator!

15a. $(\sqrt{16})(\sqrt{16})$ **b.** $\sqrt{29}\,(\sqrt{29})$ **c.** $\sqrt{x}\,(\sqrt{x})$

16a. $(\sqrt{7})^2$ **b.** $(\sqrt[3]{20})^3$ **c.** $(\sqrt[3]{4})(\sqrt[3]{4})(\sqrt[3]{4})$

17a. $\dfrac{6}{\sqrt{6}}$ **b.** $\dfrac{-15}{\sqrt{15}}$ **c.** $\dfrac{2m}{\sqrt{m}}$

18a. $(\sqrt{2b})(\sqrt{2b})$ **b.** $(3\sqrt{a})(3\sqrt{a})$ **c.** $(2\sqrt[3]{b})^3$

For Problems 19-20, put each set of numbers in order from smallest to largest. Try not to use a calculator.

19. $\dfrac{5}{4},\ 2,\ \sqrt{8},\ 2.3$ **20.** $2\sqrt{3},\ 3,\ \dfrac{23}{6},\ \sqrt{6}$

In Problems 21-23, evaluate the expression for the given values of x. Round your answers to three decimal places if necessary.

21. $\sqrt{x^2-4}$ $\quad x=3,\ \sqrt{5},\ -2$

22. $\sqrt{x}-\sqrt{x+3}$ $\quad x=1,\ 4,\ 100$

23. $2x^2-4x$ $\quad x=-3,\ \sqrt{2},\ \frac{1}{4}$

Applications

24. The distance m in miles that you can see on a clear day from a height of h miles is given by the formula $m=89.4\sqrt{h}$. How far can you see from an airplane flying at an altitude of 4.7 miles?

25. The period of a pendulum (the time it takes to complete one full swing) is given in seconds by the formula $T=6.28\sqrt{\frac{L}{32}}$, where L is the length of the pendulum in feet. What is the period of the Foucault pendulum in the United Nations Headquarters in New York, whose length is 75 feet?

26. What is a rational number? An irrational number? Give examples of each.

27. Give three examples of square roots that represent rational numbers, and three that are irrational.

28. Identify each number as rational or irrational.

a. $\sqrt{6}$ **b.** $\frac{-5}{3}$ **c.** $\sqrt{16}$ **d.** $\sqrt{\frac{5}{9}}$ **e.** 6.008 **f.** $3.\overline{23}$

In Problems 29-32, decide whether the two expressions are equivalent by completing the table.

29. Does $(a+b)^2=a^2+b^2$?

a	b	$a+b$	$(a+b)^2$	a^2	b^2	a^2+b^2
2	3					
3	4					
1	5					
−2	6					

30. Does $\sqrt{a^2+b^2}=a+b$?

a	b	$a+b$	a^2	b^2	a^2+b^2	$\sqrt{a^2+b^2}$
3	4					
2	5					
1	6					
−2	−3					

31. Does $\sqrt{a+b} = \sqrt{a} + \sqrt{b}$?

a	b	$a+b$	$\sqrt{a+b}$	$\sqrt{a}$	$\sqrt{b}$	$\sqrt{a}+\sqrt{b}$
2	7					
4	9					
1	5					
9	16					

32. Does $(\sqrt{a} + \sqrt{b})^2 = a + b$?

a	b	$a+b$	$\sqrt{a}$	$\sqrt{b}$	$\sqrt{a}+\sqrt{b}$	$(\sqrt{a}+\sqrt{b})^2$
4	9					
1	4					
3	5					
6	10					

For Problems 33-34,

a. Write an expression for the volume of each box.

b. Write an expression for the surface area of each box. (**Hint:** Each box has six faces: top and bottom, front and back, left and right. Find the areas of all six faces, then add them.)

33.

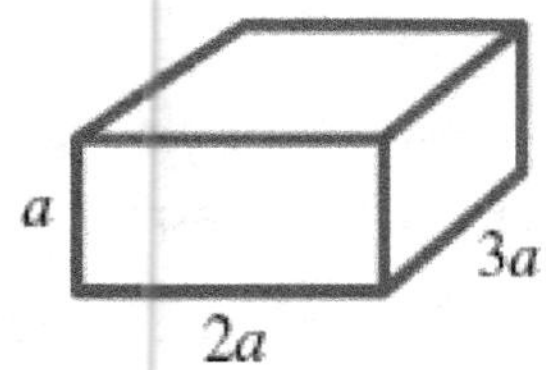

34.

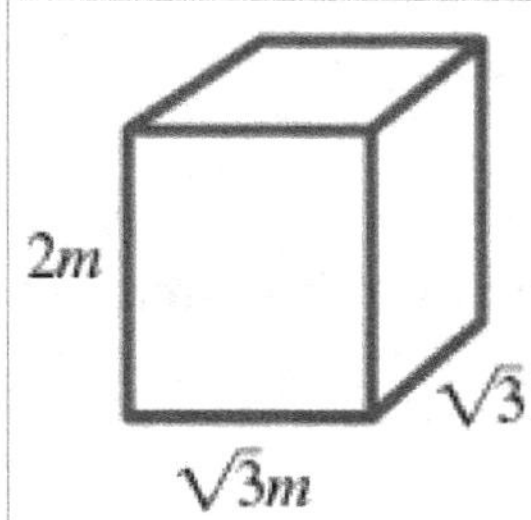

5.3 Quadratic Equations

So far you have learned how to solve linear equations. In linear equations, the variable cannot have any exponent other than 1, and for this reason such equations are often called **first-degree**. In this chapter we'll consider second-degree equations, or **quadratic equations**.

> **Quadratic Equations**
>
> A **quadratic equation** can be written in the **standard form**
>
> $$ax^2 + bx + c = 0$$
>
> where a, b, and c are constants and a is not zero.

Example 1 Some examples of quadratic equations are

$$2x^2 + 5x - 3 = 0, \qquad 7t - t^2 = 0, \qquad \text{and} \qquad 3w^2 = 16$$

The first equation is already written in standard form, with $a = 2$, $b = 5$, and $c = 3$. In standard form, the other two equations are

$$-t^2 + 7t + 0 = 0, \qquad \text{so } a = -1,\ b = 7, \text{ and } c = 0$$
$$3w^2 + 0w - 16 = 0, \qquad \text{so } a = 3,\ b = 0, \text{ and } c = -16$$

The numbers a, b, and c are called **parameters**. They are the coefficients of, respectively, the quadratic term (the x^2-term), the linear term (the x-term), and the constant term.

Caution! The x^2-term and the x-term in a quadratic expression are not like terms, so they cannot be combined.

RQ1. What is a quadratic equation? Give an example.

Graphs of Quadratic Equations

Quadratic equations appear in a variety of applications. A quadratic equation in two variables has the form

$$y = ax^2 + bx + c, \quad a \neq 0$$

Its graph is not a straight line, but a curve called a **parabola**. The simplest, or basic, parabola is the graph of $y = x^2$, shown in the figure.

x	−3	−2	−1	0	1	2	3
y	9	4	1	0	1	4	9

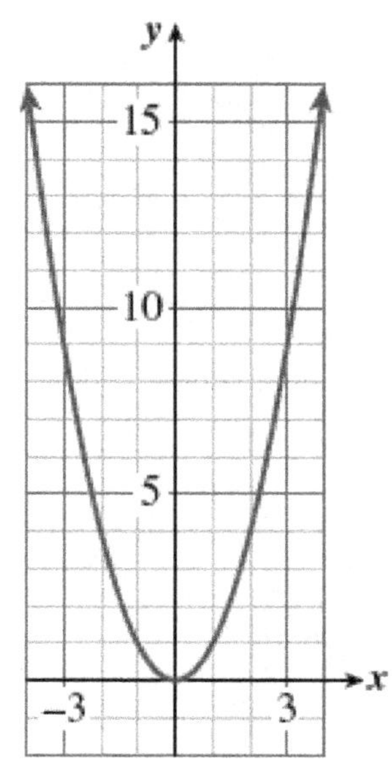

Look Closer: In the equation for the basic parabola, $y = x^2$, we have $a = 1$, $b = 0$, and $c = 0$. If the parameters have different values, the appearance or location of the graph will be altered. Example 2 illustrates a parabola that is shifted downward by 4 units.

Example 2 Graph $y = x^2 - 4$.

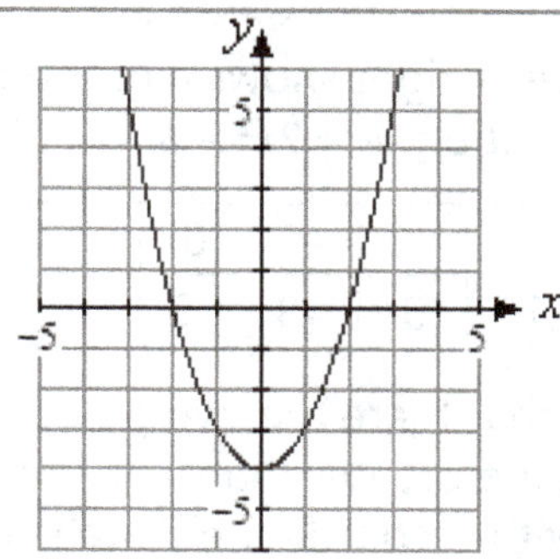

Solution We make a table of values and plot the points. The graph is shown at right.

x	−3	−2	−1	0	1	2	3
y	5	0	−3	−4	−3	0	5

Note that in this equation, $c = -4$.

RQ2. The basic parabola is the graph of what equation?

Solving Quadratic Equations

Consider a simple quadratic equation,

$$x^2 = 16$$

This equation tells us that x is a number whose square is 16. Thus, x must be a square root of 16. Now, 16 has two square roots, 4 and –4, so these are the solutions of the equation.

We can think of the solution process in the following way. Because x is squared in the equation, we perform the opposite operation, or take square roots, in order to solve for x.

$x^2 = 16$ **Take square roots of both sides.**

$x = \pm\sqrt{16}$ **Simplify.**

$x = \pm 4$

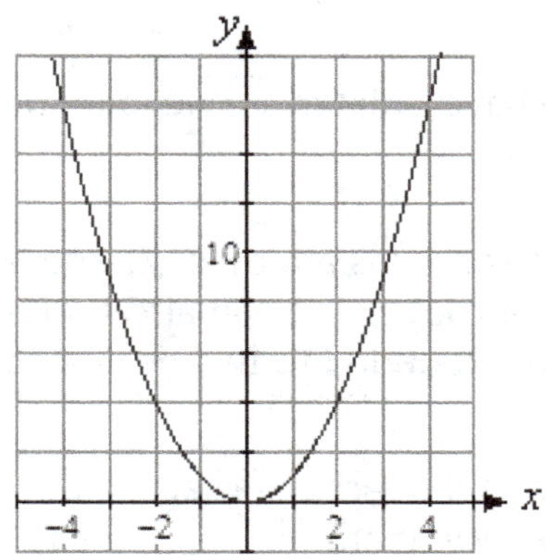

Look Closer: Remember that every positive number has two square roots, so the equation has two solutions, $x = 4$ and $x = -4$. We can also see the two solutions on a graph. The figure at right shows the graph of $y = x^2$ and a horizontal line at $y = 16$. There are two points on the graph with y-coordinate 16, and their x-coordinates are 4 and –4.

Example 3 Solve each equation without using a calculator.

a. $t^2 = 10{,}000$ **b.** $b^2 = \dfrac{4}{9}$ **c.** $h^2 = 0.04$ **d.** $m^2 = (-17)^2$

Solutions For each equation,we find the square root of the right side. (That is, find a number whose square gives the right side.) Remember that every positive number has two square roots.

a. Because $100^2 = 10{,}000,\ t = \pm 100$

b. Because $\left(\frac{2}{3}\right)^2 = \frac{4}{9},\ b = \pm\frac{2}{3}$

c. Because $0.2^2 = 0.04,\ h = \pm 0.2$

d. Because $(-17)^2 = (17)^2,\ m = \pm 17$

RQ3. Why do the equations in Example 3 have two solutions?

Extraction of Roots

The figure at right shows a graph of the quadratic equation $y = 2x^2 - 5$.

x	−3	−2	−1	0	1	2	3
y	13	3	−3	−5	−3	3	13

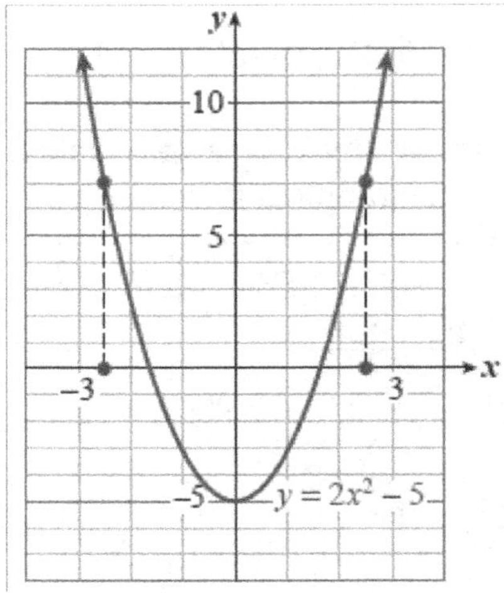

You can see that there are two points on the graph for each y-value greater than −5. For example, the two points with y-coordinate 7 are shown. To solve the quadratic equation

$$2x^2 - 5 = 7$$

we need only find the x-coordinates of these points. From the graph, the solutions appear to be about 2.5 and −2.5.

How can we solve this equation algebraically? We use the fact that taking square roots is the opposite of squaring.

Example 4 Solve the equation $2x^2 - 5 = 7$ algebraically.

Solution We first solve for x^2 as follows.

$$2x^2 - 5 = 7 \qquad \text{Add 5 to both sides.}$$
$$2x^2 = 12 \qquad \text{Divide both sides by 2.}$$
$$x^2 = 6$$

Once we have isolated x^2, we take the square root of each side to find

$$x = \pm\sqrt{6}$$

The exact solutions are thus $\sqrt{6}$ and $-\sqrt{6}$. (Don't forget that every positive number has two square roots.) We can also find decimal approximations for the solutions using a calculator. Rounded to two decimal places, the solutions are 2.45 and −2.45.

RQ4a. What are the exact solutions to the equation $2x^2 - 5 = 7$?
b. What are the approximate values of those solutions, rounded to hundredths?

This method for solving quadratic equations is called **extraction of roots**. The method applies to quadratic equation of the form $ax^2 + c = 0$, where the linear term bx is missing.

Extraction of Roots

To solve a quadratic equation of the form

$$ax^2 + c = 0$$

1. Isolate x^2 on one side of the equation.
2. Take the square root of each side.

RQ5. Explain how to solve the equation $ax^2 + c = 0$.
RQ6. Why does the equation $ax^2 + c = 0$ have two solutions?

Look Closer: In the next Example we compare ***evaluating*** a quadratic expression and ***solving*** a quadratic equation.

Example 5 A cat falls off a tree branch 20 feet above the ground. Its height t seconds later is given by $h = 20 - 16t^2$.

a. What is the height of the cat 0.5 second later?

b. How long does the cat have to get in position to land on its feet before it reaches the ground?

Solutions a. In this question, we are given the value of t and asked to find the corresponding value of h. To do this we evaluate the formula for $t = 0.5$. We substitute 0.5 for t into the formula, and simplify.

$$\begin{aligned} h &= 20 - 16(0.5)^2 && \textbf{Compute the power.} \\ &= 20 - 16(0.25) && \textbf{Multiply, then subtract.} \\ &= 20 - 4 = 16 \end{aligned}$$

The cat is 16 feet above the ground after 0.5 second. You can also use your calculator to simplify the expression for h by entering

20 [−] 16 [×] 0.5 [x^2] [**ENTER**]

b. We would like to find the value of t when the height, h, is known. We substitute $h = 0$ into the equation to obtain

$$0 = 20 - 16t^2$$

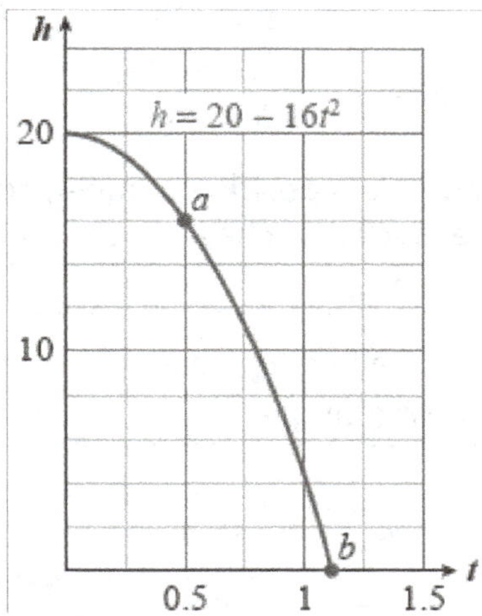

To solve this equation we use extraction of roots. We first isolate t^2 on one side of the equation.

$$\begin{aligned} 16t^2 &= 20 && \textbf{Divide by 16.} \\ t^2 &= \frac{20}{16} = 1.25 \end{aligned}$$

Next, we take the square root of both sides of the equation to find

$$t = \pm\sqrt{1.25} \approx \pm 1.118$$

Only the positive solution makes sense here, so the cat has approximately 1.12 seconds to be in position for landing. A graph of the cat's height after t seconds is shown at right. The points corresponding to parts (a) and (b) are labeled.

RQ7. In Example 5a, did we solve an equation or evaluate an expression?

The Pythagorean Theorem

A familiar example of a quadratic equation is the **Pythagorean theorem.** If we know the lengths of any two sides of a right triangle, we can find the length of the third side.

A triangle in which one of the angles is a right angle, or 90°, is called a **right triangle**. The side opposite the right angle is the longest side of the triangle, and is called the **hypotenuse**. The other two sides of the triangle are called the **legs**. We use the variable c for the length of the hypotenuse, and the lengths of the legs are denoted by a and b (it doesn't matter which is which).

Pythagorean Theorem

If c stands for the length of the hypotenuse of a right triangle, and the lengths of the two legs are represented by a and b, then

$$a^2 + b^2 = c^2$$

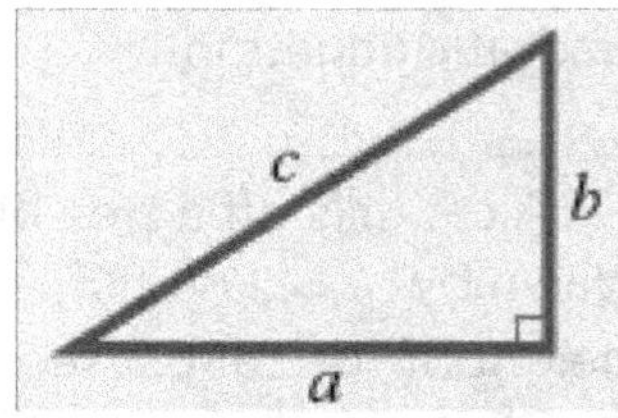

Example 6 A 25-foot ladder is placed against a wall so that its foot is 7 feet from the base of the wall. How far up the wall does the ladder reach?

Solution We make a sketch of the situation, as shown in the figure, and label the dimensions. We'll call the unknown height h. The ladder forms the hypotenuse of a right triangle, so we can apply the Pythagorean theorem, substituting 25 for c, 7 for b, and h for a.

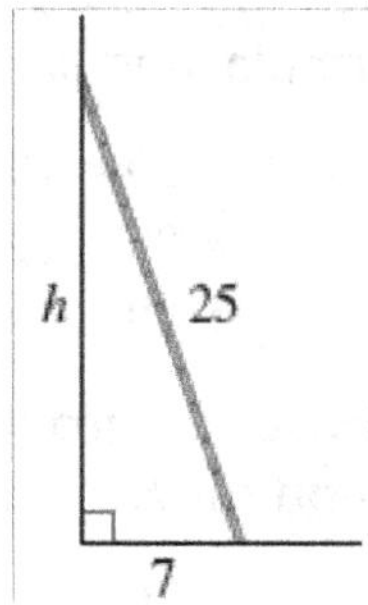

$$a^2 + b^2 = c^2$$
$$h^2 + 7^2 = 25^2$$

We solve this equation by extraction of roots:

$$h^2 + 49 = 625 \qquad \text{Subtract 49 from both sides.}$$
$$h^2 = 576 \qquad \text{Extract roots.}$$
$$h = \pm\sqrt{576} \qquad \text{Simplify the radical.}$$
$$h = \pm 24$$

The height must be a positive number, so the solution -24 does not make sense for this problem. The ladder reaches 24 feet up the wall.

Caution! The Pythagorean theorem holds only for right triangles; it is not true for any other kind of triangle.

RQ8. What is a right triangle, and what is the hypotenuse?
RQ9. What do we use the Pythagorean theorem for?

Geometric Formulas

We can use extraction of roots to solve problems involving volume and surface area.

Example 7 The volume of a can of soup is 582 cubic centimeters, and its height is 10.5 centimeters. What is its radius?

Solution The volume of a cylinder is given by the formula $V = \pi r^2 h$. We substitute 582 for V and 10.5 for h, then solve for r.

$$582 = \pi r^2(10.5) \qquad \text{Divide both sides by } 10.5\pi.$$
$$17.643 = r^2 \qquad \text{Take square roots.}$$
$$4.2 = r$$

The radius of the can is 4.2 centimeters.

RQ10. What are the units of volume in Example 7?

Solving Formulas

Extraction of roots can be used to solve some formulas with a quadratic term.

Example 8 The formula $V = \frac{1}{3}\pi r^2 h$ gives the volume of a cone in terms of its height and radius. Solve the formula for r in terms of V and h.

Solution Because the variable we want is squared, we use extraction of roots. First, we multiply both sides by 3 to clear the fraction.

$$3V = \pi r^2 h \qquad \textbf{Divide both sides by } \pi h.$$

$$\frac{3V}{\pi h} = r^2 \qquad \textbf{Take square roots.}$$

$$\pm\sqrt{\frac{3V}{\pi h}} = r$$

Because the radius of a cone must be a positive number, we use only the positive square root: $r = \sqrt{\frac{3V}{\pi h}}$.

Skills Warm-UP

a. Solve the equation.
b. Write the equation in the form $ax + b = 0$.
c. Graph the equation $y = ax + b$.
d. Find the x-intercept of your graph. Compare with your answer to part (a).

1. $2x + 5 = 11$

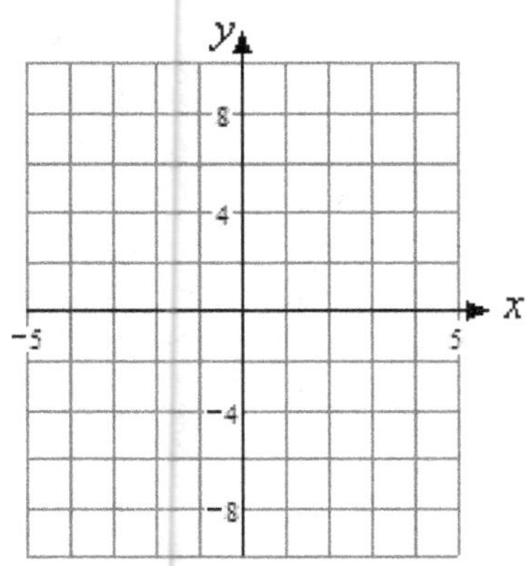

2. $2x - 3 = 5x + 9$

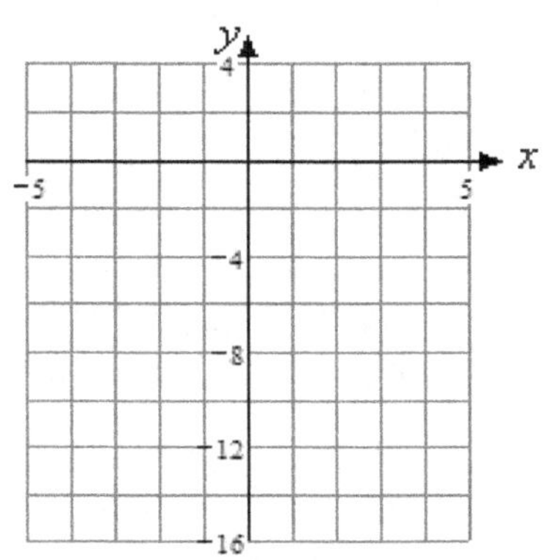

3. $0.7x + 0.2(100 - x) = 0.3(100)$

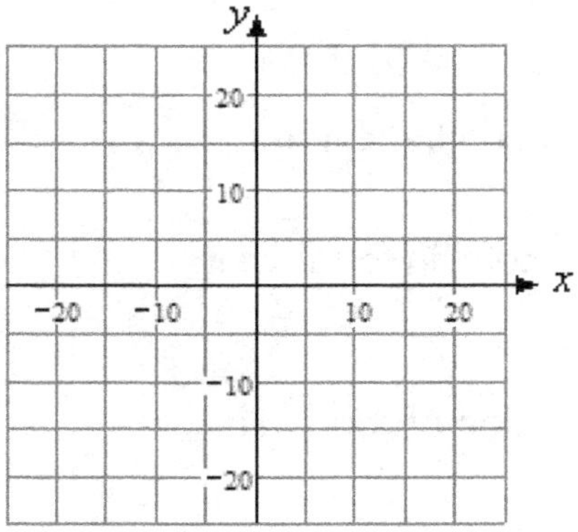

4. $4(7 - 4x) = -2(6x - 5) - 6$

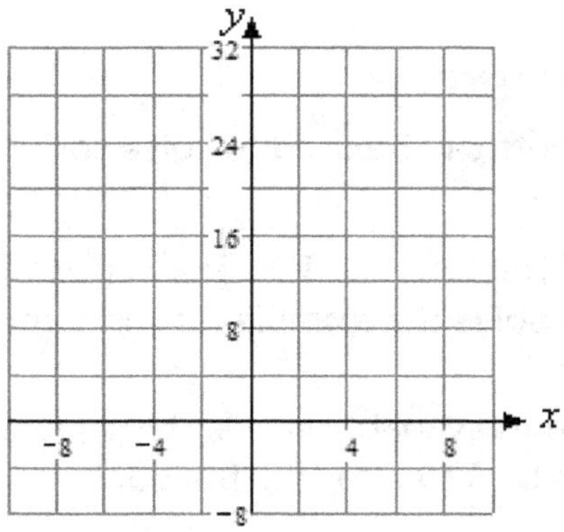

Answers: **1.** $x = 3$ **2.** $x = -4$ **3.** $x = 20$ **4.** $x = 6$

Homework 5.3

Skills Practice

1. Which of the following are quadratic equations?

a. $1 - 4x^2 = 2x$ **b.** $6a - 2a^3 = a^2$ **c.** $2b - 3 = 4b - b^2$

For Problems 2-5, complete the table of values and graph the quadratic equation.

2. $y = x^2 + 4$

x	−3	−2	−1	0	1	2	3
y							

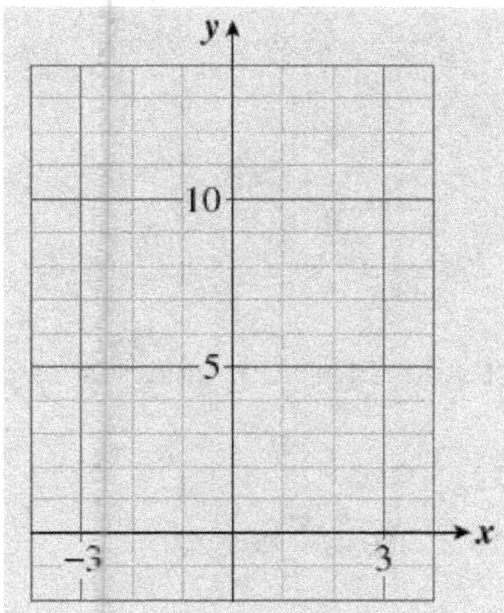

3. $y = x^2 - 4$

x	−3	−2	−1	0	1	2	3
y							

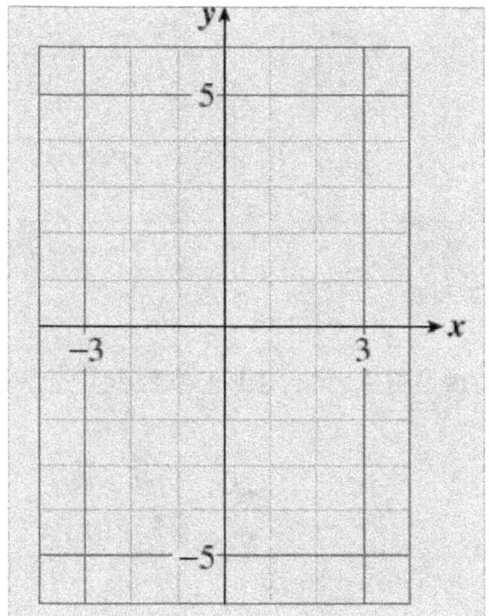

4. $y = 4 - x^2$

x	−3	−2	−1	0	1	2	3
y							

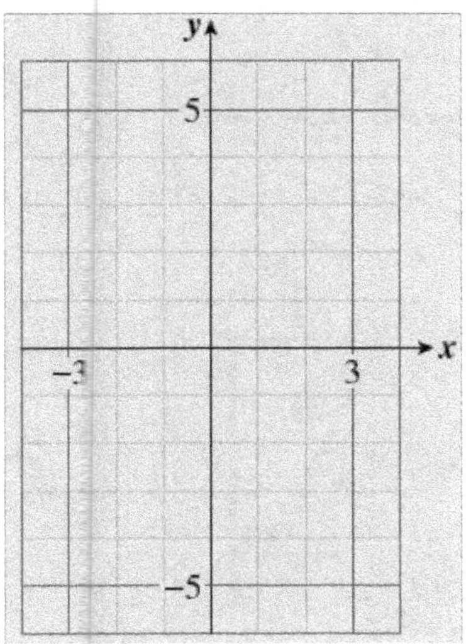

5. $y = \dfrac{x^2}{4}$

x	−4	−2	−1	0	1	2	4
y							

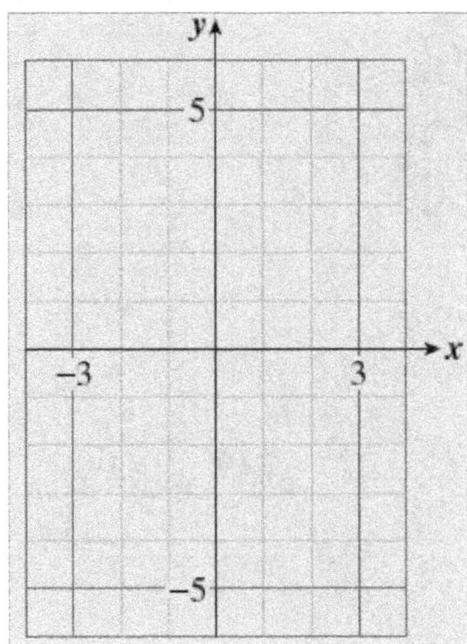

For Problems 6-8,

a. Graph the parabola in part (a).

b. Use the graph to solve the equation in part (b).

6. **a.** $y = \frac{1}{2}x^2$
b. $\frac{1}{2}x^2 = 8$

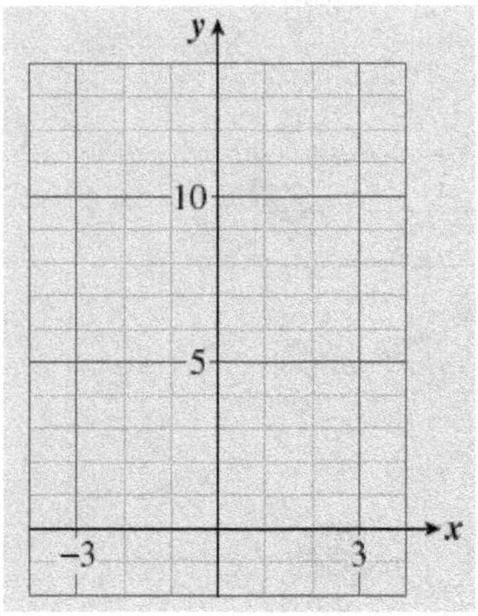

7. **a.** $y = x^2 + 2$
b. $x^2 + 2 = 11$

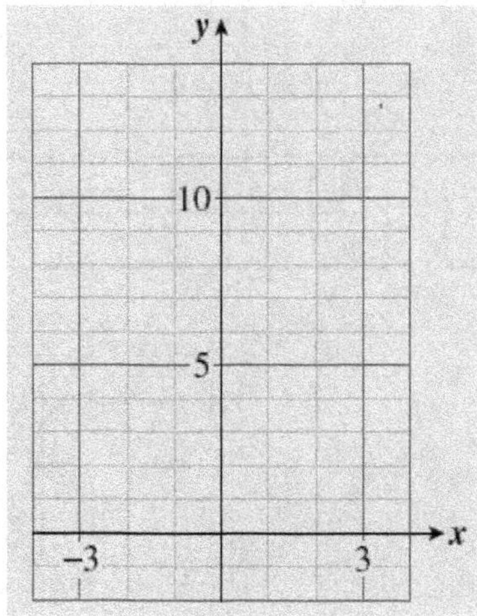

8. **a.** $y = 6 - 2x^2$
b. $6 - 2x^2 = 4$

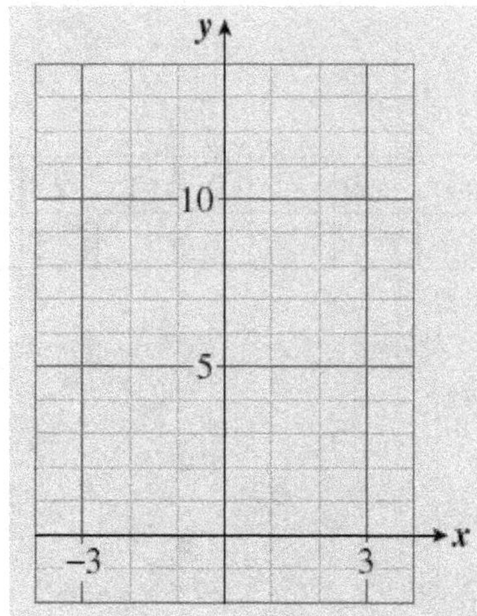

For Problems 9-14, solve by extracting roots. Give exact values for your answers.

9. $x^2 = 121$

10. $98 = 2a^2$

11. $9x^2 = 25$

12. $3g^2 - 54 = 0$

13. $\dfrac{2x^2}{3} = 4$

14. $400 + \dfrac{k^2}{4} = 625$

For Problems 15-20, solve by extracting roots. Round your answers to two decimal places.

15. $2.4m^2 = 126$

16. $55 - 3z^2 = 7$

17. $2x^2 - 200 = x^2 + 25$

18. $3t^2 - 16 = 16t^2$

19. $1.5x^2 = 0.7x^2 + 26.2$

20. $5x^2 - 97 = 3.2x^2 - 38$

Find the unknown side or sides for the right triangles in Problems 21-25.

21.

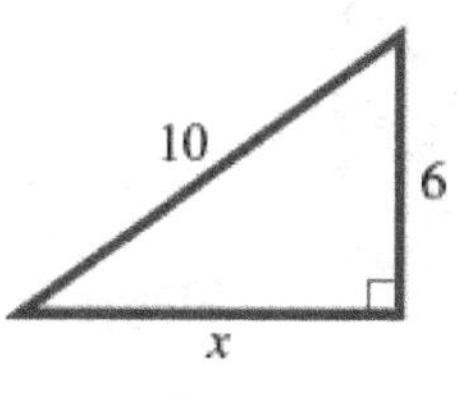

22.

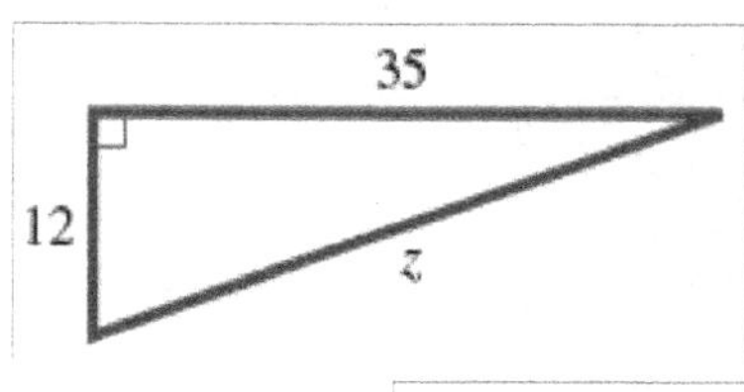

23.

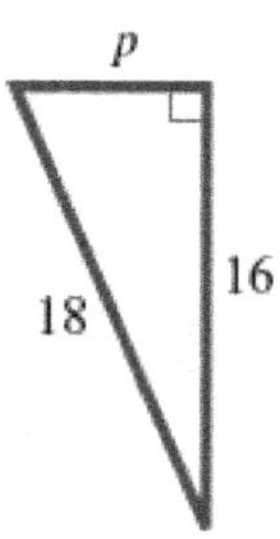

24.

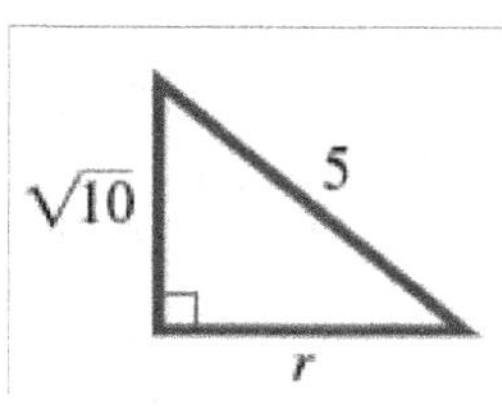

25.

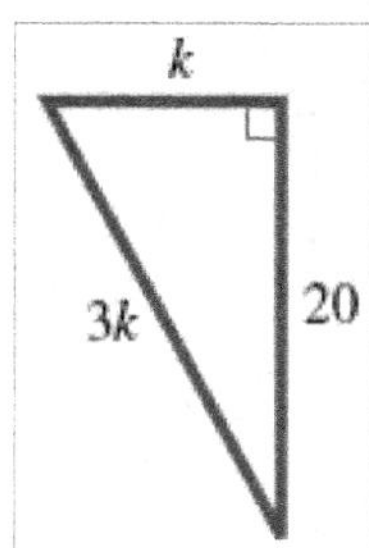

For Problems 26-28, decide whether a triangle with the given sides is a right triangle.

26. 9 in, 16 in, 25 in

27. 5 m, 12 m, 13 m

28. 5^2 ft, 8^2 ft, 13^2 ft

For Problems 29-32, explain why the equation is an incorrect application of the Pythagorean theorem for the figure.

29. 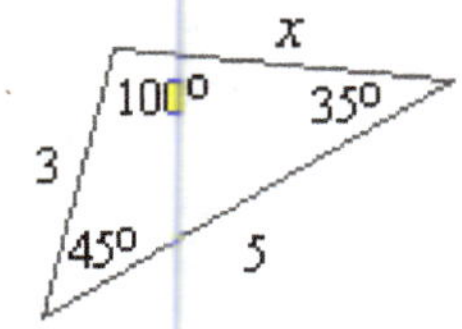

$3^2 + x^2 = 5^2$

30. 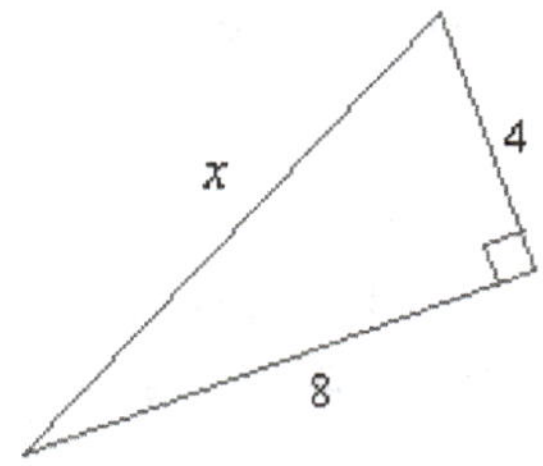

$8 + 4 = x$

31. 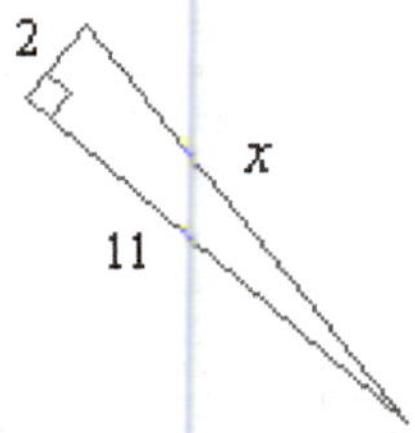

$x^2 + 2^2 = 11^2$

32. 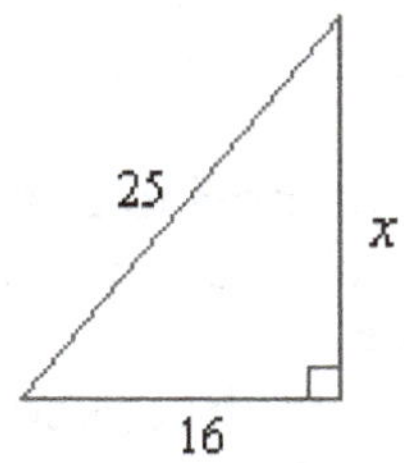

$x^2 + 16 = 25$

For Problems 33-36, solve the formula for the indicated variable.

33. $A = 4\pi r^2$ for r

34. $F = \frac{1}{2}mv^2$ for v

35. $d = 6 + kv^2$ for v

36. $h = 100 - \frac{1}{2}gt^2$ for t

Applications

37. At the Custom Pizza shop you can buy their special smoked chicken pizza in any size you like. The cost, C, of the pizza in dollars is given by the equation

$$C = \frac{1}{4}r^2$$

where r is the radius of the pizza.

a. Complete the table of values and graph the equation on the grid at right.

r	C
0	
1	
2	
4	
6	
9	
10	
14	

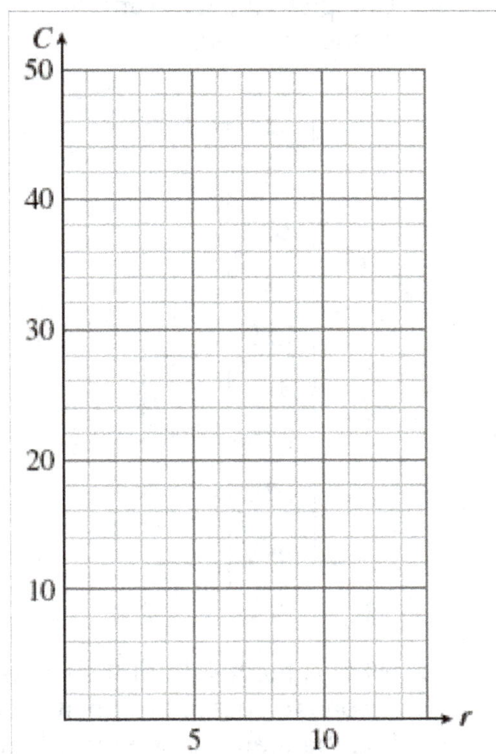

b. How much does a pizza of radius 3 inches cost? Locate this point on your graph.

c. Use your graph to find out how big a pizza you can buy for \$16.

d. What does the point $(6, 9)$ tell you about pizzas?

38. The faster a car moves, the more difficult it is to stop. The distance, d, in meters, required to stop a car traveling at velocity v, in kilometers per hour, is given by

$$d = 0.005v^2$$

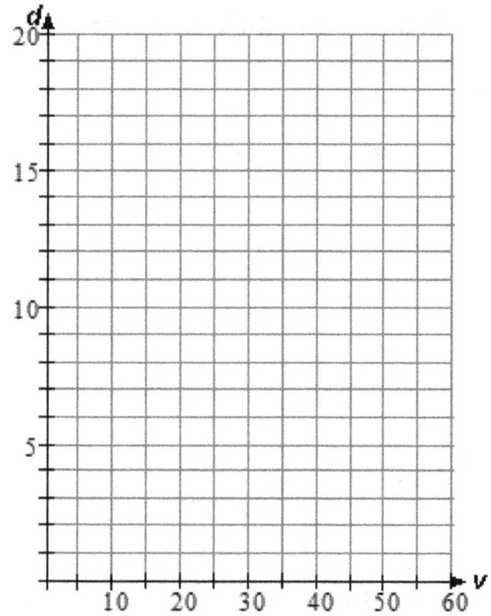

a. Complete the table of values and graph the equation on the grid above right.

Velocity (kph)	5	10	15	20	40	60
Distance (meters)						

b. What distance does a car moving at 40 kilometers per hour require to stop? Locate this point on your graph.

c. Use your graph to estimate the speed of a car that stopped in 12 meters.

d. What does the point $(20, 2)$ tell you about stopping distances?

39. If you drop a stone from a bridge 100 feet above the water, the height h of the stone t seconds after you drop it is given in feet by

$$h = 100 - 16t^2$$

a. Complete the table and sketch a graph of the equation.

t	h
0	
0.5	
1	
1.25	
1.5	
1.75	
2	
2.25	
2.5	

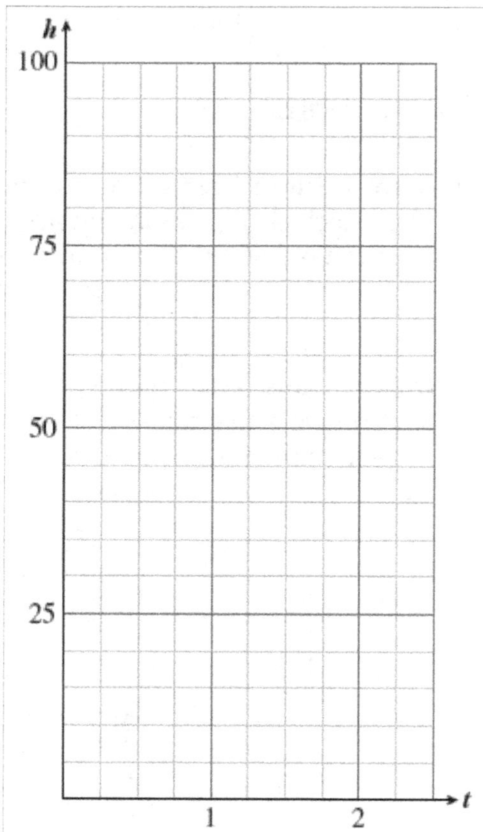

b. What is the height of the stone after 2 seconds? Locate this point on your graph.

c. Use the graph to find out when the height of the stone is 75 feet.

d. Write an equation you could solve to answer the question in part (c).

For Problems 40-41, use the Pythagorean theorem. Give exact answers, and then approximate values rounded to thousandths.

40. a. Find the height of an equilateral triangle of side 6 feet, as shown at below left. (**Hint:** The altitude divides the base into two segments of equal length.)

b. Find the area of the triangle in part (a).

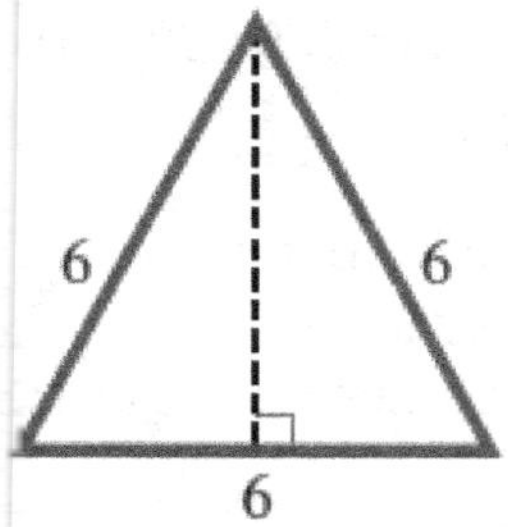

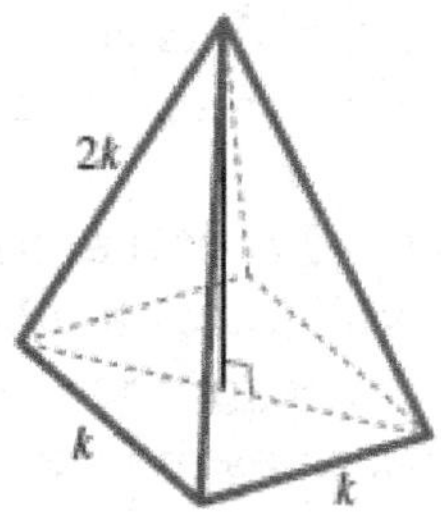

41. a. The pyramid shown above right has a square base. If $k = 2$ centimeters, find the length of the diagonal of the base.

b. Find the height of the pyramid. (**Hint:** The altitude of the pyramid divides the diagonal of the base into two segments of equal length. Use your answer to part (a).)

c. Find the volume of the pyramid.

For Problems 42-47, draw a sketch, write an equation, and solve.

42. Juliet's window is 24 feet above the ground, and there is a 10-foot moat (a ditch filled with water) at the base of the wall. How long a ladder will Romeo need to reach the window?

43. A baseball diamond is a square whose sides are 90 feet long. Find the straight-line distance from home plate to second base.

44. The light house is 5 miles east of Gravelly Point. The marina is 7 miles south of Gravelly Point. How far is it from the light house to the marina?

45. Marlene visited the Quetzalcoatl pyramid near Mexico City last summer. She measured the base and found it is about 1400 feet on each side. She unrolled a ball of string as she climbed the face of the pyramid, and it was about 722 feet to the top. How tall is the pyramid?

46. A surveyor would like to know the distance across a lake. She picks a spot P on a line perpendicular to the width of the lake, and measures the two distances shown in the figure below left. How wide is the lake?

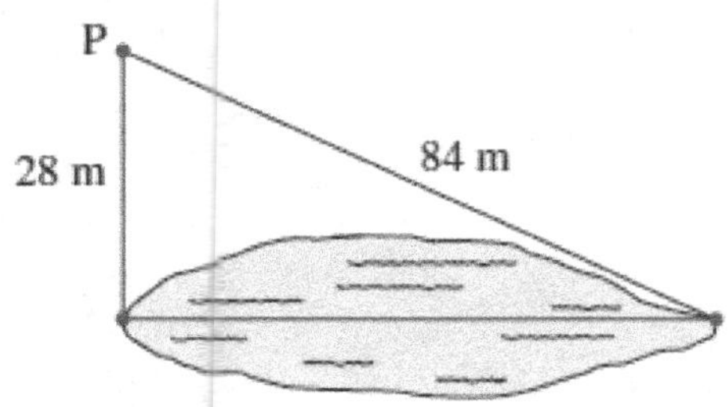

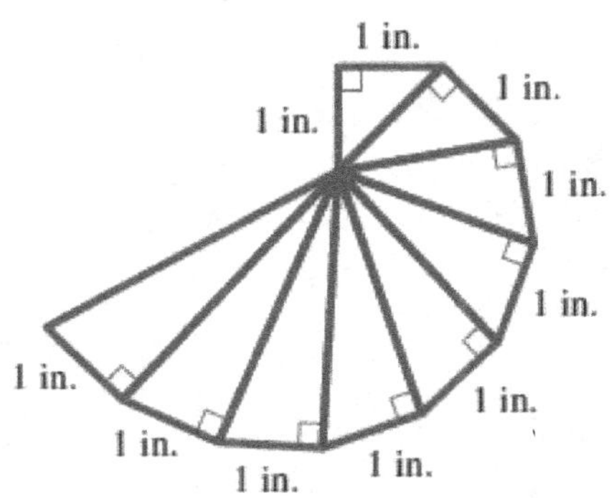

47. Find the length of each hypotenuse in the spiral figure shown above right.

■ In Problems 48-50, use the formulas for volume and surface area.

48. The famous waffle cone at Zanner's Ice Cream holds 500 cubic centimeters (a little over 1 pint) of ice cream. The cone is about 16 centimeters tall. What is the radius of the waffle cone at its top?

49. The largest monument ever constructed is the Quetzalcóatl pyramid southeast of Mexico City. It is 177 feet tall, and its volume is approximately 116,500,000 cubic feet. Approximately how long is one side of its square base?

50. A large bottle of shampoo is 20 centimeters tall and cylindrical in shape.

a. Write a formula for the volume of the bottle in terms of its radius.

b. Complete the table of values for the volume equation. If you halve the radius of the bottle, by what factor does the volume decrease?

r	1	2	3	4	5	6	7	8
V								

c. What radius should the bottle have if it must hold 240 milliliters of shampoo? (A milliliter is equal to one cubic centimeter.)

d. Use your calculator to graph the volume equation. (Use the table values to help you choose an appropriate window.) Locate the point on the graph that corresponds to the bottle in part (c).

5.4 Products of Binomials

In Section 5.1, we learned to simplify the sum of two algebraic expressions by combining like terms. In this section we see how to multiply algebraic expressions.

Products of Variables

Recall that we can simplify a product such as $3(2x)$ because

$$3(2x) = 2x + 2x + 2x = 6x$$

Three terms

This calculation is actually an application of the associative property:

$$3(2x) = (3 \cdot 2)x = 6x$$

In a similar way, we can simplify the product $(3b)(4b)$ by applying the commutative and associative properties:

$$(3b)(4b) = 3 \cdot b \cdot 4 \cdot b = 3 \cdot 4 \cdot b \cdot b = (3 \cdot 4) \cdot (b \cdot b) = 12b^2$$

Look Closer: You can convince yourself that $(3b)(4b)$ is equivalent to $12b^2$ by substituting some values for b; for example, if $b = 2$, then

$$(3 \cdot 2)(4 \cdot 2) = (6)(8) = 48, \quad \text{and} \quad 12(2)^2 = 12 \cdot 4 = 48$$

The commutative and associative properties tell us that we can multiply the factors of a product in any order.

Example 1 Simplify the product or power.

a. $(5a)(-3a)$ **b.** $(2x)^3$ **c.** $(xy^2)(4x^2)$

Solutions a. We apply the commutative property:

$$(5a)(-3a) = 5(-3) \cdot a \cdot a = -15a^2$$

b. To cube an expression means to multiply three copies of the expression together:

$$(2x)^3 = (2x)(2x)(2x) = 2 \cdot 2 \cdot 2 \cdot x \cdot x \cdot x = 8x^3$$

c. We rearrange the factors to group each variable together:

$$(xy^2)(4x^2) = x \cdot y \cdot y \cdot 4 \cdot x \cdot x = 4 \cdot x \cdot x \cdot x \cdot y \cdot y = 4x^3y^2$$

Caution! When we add like terms, we do not change the variable in the terms; we combine the coefficients. For example,

$$3a + 2a = 5a$$

When we multiply expressions, we multiply the coefficients and we multiply the variables:

$$3a(2a) = 3(2) \cdot a \cdot a = 6a^2$$

RQ1. Explain the difference between $5x - 2x$ and $5x(-2x)$.

Visualizing the Distributive Law

We can use the areas of rectangles to model products of algebraic expressions. Recall that we find the area of a rectangle by multiplying its length times its width, $A = lw$.

Example 2a. Represent the product $2x(x+9)$ as the area of a rectangle.
b. Use the distributive law to find the product.

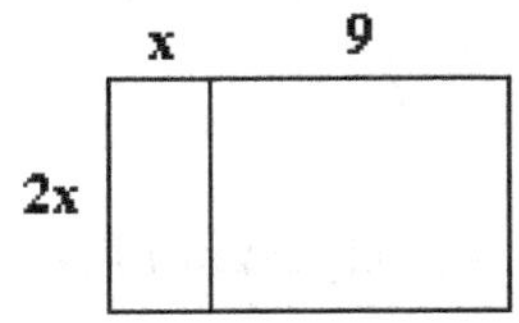

Solutions a. We sketch a rectangle whose width is $2x$ and whose length is $x+9$.
b. The area of the rectangle is the sum of the areas of the two smaller rectangles:

$$2x(x+9) = 2x \cdot x + 2x \cdot 9 = 2x^2 + 18x$$

RQ2. State the distributive law, and explain what it means.

At this stage it will be helpful to introduce some terminology.

- An algebraic expression with only one term, such as $2x^3$, is called a **monomial**.
- An expression with two terms, such as $x^2 - 16$, is called a **binomial.**
- An expression with three terms is a **trinomial**.
- The expression $ax^2 + bx + c$ is thus called a **quadratic trinomial**, because it involves the square of the variable.

Caution! Notice the difference between $(3a)(2a)$ and $3a(2+a)$:

- $(3a)(2a)$ is the product of two monomials, and we use the commutative property to simplify it:

$$(3a)(2a) = 3 \cdot a \cdot 2 \cdot a = 3 \cdot 2 \cdot a \cdot a = 6a^2$$

- $3a(2+a)$ is the product of a monomial and a binomial, and we use the distributive law to simplify it:

$$3a(2+a) = 3a(2) + 3a(a) = 6a + 3a^2$$

Multiplying Binomials

Consider the rectangle shown at right. As you can see, it is divided into four smaller rectangles. You can verify that we get the same answer when we compute its area in two different ways: We can add up the areas of the four smaller rectangles, or we can find the length and width of the entire large rectangle and then find their product:

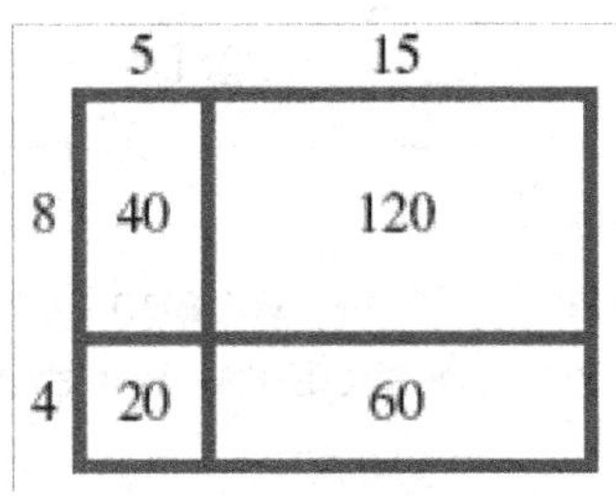

Sum of four sub-rectangles:

Area $= 8(5) + 8(15) + 4(5) + 4(15)$
$= 40 + 120 + 20 + 60$
$= 240$

One large rectangle:

Area $= (8+4)(5+15)$
$= (12)(20)$
$= 240$

RQ3. Explain the terms monomial, binomial, and trinomial.

Example 3 The City Council plans to install a 10-foot by 30-foot reflecting pool in front of City Hall. When the cost estimate comes in, they realize that they can afford to enlarge the pool. They decide to increase both the length and the width by x feet. Write an equation for the area, A, of the pool in terms of x.

Solution Look at the drawing of the pool in the figure below left. Both dimensions of the original pool have been increased by x. The area of the increased pool is thus

$$A = (30 + x)(10 + x)$$

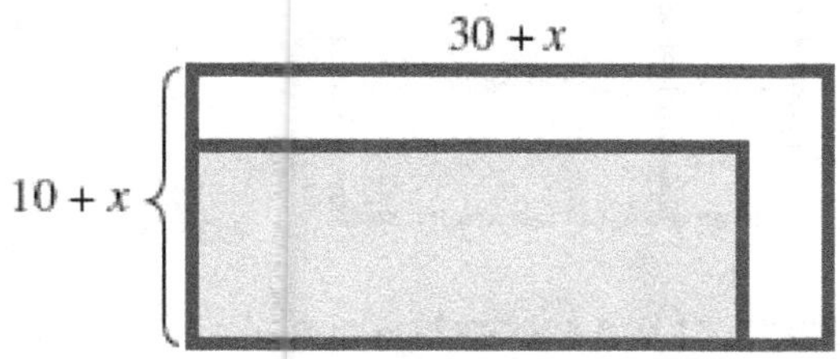

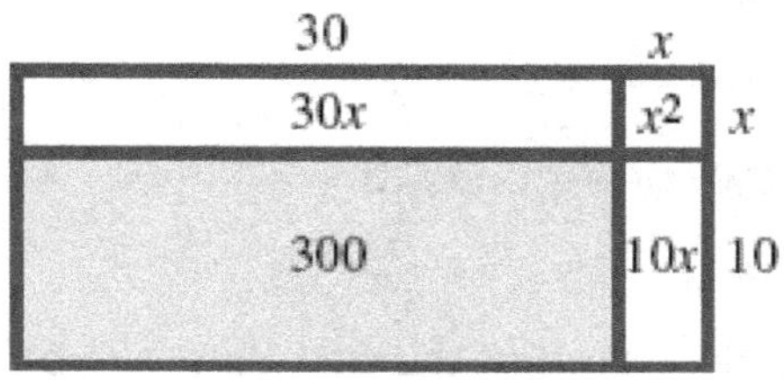

In the figure on the right, we partition the new pool into four sub-rectangles and compute the area of each. With this method we get the following expression for the area of the new pool.

$$\begin{aligned} A &= x^2 + 30x + 10x + 300 \qquad \textbf{Combine like terms.} \\ &= x^2 + 40x + 300 \end{aligned}$$

The two expressions for the area are equivalent. Thus,

$$(30 + x)(10 + x) = x^2 + 40x + 300$$

Look Ahead: Our goal in this Lesson is to understand products of binomials. We can use rectangles to illustrate, or model, the product of two binomials. The rectangles do not have to be drawn to scale; they are merely tools for visualizing the operation. With a small stretch of the imagination, we can use rectangles to represent negative numbers as well. We follow the rules for adding and multiplying negative numbers.

Example 4a. Use a rectangle to represent the product $(x - 4)(x + 6)$.
b. Write the product as a quadratic trinomial.

Solutions a. We let the first factor, $(x - 4)$, represent the width of the rectangle, and the second factor, $(x + 6)$, represent its length.
b. We find the area of each sub-rectangle, as shown in the figure. Then we add the areas together.

$$\begin{aligned} \text{Area} &= x^2 + 6x - 4x - 24 \\ &= x^2 + 2x - 24 \end{aligned}$$

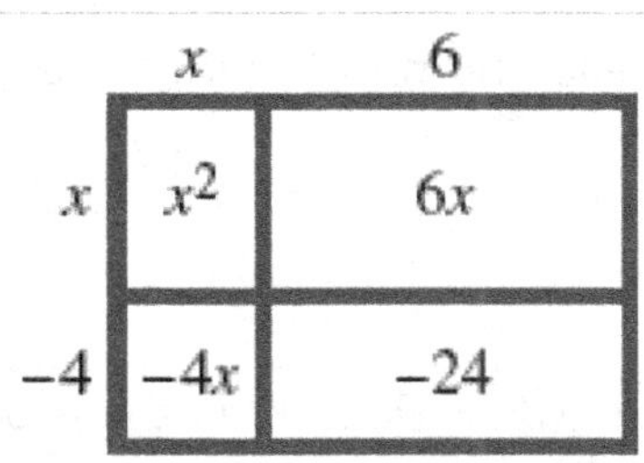

We say that $(x + 6)(x - 4)$ is the **factored form** of the product, and $x^2 + 2x - 24$ is the **expanded form**.

RQ4. When we use a rectangle to model the product of two binomials, what do the two binomials reresent? What does their product represent?

The Four Terms in a Binomial Product

Using a rectangle to multiply binomials illustrates how the distributive law works. Analyzing the rectangle method will help us in the next Lesson, when we reverse the process to **factor** a quadratic trinomial. Let us take a closer look at the example above.

Look Closer: In Example 4 we computed the product $(x-4)(x+6)$. The top row of the rectangle corresponds to

$$x(x+6) = x^2 + 6x$$

and the bottom row corresponds to

$$-4(x+6) = -4x - 24$$

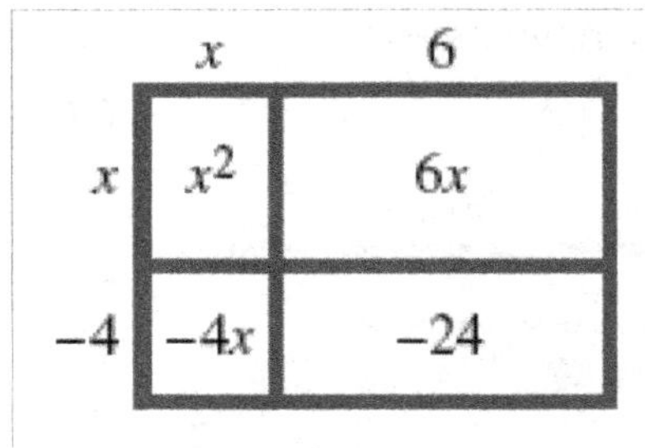

Thus, we multiply each term of the first binomial by each term of the second binomial, resulting in four multiplications in all:

$$(x-4)(x+6) = x^2 + 6x - 4x - 24$$

Each term of the product corresponds to the area of one of the four sub-rectangles, as shown below.

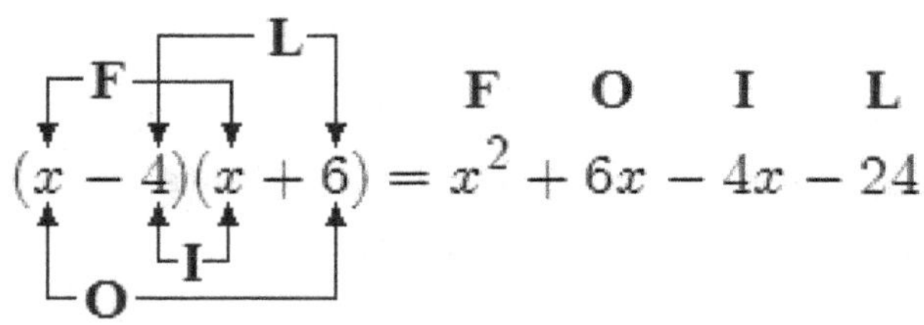

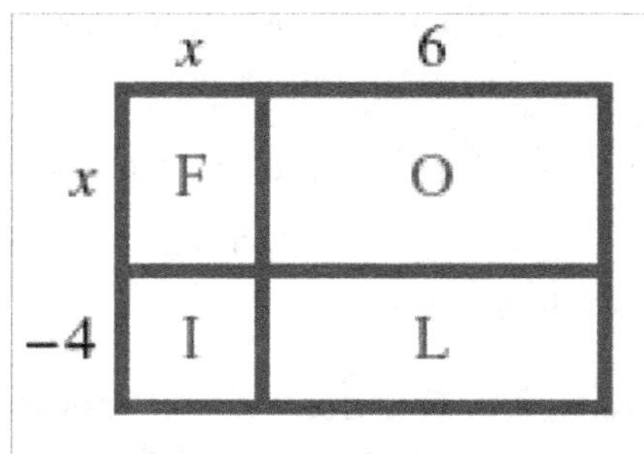

The letters **F**, **O**, **I**, **L** indicate the four steps in computing the product:

1. **F** stands for the product of the **First** terms in each binomial.
2. **O** stands for the product of the **Outer** terms.
3. **I** stands for the product of the **Inner** terms.
4. **L** stands for the product of the **Last** terms.

Note how each term of the trinomial product arises from the binomial factors:

$$(x-4)(x+6) = \underset{\textbf{F}}{x^2} \quad \underset{\textbf{O+I}}{+\,2x} \quad \underset{\textbf{L}}{-\,24}$$

- The **quadratic** term in the product comes from the **First** terms.
- The **linear** or x-term of the product is the sum of **Outer** and **Inner.**
- The **constant** term of the product comes from the **Last** terms.

RQ5. What does the acronym FOIL stand for?

Example 5 What is the constant term of the product $(x+7)(x+9)$? What is the linear term?

Solution The constant term comes from the product of the **Last** terms in each binomial: $7 \cdot 9 = 63$. The linear term comes from the sum of **Outer + Inner**.

$$9x + 7x = 16x$$

Square of a Binomial

Recall that to square a number means to multiply the number by itself. The same is true for squaring an algebraic expression. In particular to square a binomial means to multiply the binomial by itself. Thus,

$$(a+b)^2 = (a+b)(a+b)$$

Example 6 Expand the product $(2x-3)^2$.

Solution We multiply the binomial by itself.

$$\begin{aligned}(2x-3)^2 &= (2x-3)(2x-3)\\ &= 4x^2 - 6x - 6x + 9 = 4x^2 - 12x + 9\end{aligned}$$

Caution! We cannot square a binomial by squaring each term separately! In Example 6, it is **not** true that

$$(2x-3)^2 = 4x^2 + 9 \qquad \leftarrow \textbf{Incorrect!}$$

We must use the distributive law to multiply the binomial times itself.

Example 7 Show that the equation $2(x-3)^2 + 8 = 0$ is quadratic, by writing it in standard form.

Solution We follow the order of operations to expand the left side. First, we compute the power, $(x-3)^2$, to get

$$\begin{aligned}(x-3)^2 &= (x-3)(x-3)\\ &= x^2 - 3x - 3x + 9 = x^2 - 6x + 9\end{aligned}$$

We replace $(x-3)^2$ by its expanded form and simplify the result.

$$\begin{aligned}2(x-3)^2 + 8 &= 2(x^2 - 6x + 9) + 8 && \text{Apply the distributive law.}\\ &= 2x^2 - 12x + 18 + 8 && \text{Combine like terms.}\\ &= 2x^2 - 12x + 26\end{aligned}$$

Thus, the original equation may be written as $2x^2 - 12x + 26 = 0$, which is a quadratic equation in standard form.

RQ6. What is wrong with this calculation?

$$13^2 = (10+3)^2 = 10^2 + 3^2 = 100 + 9 = 109 \qquad \leftarrow \textbf{Incorrect!}$$

Skills Warm-Up

Find the area and perimeter of each figure.

1. 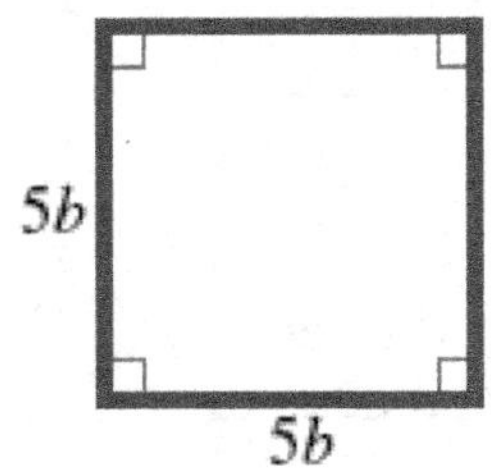

2. 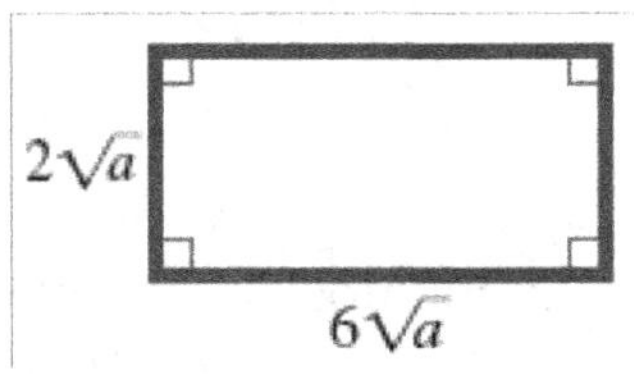

3. 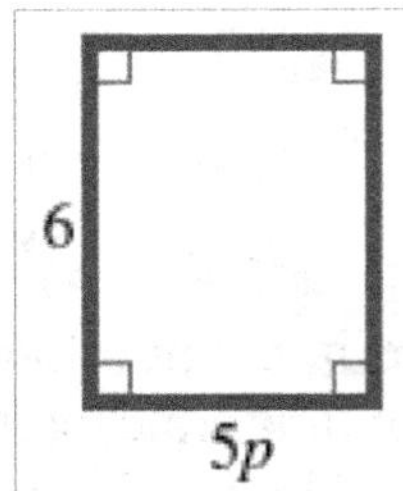

4. 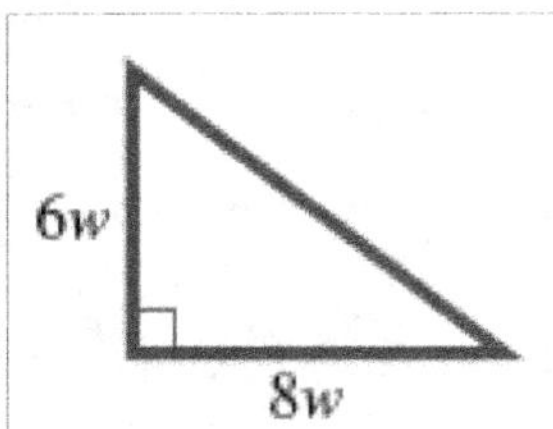

5. 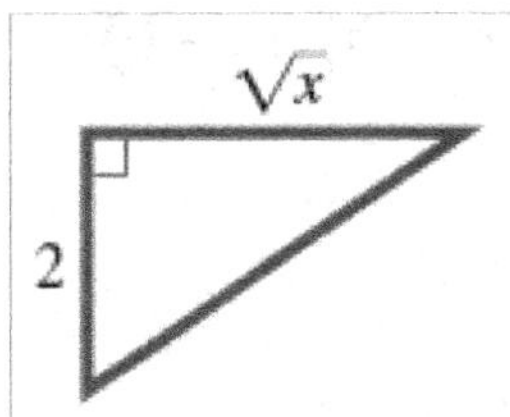

6. 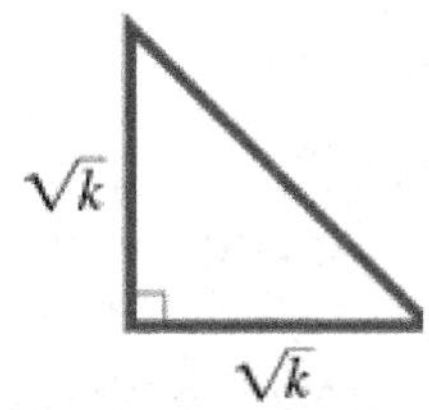

Answers: **1.** $25b^2,\ 20b$ **2.** $12a,\ 16\sqrt{a}$ **3.** $30p,\ 12+10p$ **4.** $48w^2,\ 24w$
5. $\sqrt{x},\ 2+\sqrt{x}+\sqrt{x+4}$ **6.** $\frac{k}{2},\ 2\sqrt{k}+\sqrt{2k}$

Homework 5.4

Skills Practice

In Problems 1-4, simplify each product or power.

1a. $3(4n)$ **b.** $3n(4n)$ **c.** $(4n)^3$
2a. $6x(-5x^2)$ **b.** $-6x^2(-5x)$ **c.** $(-5x)^2$
3a. $(4p)(-4p)$ **b.** $-(4p)^2$ **c.** $-4(-p)^2$

In Problems 4-6, simplify each expression.

4a. $-8x(5t)$ **b.** $-8x(5+t)$ **c.** $-8x(-5-t)$
5a. $3n(-4n)$ **b.** $3n-4n$ **c.** $(3n-4)n$
6a. $2x(-5x^2)$ **b.** $2(x-5x^2)$ **c.** $2x-(5x)^2$

In Problems 7-8,

a. Write a product (length × width) for the area of each rectangle.
b. Use the distributive law to compute the product.

7.

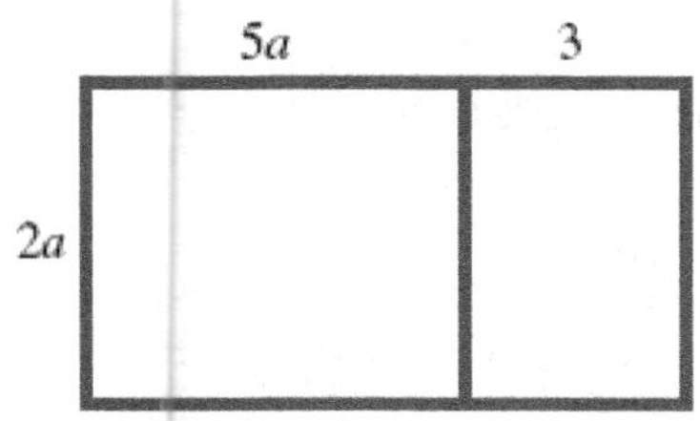

8.

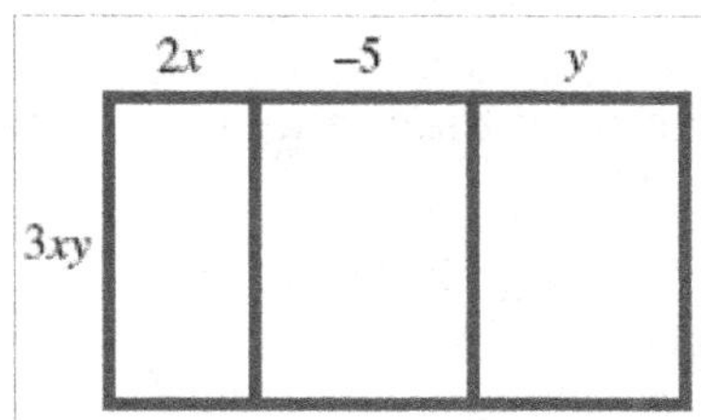

In Problems 9-14, compute the product.

9. $-2b(6b-2)$ **10.** $(6a-5)(3a)$ **11.** $3v(5v-2v^2)$
12. $-4x^2(2x+3y)$ **13.** $(y^3+3y-2)(2y)$ **14.** $-xy(2x^2-xy+3y^2)$

For Problems 15-18, fill in the missing algebraic expression.

15. $4x(2x+\textbf{?})=8x^2-12x$ **16.** $\textbf{?}(4m-9)=-12m^2+27m$
17. $6h(\textbf{?}+\textbf{?})=24h-18h^2$ **18.** $\textbf{?}(a^2-\textbf{?}+5)=2a^3-6a^2+10a$

In Problems 19-22, simplify.

19. $2a(x+3)-3a(x-3)$ **20.** $2x(3-x)+2(x^2+1)-2x$
21. $ax(x^2+2x-3)-a(x^3+2x^2)$ **22.** $3ab^2(2+3a)-2ab(3ab+2b)$

In Problems 23-24, write two different expressions for the area of each rectangle:

a. as the sum of four small areas,
b. as one large rectangle, using the formula Area = length × width.

23.

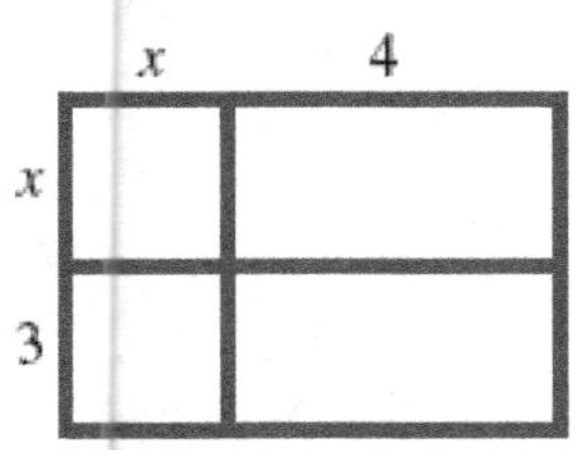

24.

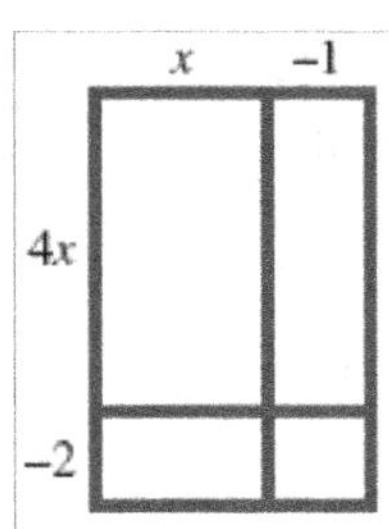

■ In Problems 25-27,
   a. Use a rectangle to represent each product.
   b. Write the product as a quadratic trinomial.

25. $(a-5)(a-3)$ **26.** $(y+1)(3y-2)$ **27.** $(5x-2)(4x+3)$

■ In Problems 28-30, use rectangles to help you multiply binomials in two variables.

28. $(x+2y)(x-y)$ **29.** $(3s+t)(2s+3t)$ **30.** $(2x-a)(x-3a)$

■ In Problems 31-34, compute the product: Multiply the binomials together first, then multiply the result by the numerical coefficient.

31. $2(3x-1)(x-3)$ **32.** $-3(x+4)(x-1)$
33. $-(4x+3)(x-2)$ **34.** $5(2x+1)(2x-1)$

Applications

■ For Problems 35-36, write and simplify algebraic expressions.

35a. For the perimeter of the triangle. **b.** For the area of the triangle.

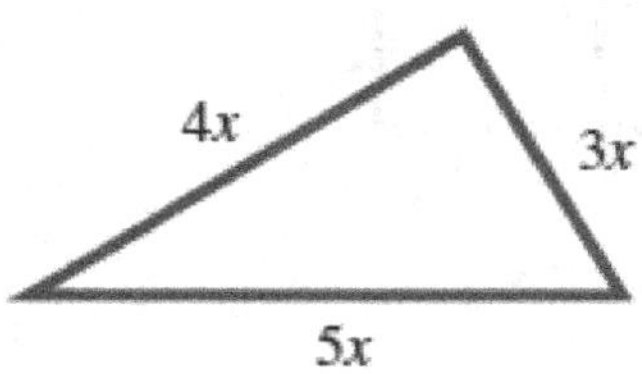

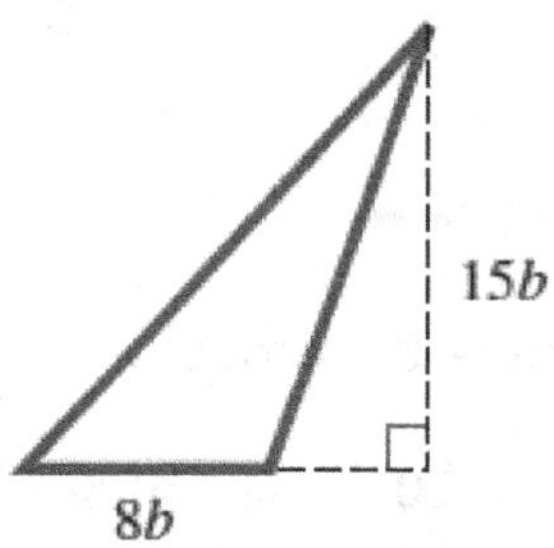

36. a. For the area of the rectangle.

b. For the perimeter of the rectangle.

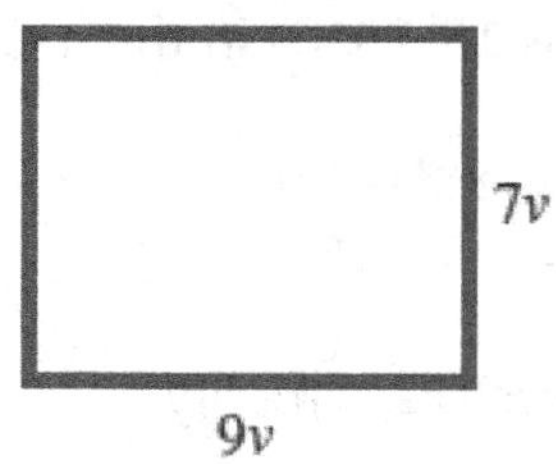

■ For Problems 37-38, write an expression for the shaded area in each combination of squares and circles.

37.

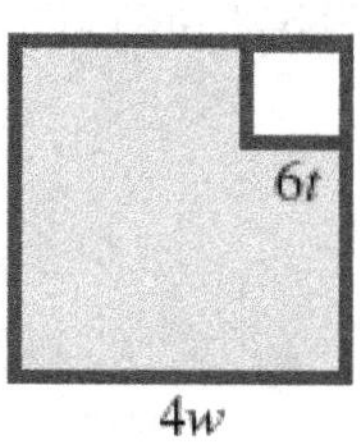

38.

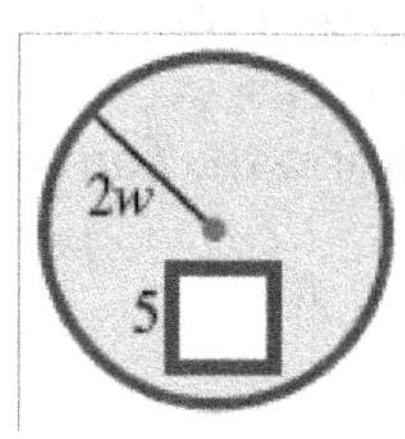

For Problems 39-40, write an algebraic expression for the volume of the figure.

39. 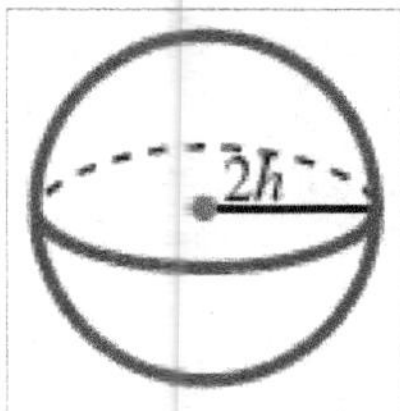

40. 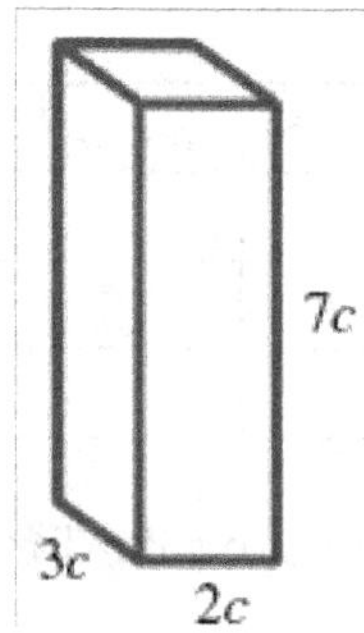

In Problems 41-42, find the product without a calculator by using rectangles.

41. 36×42

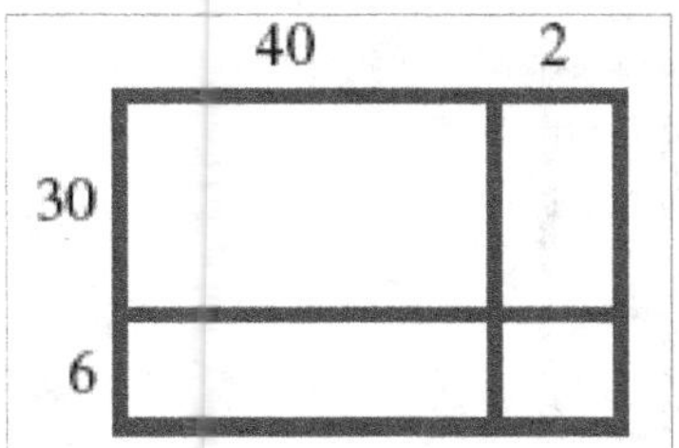

42. 82×16

(Make your own drawing)

In Problems 43-44,

a. Find the linear term in the product.

b. Shade the sub-rectangles that correspond to the linear term.

43. $(x+6)(x-9)$

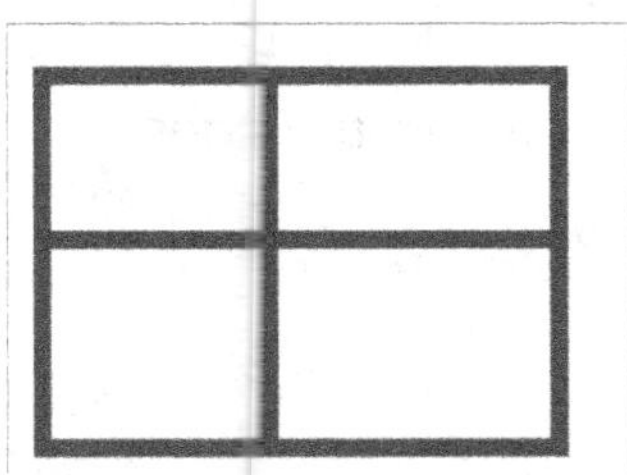

44. $(2x-5)(x+4)$

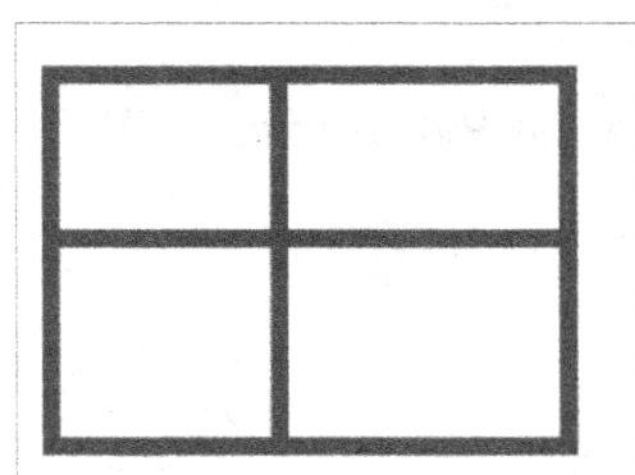

In Problems 45-47, compute the product. What do you notice? Explain why this happens.

45. $(x+3)(x-3)$

46. $(x-2a)(x+2a)$

47. $(3x+1)(3x-1)$

In Problems 48-50, compute the product.

48. $(w+4)(w+4)$

49. $(z-6)(z-6)$

50. $(3a-2c)(3a-2c)$

51. a. Complete the table below.
b. Decide whether the two expressions, $(a-b)^2$ and a^2-b^2, are equivalent.

a	b	$a-b$	$(a-b)^2$	a^2	b^2	a^2-b^2
5	3					
2	6					
−4	−3					

52. Explain why $(x-y)^2$ cannot be simplified to x^2-y^2.

In Problems 53-54, write the area of the square in two different ways:
a. as the sum of four smaller areas,
b. as one large square, using the formula Area = (length)^2.

53. 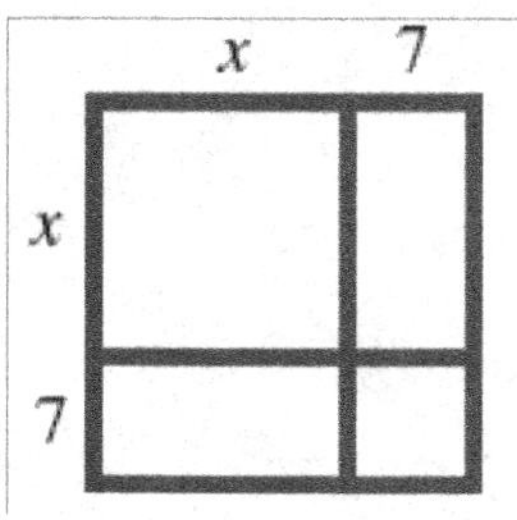

54. 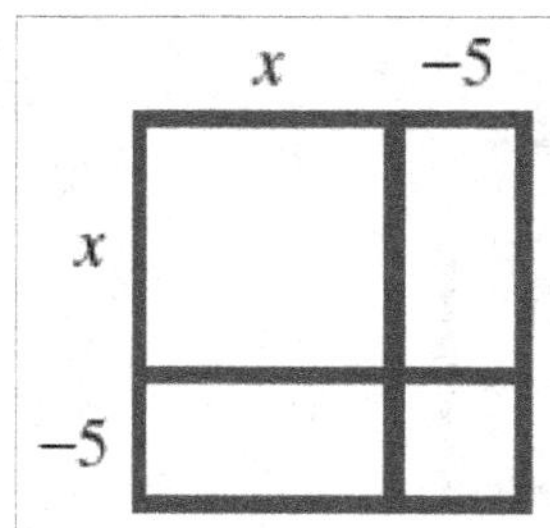

55. Is $(x+4)^2$ equivalent to x^2+4^2 ? Explain why or why not, and give a numerical example to justify your answer.

In Problems 56-58, compute the product.

56. $(x-2)^2$ **57.** $(2x+1)^2$ **58.** $(3x-4y)^2$

In Problems 59-60, use the Pythagorean theorem to write an equation about the sides of the right triangle.

59. 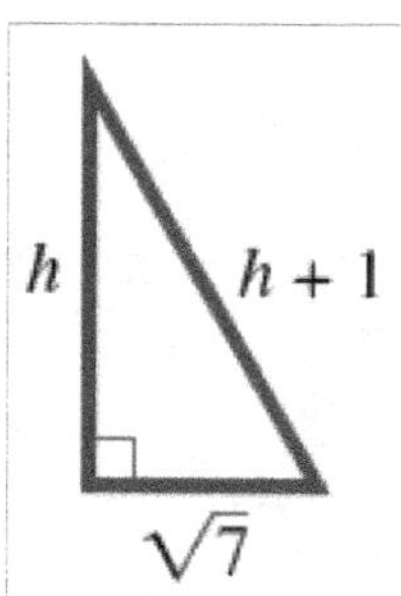

60. 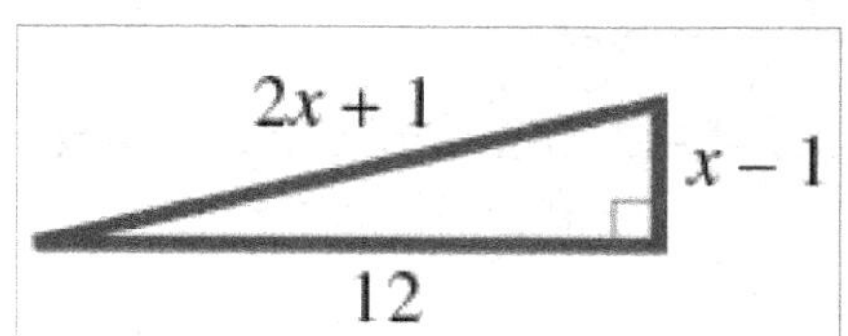

5.5 Solving Quadratic Equations

Revenue is the amount of money a company takes in from selling a product. To find the total revenue from the sale of a product, we multiply the price of one item by the number of items sold.

Revenue = (price per item) · (number of items sold)

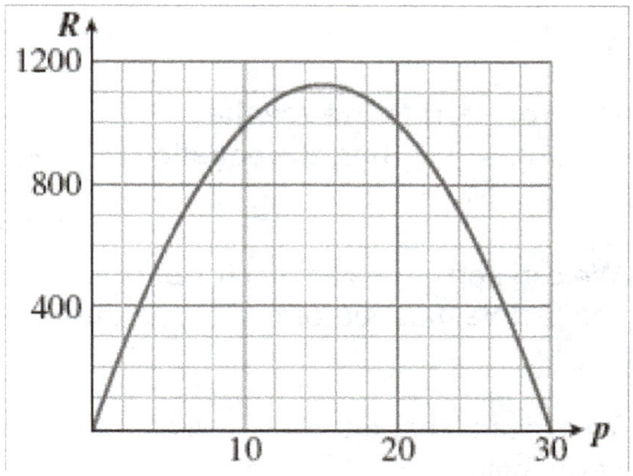

In Section 5.4 we found a quadratic equation for the revenue Raingear earns by selling umbrellas at a price of p dollars each. The equation was

$$R = p(150 - 5p) = 150p - 5p^2$$

The graph of this equation is shown at right.

If Raingear charges $p = 0$ dollars for each umbrella, they will not earn any revenue, so $R = 0$. You can see this fact in the graph, which passes through the point $p = 0$, $R = 0$. But there is another point on the graph where $R = 0$; it has p-coordinate 30. If Raingear charges 30 dollars per umbrella, they will not earn any revenue. (Why does this happen?)

RQ1. State a formula for calculating revenue.

The Zero Factor Principle

Look again at the factored form of the equation for revenue,

$$R = p\,(150 - 5p)$$

The first factor is p, and when $p = \mathbf{0}$ we have

$$R = \mathbf{0}\,(150 - 5 \cdot \mathbf{0}\,) = 0$$

so the revenue is zero as well. The second factor, $150 - 5p$, is zero when $p = \mathbf{30}$. If Raingear charges \$30 per umbrella, its revenue is

$$R = \mathbf{30}\,(150 - 5 \cdot \mathbf{30}) = 0$$

(Recall that $150 - 5p$ represents the number of umbrellas that Raingear will sell at a price of p dollars each. Apparently, no one will buy a \$30 umbrella.)

Both of these values, $p = 0$ and $p = 30$, are solutions of the quadratic equation

$$p\,(150 - 5p) = 0$$

A quadratic equation has a solution when either of its factors is equal to zero. This is a consequence of the following fact about numbers.

Zero-Factor Principle

If the product of two numbers is zero, then one (or both) of the numbers must be zero. Using symbols,

If $ab = 0$, then either $a = 0$ or $b = 0$.

The zero-factor principle also applies to algebraic expressions: If the product of two factors is zero, then at least one of the factors must be zero.

Example 1a. Compute the product $(x+5)(x-4)$.
b. Solve the equation $x^2+x-20=0$.

Solutions a. We use the FOIL method to apply the distributive law to the product.

$$(x+5)(x-4) = x^2 - 4x + 5x - 20$$
$$= x^2 + x - 20$$

b. From part (a) we see that x^2+x-20 can also be written in factored form as $(x+5)(x-4)$. Thus, the equation $x^2+x-20=0$ is equivalent to

$$(x+5)(x-4)=0$$

We can apply the zero-factor principle to this equation: If the product is zero, then one of the factors must be zero. Thus, either

$$x+5=0 \quad \text{or} \quad x-4=0$$

This means that either $x=-5$ or $x=4$. These are the solutions of the equation, as we verify below.
Check: Substitute each value into the original equation.

$x=-5:$ $(-5)^2+(-5)-20=25-5-20=0$
$x=4:$ $(4)^2+(4)-20=16+4-20=0$

RQ2. If you multiply two numbers together and the product is zero, what can you say about the original numbers?

Factoring

In order to use the zero-factor principle to solve quadratic equations, we must be able to write quadratic expressions in factored form. This process is called **factoring**, and it is the reverse of multiplying factors together. Here are some examples comparing multiplying and factoring.

Multiplying	Factoring
1. $2\cdot 3=6$	**1.** $6=2\cdot 3$
2. $3(2x-5)=6x-15$	**2.** $6x-15=3(2x-5)$
3. $(x-5)(x+2)=x^2-3x-10$	**3.** $x^2-3x-10=(x-5)(x+2)$

The first type of factoring we consider is the reverse of the distributive law. Consider the expression $6x-15$. The expression has two terms, $6x$ and -15, and each term is divisible by 3:

$$6x = \mathbf{3}\cdot 2x \qquad \text{and} \qquad -15 = \mathbf{3}\cdot -5$$

We say that 3 is a **common factor** for the expression $6x-15$ because it is a factor of each term. Using the distributive law, we can write

$$6x-15 = \mathbf{3}\cdot 2x - \mathbf{3}\cdot 15 = \mathbf{3}(2x-15)$$

We have factored out a common factor, namely 3, from $6x-15$.

RQ3. What does it mean to factor an expression?
RQ4. Factoring is the reverse process for which operation?

■ **Example 3** Factor: $36t^2 - 63t$

Solution We look for the largest common factor for the two terms. Both coefficients are divisible by 9, and t divides evenly into both terms. Thus, we factor out $\mathbf{9t}$ and write the expression as

$$36t^2 - 63t = \mathbf{9t}(\,?\, - \,?\,)$$

You may be able to see the missing factors at a glance, but if not, you can divide each term of the original expression by the common factor, as follows:

$$\frac{36t^2}{9t} = \mathbf{4t} \qquad \text{and} \qquad \frac{63t}{9t} = \mathbf{7}$$

Fill in these factors for the question marks to obtain the factored form,

$$36t^2 - 63t = \mathbf{9t}(\mathbf{4t} - \mathbf{7})$$ ■

You can always check your factorization by computing the product; it should be the same as the original expression.

RQ5. Explain the meaning of "common factor."

Solving Quadratic Equations by Factoring

Now let's see how to solve a quadratic equation using factoring.

First, we must arrange the terms so that **one side of the equation is zero**; the zero-factor principle applies only to zero products.

■ **Example 4** Solve $2x^2 = 4x$

Solution First, we write the equation in standard form by subtracting $4x$ from both sides.

$$2x^2 - 4x = 0$$

Next, we factor the left side of the equation. We can take out a common factor of $2x$.

$$2x(x - 2) = 0$$

Now, we apply the zero-factor principle: In order for the product to be zero, one of the two factors must be zero. We set each factor equal to zero, and solve for x.

$$\begin{aligned} 2x &= 0 \quad &\text{or}\quad x - 2 &= 0 \\ x &= 0 \quad &\text{or}\quad x &= 2 \end{aligned}$$

Thus, the two solutions are $x = 0$ and $x = 2$. You should check that both of these values satisfy the original equation. ■

RQ6. What is the first step in solving a quadratic equation by factoring?

Caution! We cannot solve the equation in Example 4 by dividing both sides by $2x$. If we do that, we get the equation $x = 2$. This is one of the solutions, but we have lost the second solution, $x = 0$. We should **never divide both sides of an equation by the variable**, because we risk losing one of the solutions. Instead, we follow the steps below to solve the equation by factoring. ■

To Solve a Quadratic Equation by Factoring:

1. Write the equation in standard form,

$$ax^2 + bx + c = 0 \qquad (a \neq 0)$$

2. Factor the left side of the equation.
3. Apply the zero-factor principle; that is, set each factor equal to zero.
4. Solve each equation to obtain two solutions.

RQ7. Why is it wrong to divide both sides of the equation $2x(x-2) = 0$ by $2x$?

Application to Graphing

Earlier we saw that the solutions of the equation

$$150p - 5p^2 = 0$$

namely, $p = 0$ and $p = 30$, are also the horizontal intercepts of the graph of

$$R = 150p - 5p^2$$

Recall that the x-intercept of the line $y = mx + b$ is the point where the graph crosses the x-axis. So, to find the x-intercepts of a line, we set $y = 0$ and solve for x. The same strategy applies to quadratic equations.

To find the x-intercepts of the graph of

$$y = ax^2 + bx + c$$

we set $y = 0$ and solve the equation

$$ax^2 + bx + c = 0$$

■ **Example 4a.** Find the x-intercepts of the graph of $y = 3x - x^2$.
b. Sketch the graph.

Solutions a. We set $y = 0$ and solve the equation

$3x - x^2 = 0$	**Factor the left side.**
$x(3 - x) = 0$	**Set each factor equal to 0.**
$x = 0 \quad 3 - x = 0$	**Solve each equation.**
$x = 0 \quad\quad x = 3$	

The x-intercepts are the points $(0, 0)$ and $(3, 0)$.

b. We make a table of values that includes the x-intercepts.

x	-1	0	1	2	3	4
y	-4	0	2	2	0	-4

We plot the points and connect them with a parabola, as shown in the figure.

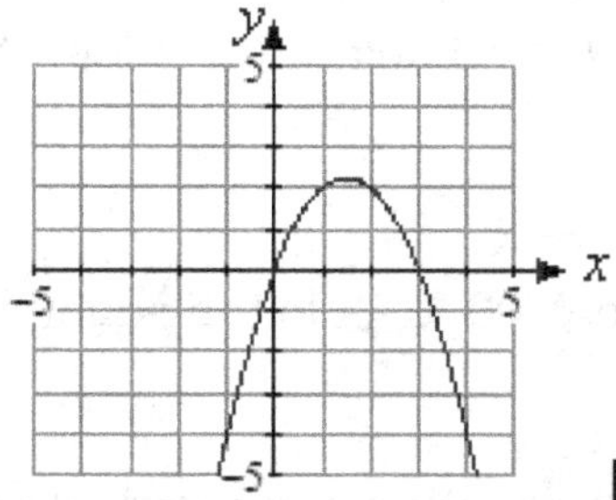

RQ8. How do we find the x-intercepts of a parabola?

More Extraction of Roots

Quadratic equations written in the form

$$a(x-p)^2 = q$$

can also be solved by extraction of roots after isolating the squared expression, $(x-p)^2$.

Example 5 Solve the equation $3(x-2)^2 = 48$.

Solution First, we isolate the perfect square, $(x-2)^2$.

$$3(x-2)^2 = 48 \qquad \text{Divide both sides by 3.}$$
$$(x-2)^2 = 16 \qquad \text{Take the square root of each side.}$$
$$x-2 = \pm\sqrt{16} = \pm 4$$

This gives us two equations for x,

$$x-2=4 \quad \text{or} \quad x-2=-4 \qquad \text{Solve each equation.}$$
$$x=6 \quad \text{or} \quad x=-2$$

The solutions are 6 and -2.

RQ9 True or false.

a. The first step in extraction of roots is to take square roots.

b. The solutions of a quadratic equation are always of the form $\pm k$.

Compound Interest

Many savings institutions offer accounts on which the interest is **compounded** annually; that is, at the end of each year the interest earned is added to the principal, and the interest for the next year is computed on this larger sum of money. After n years, the amount of money in the account is given by the formula

$$\boldsymbol{A = P\,(1+r)^n}$$

where P is the original principal and r is the interest rate, expressed as a decimal.

Example 6 Carmella invests $3000 in an account that pays an interest rate r compounded annually.

a. Write an expression for the amount of money in Carmella's account after two years.

b. What interest rate would be necessary for Carmella's account to grow to $3500 in two years?

Solutions a. We use the formula above with $P = 3000$ and $n = 2$. Carmella's account balance will be

$$A = 3000(1 + r)^2$$

b. We substitute 3500 for A in the equation.

$$3500 = 3000(1 + r)^2$$

This is a quadratic equation in the variable r, which we can solve by extraction of roots. First, we isolate the perfect square.

$$3500 = 3000(1 + r)^2 \quad \text{Divide both sides by 3000.}$$
$$1.1\overline{6} = (1 + r)^2 \quad \text{Take the square root of both sides.}$$
$$\pm 1.0801 \approx 1 + r \quad \text{Subtract 1 from both sides.}$$
$$r \approx 0.0801 \text{ or } r \approx -2.0801$$

Because the interest rate must be a positive number, we discard the negative solution. Carmella needs an account with interest rate $r \approx 0.0801$, or over 8%, in order to have an account balance of \$3500 in two years. ■

The same formula that we use for compound interest also applies to calculating the effects of **inflation**. For instance, if there is a steady inflation rate of 4% per year, then two years from now an item that costs \$100 now will cost

$$\begin{aligned} A &= P(1 + r)^2 \\ &= 100(1 + 0.04)^2 = \$108.16 \end{aligned}$$

RQ10. What is the formula for compound interest? What does each variable mean?

Skills Warm-Up

■ Mental Exercise: Evaluate each expression for the given values of the variable. Try not to use pencil, paper, or calculator.

1. $2x(x - 3), \quad x = 0, 1, 2, 3$

2. $(x + 1)(x - 6), \quad x = -1, 3, 6, 9$

3. $2(x + 2)(x + 4), \quad x = -4, -2, 0, 2$

4. $3n(2n - 1), \quad n = -1, 0, \frac{1}{2}, 1$

Answers: **1.** 0, –4, –4, 0 **2.** 0, –12, 0, 30 **3.** 0, 0, 16, 48 **4.** 9, 0, 0, 3

Homework 5.5

Skills Practice

For Problems 1-3, apply the zero-factor principle to find the solutions.

1. $(x+1)(x-4)=0$ **2.** $0=p(p+7)$ **3.** $(2v+3)(4v-1)=0$

For Problems 4-5,

a. Verify that the factored form for each expression is correct.

b. Use the factored form to find the solutions of the original equation.

c. Check algebraically that your answers to part (b) actually are solutions of the original equation.

4. $x^2-6x-27=0$
factored form:
$(x-9)(x+3)=0$

5. $x^2-81=0$
factored form:
$(x-9)(x+9)=0$

For Problems 6-8, find the largest common factor for the monomials.

6. $8x^2,\ 12x$ **7.** $6a^2b,\ 9ab^2$ **8.** $10w^2,\ 15w$

In Problems 9-12, fill in the missing algebraic expressions.

9. $15a^2-12a=3a(5a+\textbf{?})$ **10.** $30t+12t^2=\textbf{?}(5+2t)$

11. $28x^2-20x=4x(\textbf{?}+\textbf{?})$ **12.** $27n^2+24n=\textbf{?}(\textbf{?}+8)$

For Problems 13-15, factor out the largest common factor.

13. $6a^2-8a$ **14.** $-18v^2-6v$ **15.** $4h-9h^2$

For Problems 16-18, factor the right side of the formula.

16. $d=k-kat$ **17.** $A=2rh-\pi r^2$ **18.** $V=\pi r^2h-\frac{1}{3}s^2h$

For Problems 19-24, solve the equation by factoring.

19. $10x^2-15x=0$ **20.** $20x^2=x$ **21.** $0=144x+3x^2$

22. $4x-6x^2=0$ **23.** $x^2+x=0$ **24.** $\frac{1}{3}x^2=\frac{2}{3}x$

For Problems 25-27,

a. Without graphing, find the x-intercepts of the graph of each equation.

b. Verify your answer by graphing the equation on a calculator.

25. $y=(x-3)(2x+5)$ **26.** $y=2x^2-6x$ **27.** $y=8x-3x^2$

For Problems 28-33, solve by extraction of roots.

28. $(x-2)^2=9$ **29.** $(2x-1)^2=16$ **30.** $4(x+2)^2=12$

31. $\left(x-\frac{1}{2}\right)^2=\frac{3}{4}$ **32.** $81\left(x+\frac{1}{3}\right)^2=1$ **33.** $3(8x-7)^2=24$

For Problems 34-36, solve for x in terms of a, b, and c.

34. $\dfrac{ax^2}{b} = c$ **35.** $(x-a)^2 = 16$ **36.** $(ax+b)^2 = 9$

For Problems 37-38, find the mistake in the steps for solution, then write a correct solution.

37.
$$4x^2 = 12x$$
$$\frac{4x^2}{4x} = \frac{12x}{4x}$$
$$x = 3$$

38.
$$9x^2 - 4 = 16$$
$$3x - 2 = 4$$
$$3x = 6$$
$$x = 2$$

Applications

39. Mitra plans to build a rectangular pen for her rabbits in the back yard. She has 48 feet of wire fence. If she builds a pen of width w feet, then the area of the pen is given in square feet by $A = 24w - w^2$.

a. Complete the table of values.

w	A
0	
4	
8	
12	
16	
20	
24	

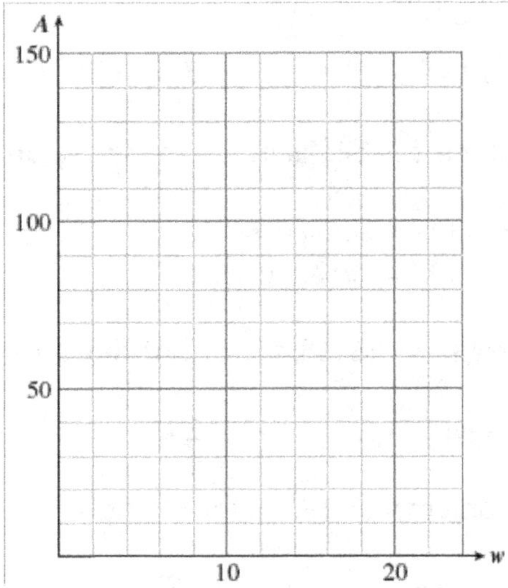

b. Sketch a graph of the equation on the grid provided.

c. Use algebra to find the w-intercepts of the graph.

d. What is the largest area that Mitra can enclose with her fence? What are the dimensions of the largest rabbit pen Mitra can build?

40. Cyril plans to invest \$5000 in a money market account paying interest compounded annually.

a. Write a formula for the balance, B, of Cyril's account after two years in terms of the interest rate, r.

b. Complete the table showing Cyril's account balance after two years for various interest rates.

r	0.02	0.04	0.06	0.08
B				

c. If Cyril would like to have \$6250 in two years, what interest rate must the account pay?

d. Use your calculator to graph the formula for Cyril's account balance. Locate the point on the graph that corresponds to the amount in part (b).

41. Two years ago Carol's living expenses were $1200 per month. This year the same items cost Carol $1400 per month. What was the annual inflation rate for the past two years?

42. A machinist wants to make a metal section of pipe that is 80 millimeters long and has an interior volume of 9000 cubic millimeters. If the pipe is 2 millimeters thick, its interior volume is given by the formula

$$V = \pi(r-2)^2h$$

where h is the length of the pipe and r is its radius. What should the radius of the pipe be?

5.6 Graphing Parabolas

All parabolas share certain features. The graph of $y = ax^2 + bx + c$ has either a highest point or a lowest point, called the **vertex** of the graph. The parabola is symmetric about a vertical line, called the **axis of symmetry**, that runs through the vertex. A parabola has a y-intercept, and it may have zero, one, or two x-intercepts. If there are two x-intercepts, they are equidistant from the axis of symmetry.

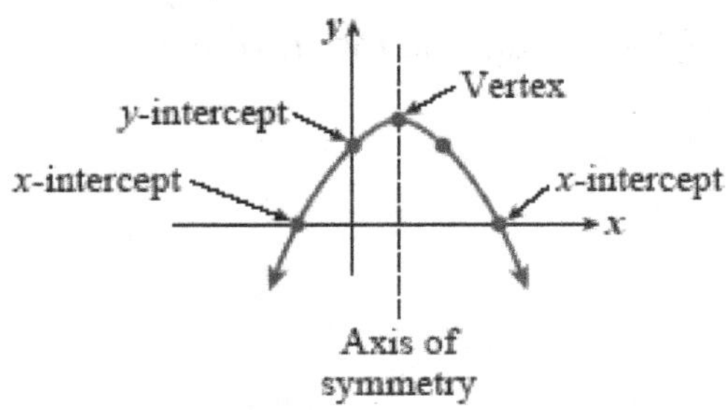

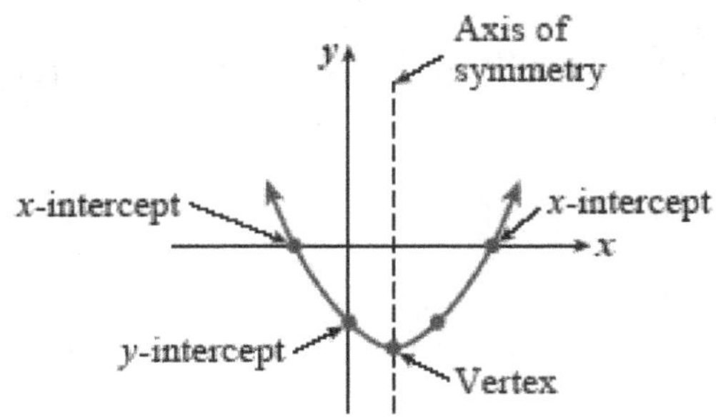

Example 1 Graph each equation and give the axis of symmetry, the vertex, and the intercepts of the graph.

a. $y = x^2 + 4$ **b.** $y = 4 - x^2$ **c.** $y = (x - 2)^2$ **d.** $y = (x + 2)^2$

Solutions We make a table of values and plot points to sketch each parabola.

a.

x	−2	−1	0	1	2
y	8	5	4	5	8

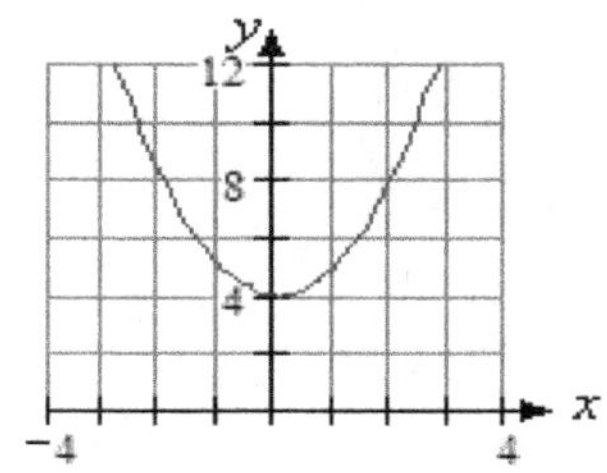

The axis of symmetry is $x = 0$, the vertex is $(0, 4)$, the y-intercept is $(0, 4)$, and there are no x-intercepts.

b.

x	−3	−1	0	1	3
y	−5	3	4	3	−5

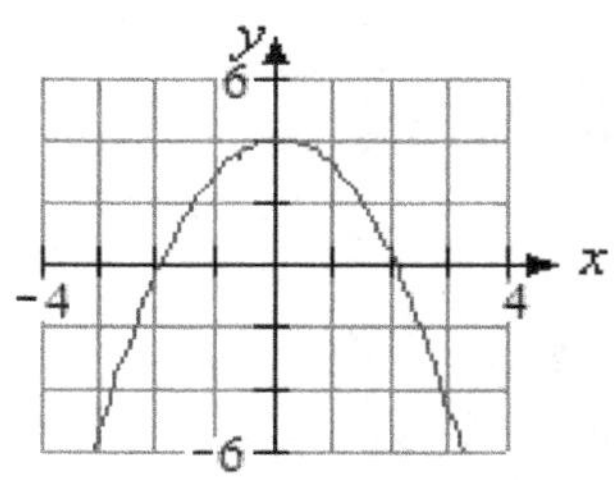

The axis of symmetry is $x = 0$, the vertex is $(0, 4)$, the y-intercept is $(0, 4)$, and the x-intercepts are $(-2, 0)$ and $(2, 0)$.

c.

x	0	1	2	3	4
y	4	1	0	1	4

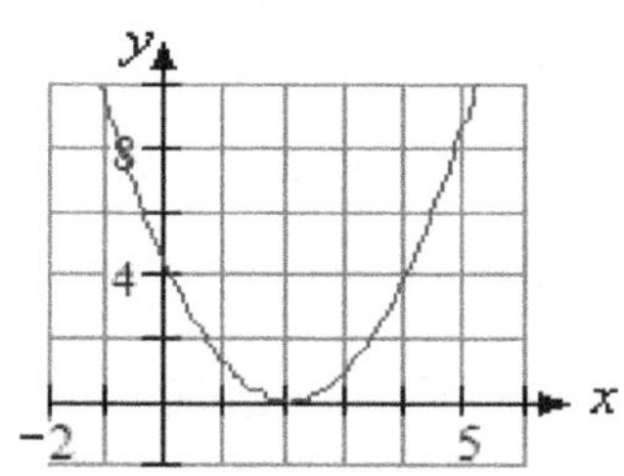

The axis of symmetry is $x = 2$, the vertex is $(2, 0)$, the y-intercept is $(0, 4)$, and the x-intercept is $(2, 0)$.

d.

x	−4	−3	−2	−1	0
y	4	1	0	1	4

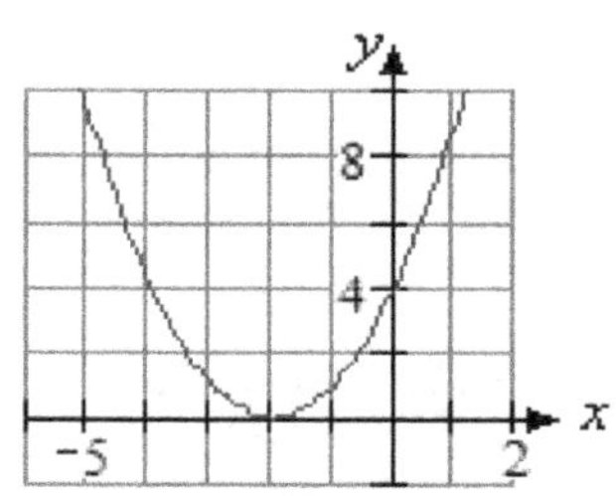

The axis of symmetry is $x = -2$, the vertex is $(-2, 0)$, the y-intercept is $(0, 4)$, and the x-intercept is $(-2, 0)$.

The values of the constants a, b, and c determine the location and orientation of the parabola. We'll begin by considering each of these constants separately.

The Graph of $y = ax^2$

First we'll investigate the effect of the constant a on the shape of the graph. Use your calculator to graph the following three equations on the same screen:

$$y = x^2$$
$$y = 3x^2$$
$$y = 0.1x^2$$

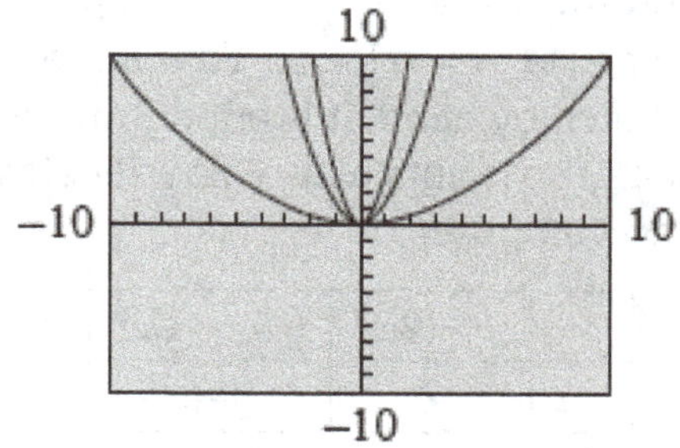

You can see that, compared to the basic parabola, the graph of $y = 3x^2$ is narrower, and the graph of $y = 0.1x^2$ is wider.

Look Closer: As x increases, the graph of $y = 3x^2$ increases faster than the basic parabola, and the graph of $y = 0.1x^2$ increases more slowly. Compare the corresponding y-values for the three graphs shown in the table.

x	$y = x^2$	$y = 3x^2$	$y = 0.1x^2$
−2	4	12	0.4
1	1	3	0.1
3	9	27	0.9

For each x-value, the points on the graph of $y = 3x^2$ are *higher* than the points on the basic parabola, while the points on the graph of $y = 0.1x^2$ are *lower*. Multiplying x^2 by a positive constant greater than 1 stretches the graph vertically, and multiplying by a positive constant less than 1 squashes the graph vertically.

What about negative values for a? Consider the graphs of

$$y = x^2$$
$$y = -2x^2$$
$$y = -0.5x^2$$

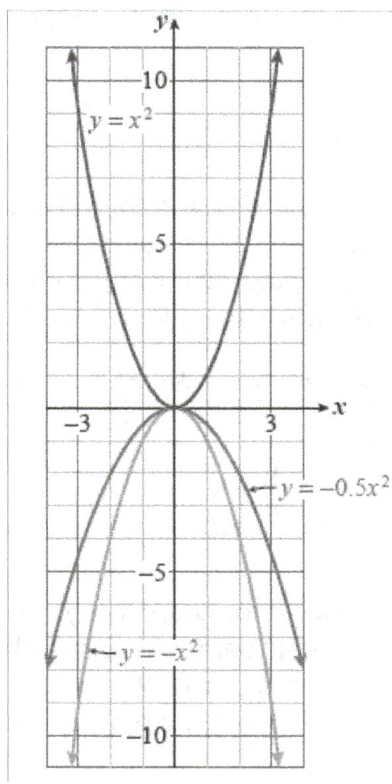

We see that multiplying x^2 by a negative constant reflects the graph about the x-axis. These parabolas open downward. In general, the graph of $y = ax^2$ opens upward if $a > 0$ and opens downward if $a < 0$. The magnitude of a determines how wide or narrow the parabola is. For equations of the form $y = ax^2$, the vertex, the x-intercepts, and the y-intercept all coincide at the origin.

The Graph of $y = ax^2$
1. opens upward if $a > 0$
2. opens downward if $a < 0$

The magnitude of $|a|$ determines how wide or narrow the parabola is.

Example 2 Sketch by hand graphs of each quadratic equation.

a. $y = 2x^2$ **b.** $y = -\frac{1}{2}x^2$

Solutions Both equations are of the form $y = ax^2$. The graph of $y = 2x^2$ opens upward because $a = 2 > 0$, and the graph of $y = -\frac{1}{2}x^2$ opens downward because $a = -\frac{1}{2} < 0$. To make a reasonable sketch by hand, it is enough to plot a few "guide points;" the points with x-coordinates 1 and -1 are easy to compute.

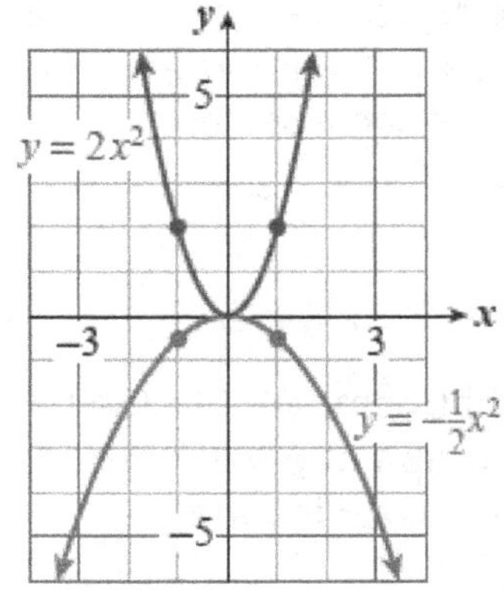

x	$y = 2x^2$	$y = -\frac{1}{2}x^2$
−1	2	$-\frac{1}{2}$
0	0	0
1	1	$-\frac{1}{2}$

We sketch parabolas through each set of guidepoints, as shown in the figure.

RQ1 True or false.
a. The vertex of a parabola lies on its axis of symmetry.
b. A parabola may have one, two, or three x-intercepts.
RQ2 True or false.
a. If $a > 1$, the graph of $y = ax^2$ is wider than the basic parabola.
b. If the y-intercept is negative, the parabola opens downward.

The Graph of $y = x^2 + c$

Next we consider the effect of the constant term, c, on the graph. Compare the graphs of

$$y = x^2$$
$$y = x^2 + 4$$
$$y = x^2 - 4$$

The graph of $y = x^2 + 4$ is shifted *upward* four units compared to the basic parabola, and the graph of $y = x^2 - 4$ is shifted *downward* four units. Look at the table, which shows the y-values for the three graphs.

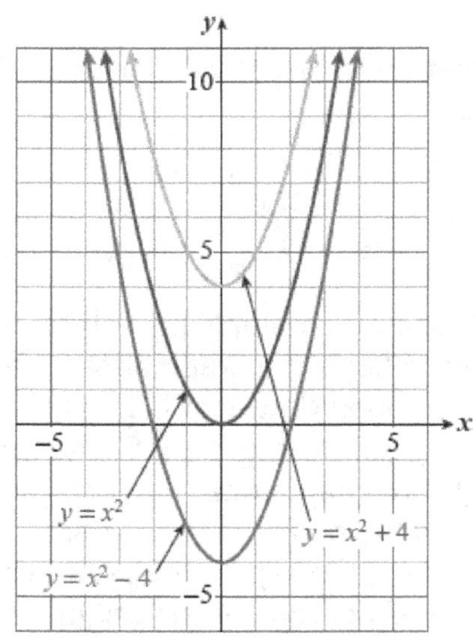

x	$y = x^2$	$y = x^2 + 4$	$y = x^2 - 4$
−1	1	5	−3
0	0	4	−4
2	4	8	0

Each point on the graph of $y = x^2 + 4$ is four units higher than the corresponding point on the basic parabola, and each point on the graph of $y = x^2 - 4$ is four units lower. In particular, the vertex of the graph of $y = x^2 + 4$ is the point $(0, 4)$, and the vertex of the graph of $y = x^2 - 4$ is the point $(0, -4)$.

The Graph of $y = x^2 + c$
1. is shifted upward by c units if $c > 0$
2. is shifted downward by c units if $c < 0$
compared to the graph of $y = x^2$.

Look Closer: We can find the x-intercepts of the graph of $y = x^2 - 4$ by setting y equal to zero and solving for x:

$$x^2 - 4 = 0$$
$$x^2 = 4 \qquad \textbf{Extract roots.}$$
$$x = 2, \qquad x = -2$$

The x-intercepts are the points $(2, 0)$ and $(-2, 0)$. The graph of $y = x^2 + 4$ has no x-intercepts, because the equation

$$0 = x^2 + 4$$

(obtained by setting $y = 0$) has no real-number solutions.

Example 3 Sketch graphs for the following quadratic equations.

a. $y = x^2 - 2$ **b.** $y = -x^2 + 4$

Solutions a. The graph of $y = x^2 - 2$ is shifted downward by two units, compared to the basic parabola. The vertex is the point $(0, -2)$, and the x-intercepts are the solutions of the equation

$$0 = x^2 - 2$$

or $\sqrt{2}$ and $-\sqrt{2}$. The graph is shown at right.

b. The graph of $y = -x^2 + 4$ opens downward and is shifted 4 units up, compared to the basic parabola. Its vertex is the point $(0, 4)$. Its x-intercepts are the solutions of the equation

$$0 = -x^2 + 4$$

or 2 and -2. You can also verify both graphs with your graphing calculator.

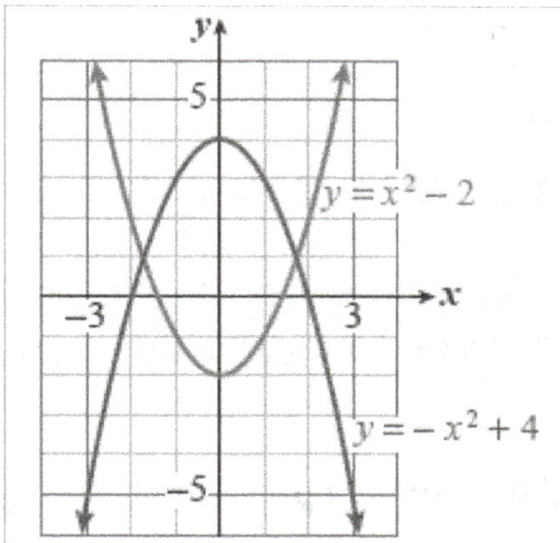

RQ3 Match each equation with the description of its graph.

a. $y = x^2 - 3$ **b.** $y = -3x^2$
c. $y = 3 - x^2$ **d.** $y = \frac{1}{3}x^2$

I. Wider than the basic parabola.
II. Shifted down 3 units from the basic parabola.
III. Opens downward, narrower than the basic parabola.
IV. Opens downward, shifted up 3 units.

The Graph of $y = ax^2 + bx$

How does the presence of a linear term, bx, affect the graph?

Example 4 Describe the graph of the equation $y = 2x^2 + 8x$

Solution The graph in the standard window is shown below. We see that the axis of symmetry for this parabola is *not* the y-axis: the graph is shifted to the left, compared to the basic parabola. We find the x-intercepts of the graph by setting y equal to zero:

$$\begin{aligned} 0 &= 2x^2 + 8x \qquad \textbf{Factor.} \\ &= 2x(x+4) \end{aligned}$$

The solutions of this equation are 0 and −4, so the x-intercepts are the points $(0, 0)$ and $(-4, 0)$.

We can find the vertex of the graph by using the symmetry of the parabola. The x-coordinate of the vertex lies exactly half-way between the x-intercepts, so we average their x-coordinates to find:

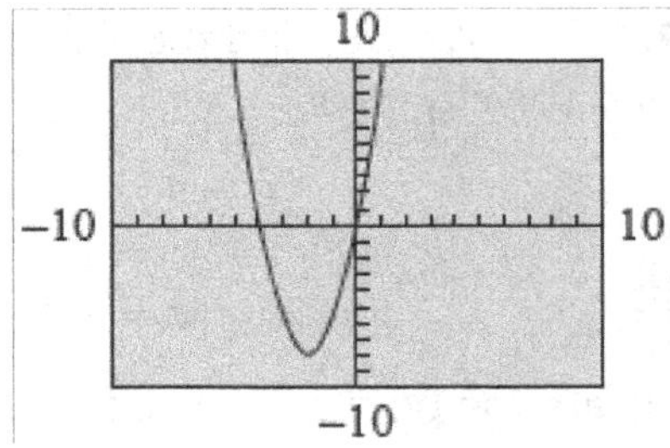

$$x_v = \frac{1}{2}[0 + (-4)] = -2$$

To find the y-coordinate of the vertex, we substitute $x = -2$ into the equation for the parabola:

$$\begin{aligned} y_v &= 2(-2)^2 + 8(-2) \\ &= 8 - 16 = -8 \end{aligned}$$

The vertex is the point $(-2, -8)$. ■

RQ4 True or false.

a. The axis of symmetry of $y = ax^2 + bx$ is shifted horizontally, compared to the basic parabola.

b. The x-coordinate of the vertex is the average of the x-intercepts.

A Formula for the Vertex

We can find a formula for the vertex of any parabola of the form

$$y = ax^2 + bx$$

First, we find the x-intercepts of the graph by setting y equal to zero and solving for x:

$$\begin{aligned} 0 &= ax^2 + bx && \textbf{Factor.} \\ &= x(ax + b) && \textbf{Set each factor equal to zero.} \\ x = 0 \quad &\text{or} \quad ax + b = 0 && \textbf{Solve for } x. \\ x = 0 \quad &\text{or} \quad x = \frac{-b}{a} \end{aligned}$$

The x-intercepts are the points $(0, 0)$ and $\left(\frac{-b}{a}, 0\right)$.

Next, we find the x-coordinate of the vertex by taking the average of the two x-intercepts.

$$x_v = \frac{1}{2}\left[0 + \left(\frac{-b}{a}\right)\right] = \frac{-b}{2a}$$

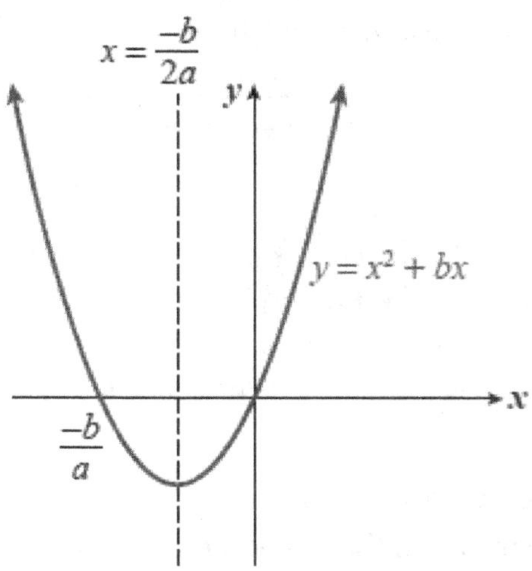

Now we have a formula for the x-coordinate of the vertex.

Vertex of a Parabola

For the graph of $y = ax^2 + bx$, the x-coordinate of the vertex is

$$x_v = \frac{-b}{2a}$$

We find the y-coordinate of the vertex by substituting its x-coordinate into the equation for the parabola.

Example 4a. Find the vertex of the graph of $y = -1.8x^2 - 16.2x$

b. Find the x-intercepts of the graph.

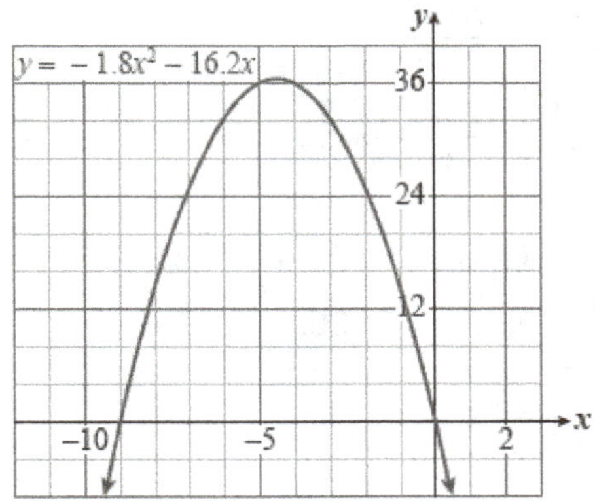

Solutions a. The x-coordinate of the vertex is

$$x_v = \frac{-b}{2a} = \frac{-(-16.2)}{2(-1.8)} = -4.5$$

To find the y-coordinate of the vertex, we evaluate y at $x = -4.5$.

$$y_v = -1.8(-4.5)^2 - 16.2(-4.5) = 36.45$$

The vertex is $(-4.5, 36.45)$.

b. To find the x-intercepts of the graph, we set $y = 0$ and solve.

$-1.8x^2 - 16.2x = 0$	**Factor.**
$-x(1.8x + 16.2) = 0$	**Set each factor equal to zero.**
$-x = 0 \quad 1.8x + 16.2 = 0$	**Solve each equation.**
$x = 0 \quad\quad x = -9$	

The x-intercepts of the graph are $(0, 0)$ and $(-9, 0)$.

RQ5 True or false.

a. The y-coordinate of the vertex is the same as the y-intercept.

b. We can use extraction of roots to solve $ax^2 + bx = 0$.

RQ6 Match each equation with its graph.

a. $y = x^2 - 2$ **b.** $y = x^2 - 2x$

c. $y = x^2 + 2x$ **d.** $y = -2x^2$

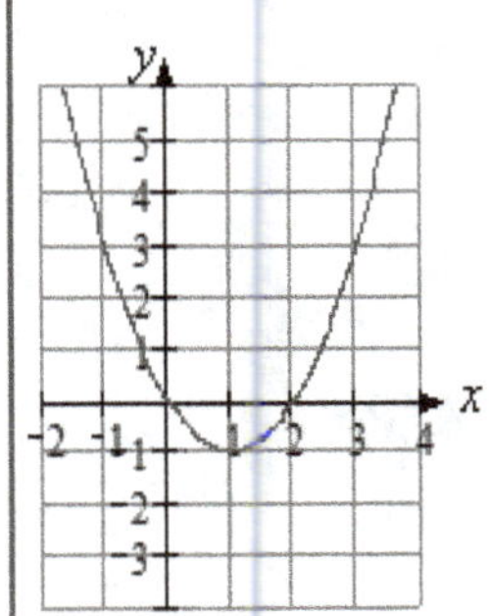

I

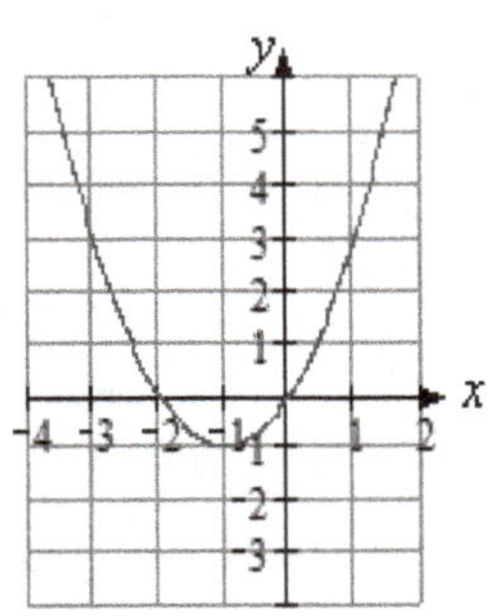

II

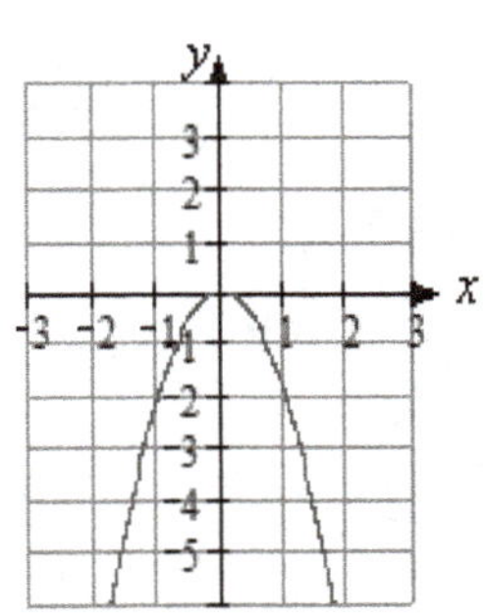

III

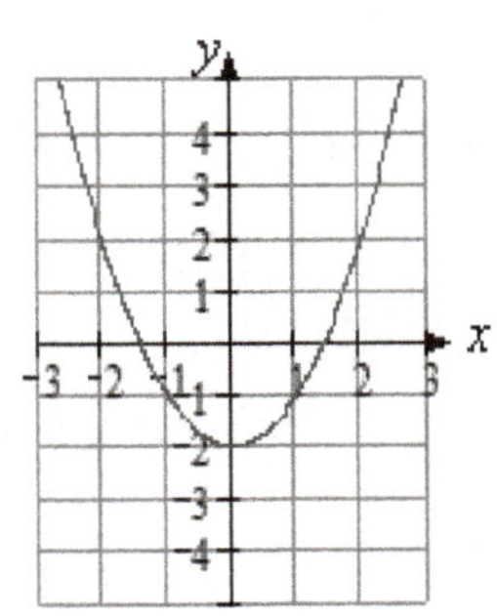

IV

Skills Warm-UP

Follow the order of operations to simplify.

1. $3(-5) - 2^3$ **2.** $6(7-4)^2$ **3.** $-6 - 2 \cdot 4^2$

4. $(5-3)^4(3-6)^3$ **5.** $4(4-4^2)$ **6.** $-3(-2)^2 - 5$

7. $-(3 \cdot 2)^2 - 5$ **8.** $3 - 4(-2)^3(-3)$

9. $\dfrac{-5^2+1}{4} + \dfrac{2(-3)^3}{-6}$ **10.** $\dfrac{2^2(-3^2)}{1-3^2} - \dfrac{7^2-6^2}{(1-3)^2}$

11. $\dfrac{3^3-3}{(3-5)^3} - \dfrac{2(-2)^3-8}{-2^2(8-2^2)}$ **12.** $3 \cdot \dfrac{5^3-(-10)^2}{3^2-3(5)} \cdot \dfrac{2^5+4}{3^2-2^2}$

Answers: **1.** -23 **2.** 54 **3.** -38 **4.** -432 **5.** -48 **6.** -17 **7.** -41
8. -93 **9.** 3 **10.** $\dfrac{5}{4}$ **11.** $\dfrac{-9}{2}$ **12.** -90

Homework 5.6

Skills Practice

■ For Problems 1-4, describe what the graph will look like compared to the basic parabola. Then sketch the graph by hand, and label the coordinates of three points on the graph.

1. $y = 2x^2$

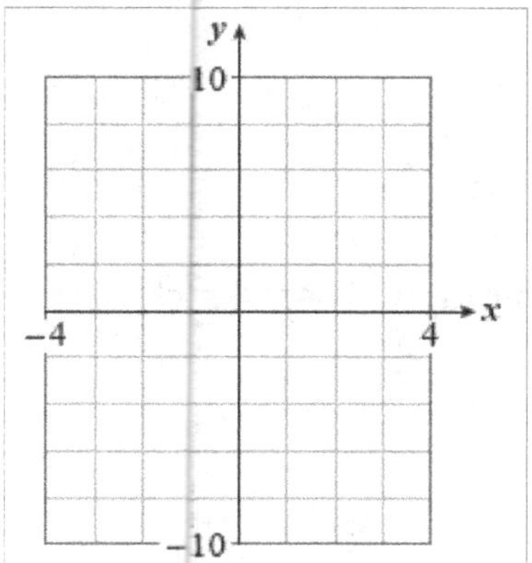

2. $y = \frac{1}{2}x^2$

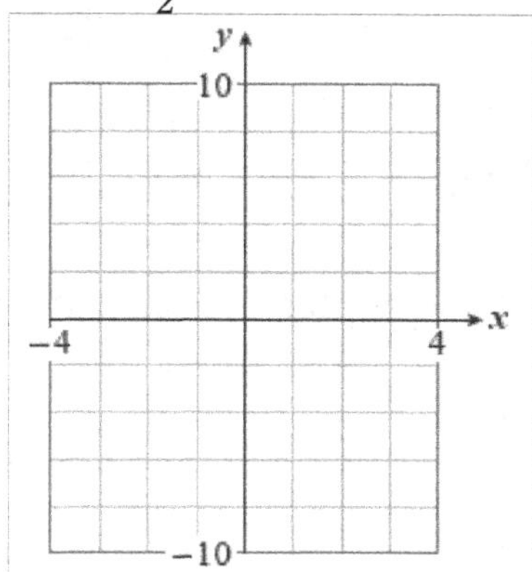

3. $y = -x^2$

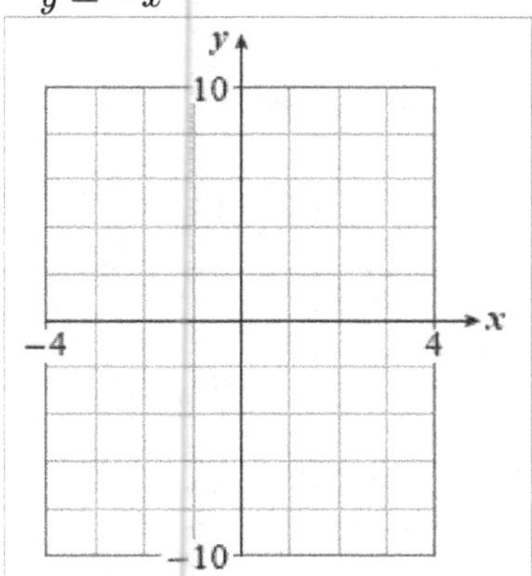

4. $y = -0.2\,x^2$

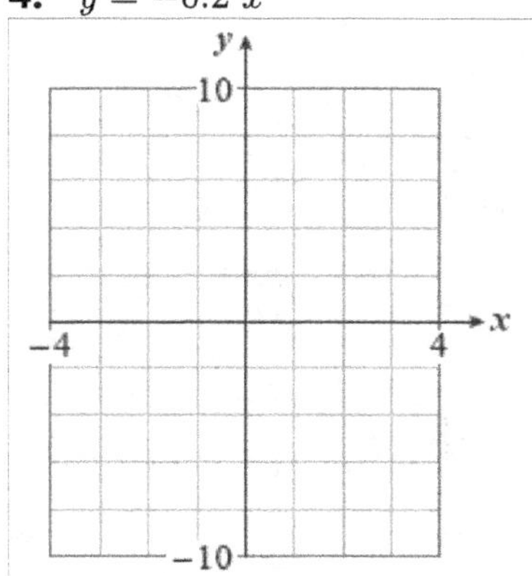

■ For Problems 5-8, describe what the graph will look like compared to the basic parabola. Then sketch the graph by hand. Label the vertex and the x-intercepts (if there are any) with their coordinates.

5. $y = x^2 + 2$

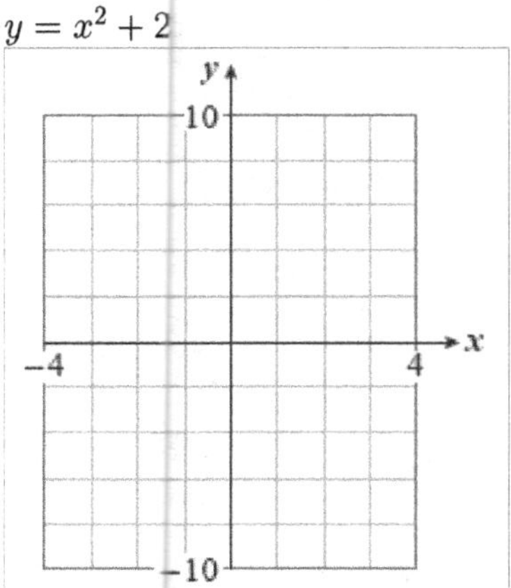

6. $y = x^2 - 1$

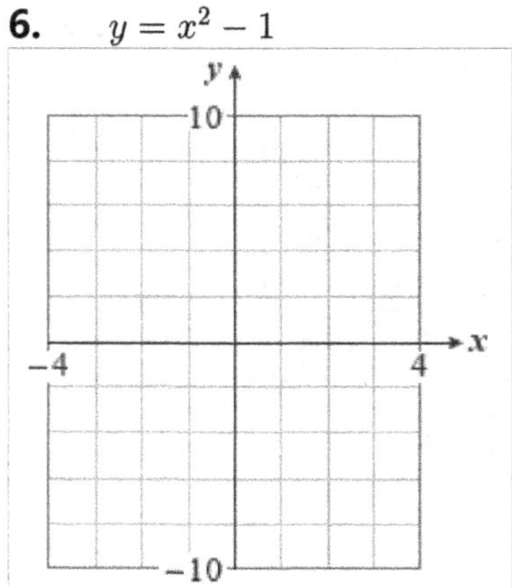

7. $y = x^2 - 5$

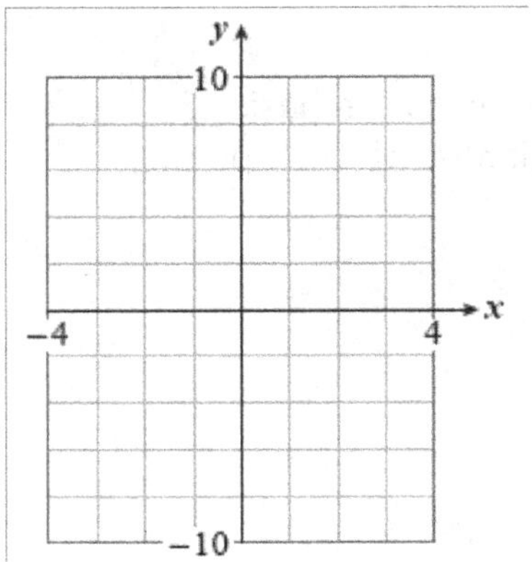

8. $y = 100 - x^2$

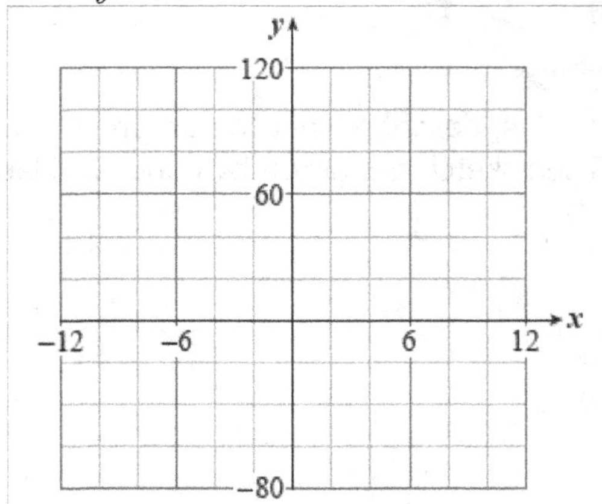

■ For Problems 9-12, find the x-intercepts and the vertex of each graph. Then sketch the graph by hand.

9. $y = x^2 - 4x$

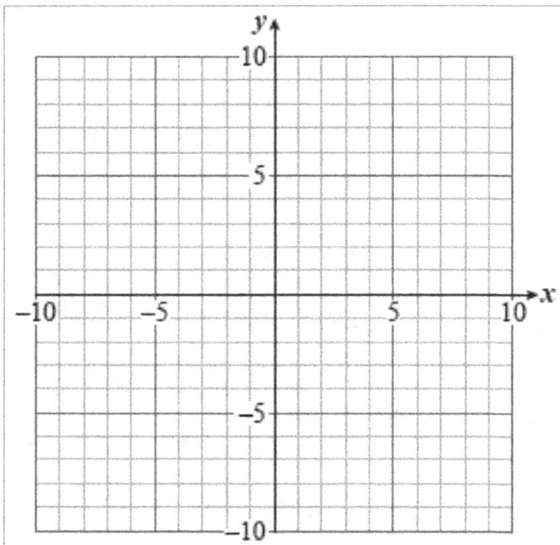

10. $y = x^2 + 2x$

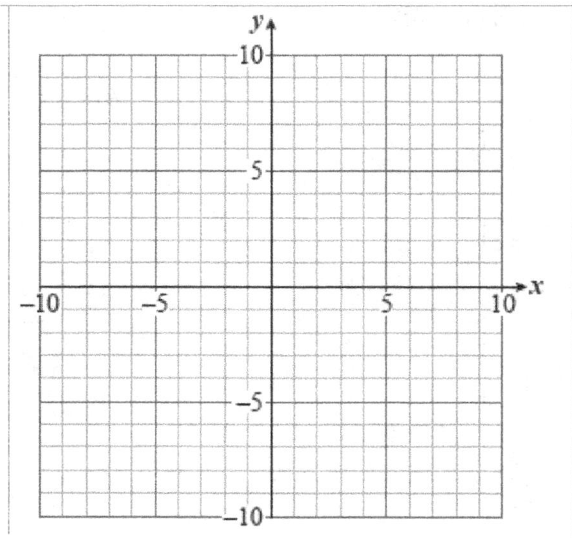

11. $y = 3x^2 + 6x$

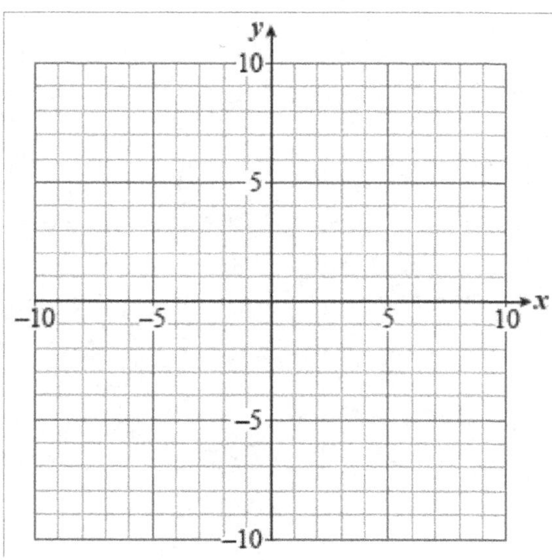

12. $y = -2x^2 + 5x$

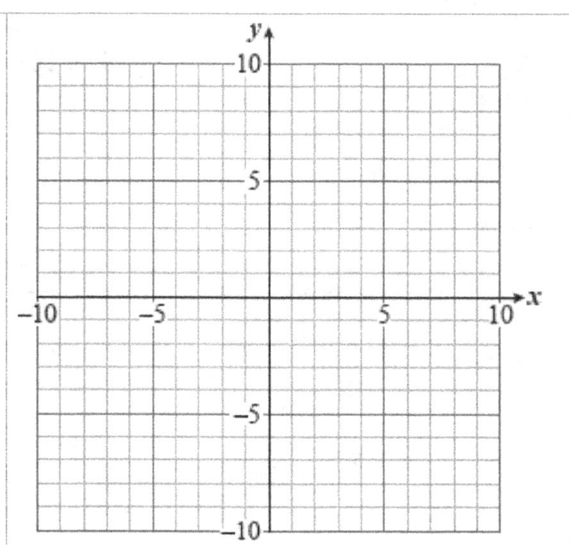

13. Match each equation with its graph. In each equation, $a > 0$.

a. $y = x^2 + a$ **b.** $y = x^2 + ax$ **c.** $y = ax^2$

d. $y = ax$ **e.** $y = x + a$ **f.** $y = x^2 - a$

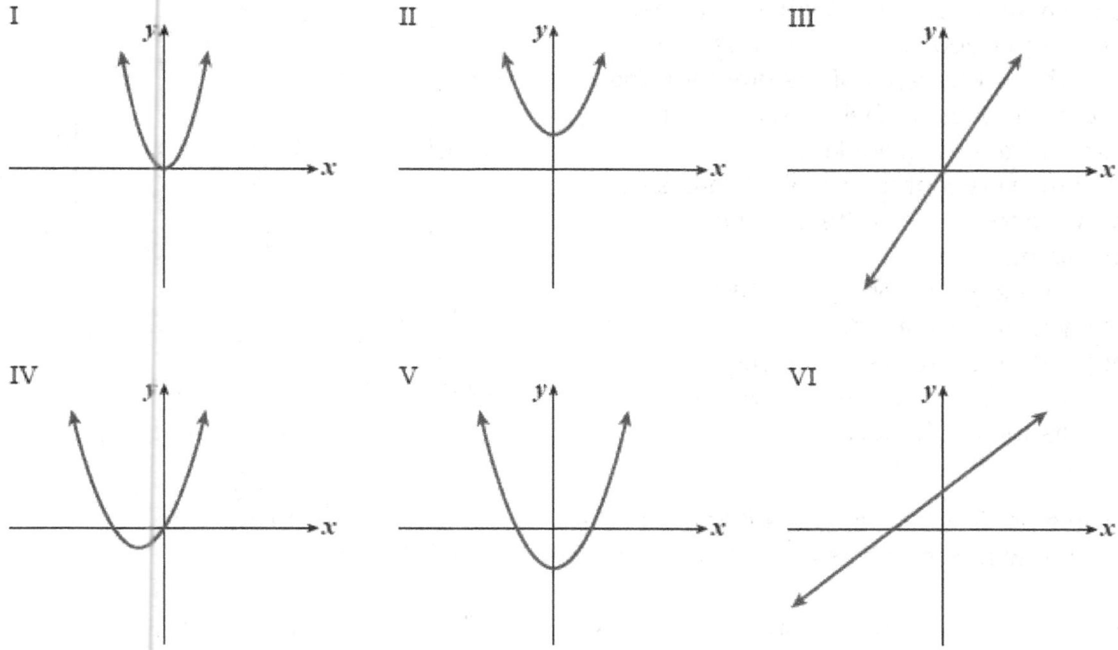

Applications

14. The revenue that you can earn by making and selling n jade bracelets is given in dollars by

$$R = -2n^2 + 80n$$

a. Factor the expression for R.

b. Find the horizontal intercepts and the vertex, and sketch the graph.

c. What is the maximum revenue that you can earn? How many bracelets should you make in order to earn that revenue?

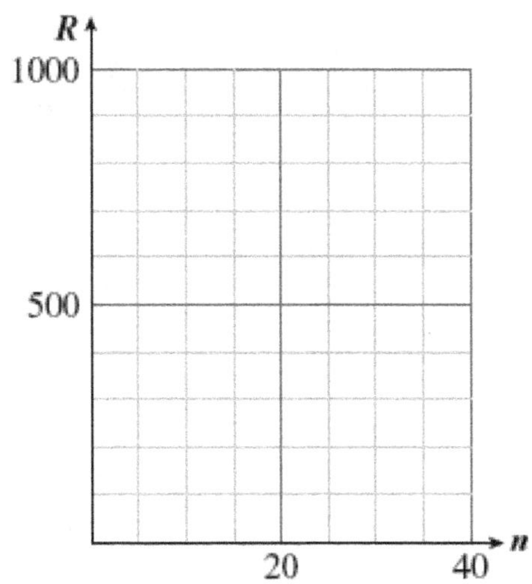

15. The height in feet of a football t seconds after being kicked from the ground is given by

$$h = -16t^2 + 80t$$

a. Factor the expression for h.

b. Find the horizontal intercepts and the vertex, and sketch the graph.

c. What is the maximum height of the football? When does it reach this height?

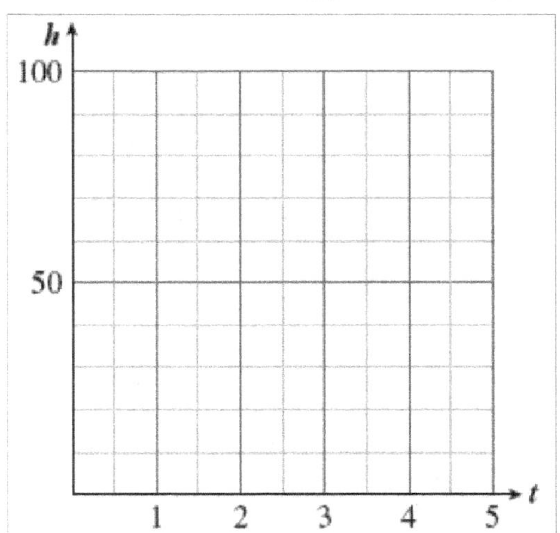

16. Commercial fishermen rely on a steady supply of fish in their area. To avoid overfishing, they adjust their harvest to the size of the population. The equation

$$y = 0.4x - 0.0001x^2$$

gives the annual rate of growth, in tons per year, of a fish population of biomass x tons.

a. Find the x-intercepts of the graph of the equation. What is their significance in terms of the fish population?

b. Find the vertex of the graph. What is its significance in the context of the problem?

c. Sketch the graph for $0 \le x \le 5000$.

d. For what values of x does the fish population decrease rather than increase? Suggest a reason why the population might decrease.

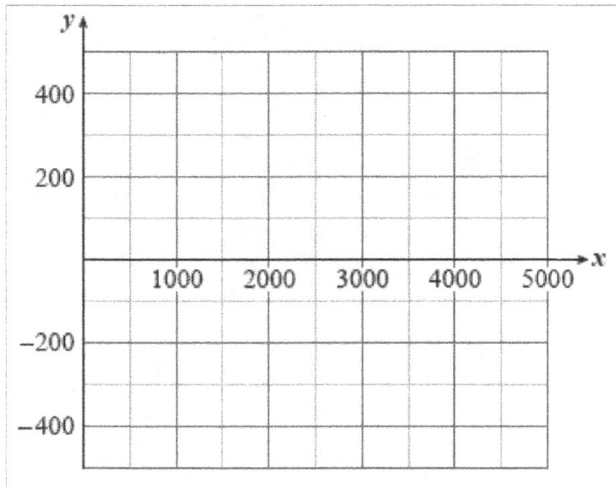

For Problems 17-18, graph the equations in the same window on your calculator. Describe how the constants in each equation affect the graph.

17a. $y = x^2$	**b.** $y = 3x^2$	**c.** $y = \frac{1}{4}x^2$	**d.** $y = -2x^2$
18a. $y = x^2 - 4x$	**b.** $y = x^2 + 4x$	**c.** $y = 4x - x^2$	**d.** $y = -x^2 - 4x$

Chapter 5 Summary and Review

Section 5.1 Exponents and Formulas

- An **exponent** tells us how many times its **base** occurs as a factor in an expression.
- To indicate that a negative number is raised to a power, we must enclose the negative number in parentheses.
- An exponent on a variable indicates repeated multiplication, while a coefficient in front of a variable indicates repeated addition.
- We can add or subtract like powers of the same variable by combining their coefficients; the exponent does not change.
- **Volume** is the amount of space contained within a three-dimensional object, and is measured in cubic units, such as cubic feet or cubic centimeters. **Surface area** is the sum of the areas of all the faces or surfaces that contain a solid, and is measured in square units.

Section 5.2 Square Roots and Cube Roots

Square Root

s is called a **square root** of a number b if $s^2 = b$

- Every positive number has two square roots, one positive and one negative. The positive square root of a number is called the **principal square root**.
- We cannot take the square root of a negative number.

Cube Root

c is called a **cube root** of a number b if $c^3 = b$

- Every number has exactly one cube root.
- A **rational number** is one that can be expressed as a quotient (or ratio) of two integers, where the denominator is not zero.

Decimal Form of a Rational Number

The decimal representation of a rational number has one of two forms.

1. The decimal representation terminates, or ends.
2. The decimal representation repeats a pattern.

- The rational and irrational numbers together make up the **real numbers**, and the number line is sometimes called the **real line**.

Order of Operations

1. Perform any operations inside parentheses, under a radical, or above or below a fraction bar.
2. Compute all indicated powers and roots.
3. Perform all multiplications and divisions in the order in which they occur from left to right.
4. Perform additions and subtractions in order from left to right.

Section 5.3 Quadratic Equations

Quadratic Equations

A **quadratic equation** can be written in the standard form

$$ax^2 + bx + c = 0$$

where a, b, and c are constants and a is not zero.

- The graph of a quadratic equation, $y = ax^2 + bx + c$, is called a **parabola**.

Extraction of Roots

To solve a quadratic equation of the form

$$ax^2 + c = 0$$

1. Isolate x^2 on one side of the equation.
2. Take the square root of each side.

Pythagorean Theorem

If c stands for the length of the hypotenuse of a right triangle, and the lengths of the two legs are represented by a and b, then

$$a^2 + b^2 = c^2$$

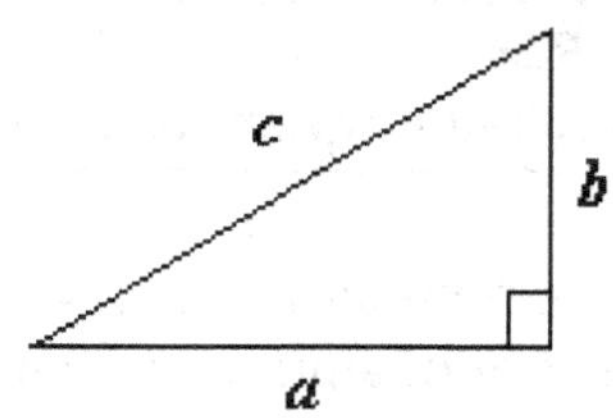

Section 5.4 Products of Binomials

- An algebraic expression with only one term is called a **monomial**. An expression with two terms is called a **binomial**, and an expression with three terms is a **trinomial**. The expression $ax^2 + bx + c$ is thus called a **quadratic trinomial**.
- We use the distributive law to expand the product of two binomials. The letters **F**, **O**, **I**, **L** indicate how each term of the product arises from the binomial factors:
  1. **F** stands for the product of the **First** terms in each binomial.
  2. **O** stands for the product of the **Outer** terms.
  3. **I** stands for the product of the **Inner** terms.
  4. **L** stands for the product of the **Last** terms.

- We cannot square a binomial by squaring each term separately; we must expand the product into its four terms.

Section 5.5 Solving Quadratic Equations

Zero-Factor Principle

If the product of two numbers is zero, then one (or both) of the numbers must be zero. Using symbols,

If $ab = 0$, then either $a = 0$ or $b = 0$.

- Writing a quadratic expression in factored form is called **factoring**, and it is the reverse of multiplying factors together.

To Solve a Quadratic Equation by Factoring:

1. Write the equation in standard form,

$$ax^2 + bx + c = 0, \qquad (a \neq 0)$$

2. Factor the left side of the equation.
3. Apply the zero-factor principle; that is, set each factor equal to zero.
4. Solve each equation to obtain two solutions.

- To find the x-intercepts of the graph of $y = ax^2 + bx + c$, we set $y = 0$ and solve the equation $ax^2 + bx + c = 0$.
- Quadratic equations of the form $a(x - p)^2 = q$ can be solved by extraction of roots after isolating the squared expression, $(x - p)^2$.
- After n years, the amount of money in an account on which interest is compounded annually is given by

$$A = P\,(1 + r)^n$$

where P is the principal and r is the interest rate, expressed as a decimal fraction.

Section 5.6 Graphing Parabolas

- The high or low point of a parabola is called the **vertex** of the graph. The parabola has **axis of symmetry** that runs through the vertex.
- The values of the constants a, b, and c determine the location and orientation of the parabola.

The Graph of $y = ax^2$

1. opens upward if $a > 0$
2. opens downward if $a < 0$

The magnitude, $|a|$, determines the width of the parabola.

The Graph of $y = x^2 + c$

1. is shifted upward by c units if $c > 0$
2. is shifted downward by $|c|$ units if $c < 0$

compared to the graph of $y = x^2$.

Vertex of a Parabola

For the graph of $y = ax^2 + bx$, the x-coordinate of the vertex is

$$x_v = \frac{-b}{2a}$$

Review Questions

Use complete sentences to answer the questions in Problems 1-20.

1. Explain the difference between -3^2 and $(-3)^2$.
2. If $s^2 = 17$, then s is called a ______________ of 17.
3. If $s^3 = 17$, then s is called a ______________ of 17.
4. What is the difference between the square of 4 and the square root of 4?
5. What is the difference between $x^2 = 9$ and $x = \sqrt{9}$?
6. What is the difference between $-\sqrt{25}$ and $\sqrt{-25}$?
7. Explain the difference between a coefficient and an exponent. Use the expressions $2x^3$ and $3x^2$ to illustrate your explanation.
8. Is the statement $\sqrt{81} = \sqrt{9} = 3$ correct? Explain why or why not.
9. When asked to state the Pythagorean theorem, a classmate says, "$a^2 + b^2 = c^2$ when the sides of a triangle are a, b, and c." Another student claims that the formula does not work for all triangles. Do you agree with either student? What necessary part or parts of the theorem are not mentioned by either of the two classmates?
10. A classmate says that $\sqrt[3]{2}$ must be a rational number because it can be written as $\frac{\sqrt[3]{2}}{1}$. Do you agree or disagree? Explain.
11. Give an example of a negative irrational number.
12. What do you get if you square $\sqrt[3]{5}$ and then multiply the result by $\sqrt[3]{5}$? Explain how to get the correct answer without using a calculator.
13. What do volume and surface area measure? What are the units of each?
14. State the zero factor principle. What is it used for?
15. Write down the steps for solving a quadratic equation by extraction of roots.
16. How can you tell from its decimal form whether a number is rational or irrational?
17. What are the four terms in the product $(a + b)(c + d)$?
18. If you know the x-intercepts of a parabola, how can you find the coordinates of its vertex?
19. Francine factored $3x^2 - 2xy + x$ and got $x(3x - 2y)$. Why is this incorrect?
20. Describe the similarities and differences between the graphs of $y = -2x^2$ and $y = x^2 - 2$.

Review Problems

For Problems 1-6, simplify.

1. a. $3(-7)^2$ b. $3^2(-7^2)$
   c. $3^2 - 7^2$ d. $(3 - 7)^2$

2. a. $\sqrt{13^2 - 5^2}$ b. $\sqrt{13^2} - \sqrt{5^2}$
   c. $\sqrt{13^2}\sqrt{5^2}$ d. $\sqrt{(13 - 5)^2}$

3. a. $\sqrt{\left(\frac{4}{5}\right)^2 + \left(\frac{3}{5}\right)^2}$ b. $\sqrt{\left(\frac{4}{5}\right)^2} + \sqrt{\left(\frac{3}{5}\right)^2}$
   c. $\sqrt{\left(\frac{4}{5}\right)^2}\sqrt{\left(\frac{3}{5}\right)^2}$ d. $\sqrt{\left(\frac{4}{5} + \frac{3}{5}\right)^2}$

4\. a. $2-\sqrt[3]{64}$ b. $2\sqrt[3]{-64}$
c. $\left(\sqrt[3]{101}\right)^3$ d. $\sqrt[3]{7}\cdot\sqrt[3]{7}\cdot\sqrt[3]{7}$

5\. a. $\left(\sqrt[3]{8}\right)^2$ b. $\sqrt[3]{8^2}$
c. $\left(\sqrt{9}\right)^3$ d. $\sqrt{9^3}$

6\. a. $(\sqrt{3x})(\sqrt{3x})$ b. $3\sqrt{x}\,(3\sqrt{x})$
c. $(5\sqrt{a})^2$ d. $(\sqrt[3]{B})^2(\sqrt[3]{B})$

For Problems 7-8, identify each number as rational, irrational or undefined.

7a. $\sqrt[3]{-5}$ b. $\sqrt{-5^2}$ c. $\sqrt{(-5)^2}$ d. $\left(\sqrt[3]{-5}\right)^2$
8a. $-3.\overline{16}$ b. -3.16 c. $-\sqrt{10}$ d. $-\pi$

For Problems 9-10, decide whether each number is rational or irrational. Then find a decimal form for each number. If it is irrational, round to hundredths.

9a. $\sqrt{300}$ b. $5+\sqrt[3]{15}$ 10a. $\sqrt[3]{512}$ b. $\dfrac{7}{\sqrt{81}}$

For Problems 11-14, simplify.

11\. $4-2\cdot 3^2$ 12. $-2-3(-3)^3-2$
13\. $\dfrac{-6-\sqrt{6^2-4(2)(4)}}{2(2)}$ 14. $18-2\sqrt[3]{\dfrac{4}{3}(48)}$

For Problems 15-16, give a decimal approximation rounded to thousandths.

15\. $-8-5\sqrt{6}$ 16. $\dfrac{3+\sqrt[3]{3}}{3}$

For Problems 17-26, evaluate the expression for the given values of the variables. Round to three decimal places any answers which are irrational numbers.

17\. $a=3$: a. $5-a^2$ b. $(5-a)^2$
18\. $a=-3$: a. $5-a^2$ b. $(5-a)^2$
19\. $w=-2$: a. $-w^2$ b. $(-w)^2$
20\. $w=-2$: a. $\sqrt{-w}$ b. $\sqrt[3]{-w}$
21\. $x=-4$: a. $3-x^2$ b. $3(-x)^2$
22\. $a=6,\ b=-2$: a. $\dfrac{3b^3}{6}-\dfrac{4-a^2}{8}$ b. $\dfrac{(a-b)^2}{ab^2}$
23\. $x=13,\ y=-5$: a. $\sqrt{x^2-y^2}$ b. $\left(\sqrt{x}\right)^2+\left(\sqrt{-y}\right)^2$
24\. $a=2,\ b=-3,\ c=1$: a. $\sqrt{b^2-4ac}$ b. $\dfrac{-b+\sqrt{b^2-4ac}}{2a}$
25\. $x=3,\ y=4,\ z=5$: a. $\sqrt[3]{x^3+y^3+z^3}$ b. $\sqrt[3]{x^3}+\sqrt[3]{y^3}+\sqrt[3]{z^3}$
26\. $m=3,\ n=2$: a. m^3+n^3 b. $(m+n)(m^2-mn+n^2)$

For Problems 27-30,

a. Find the unknown side or sides for each right triangle.
b. Find the perimeter of the triangle.
c. Find the area of the triangle.

27. 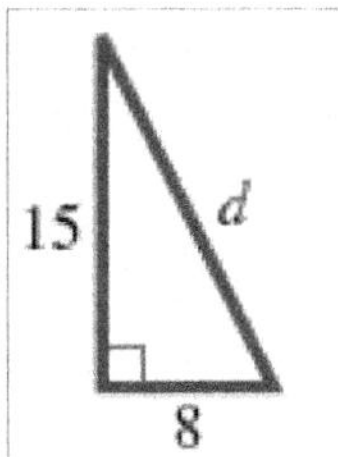

28. 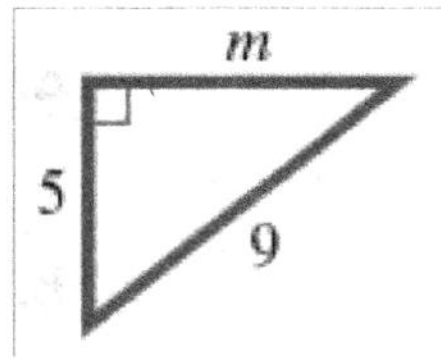

29. 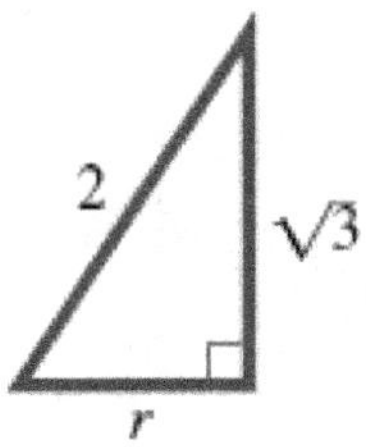

30. 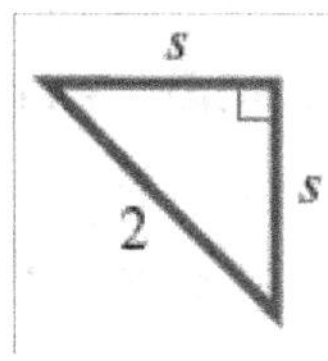

For Problems 31-32, find the volume and surface area of the box.

31. 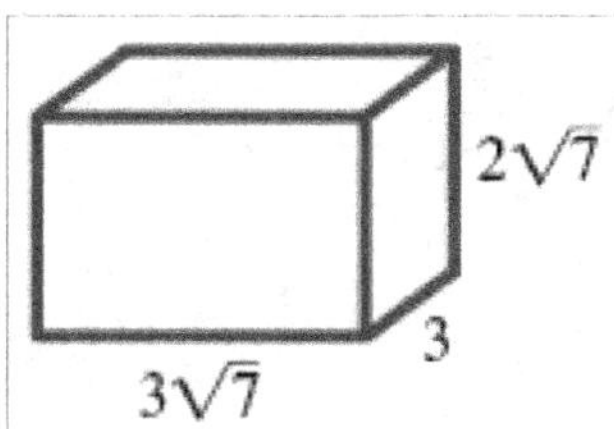

32. 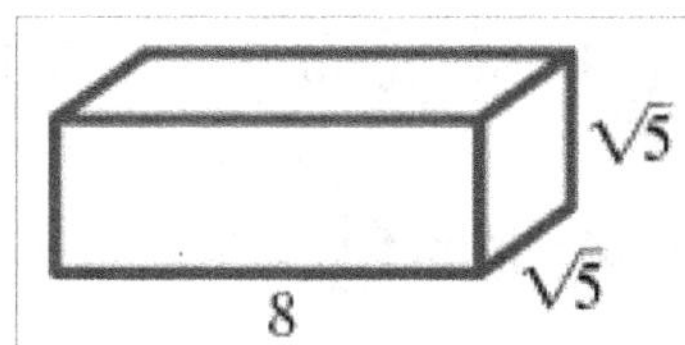

For Problems 33-34, write an algebraic expression for the volume of the figure.

33. 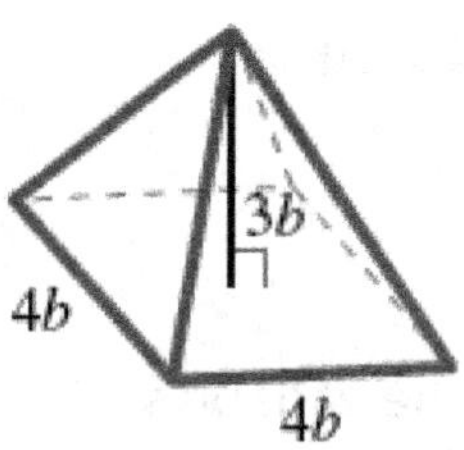

34. 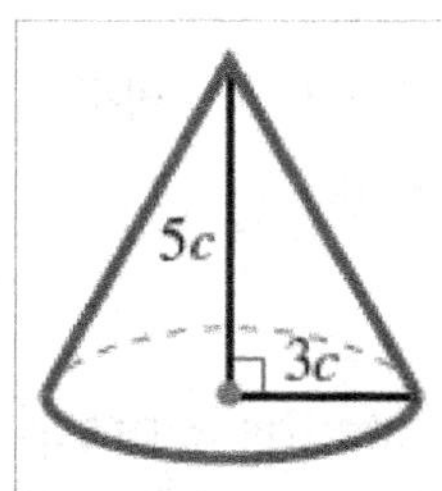

For Problems 35-36, write expressions for the volume and surface area of the figure.

35. 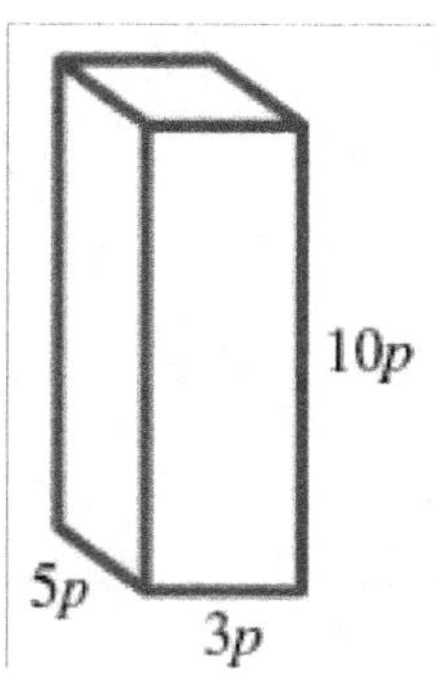

36. 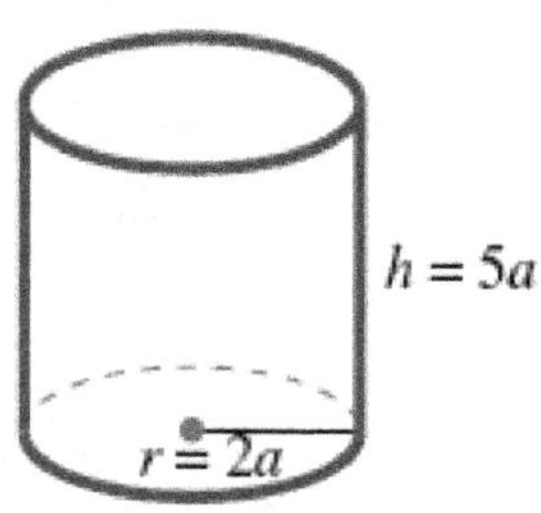

For Problems 37-46, solve by extraction of roots. Round to thousandths any answers that are irrational numbers.

37. $9x^2 - 4 = 0$

38. $3 - 2t^2 = 9$

39. $5y^2 - 12 = 2y^2$

40. $q^2 - 3 = 5 - q^2$

41. $9k^2 + 21 = 25$

42. $6a^2 + 3 = 4a^2 + 19$

43. $3(x+4)^2 = 60$

44. $(3.5 - 0.2x)^2 = 1.44$

45. $(2x-5)^2 = 9$

46. $(7x-1)^2 = 15$

For Problems 47-54, solve the formula for the indicated variable.

47. $s = vt + \frac{1}{2}at^2$ for v

48. $A = \frac{h}{2}(b+c)$ for c

49. $V = \frac{s^2h}{3}$ for s

50. $A = \frac{\pi d^2}{4}$ for d

51. $S = \frac{n}{2}(a+f)$ for a

52. $s = vt + \frac{1}{2}at^2$ for a

53. $C = bh^2r$ for h

54. $G = \frac{np}{r^2}$ for r

55. Lewis invested \$2000 in an account that compounds interest annually. He made no deposits or withdrawals after that. Two years later he closed the account, withdrawing \$2464.20. What interest rate did Lewis earn?

56. Earl borrowed \$5500 from his uncle for 2 years with interest compounded annually. At the end of two years he owed his uncle \$6474.74. What was the interest rate on the loan?

57. The distance, d, that a penny will fall in t seconds is given in feet by

$$d = 16t^2$$

a. If you drop a penny from a height of 144 feet, how long will it take to reach the ground?
b. How far will the penny fall in $1\frac{1}{2}$ seconds?

58. a. How much water would you need to fill a spherical tank of radius 3 feet?
b. A biosphere must contain 7000 cubic inches of space. What is the radius of the biosphere?

For Problems 59-66, draw a sketch, write an equation, and solve. Round your answer to the nearest tenth.

59. Francine walks along the edge of a rectangular yard from the northwest corner to the southeast corner. Delbert walks diagonally across the same yard. If the yard is 30 yards long and 20 yards wide, how much farther does Francine walk than Delbert?

60. Rani kayaks due west for 2500 meters, then due south for another 1000 meters. If she now kayaks directly back to her starting point, how long is her total trip?

61. a. The volume of a spherical communications satellite is 65.45 cubic feet. What is the radius of the satellite?
b. What is the surface area of the satellite in part (a)?

62. a. Francine's coffee cup is shaped like a cylinder 10 centimeters tall with a radius of 3.36 centimeters. What is its volume in cubic centimeters?

b. One fluid ounce is about 29.56 cubic centimeters. Find the volume of Francine's coffee cup in ounces, rounded to the nearest ounce.

c. Delbert's coffee cup is also 10 centimeters tall, but its volume is 24 fluid ounces. What is its radius?

63. A rectangle is 7 meters wide and 9 meters long. What is the length of its diagonal?

64. The size of a TV screen is the length of its diagonal. What is the size of a TV whose screen measures 12 inches by 16 inches?

65. A 26-meter guy wire is attached to a radio antenna for support. One end of the wire is attached to the antenna 24 meters above the ground, and the other is attached to an iron ring in a cement slab on the ground. How far is the ring from the base of the antenna?

66. Find the height of an equilateral triangle whose sides are each 8 centimeters long.

■ For Problems 67-70, compute each product.

67. $-5x(4-3x)$ **68.** $(3xy-4x^2+2)(-2xy)$

69. $3xy(2x-4-y)$ **70.** $-2a(-a^2-2a+4)$

■ For Problems 71-72, simplify if possible.

71a. $6t^2-8t^2$ **b.** $6t-8t^2$ **c.** $6t(-8t^2)$

72a. w^2+w^2 **b.** $-w-w$ **c.** w^2-w

■ For Problems 73-76, simplify.

73. $6a(2a-1)-(3a^2-3a)$ **74.** $4b-2(3-b^2)-b(b-3)$

75. $5a(2a-3)-4(3a^2-2)+6a$ **76.** $2x(x-3y)-3y(x-3y)$

■ For Problems 77-78, fill in the missing algebraic expressions.

77. $-2n(3n+\mathbf{?})=-6n^2-2n$ **78.** $5b(6+\mathbf{?})=30b-35b^2$

■ For Problems 79-80, write the area of the rectangle in two different ways:

a. as the sum of four smaller areas,

b. as one large rectangle, using the formula Area = length × width.

79.

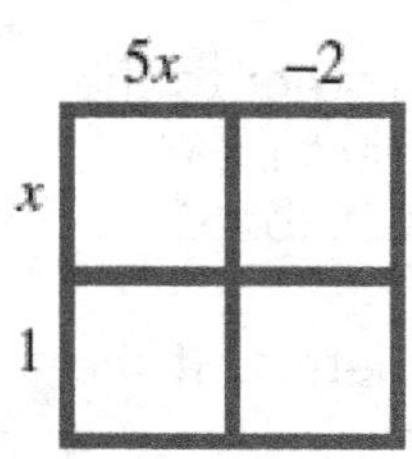

80.

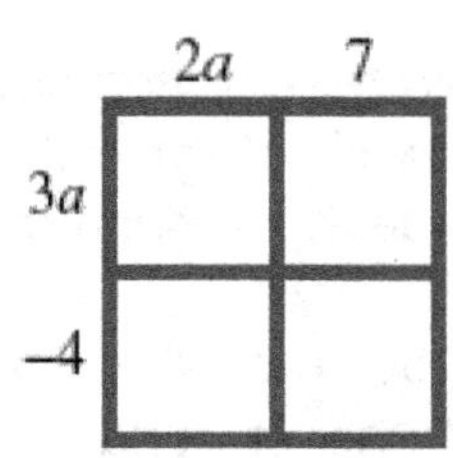

■ For Problems 81-92, compute the product.

81. $(x-2)(x+4)$
82. $(2y+1)(3y-2)$
83. $(5a+1)(5a-1)$
84. $(n-7)(n-7)$
85. $(2q+5)(2q+5)$
86. $(3c-8)(3c+8)$
87. $(u-5)(u-2)$
88. $(3r+2)(r-4)$
89. $(a+6)^2$
90. $(6y+5)(6y-5)$
91. $(2a-5c)(3a+2c)$
92. $-3(x-4)(2x+5)$

93. If you drop a stone from a bridge 100 feet above the water, the height h of the stone t seconds after you drop it is given in feet by

$$h = 100 - 16t^2$$

a. Complete the table.

t	h
0	
0.5	
1	
1.25	
1.5	
1.75	
2	
2.25	
2.5	

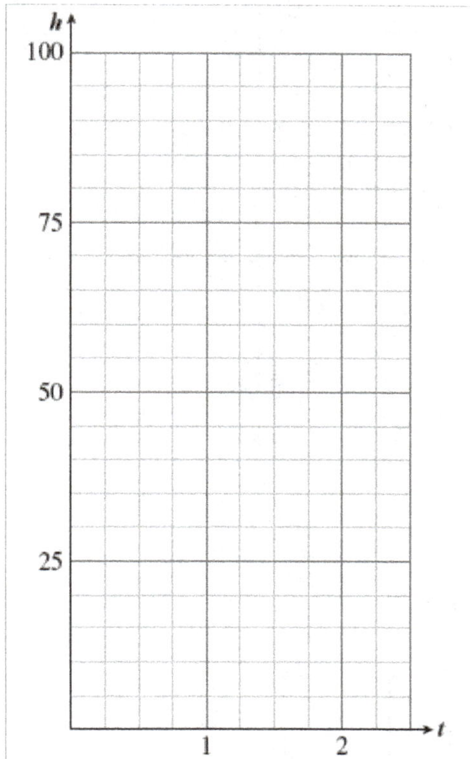

b. Sketch the graph of the equation.
c. What is the height of the stone after 2 seconds? Verify your answer algebraically.
d. When is the height of the stone 75 feet? Write an algebraic equation you can solve to verify your answer.

94. Sportsworld sells $180 - 3p$ pairs of their name-brand running shoes per week when they charge p dollars per pair.
a. Write an equation for Sportsworld's revenue, R, in terms of p.
b. Fill in the table, and graph your equation.

p	R
0	
10	
20	
30	
40	
50	
60	

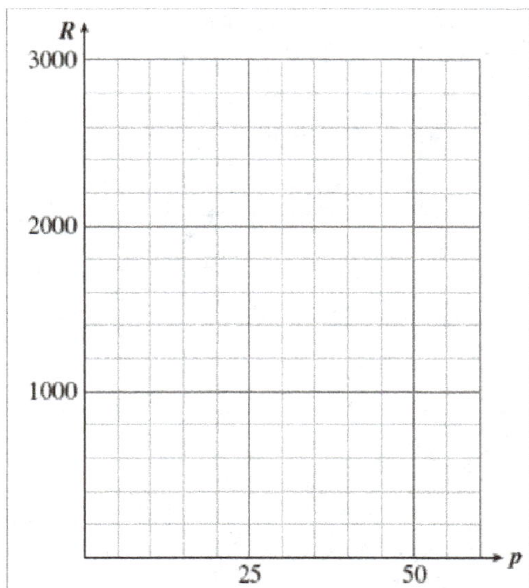

c. At what price(s) will Sportsworld's revenue be zero?
d. At what price will Sportsworld's revenue be maximum? What is their maximum revenue?

For Problems 95-98,

a. Find the coordinates of the vertex and the intercepts.

b. Sketch the graph.

95. $y = \frac{1}{2}x^2$

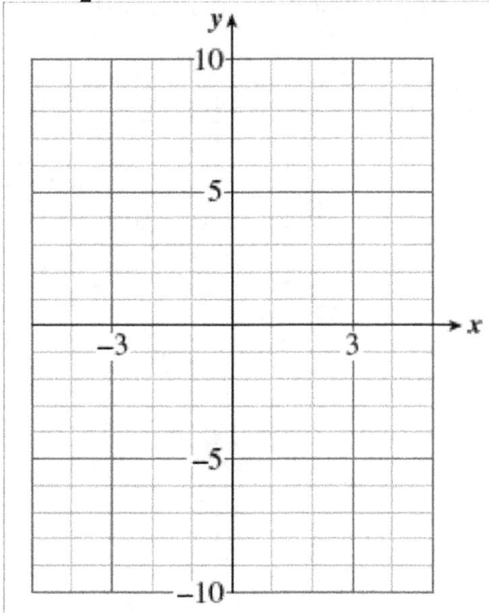

96. $y = x^2 - 4$

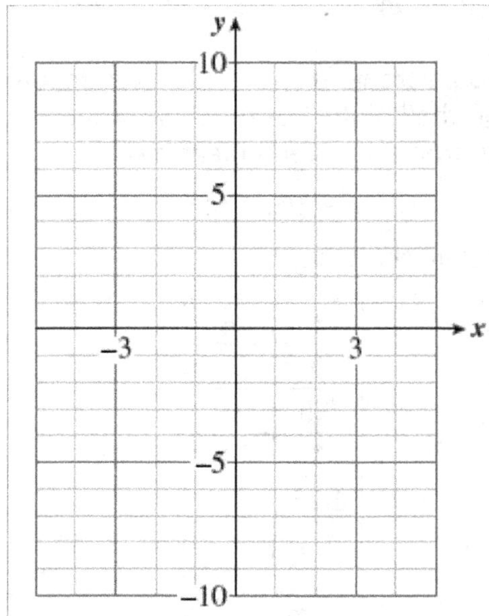

97. $y = x^2 - 9x$

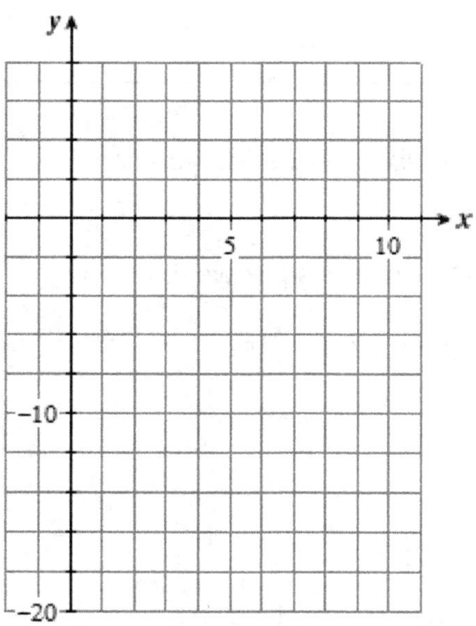

98. $y = -2x^2 - 4x$

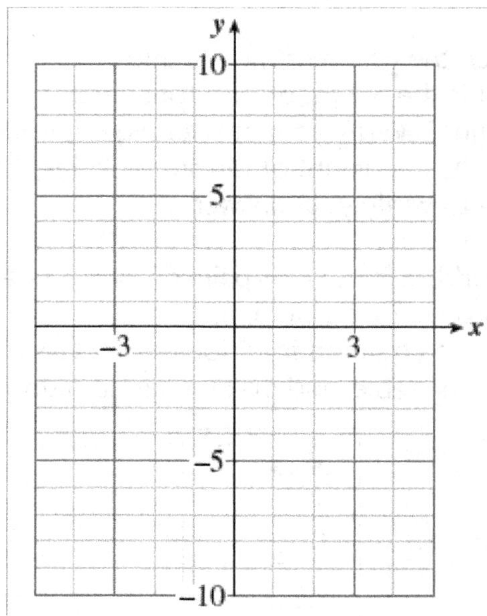

99. The height, h, of an object t seconds after being thrown from ground level is given by

$$h = v_0 t - \frac{1}{2} g t^2$$

where v_0 is its starting velocity and g is a constant that depends on gravity. On the moon, the value of g is approximately 5.6. Suppose you hit a golf ball on the moon with an upwards velocity 100 feet per second.

a. Write an equation for the height of the golf ball t seconds after you hit it.

b. Graph your equation in the window

$$\text{Xmin} = 0 \quad \text{Xmax} = 47$$
$$\text{Ymin} = 0 \quad \text{Ymax} = 1000$$

c. Use the **Trace** key to estimate the maximum height the golf ball reaches.

d. Use your equation to calculate when the golf ball will reach a height of 880 feet.

100. An acrobat is catapulted into the air from a springboard at ground level. His height h in meters is given by the formula

$$h = -4.9t^2 + 14.7t$$

where t is the time in seconds from launch. Use your calculator to graph the acrobat's height versus time. Set the **Window** values on your calculator to

$$\text{Xmin} = 0 \quad \text{Xmax} = 4.7$$
$$\text{Ymin} = 0 \quad \text{Ymax} = 12$$

a. Use the **Trace** key to find the coordinates of the highest point on the graph. When does the acrobat reach his maximum height, and what is that height?

b. Use the formula to find the height of the acrobat after 2.4 seconds.

c. Use the **Trace** key to verify your answer to part (b). Find another time when the acrobat is at the same height.

d. Use the formula to find two times when the acrobat is at a height of 6.125 meters. Verify your answers on the graph.

e. What are the coordinates of the horizontal intercepts of your graph? What do these points have to do with the acrobat?

101. What is the smallest positive integer (other than 1) that is both a perfect square and a perfect cube?

102. Find the cube root of 27^2 without using a calculator.

103. Use the given areas to find the area of the large square shown below.

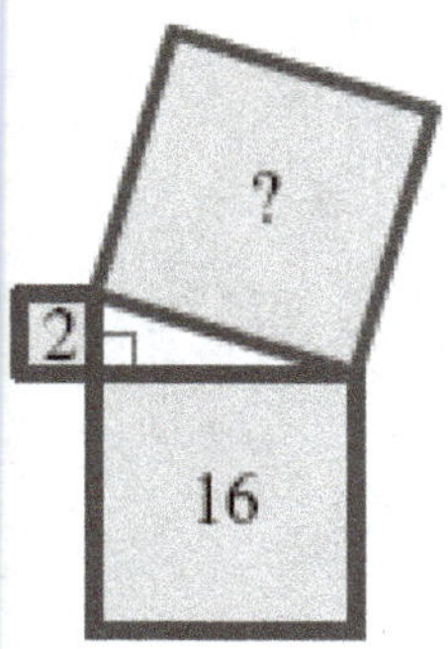

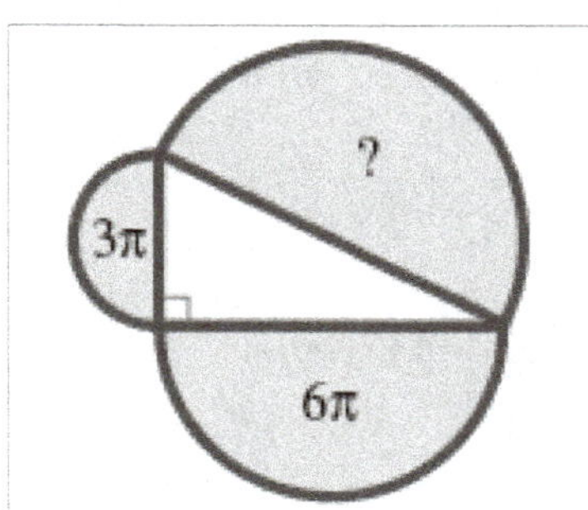

104. Use the given areas to find the area of the large semicircle shown above.

Chapter 6 Quadratic Models

Lesson 1 Factoring Quadratic Trinomials

- Factor quadratic trinomials
- Factor the difference of two squares
- Solve quadratic equations by factoring

Lesson 2 Quadratic Models

- Read and interpret points on the graph of a quadratic equation
- Solve quadratic equations
- Find a quadratic equation with given solutions
- Use the connections between factors, solutions, and x-intercepts

Lesson 3 Completing the Square

- Solve a quadratic equation by completing the square
- Find the vertex of a parabola
- Sketch the graph of a quadratic equation

Lesson 4 Quadratic Formula

- Solve a quadratic equation by the quadratic formula
- Solve a formula for one variable in terms of the others
- Sketch the graph of a quadratic equation
- Use the discriminant

Lesson 5 Problem Solving

- Find the maximum or minimum value of a quadratic model
- Write a quadratic equation in vertex form
- Use the vertex form to find a quadratic equation

Lesson 6 Quadratic Curve Fitting

- Use a system of equations to fit a parabola through three points
- Use a calculator for quadratic regression
- Evaluate the fit of a regression equation

6.1 Factoring Quadratic Trinomials

In Section 5.5 we learned to factor quadratic expressions of the form $ax^2 + bx$. Now we consider quadratic trinomials of the form $x^2 + bx + c$. Many of these can be factored into the product of two binomials. For example,

$$x^2 + 7x + 12 = (x + 3)(x + 4)$$

How can we come up with the factors of the quadratic trinomial? Let's recall how the product is obtained. We apply the distributive law and multiply each term of the first binomial by each term of the second binomial. We use the word "FOIL" to label the four terms of the product: First, Outside, Inside, and Last.

$$\begin{array}{c} (x+3)(x+4) = x^2 + 4x + 3x + 12 \\ \qquad\qquad\quad \textbf{F} \quad\ \ \textbf{O+I} \quad\ \ \textbf{L} \end{array}$$

	x	4
x	x^2	$4x$
3	$3x$	12

RQ1. Where does the constant term of the trinomial appear in the product box?
RQ2. Where does the linear term appear?

We make several observations about this product.

1. The quadratic term, x^2, comes from the product of the **F**irst terms in each binomial.
2. The constant term, 12, is the product of the **L**ast terms.
3. The middle term, $7x$, is the sum of two terms, the product of the **I**nside terms and the product of the **O**utside terms.

Look Closer: We can see that the first term of each factor must be x, so we only need to fill in the blanks below with the correct numbers:

$$x^2 + 7x + 12 = (x + \_\_\_\_\_)(x + \_\_\_\_\_)$$

If the two numbers are p and q, then

$$\begin{aligned}(x+p)(x+q) &= x^2 + qx + px + pq \\ &= x^2 + (\boldsymbol{p+q})x + \boldsymbol{pq}\end{aligned}$$

We see that the constant term of the trinomial is the product pq, and the linear term of the trinomial has as coefficient the sum $p + q$.

Thus, to factor $x^2 + 7x + 12$, we look for two integers p and q whose sum is 7 and whose product is 12. With a little trial and error, we discover that the two integers are 3 and 4. So the factored form is

$$x^2 + 7x + 12 = (x + 3)(x + 4)$$

To Factor a Quadratic Trinomial

To factor $x^2 + bx + c$, we look for two numbers p and q so that

$$\boldsymbol{pq = c} \text{ and } \boldsymbol{p + q = b}$$

Example 1 Factor: $a^2 + 13a + 40$

Solution The factored form looks like

$$a^2 + 13a + 40 = (a + p)(a + q)$$

where $pq = 40$ and $p + q = 13$. To find the numbers p and q, it may help to list all possible pairs of numbers whose product is 40. Then we can check each pair of numbers to see which pair has a sum of 13.

p	q	$p+q$
1	40	41
2	20	22
4	10	14
5	**8**	**13**

We see that $5 \cdot 8 = 40$ and $5 + 8 - 13$, so the correct factorization is

$$a^2 + 13a + 40 = (a + 5)(a + 8)$$

We can also write the factored form as $(a + 8)(a + 5)$; the order of the factors does not matter. Aside from rearranging the factors, there is only one correct factorization for these quadratic trinomials. ■

Look Ahead: In the Activities we'll see how to factor trinomials with negative coefficients. Here are the results:

Sign Patterns for Factoring Quadratic Trinomials

Assume that b, c, p, and q are positive integers.

1. $x^2 + bx + c = (x + p)(x + q)$

   If all the coefficients of the trinomial are positive, then both p and q are positive.

2. $x^2 - bx + c = (x - p)(x - q)$

   If the linear term of the trinomial is negative and the other two terms positive, then p and q are both negative.

3. $x^2 \pm bx - c = (x + p)(x - q)$

   If the constant term of the trinomial is negative, then p and q have opposite signs.

Example 2 Factor $x^2 + 3x - 18$

Solution We look for two integers p and q for which $pq = -18$ and $p + q = 3$. Because pq is negative, p and q must have opposite signs. Because $p + q$ is positive, the choices are

-1 and 18, -2 and 9, or -3 and 6

Because $-3 + 6 = 3$, the correct factorization is

$$x^2 + 3x - 18 = (x - 3)(x + 6)$$

You can check the factorization by multiplying the two binomials to see that their product is in fact $x^2 + 3x - 18$. ■

Solving Quadratic Equations

Now that we can factor quadratic trinomials, we can use the zero-factor principle to solve many quadratic equations.

■ **Example 3a.** Solve the equation $x^2 + 3x = 18$
b. Solve the equation $2x^2 + 6x = 36$

Solutions a. Recall the steps for solving quadratic equations, given on page 294 . We must have the right side equal to zero if we want to apply the zero-factor principle, so we begin by subtracting 18 from both sides.

$$x^2 + 3x - 18 = 0 \qquad \text{Factor the left side.}$$

$$(x+6)(x-3) = 0 \qquad \text{Apply the zero-factor principle: set each factor equal to zero.}$$

$$x + 6 = 0 \qquad x - 3 = 0 \qquad \text{Solve each equation.}$$

$$x = -6 \qquad x = 3$$

The solutions are −6 and 3. You should check that each of these solutions satisfies the original equation.
b. You may notice that each term of the equation in part (b) is twice the corresponding term of the equation in part (a). We can factor out a common factor of 2 from the left side of the standard form of this equation.

$$2x^2 + 6x - 36 = 0 \qquad \text{Factor out 2 from the left side.}$$

$$2(x^2 + 3x - 18) = 0$$

Then we can divide both sides of the equation by 2.

$$\frac{2(x^2 + 3x - 18)}{\mathbf{2}} = \frac{0}{\mathbf{2}}$$

$$x^2 + 3x - 18 = 0$$

This new equation is the same as the equation in part (a). Thus, the solutions of the equation in part (b) are the same as the solutions in part (a), namely, −6 and 3. ■

RQ3. How did the factor of 2 in Example 3b affect the solutions of the equation?

Applications

Here is an example of how quadratic equations arise in applications.

■ **Example 4** The size of a rectangular computer monitor screen is the length of its diagonal. If the length of the screen should be 3 inches greater than its width, what are the dimensions of a 15-inch monitor?

Solution The diagonal of the monitor is 15 inches. We express the dimensions of the screen in terms of a single variable:

Width of screen: w
Length of screen: $w+3$

We use the Pythagorean theorem to write an equation. (It may be helpful to sketch and label a diagram of the monitor screen.)

$$w^2+(w+3)^2=15^2$$

To solve the equation, we must write it in standard form, so we begin by simplifying the left side. First we expand the square of the binomial, $(w+3)^2$.

$w^2+w^2+6w+9=225$	**Combine like terms.**
	Subtract 225 from both sides.
$2w^2+6w-216=0$	**Factor 2 from the left side.**
$2(w^2+3w-108)=0$	**Divide both sides by 2.**
$w^2+3w-108=0$	**Factor the quadratic expression.**
$(w-9)(w+12)=0$	**Set each factor equal to zero.**
$w-9=0$ or $w+12=0$	**Solve each equation.**
$w=9$ or $w=-12$	

Because the width of the screen cannot be a negative number, we discard the solution $w=-12$. Thus, the width is 9 inches and the length is $9+3=12$ inches. ■

More Factoring

What if the coefficient of x^2 is not 1? If the coefficients are small integers, we can often factor a quadratic expression by the guess-and-check method.

■ **Example 5a.** Factor $2x^2-7x+3$ into a product of two binomials.
b. Solve the equation $2x^2-7x+3=0$

Solutions a. We begin by factoring the quadratic term, $2x^2$, which can only be factored as x times $2x$, so we can fill in the First terms in each binomial.

$$2x^2-7x+3=(2x\_\_\_\_)(x\_\_\_\_)$$

Next, we factor the constant term, 3, which can only be factored as 3 times 1. Because the linear term is negative, we make both factors negative: $3=-3(-1)$. Finally, we have to decide which number appears as the Last term in each binomial. We check **O + I** for each possibility.

$$(2x-3)(x-1) \quad \boldsymbol{O+I=-2x-3x=-5x}$$
$$(2x-1)(x-3) \quad \boldsymbol{O+I=-6x-x=-7x}$$

The second choice gives the correct middle term, so the factorization is

$$2x^2-7x+3=(2x-1)(x-3)$$

b. After we have factored the quadratic trinomial, we apply the Zero-Factor principle and set each factor equal to zero.

$2x-1=0$	$x-3=0$	**Solve each equation.**
$x=\frac{1}{2}$	$x=3$	

The solutions are $x=\frac{1}{2}$ and $x=3$. ■

Caution! The two parts of Example 5 are different. In part (a), we were asked to *factor* a trinomial. The answer was the factored form, $(2x-1)(x-3)$. In part (b), we were asked to *solve* an equation. The answer was the solutions of the equation, $\frac{1}{2}$ and 3. ■

RQ4. What is the difference between factoring a quadratic trinomial and solving a quadratic equation?

In a quadratic equation, the coefficient of the x^2-term is sometimes called the **lead coefficient**, because the quadratic term comes first in the standard form of the equation. In the optional portion of this section, we explore more techniques for factoring quadratic trinomials whose lead coefficient is not 1.

Difference of Squares

Consider the following product of binomials:

$$(x-4)(x+4) = x^2 + 4x - 4x - 16 = x^2 - 16$$

The linear terms, $4x$ and $-4x$, cancel, and we are left with the difference of two squares. Of course, we can reverse the operation and factor the expression $x^2 - 16$ as

$$x^2 - 16 = (x-4)(x+4)$$

The difference of two squared terms can always be factored in this way. All we need to do is identify the square root of each term.

Difference of Squares

We can factor a quadratic expression that is the difference of two squares.

$$\boldsymbol{x^2 - p^2 = (x-p)(x+p)}$$

■ **Example 6** Factor.

a. $w^2 - 49$

b. $9t^2 - 1$

Solutions a. Because 49 is the square of 7, we have

$$w^2 - 49 = (w-7)(w+7)$$

b. We see that $9t^2$ is the square of $3t$, and 1 is the square of 1, so

$$9t^2 - 1 = (3t-1)(3t+1)$$ ■

Caution! We cannot factor the sum of two squares. For example, by trying any reasonable possibilities, you can verify that $x^2 + 16$ cannot be factored. ■

RQ5. Factor each expression if possible:

a. $x^2 - 64$ **b.** $x^2 - 64x$ **c.** $x^2 + 64$ **d.** $x^2 + 64x$

Optional Extension: A Property of Binomial Products

If we cannot factor a quadratic trinomial $ax^2 + bx + c$ by guess-and-check, we will need a property of the products of binomials, represented as areas of rectangles. Remember that we use the two binomials to represent the length and width of a rectangle, then partition the rectangle into four smaller rectangles. The product appears as the sum of the areas of the sub-rectangles.

Example 7a. Compute the product $(3t+2)(t+3)$ using the area of a rectangle.
b. Verify that the products of the diagonal entries are equal.

Solutions a. We construct a rectangle whose sides are represented by $3t+2$ and $t+3$, as shown at right. We see that the product of the two binomials is

$$3t^2 + 9t + 2t + 6 = 3t^2 + 11t + 6$$

	t	3
3t	$3t^2$	9t
2	2t	6

b. Now let's compute the product of the expressions along each diagonal of the rectangle:

$$3t^2 \cdot 6 = 18t^2$$
$$9t \cdot 2t = 18t^2$$

The products of the diagonal entries are equal. This is not surprising when you think about it, because each diagonal product is the product of all four terms of the binomials, namely $3t$, 2, t, and 3, just multiplied in a different order. You can see where the diagonal entries came from in our example:

$$18t^2 = 3t^2 \cdot 6 = 3t \cdot t \cdot 2 \cdot 3$$
$$18t^2 = 9t \cdot 2t = 3t \cdot 3 \cdot 2 \cdot t$$

When we represent the product of two binomials by the area of a rectangle, **the products of the entries on the two diagonals are equal.**

Factoring Quadratic Trinomials, $ax^2 + bx + c$

Now suppose that we'd like to factor the trinomial

$$3x^2 + 10x + 8$$

$3x^2$	
	8

We'll try to reverse the steps for multiplication. Instead of starting with the factors on the outside of the rectangle, we begin by filling in the areas of the sub-rectangles. We know that the quadratic term, $3x^2$, goes in the upper left,
and the constant term, 8, goes in the lower right, as shown in the figure.

What about the other two entries? We know that their sum must be $10x$, but we don't know what expressions go in each! This is where we need our observation that the products of the entries on the two diagonals are equal.

We compute the product of the entries on the first diagonal:

$$D = 3x^2 \cdot 8 = \mathbf{24x^2}$$

The product of the entries on the other diagonal must also be $24x^2$. We now know two things about those entries:

1. Their product is $24x^2$ and
2. Their sum is $10x$

To find the two unknown entries,we list all the ways to factor $D = 24x^2$, then choose the factors whose sum is $10x$.

Factors of $D = 24x^2$		Sum of Factors
x	$24x$	$x + 24x = 25x$
$2x$	$12x$	$2x + 12x = 14x$
$3x$	$8x$	$3x + 8x = 11x$
$\mathbf{4x}$	$\mathbf{6x}$	$\mathbf{4x + 6x = 10x}$

The last pair of factors, $4x$ and $6x$, has a sum of $10x$. These are the expressions we enter in the remaining sub-rectangles. (It doesn't matter which one goes in which spot.) We now have all the sub-rectangles filled in, as shown at right.

$3x^2$	$4x$
$6x$	8

Finally, we work backwards to discover what length and width produce the areas of the four subrectangles. We factor each row of the rectangle, and write the factors on the outside. Start with the top row, factoring out x and writing the result, $3x + 4$, at the top, as shown in the next figure.

We get the same result when we factor 2 from the bottom row. The final rectangle is shown at right, and the factors of $3x^2 + 10x + 8$ appear as the length and width of the rectangle.

	$3x$	4
x	$3x^2$	$4x$
2	$6x$	8

Our factorization is thus

$$3x^2 + 10x + 8 = (x + 2)(3x + 4)$$

Here is a summary of our factoring method.

To Factor $ax^2 + bx + c$ Using Rectangles:

1. Write the quadratic term ax^2 in the upper left sub-rectangle, and the constant term c in the lower right.
2. Multiply these two terms to find the diagonal product, D.
3. List all possible factors px and qx of D, and choose the pair whose sum is the linear term, bx, of the quadratic trinomial.
4. Write the factors px and qx in the remaining subrectangles.
5. Factor each row of the rectangle, writing the factors on the outside. These are the factors of the quadratic trinomial.

Example 8 Factor $2x^2 - 11x + 15$.

Solution Step 1 We enter $2x^2$ and 15 on the diagonal of the rectangle, as shown in the figure.

$2x^2$	
	15

Step 2 We compute the diagonal product:

$$D = 2x^2 \cdot 15 = 30x^2$$

Step 3 We list all possible factors of D, and compute the sum of each pair of factors. (Note that both factors must be negative.)

Factors of $D = 30x^2$		Sum of Factors
$-x$	$-30x$	$-x - 30x = -31x$
$-2x$	$-15x$	$-2x - 15x = -17x$
$-3x$	$-10x$	$-3x - 10x = -13x$
$\mathbf{-5x}$	$\mathbf{-6x}$	$\mathbf{-5x - 6x = -11x}$

The correct factors are $-5x$ and $-6x$.

Step 4 We enter the factors $-5x$ and $-6x$ into the rectangle.

$2x^2$	$-6x$
$-5x$	15

Step 5 We factor $2x$ from the top row of the rectangle, and write the result, $x - 3$, at the top, as shown below.

	x	-3
$2x$	$2x^2$	$-6x$
	$-5x$	15

	x	-3
$2x$	$2x^2$	$-6x$
-5	$-5x$	15

Finally, we factor $x - 3$ from the bottom row, and write the result, -5, on the left. The correct factorization is

$$2x^2 - 11x + 15 = (2x - 5)(x - 3)$$

■

■ **Example 9a.** Factor $5x^2 + 2x - 3$

b. Solve $5x^2 + 2x - 3 = 0$

Solutions a. There is only one way to factor $5x^2$ and one way to factor -3, so there are only two possibilities for the factorization. You can check that

$$5x^2 + 2x - 3 = (5x - 3)(x + 1)$$

b. The equation is already in standard form, so we factor the left side as we did in part (a) and apply the zero-factor principle.

$$5x^2 + 2x - 3 = 0$$

$$(5x - 3)(x + 1) = 0 \qquad \textbf{Set each factor equal to 0.}$$

$$5x - 3 = 0 \quad x + 1 = 0 \qquad \textbf{Solve each equation.}$$

$$x = \frac{3}{5} \qquad x = -1$$

The solutions are -1 and $\frac{3}{5}$.

■

Skills Warm-Up

■ **a.** Find the linear term of each product mentally.
b. Find the constant term.

1. $(x + 1)(x + 3)$ **2.** $(a - 6)(a - 3)$ **3.** $(p + 7)(p - 4)$

4. $(q - 8)(q + 2)$ **5.** $(3w + 4)(w + 2)$ **6.** $(2z - 5)(3z + 2)$

Answers: **1a.** $4x$ **b.** 3 **2a.** $-9a$ **b.** 18 **3a.** $3p$ **b.** -28 **4a.** $-6q$ **b.** -16 **5a.** $10w$ **b.** 8 **6a.** $-11z$ **b.** -10

Homework 6.1

Skills Practice

For Problems 1-3, the area of the rectangle represents a product.

a. Fill in the missing expressions.

b. Write the area as a product of binomials, then compute the product.

1.

	x	?
x	x^2	$6x$
?	$5x$	30

2.

	x	-3
x	x^2	$-3x$
?	?	27

3.

	x	?
?	x^2	$2x$
?	?	16

4.

	?	?
?	x^2	$8x$
?	$-4x$	?

For Problems 5-10, factor each quadratic trinomial.

5. $n^2 + 10n + 16$ **6.** $h^2 + 26h + 48$ **7.** $a^2 - 8a + 12$

8. $t^2 - 15t + 36$ **9.** $x^2 - 3x - 10$ **10.** $a^2 + 8a - 20$

For Problems 11-16, factor if possible. If the trinomial cannot be factored, say so.

11. $x^2 - 17x + 30$ **12.** $x^2 + 4x + 2$ **13.** $y^2 - 44y - 45$

14. $t^2 - 9t - 20$ **15.** $q^2 - 5q - 6$ **16.** $n^2 - 5n + 6$

For Problems 17-19, factor the expression as a product of binomials.

17. $x^2 - 9$ **18.** $4 - w^2$ **19.** $-121 + b^2$

For Problems 20-22, factor the quadratic trinomial.

20. $2x^2 + 11x + 5$ **21.** $5t^2 + 7t + 2$ **22.** $3x^2 - 8x + 5$

For Problems 23-28, solve the quadratic equation.

23. $x^2 + 3x - 10 = 0$ **24.** $t^2 + t = 42$

25. $2x^2 - 10x = 12$ **26.** $0 = n^2 - 14n + 49$

27. $5q^2 = 10q$ **28.** $3z^2 + 2 = 5z$

Applications

29. Icon Industries produces all kinds of electronic equipment. The cost, C, of producing a piece of equipment depends on the number of hours, t, it takes to build it, where

$$C = 8t^2 - 32t - 16$$

How many hours does it take to build a transformer if it costs $80 to produce?

30. A competitive diver executes a double somersault with a twist from a platform. Her height above the water t seconds after leaving the board is given in feet by

$$h = -16t^2 + 8t + 24$$

How long does she have to complete her somersault before entering the water?

For Problems 31-33,

a. Draw and label a sketch to illustrate the problem,
b. Write an equation for the problem,
c. Solve your equation and complete the problem.

31. The area of a computer circuit board must be 60 square centimeters. The length of the circuit board should be 2 centimeters shorter than twice its width. Find the dimensions of the circuit board.

32. A paper airplane in the shape of a triangle is 40 square inches in area. Its base is 11 inches longer than its altitude. Find the base and altitude of the triangle.

33. Steve's boat locker is 2 feet longer than twice its width. Find the dimensions of the locker if the 13-foot mast of Steve's boat will just fit diagonally across the floor of the locker.

For Problems 34-35, solve the proportion.

34. $\dfrac{p-3}{2} = \dfrac{7}{p+2}$

35. $\dfrac{m-4}{2m+1} = \dfrac{m+1}{2m-1}$

For Problems 97-106: Mental exercise: Find the other factor without using pencil, paper, or calculator.

36. $x^2 + 15x + 56 = (x+7)(\_\_\_\_\_\_)$

37. $b^2 + 8b - 240 = (b-12)(\_\_\_\_\_\_)$

38. $n^2 - 97n - 300 = (n-100)(\_\_\_\_\_\_)$

39. $3u^2 - 17u - 6 = (u-6)(\_\_\_\_\_\_)$

40. $2t^2 - 21t + 54 = (t-6)(\_\_\_\_\_\_)$

Extension

For Problems 41-46, use a rectangle to factor the quadratic trinomial.

41. $2x^2 - 13x + 18$

42. $5x^2 + 16x - 16$

43. $6h^2 + 7h + 2$

44. $9n^2 - 8n - 1$

45. $6t^2 - 5t - 25$

46. $5x^2 - 14x - 24$

For Problems 47-49, arrange in standard form, then factor.

47. $9x^2 - 8 - 21x$

48. $-5 - 2z + 16z^2$

49. $23a + 4a^2 - 6$

For Problems 50-53, solve the equation.

50. $3n^2 - n = 4$

51. $11t = 6t^2 + 3$

52. $1 = 4y - 4y^2$

53. $12z^2 + 26z = 10$

For Problems 54-55, use the given areas to find the length and width of the rectangle.

54.

$6x^2$	$9x$
$10x$	15

55.

$12m^2$	$-10m$
$30m$	-25

6.2 Quadratic Models

Suppose a baseball player pops up, that is, hits the baseball straight up into the air. The height, h, of the baseball t seconds after it leaves the bat can be calculated using a formula from physics.

This formula takes into account the initial speed of the ball and its height when it was hit. The formula for the height of the ball (in feet) is

$$h = -16t^2 + 64t + 4$$

We evaluate the formula to make a table of values for the height of the baseball.

t	0	1	2	3	4
h	4	52	68	52	4

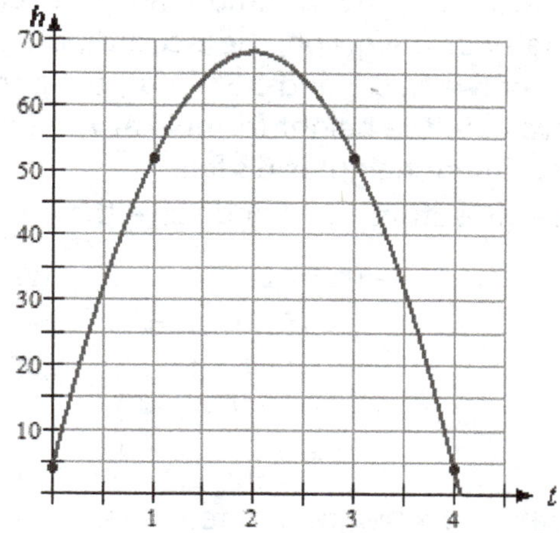

Then we graph the height of the baseball versus time by plotting data points from the table, and connecting the points with a smooth curve.

Now we'll use the model to answer some questions about the baseball.

Example 1a. Find the height of the baseball after $\frac{1}{2}$ second. Locate the point on the graph that corresponds to your answer. Use the graph to find another time when the baseball is at the same height.

b. Find two times when the baseball is at a height of 64 feet.

Solutions a. We evaluate the formula for height at $t = \frac{1}{2}$.

$$h = -16\left(\frac{1}{2}\right)^2 + 64\left(\frac{1}{2}\right) + 4 = 32$$

After $\frac{1}{2}$ second, the ball is at a height of 32 feet. We locate another point on the graph with h-coordinate 32, and read its t-coordinate. The ball is 32 feet high again (on the way down) after about 3.5 seconds.

b. We substitute 64 for h and solve the equation

$-16t^2 + 64t + 4 = 64$	**Write the equation in standard form.**
$-16t^2 + 64t - 60 = 0$	**Factor out -4, then divide by -4.**
$-4(4t^2 - 16t + 15) = 0$	**Factor the left side.**
$(2t - 5)(2t - 3) = 0$	**Apply the zero-factor principle.**
$2t - 5 = 0 \quad 2t - 3 = 0$	**Solve each equation.**
$t = \frac{5}{2} \quad t = \frac{3}{2}$	

The ball is at a height of 64 feet at 1.5 seconds and at 2.5 seconds. We can verify these values on the graph.

Look Closer: Notice how the symmetry of the parabola is illustrated on the graph in Example 1: Points with the same h-coordinate are located at equal distances from the axis of the parabola.

Example 2a. When does the baseball reach its maximum height, and what is that height?

b. Suppose the catcher catches the baseball at a height of 4 feet, before it strikes the ground. At what time was the ball caught?

Solutions a. The maximum height occurs at the vertex of the parabola. Because of the symmetry of the graph, the t-coordinate of the vertex must be halfway between 1.5 and 2.5, or between 0.5 and 3.5. Thus, the baseball reaches its highest point at 2 seconds. We can evaluate the height formula at $t = 2$ or read the h-coordinate of the vertex to find that the maximum height is 68 feet.

b. We substitute $h = 4$ and solve the equation

$-16t^2 + 64t + 4 = 4$	**Write the equation in standard form.**
$-16t^2 + 64t = 0$	**Factor the left side.**
$-16t(t-4) = 0$	**Apply the zero-factor principle.**
$-16t = 0 \quad t - 4 = 0$	**Solve each equation.**
$t = 0 \quad t = 4$	

The ball is at a height of 4 feet again 4 seconds after it was hit. Instead of using algebra to find the solution, we can use the symmetry of the graph: the points at $t = 0$ and $t = 4$ lie at equal distances from the axis of symmetry, $x = 2$.

RQ1. If we know two points on a parabola with the same y-coordinate, how can we find the axis of symmetry?

Using the Table Feature on a Graphing Calculator

We can use a graphing calculator to make a table of values for an equation in two variables.

Example 3 Make a table for the quadratic model from Example 1:

$$h = -16t^2 + 64t + 4$$

Solution Step 1 First, we enter the equation: we press the [Y =] key and clear out any other definitions, then type in $\mathbf{Y_1} = -16\mathbf{X}^2 + 64\mathbf{X} + 4$.

Step 2 Next, we choose x-values for the table: we press [2nd] [WINDOW] to access the **TblSet** (Table Setup) menu and set it to look like Figure (a). This setting gives us an initial x-value of 0 (**TblStart** = **0**), and an increment of one unit in the x-values (**ΔTbl** = **1**).

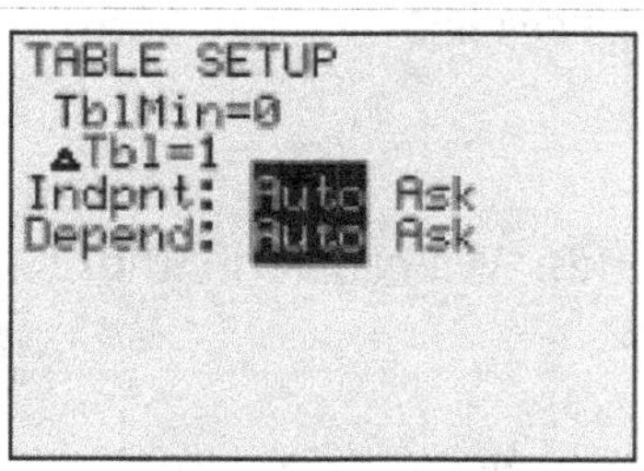

a.

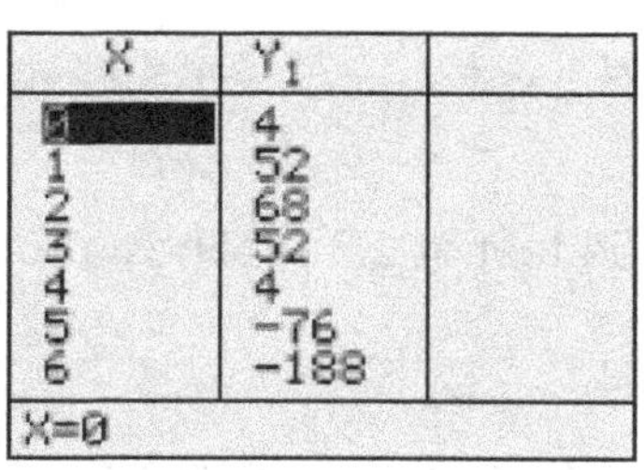

b.

Step 3 Finally, we press [2nd] [GRAPH] to see the table of values, as shown in Figure (b). We use the [△] and [▽] keys to scroll up and down the table.

Factors and x-Intercepts

There is a connection between:
1. the **factors** of $ax^2 + bx + c$,
2. the **solutions** of the quadratic equation $ax^2 + bx + c = 0$, and
3. the **x-intercepts** of the graph of $y = ax^2 + bx + c$.

Example 4a. Factor $2x^2 - 5x - 7$

b. Solve $2x^2 - 5x - 7 = 0$

c. Find the x-intercepts of the graph of $y = 2x^2 - 5x - 7$

Solutions a. Because $2x^2$ must be factored as $2x$ times x, and 7 can only be factored as 7 times 1, the possibilities for the factored form are

$$(2x \quad 7)(x \quad 1) \quad \text{or} \quad (2x \quad 1)(x \quad 7)$$

By checking **O+I** for each and inserting the correct signs, we find that

$$2x^2 - 5x - 7 = (2x - 7)(x + 1)$$

b. We factor the left side as in part (a) to get

$$(2x - 7)(x + 1) = 0 \qquad \textbf{Set each factor equal to 0.}$$
$$2x - 7 = 0 \quad x + 1 = 0 \qquad \textbf{Solve each equation.}$$
$$x = \frac{7}{2} \quad x = -1$$

The solutions are $\frac{7}{2}$ and -1.

c. To find the x-intercepts of the graph, we set $y = 0$ and solve for x.

$$0 = 2x^2 - 5x - 7$$

But this is the same equation we solved in part (b), so the x-intercepts are $(-1, 0)$ and $(\frac{7}{2}, 0)$. We can graph the equation in the window

Xmin $= -4.7$ Xmax $= 4.7$
Ymin $= -10$ Ymax $= 10$

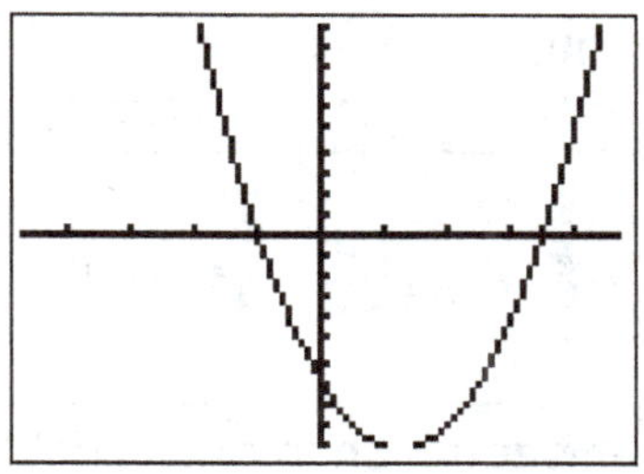

and use the **Trace** to locate the x-intercepts.

Look Closer: In Example 4, notice that each factor of $2x^2 - 5x - 7$ yields a solution of the equation $2x^2 - 5x - 7 = 0$, and each solution of $2x^2 - 5x - 7 = 0$ is an x-intercept of the graph of $y = 2x^2 - 5x - 7$.

RQ2. How are the factors of $ax^2 + bx + c$ related to the solutions of $ax^2 + bx + c = 0$?

RQ3. How are the solutions of $ax^2 + bx + c = 0$ related to the x-intercepts of the graph of $y = ax^2 + bx + c$?

To solve a quadratic equation by factoring, we must first write the equation so that one side of the equation is zero.

Example 4 Solve $3x(x+1) = 2x+2$

Solution First, we write the equation in standard form.

$$\begin{aligned} 3x(x+1) &= 2x+2 && \text{Apply the distributive law to the left side.} \\ 3x^2+3x &= 2x+2 && \text{Subtract } 2x+2 \text{ from both sides.} \\ 3x^2+x-2 &= 0 \end{aligned}$$

Now we can factor the left side to obtain

$$\begin{aligned} (3x-2)(x+1) &= 0 && \text{Apply the zero-factor principle.} \\ 3x-2=0 \quad & x+1=0 && \text{Solve each equation.} \\ x=\frac{2}{3} \quad & x=-1 \end{aligned}$$

The solutions are $\frac{2}{3}$ and -1.

Caution! When we apply the zero-factor principle, one side of the equation must be zero. For example, to solve the equation

$$(x-2)(x-4) = 15$$

it is incorrect to set each factor equal to 15! (There are many ways that the product of two numbers can equal 15; it is not necessary that one of the numbers be 15.) We must first simplify the left side and write the equation in standard form. (The correct solutions are 7 and -1; make sure you can find these solutions.)

RQ4 Fill in the blanks.
a. If the sum of two numbers is zero, the numbers must be ________.
b. If a fraction equals zero, the ____________ must be zero.
c. If the product of two numbers is zero, one of the numbers must ____.
d. If the divisor in a quotient is zero, the quotient is __________.

Solutions of Quadratic Equations

A quadratic equation in one variable always has two solutions.

However, in some cases, the solutions may be equal. For example, the equation

$$x^2-2x+1=0$$

can be solved by factoring as follows:

$$\begin{aligned} (x-1)(x-1) &= 0 && \text{Apply the zero-factor principle.} \\ x-1=0 \quad \text{or} \quad & x-1=0 \end{aligned}$$

Both of these equations have solution 1. We say that 1 is a solution **of multiplicity two**, meaning that it occurs twice as a solution of the quadratic equation.

RQ5. What is a solution of multiplicity two?

Because of the zero-factor principle, the solutions of the quadratic equation

$$(x-r_1)(x-r_2)=0$$

are r_1 and r_2. If we know the two solutions of a quadratic equation, we can work backwards to reconstruct the equation in factored form. By multiplying together the two factors, we can then write the equation in standard form.

Example 5 Find a quadratic equation whose solutions are $\frac{1}{2}$ and -3.

Solution The quadratic equation is

$$\left(x - \frac{1}{2}\right)[x - (-3)] = 0, \quad \text{or} \quad \left(x - \frac{1}{2}\right)(x + 3) = 0$$

To write the equation in standard form, we multiply the factors together.

$$x^2 + \frac{5}{2}x - \frac{3}{2} = 0$$

We can find an equation with integer coefficients if we clear the equation of fractions. Multiply both sides by 2:

$$\mathbf{2}(x^2 + \frac{5}{2}x - \frac{3}{2}) = \mathbf{2}(0)$$

$$2x^2 + 5x - 3 = 0$$

You should check that $\frac{1}{2}$ and -3 are in fact solutions of this last equation. Multiplying by a constant factor does not change the equation's solutions.

Skills Warm-Up

Write expressions for the area and perimeter of a rectangle with the dimensions given.

1. 12 cm and $x + 4$ cm

2. w in and $20 - w$ in

a. Write an expression for the length of the rectangle.
b. Write an expression for the area of the rectangle.

3. Perimeter $= 30$ ft, width $= 6$ ft

4. Perimeter $= 30$ ft, width $= x$ ft

Answers: 1. $A = 12x + 48$ sq cm, $P = 2x + 32$ cm **2.** $A = 20w - w^2$ sq in, $P = 40$ in
3a. $l = 9$ ft **b.** $A = 54$ sq ft **4a.** $l = 15 - x$ ft **b.** $A = 15x - x^2$ sq ft

Homework 6.2

Skills Practice

1. **a.** Give the dimensions of two different rectangles with perimeter 60 meters. Compute the areas of the two rectangles.
   b. A rectangle has a perimeter of 60 meters. If the length of the rectangle is x meters, write an expression for its width.
   c. Write an expression for the area of the rectangle.

2. Francine stands at the top of a 300-foot cliff and throws a diamond ring upward with a velocity of 20 feet per second. The height of the ring above the ground t seconds later is given by the equation

$$h = -16t^2 + 20t + 300$$

   where h is in feet.
   a. Use your calculator to make a table of values for the height equation, with increments of 0.5 second.
   b. Graph the height equation on your calculator. Use your table of values to help you choose appropriate **Window** settings.
   c. What is the highest altitude the ring reaches? When does it reach that height? Use the **Trace** feature to find approximate answers first. Then use the **Table** feature to improve your estimate.
   d. When does the ring pass Francine on its way down? (Francine is standing at a height of 300 feet.) Use the **intersect** command.
   e. How long will it take the ring to hit the ground at the bottom of the cliff?

■ For Problems 3-6, solve.

3. $x(2x - 3) = -1$
4. $t(t - 3) = 2(t - 3)$
5. $z(3z + 2) = (z + 2)^2$
6. $(v + 2)(v - 5) = 8$

■ For Problems 7-8, use a graphing calculator to graph the equations. Use the standard graphing window. What do you notice about the x-intercepts? Try to generalize your observation, and test your idea with a few examples.

7. **a.** $y = x^2 - x - 20$
   b. $y = 2(x^2 - x - 20)$
   c. $y = 0.5(x^2 - x - 20)$
8. **a.** $y = x^2 + 6x - 16$
   b. $y = -2(x^2 + 6x - 16)$
   c. $y = -0.1(x^2 + 6x - 16)$

■ For Problems 9-12, write a quadratic equation whose solutions are given. The equation should be in standard form with integer coefficients.

9. -2 and 1
10. 0 and -5
11. -3 and $\frac{1}{2}$
12. $\frac{-1}{4}$ and $\frac{3}{2}$

■ For Problems 13-14, graph the equation on a graphing calculator. (Use the **ZInteger** setting.) Locate the x-intercepts of the graph. Use the x-intercepts to write the quadratic expression in factored form.

13. $y = 0.1(x^2 - 3x - 270)$
14. $y = -0.08(x^2 + 14x - 576)$

Applications

15. Suppose you have 36 feet of rope and you want to enclose a rectangular display area against one wall of an exhibit hall. The area enclosed depends upon the dimensions of the rectangle you make. Because the wall makes one side of the rectangle, the length of the rope accounts for only three sides. Thus

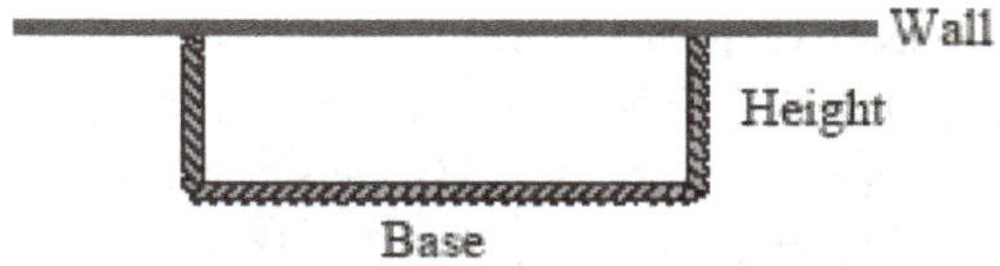

$$\textit{Base} + 2\,(\textit{Height}) = 36$$

a. Complete the table showing the base and the area of the rectangle for various heights.

Height	Base	Area	Height	Base	Area
1	34	34	10		
2	32	64	11		
3			12		
4			13		
5			14		
6			15		
7			16		
8			17		
9			18		

b. Make a graph with Height on the horizontal axis and Area on the vertical axis. Draw a smooth curve through your data points.

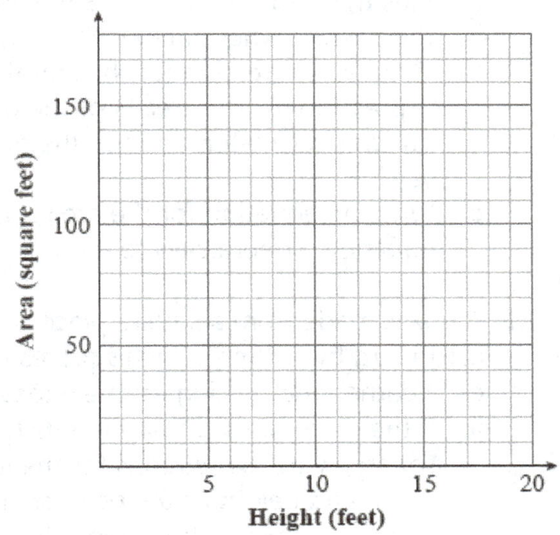

c. What is the area of the largest rectangle you can enclose in this way? What are its dimensions? Label the point on your graph that corresponds to this rectangle with the letter M.

d. Let x stand for the height of a rectangle, and write algebraic expressions for the base and the area of the rectangle.

e. Enter your algebraic expression for the area in your calculator, and use the **Table** feature to verify the entries in your table in part (a).

f. Graph your formula for area on your graphing calculator. Use your table of values and your hand-drawn graph to help you choose appropriate **Window** settings.

g. Use the **intersect** command to find the height of the rectangle whose area is 149.5 square feet.

■ If an object is thrown into the air from a height s_0 above the ground with an initial velocity v_0, then its height t seconds later is given by the formula

$$h = -\frac{1}{2}\,gt^2 + v_0t + s_0$$

where g is a constant that measures the force of gravity. Use this formula to answer Problem 16.

16. A tennis ball is thrown into the air with an initial velocity of 16 feet per second from a height of 8 feet. The value of g is 32.

a. Write a quadratic equation that gives the height of the tennis ball at time t.

b. Find the height of the tennis ball at $t = \frac{1}{2}$ second and at $t = 1$ second.

c. Write and solve an equation to answer the question: At what time is the tennis ball 11 feet high?

d. Use the **Table** feature on your calculator to verify your answers to parts (b) and (c). (What value of **ΔTbl** is useful for this problem?)

e. Graph your equation for the tennis ball's height from part (a) on your calculator. Use your table to help you choose an appropriate window.

f. If nobody hits the tennis ball, approximately how long will it be in the air?

17. A rancher has 360 yards of fence to enclose a rectangular pasture. If the pasture should be 8000 square yards in area, what should its dimensions be? We'll use three methods to solve this problem: a table of values, a graph, and an algebraic equation.

a. Make a table by hand that shows the areas of pastures of various widths, as shown below.

Width	Length	Area
10	170	1700
20	160	3200
⋮	⋮	⋮

(To find the length of each pasture, ask yourself: What is the sum of the length plus the width if there are 360 yards of fence?) Continue the table until you find the pasture whose area is 8000 square yards.

b. Write an expression for the length of the pasture if its width is x. Next write an expression for the area A of the pasture if its width is x. Graph the equation for A on your calculator, and use the graph to find the pasture of area 8000 square yards.

c. Write an equation for the area A of the pasture in terms of its width x. Solve your equation algebraically for $A = 8000$. Explain why there are two solutions.

18. A box is made from a square piece of cardboard by cutting 2-inch squares from each corner and turning up the edges.

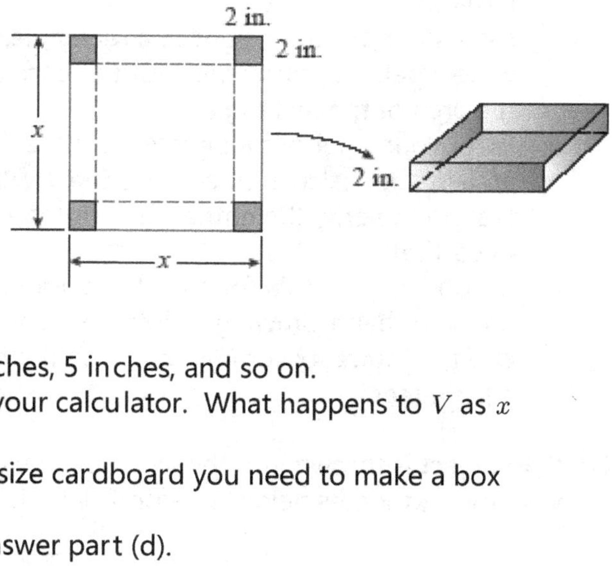

a. If the piece of cardboard is x inches square, write expressions for the length, width and height of the box. Then write an expression for the volume V of the box in terms of x.

b. Use your calculator to make a table of values showing the volumes of boxes made from cardboard squares of side 4 inches, 5 inches, and so on.

c. Graph your expression for the volume on your calculator. What happens to V as x increases?

d. Use your table or your graph to find what size cardboard you need to make a box with volume 50 cubic inches.

e. Write and solve a quadratic equation to answer part (d).

6.3 Completing the Square

So far we know two methods for solving quadratic equations: **factoring** and **extraction of roots**. The equation $2x^2+3x-1=0$ cannot be solved by either of these methods. (Do you see why?) In this lesson we learn a method that will solve *any* quadratic equation.

Squares of Binomials

We use extraction of roots to solve equations of the form

$$a(x-p)^2=q$$

where the left side of the equation includes the square of a binomial, or a **perfect square**. It turns out that we can write any quadratic equation in this form.

Study the following squares of binomials.

Square of Binomial, $(x+p)^2$		p	$2p$	p^2
1. $(x+5)^2=x^2+10x+25$		5	$2(5)=10$	$5^2=25$
2. $(x-3)^2=x^2-6x+9$		-3	$2(-3)=-6$	$(-3)^2=9$
3. $(x-12)^2=x^2-24x+144$		-12	$2(-12)=-24$	$(-12)^2=144$

In each case the square of the binomial is a special case of a quadratic trinomial,

$$(x+p)^2=x^2+2px+p^2$$

The coefficient of the linear term, $2p$, is twice the constant in the binomial, and the constant term of the trinomial, p^2, is its square.

We would like to reverse the process and write a quadratic trinomial as the square of a binomial. For example, what constant term can we add to

$$x^2-16x$$

to produce a perfect square trinomial? Compare the expression to the formula above:

$$\begin{aligned} x^2+\mathbf{2p}x+\ p^2\ &=(x+p)^2 \\ x^2-\mathbf{16}x+\ \underline{?}\ &=(x+\underline{?})^2 \end{aligned}$$

We see that $2p=-16$, so $p=\frac{1}{2}(-16)=-8$, and $p^2=(-8)^2=\mathbf{64}$. We substitute these values for p^2 and p into the equation to find

$$x^2-16x+\mathbf{64}=(x-\mathbf{8})^2$$

In the resulting trinomial, the constant term is equal to *the square of one-half the coefficient of x*. In other words, we can find the constant term by taking one-half the coefficient of x, and then squaring the result. Adding a constant term obtained in this way is called **completing the square.**

Example 1 Complete the square by adding an appropriate constant, and write the result as the square of a binomial.

a. x^2-12x+ ____ **b.** x^2+5x+ ____

Solutions a. One-half of -12 is -6, so the constant term is $(-6)^2$ or 36. We add 36 to obtain

$$x^2 - 12x + \mathbf{36} = (x-6)^2 \qquad \begin{aligned} p &= \tfrac{1}{2}(-12) = -6 \\ p^2 &= (-6)^2 = 36 \end{aligned}$$

b. One-half of 5 is $\dfrac{5}{2}$, so the constant term is $\left(\dfrac{5}{2}\right)^2$, or $\dfrac{25}{4}$. We add $\dfrac{25}{4}$ to obtain

$$x^2 + 5x + \mathbf{\frac{25}{4}} = \left(x + \frac{5}{2}\right)^2 \qquad \begin{aligned} p &= \tfrac{1}{2}(5) = \tfrac{5}{2} \\ p^2 &= \left(\tfrac{5}{2}\right)^2 = \tfrac{25}{4} \end{aligned}$$

RQ1 True or false.
a. Every quadratic equation can be solved by factoring.
b. Every expression of the form $x^2 + bx$ can be turned into a perfect square by adding an appropriate constant.
c. The coefficient of the linear term in $(x+p)^2$ is twice the constant term.
d. To *complete the square* means to square the expression.

Look Closer: You may find it helpful to visualize completing the square geometrically. We think of the expression $x^2 + 2px$ as the area of a rectangle with dimensions x and $x + 2p$. For example:

(a) The rectangle with length $x + 10$ and width x has area $x(x+10) = x^2 + 10x$.
(b) We would like to cut the rectangle into pieces and rearrange them to make a square.
(c) We move half of the x-term so that each side of the square has length $x + 5$ (note that $p = \frac{1}{2}(10) = 5$).
(d) We see that the missing corner piece has area $p^2 = 5^2 = 25$.

a. 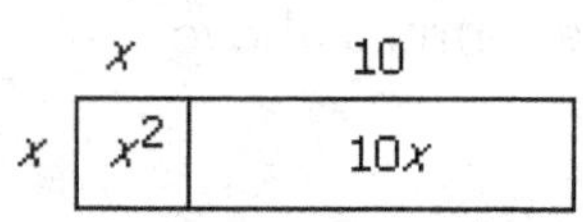

Represent $x^2 + 10x$ as the area of a rectangle

b. 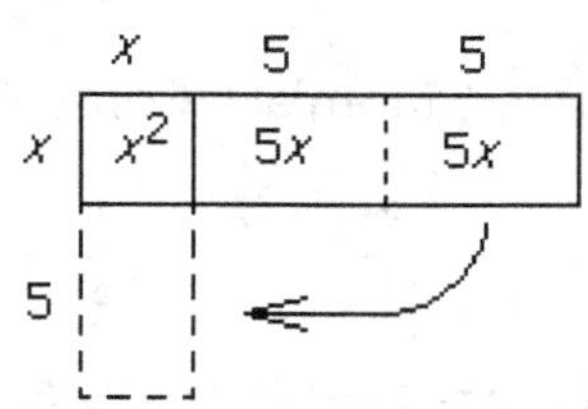

Take $\frac{1}{2}$ of the linear term, and rearrange the pieces

c. 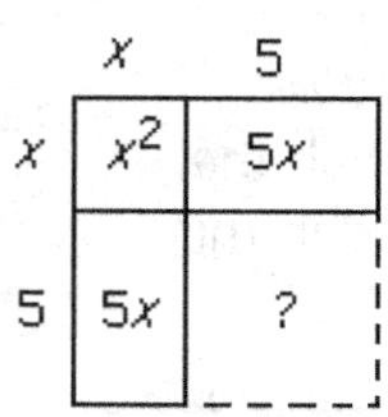

What is the area of the missing piece?

d. 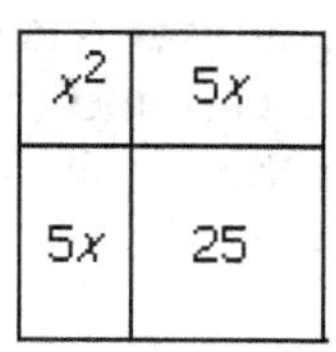

We must add $5^2 = 25$:
$x^2 + 10x + \mathbf{25} = (x+5)^2$

Solving Quadratic Equations by Completing the Square

Now we can use completing the square to solve quadratic equations. Consider the equation

$$x^2 - 6x - 7 = 0$$

Step 1 To begin, we move the constant term to the other side of the equation to get

$$x^2 - 6x \quad \underline{\qquad} \quad = 7$$

Step 2 Next, we complete the square on the left. Because

$$p = \frac{1}{2}(-6) = \mathbf{-3} \quad \text{and} \quad p^2 = (-3)^2 = \mathbf{9}$$

we add 9 to both sides of our equation to get

$$x^2 - 6x + \mathbf{9} = 7 + \mathbf{9}$$

Step 3 The left side of the equation is now the square of a binomial, namely $(x - \mathbf{3})^2$. We write the left side in its square form, and simplify the right side, to get

$$(x - \mathbf{3})^2 = 16$$

Step 4 We can now use extraction of roots to find the solutions. Taking square roots of both sides, we get

$$x - 3 = 4 \quad \text{or} \quad x - 3 = -4 \qquad \textbf{Solve each equation.}$$
$$x = 7 \quad \text{or} \quad x = -1$$

The solutions are 7 and −1.

Look Closer: You can check that this equation is equivalent to the original one; if you expand the left side and collect like terms you will return to the original equation.

$$(x - 3)^2 = 16 \qquad \textbf{Expand the square.}$$
$$x^2 - 6x + 9 = 16 \qquad \textbf{Subtract 16 from both sides.}$$
$$x^2 - 6x - 7 = 0$$

The graph of $y = x^2 - 6x - 7$ is shown at right. The x-intercepts of the graph are $x = 7$ and $x = -1$, and the axis of symmetry is halfway between the intercepts, at $x = 3$.

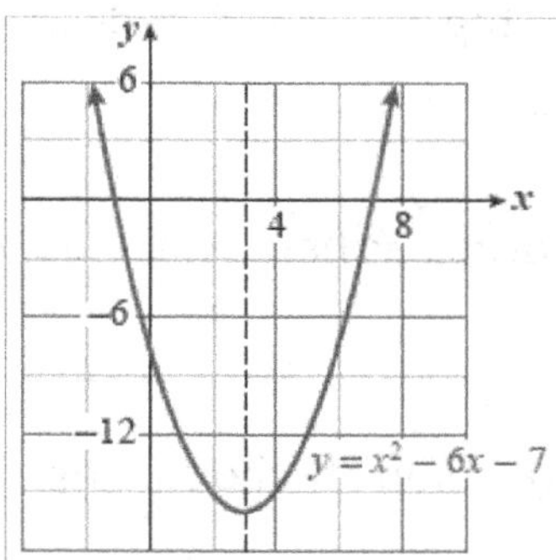

We can also solve the equation $x^2 - 6x - 7 = 0$ by factoring instead of completing the square. Of course, we get the same solutions by either method. In Example 2 we solve an equation that cannot be solved by factoring.

Example 2 Solve $x^2 - 4x - 3 = 0$ by completing the square.

Solution Step 1 We write the equation with the constant term on the right side:

$$x^2 - 4x \quad \underline{\qquad} \quad = 3$$

Step 2 We complete the square on the left side. The coefficient of x is −4, so

$$p = \frac{1}{2}(-4) = \mathbf{-2} \quad \text{and} \quad p^2 = (-2)^2 = \mathbf{4}$$

We add 4 to both sides of the equation.

$$x^2 - 4x + \mathbf{4} = 3 + \mathbf{4}$$

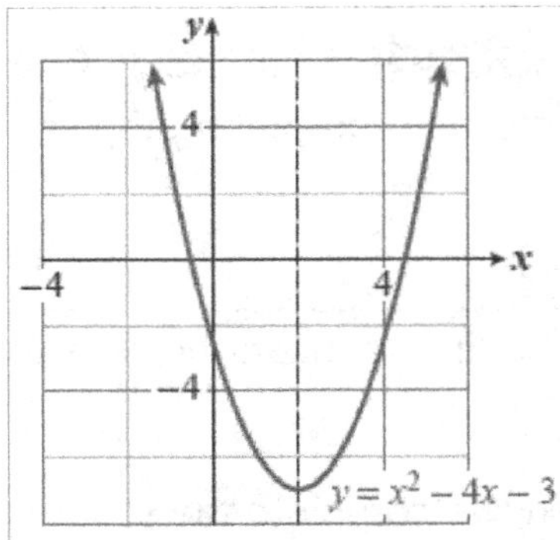

Step 3 We write the left side as the square of a binomial, and combine terms on the right side:

$$(x-2)^2 = 7$$

Step 4 Finally, we use extraction of roots to obtain

$$x - 2 = \sqrt{7} \quad \text{or} \quad x - 2 = -\sqrt{7}$$
$$x = 2 + \sqrt{7} \qquad\qquad x = 2 - \sqrt{7}$$

These are the exact values of the solutions. We can use a calculator to find decimal approximations for each solution:

$$2 + \sqrt{7} \approx 4.646 \quad \text{and} \quad 2 - \sqrt{7} \approx -0.646$$

The graph of $y = x^2 - 4x - 3$ is shown in the figure. ■

RQ2 Put the steps in the correct order:
a. Add p^2 to both sides of the equation.
b. Use extraction of roots.
c. Write the left side as a perfect square.
d. Write the equation with the constant term on the right side.

The General Case: $ax^2 + bx + c = 0,\ a \neq 1$

Our method for completing the square only works if the coefficient of x^2 is 1. If we want to solve a quadratic equation whose lead coefficient is not 1, we must divide each term of the equation by the x^2-coefficient.

■ **Example 3** Solve $2x^2 - 6x - 5 = 0$

Solution Step 1 We must divide each term of the equation by 2:

$$x^2 - 3x - \frac{5}{2} = 0$$

Now we proceed as before. We write the equation with the constant on the right side.

$$x^2 - 3x \quad \underline{\qquad} \quad = \frac{5}{2} \qquad \textbf{Add } \frac{5}{2} \textbf{ to both sides.}$$

Step 2 We complete the square:

$$p = \frac{1}{2}(-3) = \frac{-3}{2}, \quad \text{and} \quad p^2 = \left(\frac{-3}{2}\right)^2 = \frac{9}{4}$$

so we add $\frac{9}{4}$ to both sides of the equation.

$$x^2 - 3x + \frac{\mathbf{9}}{\mathbf{4}} = \frac{5}{2} + \frac{\mathbf{9}}{\mathbf{4}}$$

Step 3 We rewrite the left side as the square of a binomial and simplify the right side to get

$$\left(x - \frac{3}{2}\right)^2 = \frac{19}{4}$$

Step 4 Finally, we extract roots and solve each equation for x.

$$x - \frac{3}{2} = \sqrt{\frac{19}{4}} \quad \text{or} \quad x - \frac{3}{2} = -\sqrt{\frac{19}{4}}$$

The solutions are $\frac{3}{2} + \sqrt{\frac{19}{4}}$ and $\frac{3}{2} - \sqrt{\frac{19}{4}}$. Using a calculator, we can find decimal approximations for the solutions: 3.679 and −0.679.

Caution! In Example 3, it is essential that we first divide each term of the equation by 2, the coefficient of x^2. The following attempt at a solution is incorrect.

$$2x^2 - 6x = 5$$
$$2x^2 - 6x + 9 = 5 + 9$$
$$(2x - 3)^2 = 14 \qquad \leftarrow \textbf{Incorrect!}$$

You can check that $(2x - 3)^2$ is not equal to $2x^2 - 6x + 9$. We have not written the left side of the equation as a perfect square, so the solutions we obtain by extracting roots will not be correct.

RQ3. What is the first step for completing the square on $ax^2 + bx + c = 0$?

To Solve a Quadratic Equation by Completing the Square:

1. Write the equation in standard form.
2. Divide both sides of the equation by the coefficient of the quadratic term, and subtract the constant term from both sides.
3. Complete the square on the left side:
   a. Multiply the coefficient of the first-degree term by one-half, then square the result.
   b. Add the value obtained in (a) to both sides of the equation.
4. Write the left side of the equation as the square of a binomial. Simplify the right side.
5. Use extraction of roots to finish the solution.

The Vertex Formula

In Section 5.6 we found a formula for the vertex of a parabola with equation $y = ax^2 + bx$. The x-coordinate of the vertex is given by

$$x_v = \frac{-b}{2a}$$

Now we'll see that the same formula holds for any parabola. Use your calculator to graph the two parabolas

(i) $y = 2x^2 + 8x$
(ii) $y = 2x^2 + 8x + 6$

in the window

Xmin $= -6$ Xmax $= 4$
Ymin $= -10$ Ymax $= 10$

The graphs are shown at right.

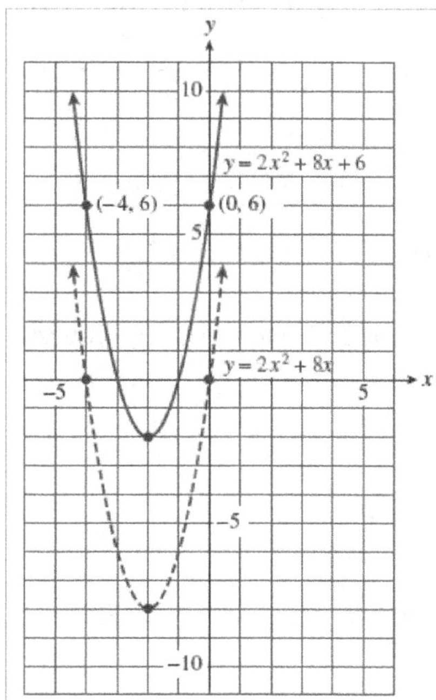

Now compare the two graphs. You should notice that:

(1) The second graph is identical to the first, except shifted upward by 6 units.
(2) For both graphs, the x-coordinate of the vertex is $x_v = -2$

We see that the x-coordinate of the vertex is not affected by an upward shift. The formula for the x-coordinate of the vertex still holds.

Vertex of a Parabola

For the graph of $y = ax^2 + bx + c$, the x-coordinate of the vertex is

$$\boldsymbol{x_v = \frac{-b}{2a}}$$

Example 4 Find the vertex of the graph of $y = -2x^2 + x + 1$

Solution For this equation, $a = -2$, $b = 1$, and $c = 1$. The x-coordinate of the vertex is given by

$$x_v = \frac{-b}{2a} = \frac{-1}{2(-2)} = \frac{1}{4}$$

To find the y-coordinate of the vertex, we substitute $x = \frac{1}{4}$ into the equation. We can do this by hand to find

$$\begin{aligned} y_v &= -2\left(\frac{1}{4}\right)^2 + \frac{1}{4} + 1 \\ &= -2\left(\frac{1}{16}\right) + \frac{4}{16} + \frac{16}{16} = \frac{18}{16} = \frac{9}{8} \end{aligned}$$

The coordinates of the vertex are $(\frac{1}{4}, \frac{9}{8})$. Or we can use the calculator to evaluate $-2x^2 + x + 1$ for $x = 0.25$. We enter

[(−)] 2 [×] 0.25 [x^2] [+] 0.25 [+] 1

and press [ENTER]. The calculator returns the y-value 1.125. Thus, the vertex is the point $(0.25, 1.125)$, which is the decimal equivalent of $(\frac{1}{4}, \frac{9}{8})$. ■

RQ4. What is the x-coordinate of the vertex of the parabola $y = ax^2 + bx + c$?

Skills Warm-Up

True or False. Explain why your answers are correct.

1. $\dfrac{15}{\sqrt{5}} = 3\sqrt{5}$ **2.** $\dfrac{7}{\sqrt[3]{7}} = \sqrt[3]{7}$ **3.** $(\sqrt{13})^3 = 13\sqrt{13}$

4. $(\sqrt[3]{3})^2 = 3$ **5.** $(\sqrt{2}m)(\sqrt{2}m) = 2m$ **6.** $(5\sqrt{x})^2 = 25x$

Answers: **1.** True **2.** False **3.** True **4.** False **5.** False **6.** True

Homework 6.3

Skills Practice

For Problems 1-4, complete the square and write the result as the square of a binomial.

1. $x^2 + 8x$ **2.** $x^2 - 7x$ **3.** $x^2 + \frac{3}{2}x$ **4.** $x^2 - \frac{4}{5}x$

For Problems 5-12, solve by completing the square.

5. $x^2 - 2x + 1 = 0$
6. $x^2 + 9x + 20 = 0$
7. $x^2 = 3 - 3x$
8. $2x^2 + 4x - 3 = 0$
9. $3x^2 + x = 4$
10. $4x^2 - 3 = 2x$
11. $3x^2 - x - 4 = 0$
12. $5x^2 + 8x = 4$

For Problems 13-16, choose the best method, then solve the equation.

13. $x^2 - 3x = 40$
14. $(x - 3)^2 = 40$
15. $3x^2 + x = 12$
16. $3x^2 + 6x = 15$

For Problems 17-9, solve by completing the square. Your answers will involve a, b, or c.

17. $x^2 + 2x + c = 0$ **18.** $x^2 + bx + 1 = 0$ **19.** $ax^2 + 2x - 4 = 0$

For Problems 20-22, solve the formula for the indicated variable.

20. $V = \pi(r - 3)^2h$, for r
21. $E = \frac{1}{2}mv^2 + mgh$, for v
22. $V = 2(s^2 + t^2)w$, for t

For Problems 23-26, solve for y in terms of x. Use whichever method seems easiest.

23. $x^2y - y^2 = 0$
24. $(2y + 3x)^2 = 9$
25. $4x^2 - 9y^2 = 36$
26. $4x^2 - 25y^2 = 0$

For Problems 27-30, find the coordinates of the vertex. Decide whether the vertex is a maximum point or a minimum point of the graph.

27. $y = x^2 + 6x - 2$
28. $y = 2 + 3x - x^2$
29. $y = \frac{1}{2}x^2 - \frac{2}{3}x + \frac{1}{3}$
30. $y = 2.3 - 7.2x - 0.8x^2$

Applications

31. a. Write an expression for the area of the square.
b. Express the area as a quadratic trinomial.
c. Divide the square into four pieces whose areas are given by the terms of your answer to (b).

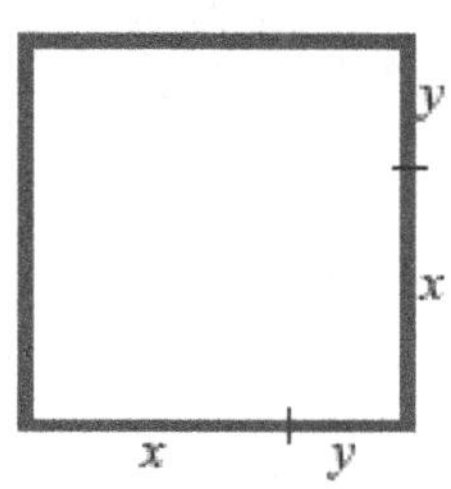

32. The sail shown is a right triangle of base and height x. It has a colored stripe along the hypotenuse and a white triangle of base and height y in the lower corner.

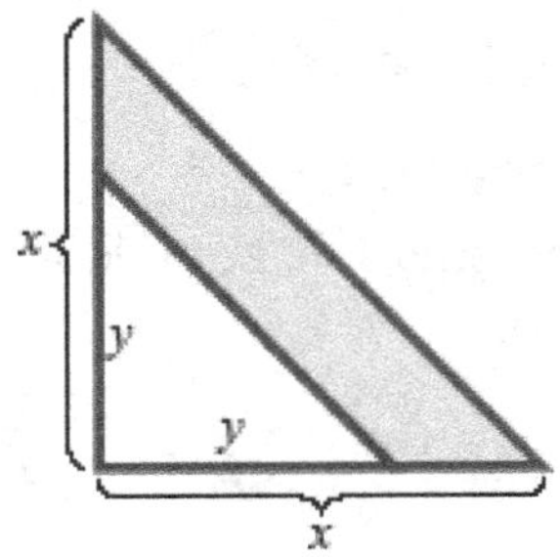

a. Write an expression for the area of the colored stripe.

b. Express the area of the stripe in factored form.

c. If the sail is 7½ feet high and the white strip is 4½ feet high, use your answer to (b) to calculate mentally the area of the stripe.

33. Complete the square to find the solutions of the equation $x^2 + bx + c = 0$. (Your answers will be expressions in b and c.)

34. Explain why the "solution" is incorrect.

$$x^2 - 6x + 8 = 0$$
$$x^2 = 6x - 8$$
$$x = \frac{6x - 8}{x}$$

For Problems 35-36,

a. Use completing the square to find the x-intercepts of the graph.

b. Find the vertex of the graph.

c. Sketch the graph.

35. $y = 4x^2 - 2x - 3$

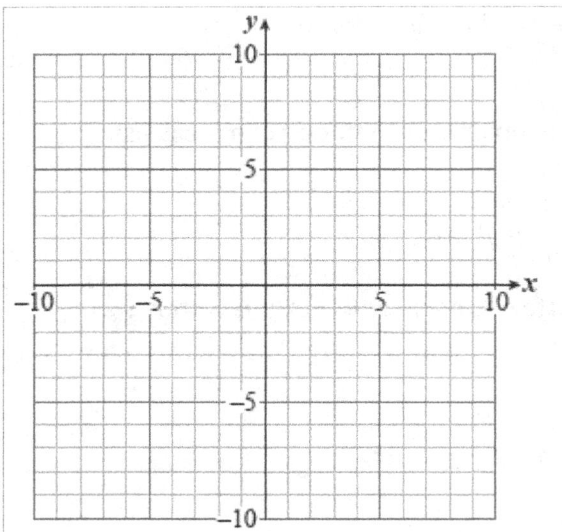

36. $y = 5x^2 + 8x - 4$

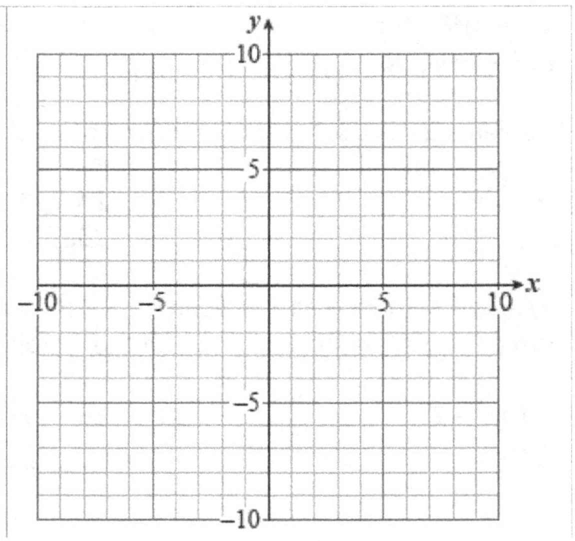

6.4 Quadratic Formula

Instead of completing the square every time we solve a new quadratic equation, we can complete the square on the general quadratic equation,

$$ax^2 + bx + c = 0, \qquad a \neq 0$$

and obtain a formula for the solutions of any quadratic equation.

The Quadratic Formula
The solutions of the equation $ax^2 + bx + c = 0,\ a \neq 0,$ are

$$\boldsymbol{x = \frac{-b \pm \sqrt{b^2 - 4ac}}{2a}}$$

This formula expresses the solutions of a quadratic equation in terms of its coefficients. The symbol $\pm$, read *plus or minus*, is used to combine the two equations

$$x = \frac{-b + \sqrt{b^2 - 4ac}}{2a} \qquad \text{and} \qquad x = \frac{-b - \sqrt{b^2 - 4ac}}{2a}$$

into a single equation.

To solve a quadratic equation using the quadratic formula, all we have to do is substitute the coefficients a, b, and c into the formula.

RQ1 Which of the following is a correct statement of the quadratic formula?

a. $x = -b \pm \dfrac{\sqrt{b^2 - 4ac}}{2a}$ **b.** $x = \dfrac{-b}{2a} \pm \dfrac{\sqrt{b^2 - 4ac}}{2a}$

c. $x = -b \pm \dfrac{b - \sqrt{4ac}}{2a}$ **d.** $x = \dfrac{-b \pm \sqrt{b^2 - ac}}{a}$

Example 1 Solve $2x^2 + 1 = 4x$

Solution We write the equation in standard form as

$$2x^2 - 4x + 1 = 0$$

We substitute 2 for a, -4 for b, and 1 for c into the quadratic formula, and simplify.

$$x = \frac{-(-4) \pm \sqrt{(-4)^2 - 4(2)(1)}}{2(2)}$$

$$= \frac{4 \pm \sqrt{8}}{4}$$

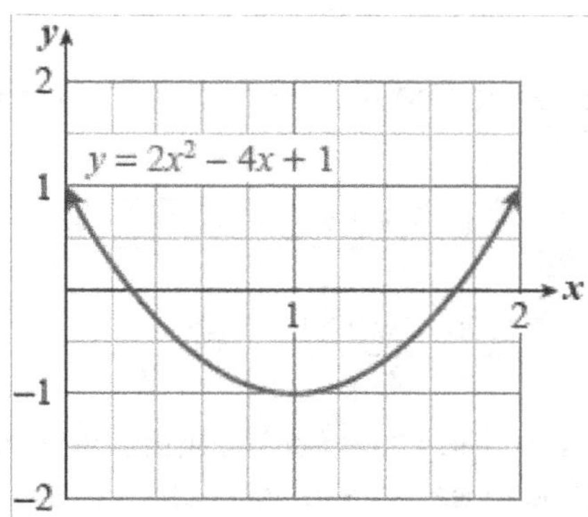

The two solutions are $\dfrac{4 + \sqrt{8}}{4}$ and $\dfrac{4 - \sqrt{8}}{4}$. Using a calculator, we find that these solutions are approximately 1.707 and 0.293. These values are the x-intercepts of the graph of $y = 2x^2 - 4x + 1$, as shown in the figure.

Solving Quadratic Equations

We have now seen four different algebraic methods for solving quadratic equations:

1. Factoring
2. Extraction of Roots
3. Completing the Square
4. Quadratic Formula

Factoring and extraction of roots are relatively fast and simple, but they don't work on all quadratic equations. The quadratic formula will work on any quadratic equation.

RQ2 Match each equation with the most efficient method of solution.

a. $6(x-4)^2 = 120$ **b.** $x^2 - 9x + 20 = 0$

c. $1.4x^2 - 3.2x + 6.5 = 0$ **d.** $x^2 - 20x = 44$

I. Factoring **II.** Extraction of roots

III. Completing the square **IV.** Quadratic formula

Example 2 The owners of a day-care center plan to enclose a divided playground against the back wall of their building, as shown in the figure. They have 300 feet of picket fence, and they would like the total area of the playground to be 6000 square feet. Can they enclose the playground with the fence they have, and if so what should the dimensions of the playground be?

Solution Suppose the width of the playground is x feet. Because there are three sections of fence along the width of the playground, that leaves $300 - 3x$ feet of fence for its length.

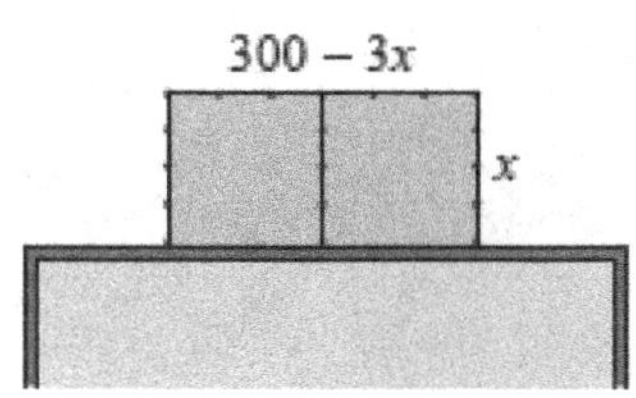

The area of the playground should be 6000 square feet, so we have the equation

$$x(300 - 3x) = 6000$$

width · length = area

This is a quadratic equation. In standard form,

$$3x^2 - 300x + 6000 = 0 \qquad \textbf{Divide each term by 3.}$$
$$x^2 - 100x + 2000 = 0$$

The left side cannot be factored, so we use the quadratic formula with $a = 1$, $b = -100$, and $c = 2000$.

$$x = \frac{-(-100) \pm \sqrt{(-100)^2 - 4(1)(2000)}}{2(1)} \qquad \textbf{Simplify the radicand.}$$

$$= \frac{100 \pm \sqrt{2000}}{2} \approx \frac{100 \pm 44.7}{2}$$

Thus, $x \approx 72.4$ or $x \approx 27.6$. Both values give solutions to the problem. To find the length of the playground in each case, we substitute x into $300 - 3x$.

If the width $= x = 72.4$ ft,
length $= 300 - 3(72.4) = 828.8$ ft

If the width $= x = 27.6$ ft,
length $= 300 - 3(27.6) = 217.2$ ft

The dimensions of the playground can be 72.4 feet by 82.8 feet, or it can be 27.6 feet by 217.2 feet.

RQ3. What should you do if you can't solve a quadratic equation by factoring?

Sometimes it is useful to solve a quadratic equation for one variable in terms of the others.

Example 3 Solve $x^2 - xy + y = 2$ for x in terms of y.

Solution We first write the equation in standard form with x as the variable:

$$x^2 - yx + (y - 2) = 0$$

Expressions in y are treated as constants with respect to x, so that $a = 1$, $b = -y$, and $c = y - 2$. We substitute these expressions into the quadratic formula.

$$x = \frac{-(-y) \pm \sqrt{(-y)^2 - 4(1)(y-2)}}{2(1)} \qquad \text{Simplify the radicand.}$$

$$= \frac{y \pm \sqrt{y^2 - 4y + 8}}{2}$$

Sketching a Parabola

If we can locate the vertex of a parabola, the x-intercepts, and the y-intercept, we can sketch a reasonably accurate graph. Recall that the graph should be symmetric about a vertical line through the vertex.

To graph the quadratic equation $y = ax^2 + bx + c$:

1. Determine whether the parabola opens upward (if $a > 0$) or downward (if $a < 0$).
2. Locate the vertex of the parabola.
   a. The x-coordinate of the vertex is $x_v = \dfrac{-b}{2a}$.
   b. Find the y-coordinate of the vertex by substituting x_v into the equation of the parabola.
3. Locate the x-intercepts (if any) by setting $y = 0$ and solving for x.
4. Locate the y-intercept by evaluating y for $x = 0$.
5. Locate the point symmetric to the y-intercept across the axis of symmetry.

Example 4 Sketch a graph of the equation $y = x^2 + 3x + 1$, showing the significant points.

Solution We follow the steps outlined above.

Step 1 Because $a = 1 > 0$, we know that the parabola opens upward.

Step 2 We compute the coordinates of the vertex:

$$x_v = \frac{-b}{2a} = \frac{-3}{2(1)} = -1.5$$

$$y_v = (-1.5)^2 + 3(-1.5) + 1 = -1.25$$

The vertex is the point $(-1.5, -1.25)$.

Step 3 We set y equal to zero and use the quadratic formula to find the x-intercepts.

$$0 = x^2 + 3x + 1$$

$$x = \frac{-3 \pm \sqrt{3^2 - 4(1)(1)}}{2(1)}$$
$$= \frac{-3 \pm \sqrt{5}}{2}$$

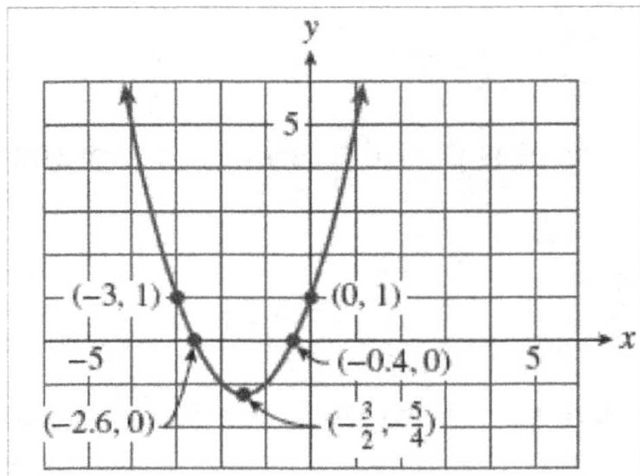

Rounding to the nearest tenth, we find that the x-intercepts are approximately $(-2.6, 0)$ and $(-0.4, 0)$.

Step 4 We substitute $x = 0$ to find the y-intercept, $(0, 1)$.

Step 5 The axis of symmetry is the vertical line $x = -1.5$, so the y-intercept lies 1.5 units to the right of the axis. There must be another point on the parabola with the same y-coordinate as the y-intercept, but 1.5 units to the left of the axis. The coordinates of this point are $(-3, 1)$. We plot the x-intercepts, the vertex, the y-intercept, and its symmetric point, and draw a parabola through them. The finished graph is shown in the figure. ■

RQ4. What points should you find in order to sketch a parabola?

Complex Numbers

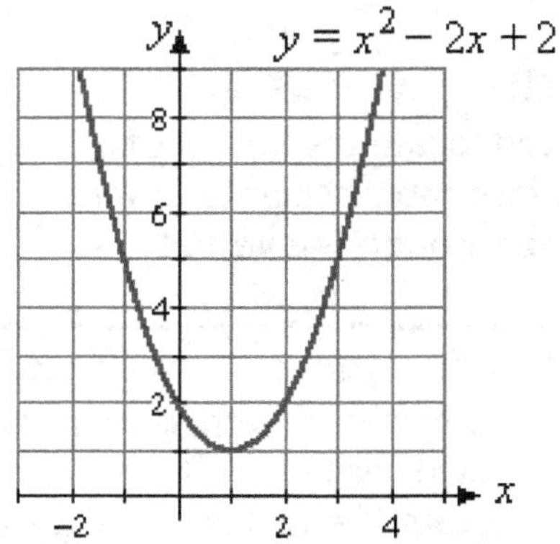

Not all quadratic equations have real solutions. For example, the graph of

$$y = x^2 - 2x + 2$$

shown at right has no x-intercepts, and the equation

$$x^2 - 2x + 2 = 0$$

has no real solutions. We can still use completing the square or the quadratic formula to solve the equation.

■ **Example 5** Solve the equation $x^2 - 2x + 2 = 0$ by using the quadratic formula.

Solution We substitute $a = 1$, $b = -2$, and $c = 2$ into the quadratic formula to get

$$x = \frac{-(-2) \pm \sqrt{(-2)^2 - 4(1)(2)}}{2(1)} = \frac{2 \pm \sqrt{-4}}{2}$$

Because $\sqrt{-4}$ is not a real number, the equation $x^2 - 2x + 2 = 0$ has no real solutions. ■

Although square roots of negative numbers like $\sqrt{-4}$ are not real numbers, they occur frequently in mathematics and its applications. Mathematicians in the sixteenth century gave them the name **imaginary numbers**, which reflected the mistrust with which they were viewed at the time. Today, however, such numbers are well understood and used routinely by scientists and engineers.

We begin by defining a new number, i, whose square is -1.

Imaginary Unit

We define the **imaginary unit**, $\boldsymbol{i}$, by

$$\boldsymbol{i^2 = -1} \quad \text{or} \quad \boldsymbol{i = \sqrt{-1}}$$

Look Closer: The letter i used in this way is not a variable, it is the name of a specific number, and hence is a constant. The square root of any negative number can be written as the product of a real number and i. For example,

$$\begin{aligned}\sqrt{-4} &= \sqrt{-1 \cdot 4} \\ &= \sqrt{-1}\ \sqrt{4} = i \cdot 2\end{aligned}$$

or $\sqrt{-4} = 2i$. Any number that is the product of i and a real number is called an imaginary number.

Imaginary Numbers

For $a > 0$,

$$\sqrt{-a} = \sqrt{-1}\sqrt{a} = i\sqrt{a}$$

Examples of imaginary numbers are

$$3i, \quad \frac{7}{8}i, \quad -38i, \quad \text{and} \quad i\sqrt{5}$$

Example 6 Write each radical as an imaginary number.

a. $\sqrt{-25}$

b. $2\sqrt{-3}$

Solutions a. $\sqrt{-25} = \sqrt{-1}\sqrt{25} = i\sqrt{25} = 5i$

b. $2\sqrt{-3} = 2\sqrt{-1}\sqrt{3} = 2i\sqrt{3}$

The square root of any negative number can be written as the product of a real number and i. For example, the two square roots of -9 are $3i$ and $-3i$.

Using this new notation, we write the solutions to Example 5, $\dfrac{2 \pm \sqrt{-4}}{2}$, as $\dfrac{2 \pm 2i}{2}$, or $1 \pm i$. The sum of a real number and an imaginary number is called a **complex number**.

Complex Numbers

A **complex number** can be written in the form $a + bi$, where a and b are real numbers.

Examples of complex numbers are

$$3 - 5i, \quad 2 + \sqrt{7}\,i, \quad \frac{4 - i}{3}, \quad 6i, \quad \text{and} \quad -9$$

In a complex number $a + bi$, a is called the **real part,** and b is called the **imaginary part**. All real numbers are also complex numbers (with imaginary part equal to zero). A complex number whose real part equals 0 is called a **pure imaginary** number.

Example 7 Solve $x^2 - 2x + 5 = 0$

Solution We use the quadratic formula to find

$$x = \frac{-(-2) \pm \sqrt{(-2)^2 - 4(1)(5)}}{2} = \frac{2 \pm \sqrt{-16}}{2}$$

If we now replace $\sqrt{-16}$ by $4i$, we have

$$x = \frac{2 \pm 4i}{2} = 1 \pm 2i$$

The solutions are $1 + 2i$ and $1 - 2i$. ■

Number of x-Intercepts

The graph of the quadratic equation $y = ax^2 + bx + c$ may have two, one, or no x-intercepts, according to the number of distinct real-valued solutions of the equation $ax^2 + bx + c = 0$.

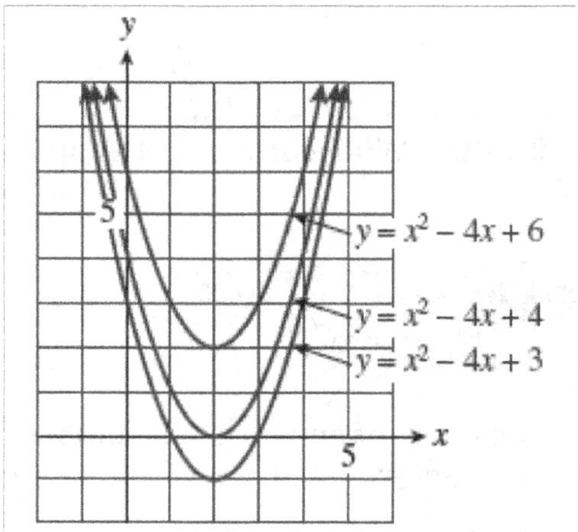

For example, the graph of

$$y = x^2 - 4x + 3$$

has two x-intercepts, because the equation

$$x^2 - 4x + 3 = 0$$

has two real-valued solutions, $x = 1$ and $x = -3$, as shown in the figure at right. The graph of

$$y = x^2 - 4x + 4$$

has only one x-intercept, because the equation

$$x^2 - 4x + 4 = 0$$

has only one (repeated) real-valued solution, $x = 2$. The graph of $y = x^2 - 4x + 6$ has no x-intercepts, because the equation $x^2 - 4x + 6 = 0$ has no real-valued solutions.

Look Closer: A closer look at the quadratic formula reveals useful information about the solutions of quadratic equations. For the three equations above, we have the following:

$y = x^2 - 4x + 3$ two x-intercepts	$y = x^2 - 4x + 4$ one x-intercept	$y = x^2 - 4x + 6$ no x-intercepts
$x = \frac{4 \pm \sqrt{(-4)^2 - 4(1)(3)}}{2}$	$x = \frac{4 \pm \sqrt{(-4)^2 - 4(1)(4)}}{2}$	$x = \frac{4 \pm \sqrt{(-4)^2 - 4(1)(6)}}{2}$
$= \frac{4 \pm \sqrt{4}}{2}$	$= \frac{4 \pm \sqrt{0}}{2}$	$= \frac{4 \pm \sqrt{-12}}{2}$
(two solutions)	(one repeated solution)	(no solutions) ●

The Discriminant

The expression $b^2 - 4ac$, which appears under the radical in the quadratic formula, is called the **discriminant** of the equation. The value of the discriminant determines the nature of the solutions of the equation.

Discriminant
The **discriminant** of a quadratic equation $ax^2 + bx + c = 0$ is

$$\boldsymbol{D = b^2 - 4ac}$$

1. If $D > 0$, the equation has two unequal real solutions.
2. If $D = 0$, the equation has one real solution of multiplicity two.
3. If $D < 0$, the equation has two complex solutions.

Example 8 Use the discriminant to determine the nature of the solutions of each equation.

a. $x^2 - x - 3 = 0$ **b.** $2x^2 + x + 1 = 0$ **c.** $x^2 - 2\sqrt{3}x + 3 = 0$

Solutions We compute the discriminant for each equation.
a. $D = b^2 - 4ac = (-1)^2 - 4(1)(-3) = 13 > 0.$ The equation has two real, unequal solutions.
b. $D = b^2 - 4ac = 1^2 - 4(2)(1) = -7 < 0.$ The equation has two complex solutions.
c. $D = b^2 - 4ac = (-2\sqrt{3})^2 - 4(1)(3) = 0.$ The equation has one real solution of multiplicity two. You can verify the conclusions above by solving each equation. ■

We can also use the discriminant to decide whether a quadratic equation with integer coefficients can be solved by factoring.

If the discriminant is a perfect square, that is, the square of an integer, the solutions are rational numbers. This in turn means that the equation can be solved by factoring.

In any case, the solutions of a quadratic equation always occur in **conjugate pairs**,

$$\frac{-b}{2a} + \frac{\sqrt{b^2 - 4ac}}{2a} \quad \text{and} \quad \frac{-b}{2a} - \frac{\sqrt{b^2 - 4ac}}{2a}$$

The only difference between the two solutions is the sign between the terms. For example, if we know that one solution of a particular quadratic equation is $3 + \sqrt{2}$, then the other solution must be $3 - \sqrt{2}$.

RQ5 True or False.
a. The discriminant is part of the quadratic formula.
b. We use the discriminant to calculate the solutions of a quadratic equation.
c. If the discriminant is negative, both x-intercepts of the graph are negative.
d. If a quadratic equation won't factor, its graph has no x-intercepts.

Skills Warm-Up

Solve each equation by the easiest method.

1. $x^2 = 36$	**2.** $x^2 - 9x = 36$	**3.** $x^2 - 9x = 0$
4. $9x^2 - 36 = 0$	**5.** $3x^2 - 24x + 36 = 0$	**6.** $3x^2 = 31x - 36$

Answers: **1.** 6, −6 **2.** 12, −3 **3.** 0, 9 **4.** 2, −2 **5.** 2,6 **6.** $\frac{4}{3}$, 9

Homework 6.4

Skills Practice

For Problems 1-6, solve by using the quadratic formula. Round your answers to three decimal places.

1. $x^2 - x - 1 = 0$
2. $y^2 + 2y = 5$
3. $3z^2 = 4.2z + 1.5$
4. $0 = x^2 - \frac{5}{3}x + \frac{1}{3}$
5. $-5.2z^2 + 176z + 1218 = 0$
6. $\frac{1}{4}t^2 + 4t - 9 = 0$

For Problems 7-8, the figure is a right triangle. Find the unknown sides.

7. 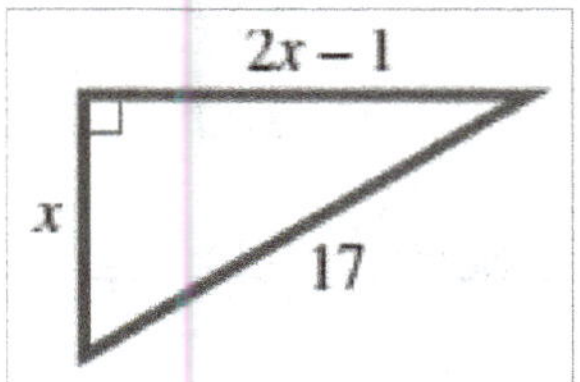

8. 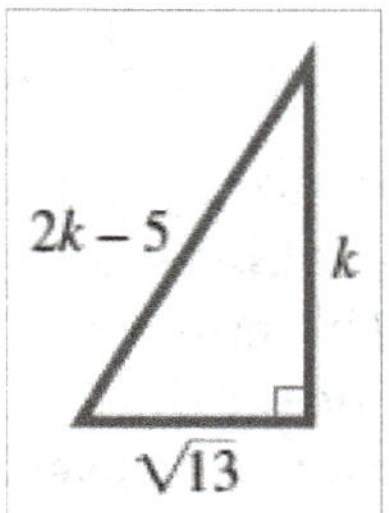

For Problems 9-11, write the complex number in the form $a + bi$, where a and b are real numbers.

9. $\sqrt{-25} - 4$
10. $\dfrac{-8 + \sqrt{-4}}{2}$
11. $\dfrac{-5 - \sqrt{-2}}{6}$

For Problems 12-13, solve. Write your answers as complex numbers.

12. $x^2 + 6x + 13 = 0$
13. $3x^2 - x + 1 = 0$

For Problems 14-16, use the discriminant to determine the nature of the solutions of each equation.

14. $3x^2 + 26 = 17x$
15. $16x^2 - 712x + 7921 = 0$
16. $65.2x = 13.2x^2 + 41.7$

For Problems 17-20, use the quadratic formula to solve the equation for the indicated variable.

17. $A = 2w^2 + 4lw$, for w
18. $h = 4t - 16t^2$, for t
19. $s = vt - \frac{1}{2}at^2$, for t
20. $3x^2 + xy + y^2 = 2$, for y

21. a. Graph the three equations $y = x^2 - 6x + 5$, $y = x^2 - 6x + 9$, $y = x^2 - 6x + 12$ in the window

Xmin $= -2$ Xmax $= 7.4$
Ymin $= -5$ Ymax $= 15$

Use the **Trace** to locate the x-intercepts of each graph.

b. Calculate the discriminant for each of the equations

$$x^2 - 6x + 5 = 0$$
$$x^2 - 6x + 9 = 0$$
$$x^2 - 6x + 12 = 0$$

What does the discriminant tell you about the solutions of the equation? How does your answer relate to the graphs in part (a)?

Applications

22. A car traveling at s miles per hour on a dry road surface will require approximately d feet to stop, where d is given by

$$d = \frac{s^2}{24} + \frac{s}{2}$$

a. Make a table showing the stopping distance, d, for speeds of 10, 20, ..., 100 miles per hour. (Use the **Table** feature of your calculator.)
b. Graph the equation for d in terms of s. Use your table values to help you choose appropriate window settings.
c. Write and solve an equation to answer the question: If a car must be able to stop in 50 feet, what is the maximum safe speed it can travel? Verify your answer on your graph.

23. A skydiver jumps out of an airplane at 11,000 feet. While she is in free-fall, her altitude in feet t seconds after jumping is given by

$$h = -16t^2 - 16t + 11{,}000$$

a. Make a table of values showing the skydiver's altitude at 5 second intervals after she jumps from the airplane. (Use the **Table** feature of your calculator.)
b. Graph the equation. Use your table of values to choose appropriate window settings.
c. If the skydiver must open her parachute at an altitude of 1000 feet, how long can she free-fall? Write and solve an equation to find the answer.
d. If the skydiver drops a marker just before she opens her parachute, how long will it take the marker to hit the ground? (Hint: The marker continues to fall according to the equation given above.)
e. Find points on your graph that correspond to your answers to parts (c) and (d).

24. A dog trainer has 100 meters of chain link fence. She wants to enclose 250 square meters in three pens of equal size as shown at right.

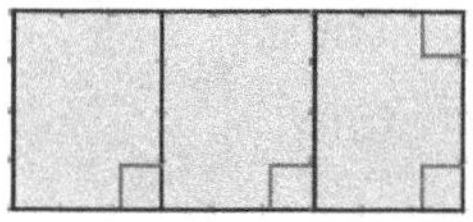

a. Let l and w represent the length and width of the entire area. Write an equation about the amount of chain link fence.
b. Solve your equation for l in terms w.
c. Write and solve an equation in w for the total area enclosed.
d. Find the dimensions of each pen.

25. When you look down from a height, say a tall building or a mountain peak, your line of sight is tangent to the earth at the horizon.

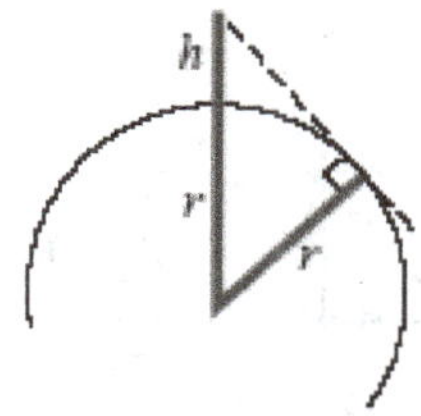

a. Suppose you are standing on top of the Petronas Tower in Kuala Lumpur, 1483 feet high. How far can you see on a clear day? (You will need to use the Pythagorean theorem, and the fact that the radius of the earth is 3960 miles. Don't forget to convert the height of the Petronas Tower to miles.)

b. How tall a building should you stand on in order to see 100 miles?

26. The volume of a cedar chest is 12,000 cubic inches. The chest is 20 inches high, and its length is 5 inches less than three times its width. Find the dimensions of the chest.

For Problems 27-31,

a. Find the coordinates of the intercepts and the vertex.

b. Sketch the graph by hand.

c. Use your calculator to verify your graph.

27. $y = -2x^2 + 7x + 4$

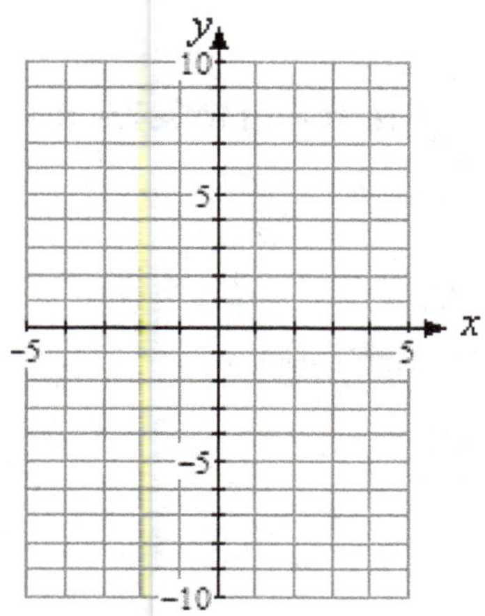

28. $y = 0.6x^2 + 0.6x - 1.2$

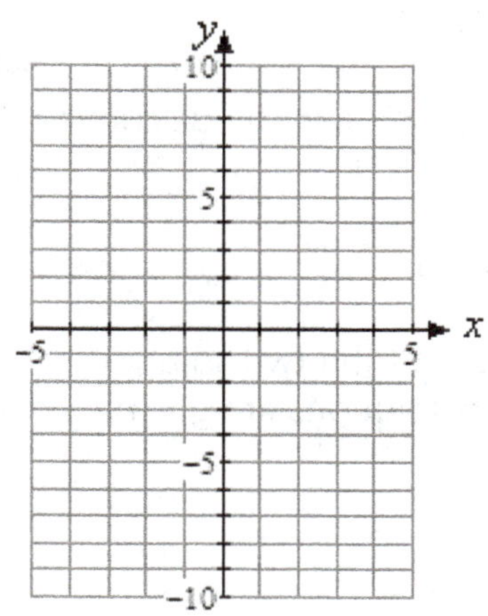

29. $y = x^2 + 4x + 7$

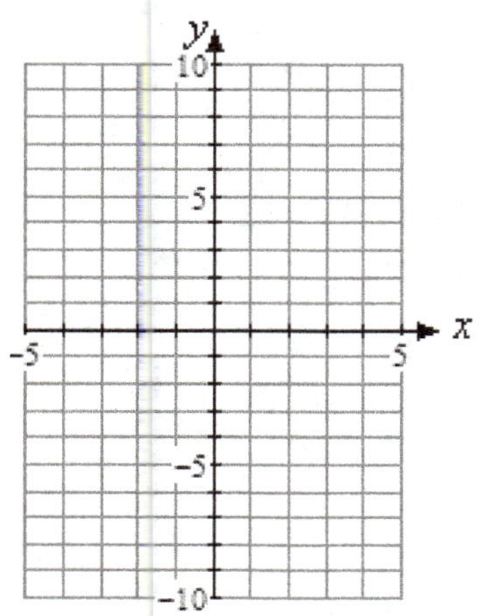

30. $y = x^2 + 2x - 1$

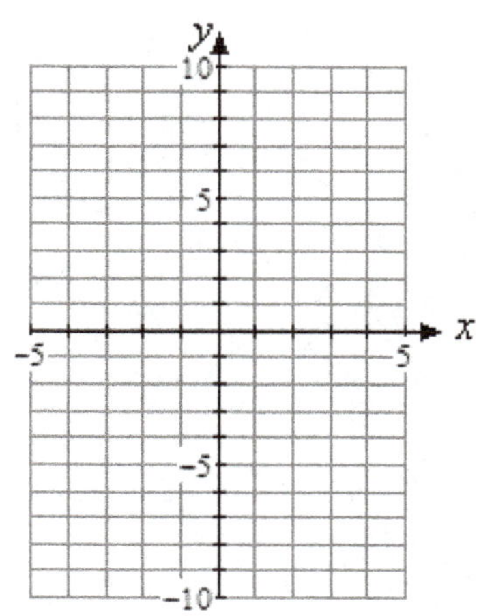

31. $y = -2x^2 + 6x - 3$

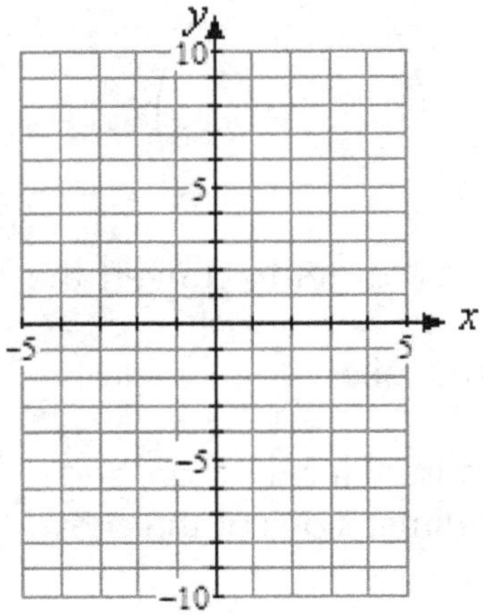

For Problems 32-33,

a. Given one solution of a quadratic equation with rational coefficients, find the other solution.

b. Write a quadratic equation that has those solutions.

32. $2 + \sqrt{5}$

33. $4 - 3\sqrt{2}$

For Problems 34-35, use the discriminant to decide if we can use factoring to solve the equation.

34. $3x^2 - 7x + 6 = 0$

35. $15x^2 - 52x - 32 = 0$

36. What is the sum of the two solutions of the quadratic equation $ax^2 + bx + c = 0$? (**Hint:** The two solutions are given by the quadratic formula.)

6.5 Problem Solving

Many quadratic models arise as the product of two variables, one of which increases while the other decreases. For example, in the Activities for Section 6.2 we looked at the areas of different rectangles with the same perimeter. The area of a rectangle is the product of its length and its width, or $A = lw$. If we require that the rectangle have a certain perimeter, then as we increase its length, we must also decrease its width.

Here is another example. How many eggs in a nest will maximize the number of baby birds that survive? The average weight of the hatchlings decreases as the number of eggs increases, but a baby bird has a better chance of surviving if it weighs more. Here are some data showing the number of eggs in a nest of a particular species and the percent of hatchlings that survived to adulthood.

Number of eggs	2	4	6	8	10	12	14	16
Percent survival	17.9	15.8	13.7	11.6	9.5	7.4	5.3	3.1
Adult birds								

If you multiply the number of eggs times the survival rate, you will find the number of birds from each nest that survive until adulthood. You should find that a nest size of 10 eggs produces the largest number of adult birds. And in fact, most nests of this species contain 10 eggs. The graph shows the number of eggs, the survival rate, and the number of adult birds fom 100 nests.

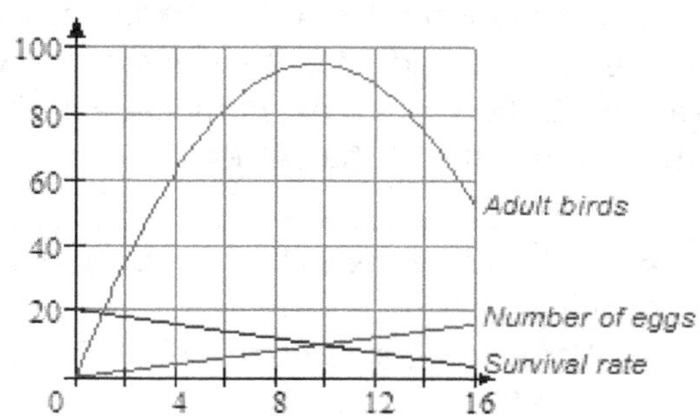

Maximum or Minimum Values

Finding the maximum or minimum value for a variable expression is a common problem in applications. For example, if you own a company that manufactures blue jeans, you might like to know how much to charge for your jeans in order to maximize your revenue. Recall the formula for the revenue from sales of an item:

Revenue = (price of one item) (number of items sold)

As you increase the price of the jeans, your revenue may increase for a while. But if you charge too much for the jeans, consumers will not buy as many pairs, and your revenue may actually start to decrease. Is there some optimum price you should charge for a pair of jeans in order to achieve the greatest revenue?

Example 1 Late Nite Blues finds that it can sell $600 - 15x$ pairs of jeans per week if they charge x dollars per pair. (Notice that as the price increases, the number of pairs of jeans sold decreases.)

a. Write an equation for the revenue in terms of the price of a pair of jeans.

b. Graph the equation.

c. How much should Late Nite Blues charge for a pair of jeans in order to maximize their revenue?

Solutions a. We use the formula for revenue to find

Revenue = (price of one item) (number of items sold)

$$R = x(600 - 15x)$$
$$R = 600x - 15x^2$$

b. We recognize the equation as quadratic, so the graph is a parabola. You can use your calculator to verify the graph shown.

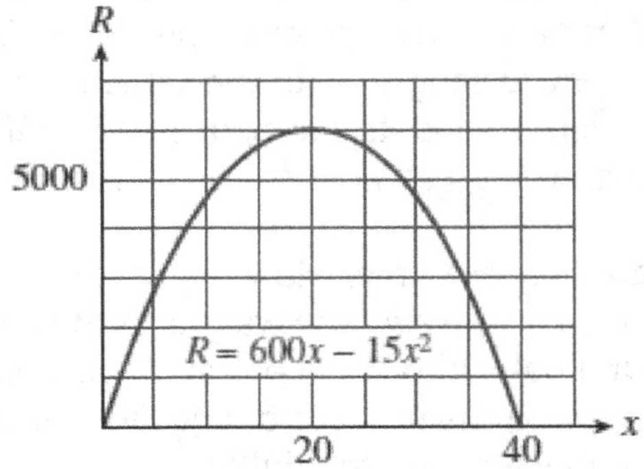

c. The maximum value of R occurs at the vertex of the parabola. Thus,

$$x_v = \frac{-b}{2a} = \frac{-600}{2(-15)} = 20$$
$$R_v = 600(20) - 15(20)^2 = 6000$$

The revenue takes on its maximum value when $x = 20$, and the maximum value is $R = 6000$. This means that Late Nite Blues should charge \$20 for a pair of jeans in order to maximize revenue at \$6000 a week. ■

If an equation relating two variables is quadratic, then the maximum or minimum value is easy to find: It is the value at the vertex. If the parabola opens downward, as in Example 1, there is a maximum value at the vertex. If the parabola opens upward, there is a minimum value at the vertex.

RQ1 To find the maximum or minimum value of a quadratic expression, we should:

a. Set $y = 0$ and solve for x. **b.** Factor the expression.
c. Use the quadratic formula. **d.** Find the vertex.

The Vertex Form for a Parabola

Consider the quadratic equation

$$y = 2(x - 3)^2 - 8$$

By expanding the squared expression and collecting like terms, we can rewrite the equation in standard form as

$$y = 2(x^2 - 6x + 9) - 8$$
$$y = 2x^2 - 12x + 10$$

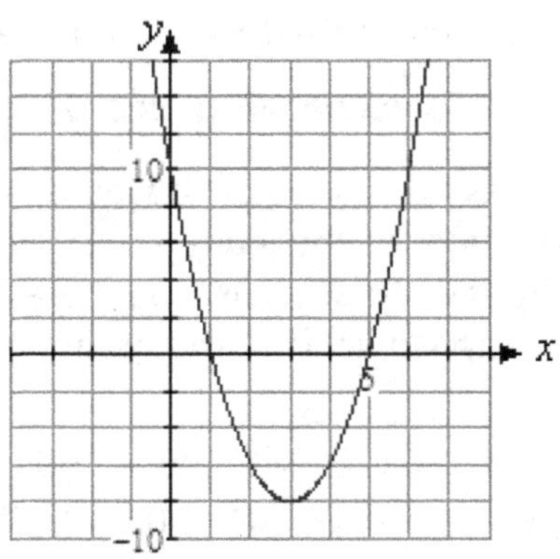

The vertex of this parabola is

$$x_v = \frac{-(-12)}{2(2)} = 3$$
$$y_v = 2(3)^2 - 12(3) + 10 = -8$$

and its graph is shown at right.

Notice that the coordinates of the vertex, $(3, -8)$, are apparent in the original equation; we don't need to do any computation to find the vertex.

$$y = 2(x - \mathbf{3})^2 - \mathbf{8}$$
$$\qquad\quad \boldsymbol{x_v} \quad \boldsymbol{y_v}$$

This equation is an example of the **vertex form** for a quadratic equation.

Vertex Form for a Quadratic Equation

A quadratic equation $y = ax^2 + bx + c,\ a \neq 0,$ can be written in the **vertex form**

$$\boldsymbol{y = a(x - x_v)^2 + y_v}$$

where the vertex of the graph is (x_v, y_v).

Example 2 Find the vertex of the graph of $y = -3(x-4)^2 + 6$. Is the vertex a maximum or a minimum point of the graph?

Solution We compare the equation to the vertex form to see that the coordinates of the vertex are $(4, 6)$. For this equation, $a = -3 < 0$, so the parabola opens downward. The vertex is the maximum point of the graph.

Look Closer: To understand why the vertex form works, substitute $x_v = 4$ into $y = -3(x-4)^2 + 6$ from Example 2 to find

$$y = -3(4-4)^2 + 6 = 6$$

which confirms that when $x = 4$, $y = 6$. Next, notice that if x is any number except 4 the expression $-3(x-4)^2$ is negative, so $y < 6$. Therefore, 6 is the maximum value for y on the graph, and $(4, 6)$ is the high point or vertex.

You can also rewrite $y = -3(x-4)^2 + 6$ in standard form and use the formula $x_v = \dfrac{-b}{2a}$ to confirm that the vertex is the point $(4, 6)$.

RQ2 Which of the following is the vertex form for a parabola with vertex $(-2, 3)$?

a. $y = 4(x-2) + 3$ **b.** $y = 4(x-2)^2 + 3$

c. $y = 4(x+2)^2 + 3$ **d.** $y = 4(x-3)^2 + 2$

Any quadratic equation in vertex form can be written in standard form by expanding, and any quadratic equation in standard form can be put into vertex form by completing the square.

Example 3 Write the equation $y = 3x^2 - 6x - 1$ in vertex form and find the vertex of its graph.

Solution We factor the lead coefficient, 3, from the variable terms, leaving a space to complete the square.

$$y = 3(x^2 - 2x \ \underline{\qquad}) - 1$$

Next, we complete the square inside parentheses. Take half the coefficient of x and square the result: $p = \frac{1}{2}(-2) = -1$, and $p^2 = (-1)^2 = 1$. We must add 1 to complete the square. However, we are really adding $3(1)$ to the right side of the equation, so we must also subtract 3 to compensate.

$$y = 3(x^2 - 2x + \mathbf{1}) - 1 - \mathbf{3}$$

The expression inside parentheses is now a perfect square, and the vertex form is

$$y = 3(x-1)^2 - 4$$

The vertex of the parabola is $(1, -4)$.

Using the Vertex Form

If we know the vertex of a parabola and one other point, we can use the vertex form to find its equation.

Example 4 When Andre practices free-throws at the park, the ball leaves his hands at a height of 7 feet, and reaches the vertex of its trajectory 10 feet away at a height of 11 feet.

a. Find a quadratic equation for the ball's trajectory.

b. Do you think Andre's free-throw would score on a regulation basketball court, where the hoop is 15 feet from the shooter and 10 feet high?

Solutions a. If Andre's feet are at the origin, then the vertex of the ball's trajectory is the point $(10, 11)$, and its y-intercept is $(0, 7)$. We start with the vertex form for a parabola.

$$y = a(x - x_v)^2 + y_v \quad \text{Substitute the coordinates of the vertex.}$$
$$y = a(x - 10)^2 + 11$$

We use the point $(0, 7)$ to find the value of a.

$$7 = a(0 - 10)^2 + 11 \quad \text{Substitute } (0,7)\text{ .}$$
$$7 = 100a + 11 \quad \text{Solve for } a.$$
$$a = -0.04$$

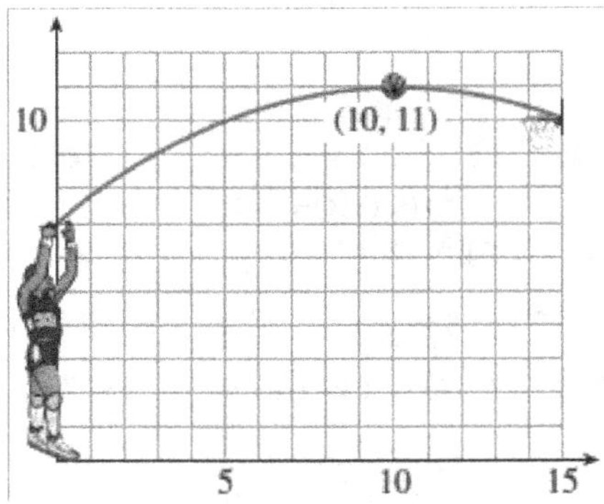

The equation of the trajectory is $y = -0.04(x - 10)^2 + 11$.

b. We'd like to know if the point $(15, 10)$ is on the trajectory of Andre's free-throw. We substitute $x = 15$ into the equation.

$$y = -0.04(15 - 10)^2 + 11$$
$$= -0.04(25) + 11 = 10$$

Andre's shot will probably hit the backboard just where the hoop attaches and bounce off, so it is unlikely that his shot will score.

Skills Warm-Up

Graph all three equations on the same grid. At what x-value does the maximum or minimum of the parabola occur?

1. $y = x$
$y = 2 - x$
$y = 2x - x^2$

2. $y = x$
$y = 3 - x$
$y = 3x - x^2$

3. $y = 1 - x$
$y = 3 + x$
$y = 3 - 2x - x^2$

4. $y = 4 - x$
$y = 2 + x$
$y = 8 + 2x - x^2$

Answers: **1.** 1 **2.** 1.5 **3.** −1 **4.** 1

Homework 6.5

Skills Practice

For Problems 1-2, graph the parabola. What is the vertex of the graph?

1. **a.** $y = (x-3)^2$
b. $y = -(x-3)^2$
c. $y = -(x-3)^2 + 4$

2. **a.** $y = (x+4)^2$
b. $y = \frac{1}{2}(x+4)^2$
c. $y = 3 + \frac{1}{2}(x+4)^2$

For Problems 3-4,

a. Find the vertex of the parabola.
b. Sketch the graph.
c. Write the equation in standard form.

3. $y = 2(x-3)^2 + 4$

4. $y = -\frac{1}{2}(x+4)^2 - 3$

5. Match each equation with one of the eight graphs shown below.

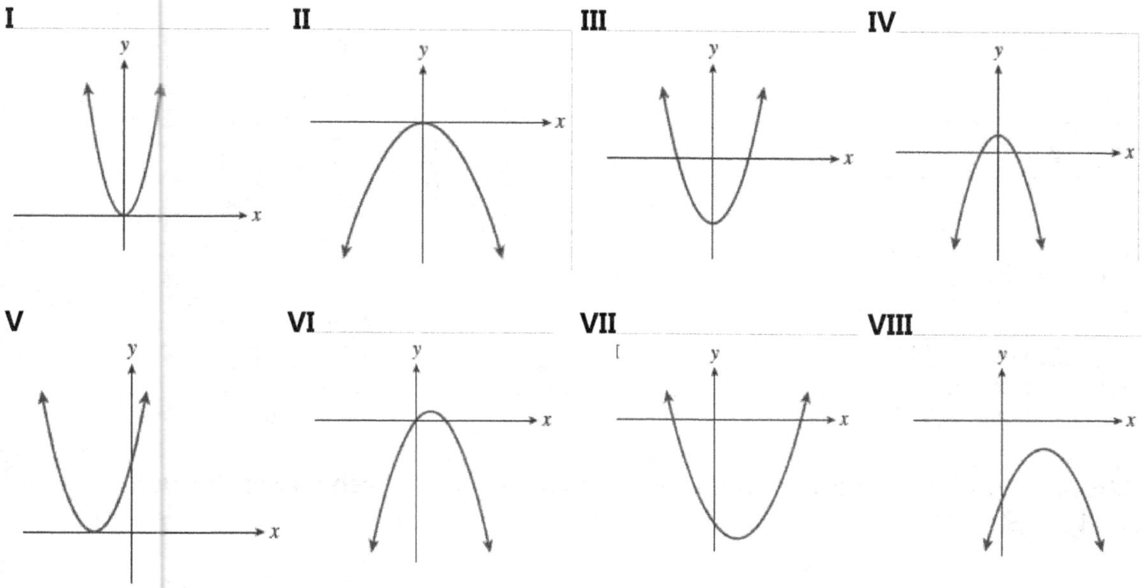

a. $y = 1 - x^2$
b. $y = (x+2)^2$
c. $y = 2x^2$
d. $y = (x-4)(x+2)$

For Problems 6-8, write the equation in the form $y = a(x-p)^2 + q$ by completing the square.

6. $y = x^2 - 4x + 7$

7. $y = 3x^2 + 6x - 2$

8. $y = -2x^2 - 8x + 3$

9. **a.** Write an equation for a parabola that has x-intercepts at $(2, 0)$ and $(-3, 0)$.
b. Write an equation for another parabola that has the same x-intercepts.

10. **a.** Write an equation for a parabola whose vertex is the point $(-2, 6)$. (Many answers are possible.)
b. Find the value of a if the y-intercept of the parabola in part (a) is 18.

11. a. Write an equation for a parabola with vertex at $(0,-3)$ and one of its x-intercepts at $(2,0)$.

b. Write an equation for a parabola with vertex at $(0,-3)$ and no x-intercepts.

For Problems 12-15, find an equation for the parabola. Use the vertex form or the factored form of the equation, whichever is more appropriate.

12.

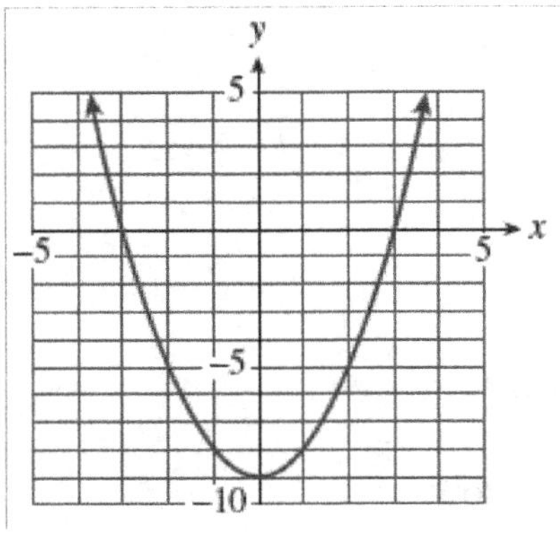

13.

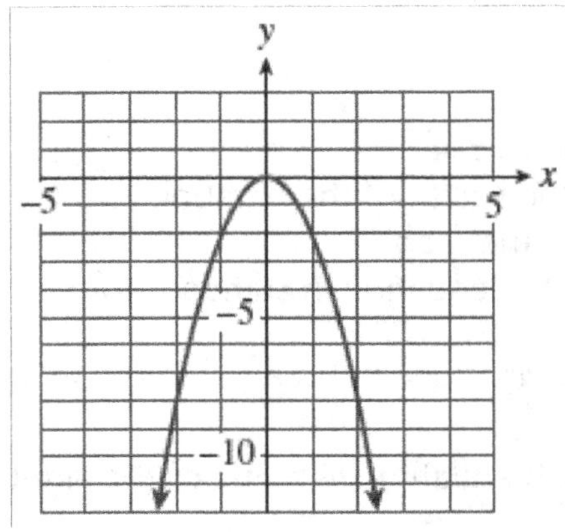

14.

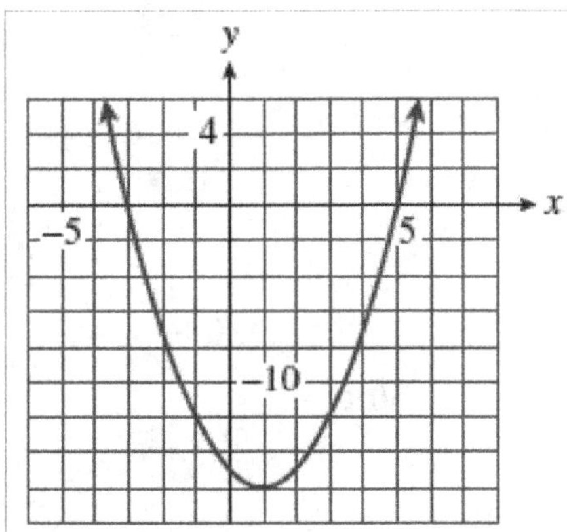

15.

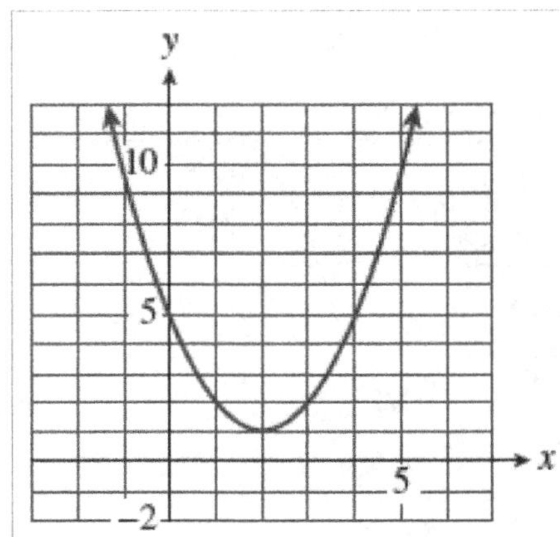

16. Find the equation for a parabola whose vertex is $(30,280)$ and which passes through the point $(20,80)$.

Applications

17. The owner of a motel has 60 rooms to rent. She finds that if she charges \$20 per room per night, all the rooms will be rented. For every \$2 that she increases the price of a room, 3 rooms will stand vacant.

a. Complete the table. The first two rows are filled in for you.

No. of price increases	Price of room	No. of rooms rented	Total revenue
0	20	60	1200
1	22	57	1254
2			
3			
4			
5			
6			
7			
8			
10			
12			
16			
20			

b. Let x stand for the number of \$2 price increases the owner makes. Write algebraic expressions for the price of a room, the number of rooms that will be rented, and the total revenue earned at that price.

c. Use your calculator to make a table of values for your algebraic expressions. Let $\mathbf{Y_1}$ stand for the price of a room, $\mathbf{Y_2}$ for the number of rooms rented, and $\mathbf{Y_3}$ for the total revenue. Verify the values you calculated in part (a).

d. Use your table to find a value of x that causes the total revenue to be zero.

e. Use your graphing calculator to graph your formula for total revenue.

f. What is the lowest price that the owner can charge for a room if she wants her revenue to exceed \$1296 per night? What is the highest price she can charge to obtain this revenue?

g. What is the maximum revenue the owner can earn in one night? How much should she charge for a room to maximize her revenue? How many rooms will she rent at that price?

18. Delbert launches a toy water rocket from ground level. Its distance above the ground t seconds after launch is given, in feet, by $d = 96t - 16t^2$. When will the rocket reach its greatest height, and what will that height be?

19. The owners of a small fruit orchard decide to produce gift baskets as a sideline. The cost per basket for producing x baskets is $C = 0.01x^2 - 2x + 120$. How many baskets should they produce in order to minimize the cost per basket? What will their total cost be at that production level?

20. As part of a collage for her art class, Sheila wants to enclose a rectangle with 100 inches of yarn.

a. Let w represent the width of the rectangle, and write an expression for its length. Then write an expression in terms of w for the area, A, of the rectangle.

b. What is the area of the largest rectangle that Sheila can enclose with 100 inches of yarn?

21. A farmer plans to fence a rectangular grazing area along a river with 300 yards of fence.

a. Write an expression for the area A of the grazing land in terms of the width w of the rectangle.

b. What is the largest area he can enclose?

22. A travel agent offers a group rate of $2400 per person for a week in London if 16 people sign up for the tour. For each additional person who signs up, the price per person is reduced by $100.

a. Let x represent the number of additional people who sign up. Write expressions for the total number of people signed up, the price per person, and the total revenue.

b. How many people must sign up for the tour in order for the travel agent to maximize her revenue?

23. In skeet shooting, the clay pigeon is launched from a height of 4 feet and reaches a maximum height of 164 feet at a distance of 80 feet from the launch site.

a. Write an equation for the height of the clay pigeon in terms of the horizontal distance it has traveled.

b. If the shooter misses the clay pigeon, how far from the launch site will it hit the ground?

24. The cables on a suspension bridge hang in the shape of parabolas. Imagine a coordinate system superimposed on a diagram of a suspension bridge, as shown at right. Each of the towers is 500 feet high, and the span between the towers is 4000 feet long. At its lowest point, the cable hangs 20 feet above the roadway.

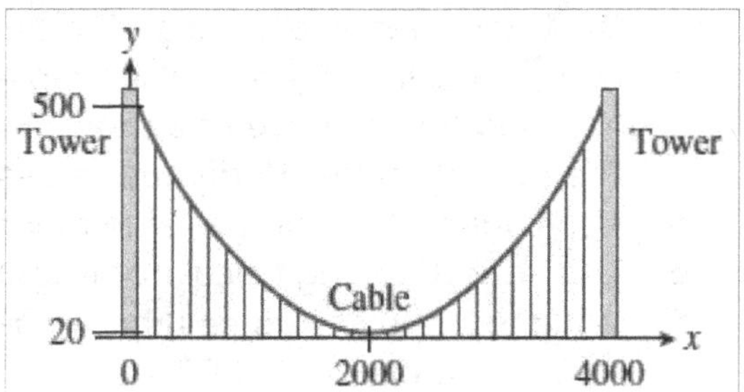

a. Find the coordinates of the vertex and one other point on the cable.

b. Use the points from part (a) to find an equation for the shape of the cable in vertex form.

6.6 Quadratic Curve Fitting

Earlier, we used linear regression to fit a line to a collection of data points. In this section we'll see how to fit a quadratic equation to a collection of data points.

Finding a Quadratic Equation through Three Points

Every linear equation can be written in the form

$$y = mx + b$$

To find a specific line we must find values for the two parameters (constants) m and b. We need two data points in order to find those two parameters. A quadratic equation, however, has three parameters, a, b, and c:

$$y = ax^2 + bx + c$$

To find these parameters we need three data points.

Example 1 Find values for a, b, and c so that the points $(1,3)$, $(3,5)$, and $(4,9)$ lie on the graph of $y = ax^2 + bx + c$.

Solution We substitute the coordinates of each of the three points into the equation of the parabola to obtain three equations:

$$\begin{aligned} 3 &= a(1)^2 + b(1) + c \\ 5 &= a(3)^2 + b(3) + c \\ 9 &= a(4)^2 + b(4) + c \end{aligned}$$

or, equivalently,

$$\begin{aligned} a + b + c &= 3 && (1) \\ 9a + 3b + c &= 5 && (2) \\ 16a + 4b + c &= 9 && (3) \end{aligned}$$

This is a system of three equations in the three unknowns a, b, and c. To solve the system, we first eliminate c. We subtract Equation (1) from Equation (2) to obtain

$$8a + 2b = 2 \qquad (4)$$

and subtract Equation (1) from Equation (3) to get

$$15a + 3b = 6 \qquad (5)$$

Next we eliminate b from Equations (4) and (5): we add -3 times Equation (4) to 2 times Equation (5) to get

$$\begin{aligned} -24a - 6b &= -6 && \textbf{-3 times Equation (4)} \\ 30a + 6b &= 12 && \textbf{2 times Equation (5)} \\ \hline 6a &= 6 \end{aligned}$$

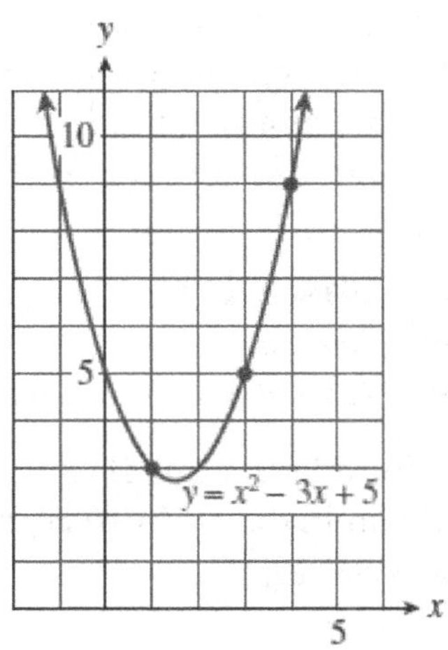

or $a = 1$. We substitute 1 for a in Equation (4) to find

$$\begin{aligned} 8(1) + 2b &= 2 && \textbf{Solve for } b. \\ b &= -3 \end{aligned}$$

Finally, we substitute -3 for b and 1 for a in Equation (1) to find

$$\begin{aligned} 1+(-3)+c &= 3 \quad \textbf{Solve for } c. \\ c &= 5 \end{aligned}$$

Thus, the equation of the parabola is

$$y = x^2 - 3x + 5$$

The parabola and the three points are shown above. ■

RQ1 Fill in the blanks. To find the equation for a parabola:

a. We can use the formula $y = a(x - x_v)^2 + y_v$ if we know the __________ and one other point.

b. We can use the formula $y = a(x - r_1)(x - r_2)$ if we know the __________ and one other point.

c. Otherwise, we must know at least _____ points on the graph.

d. In that case, we use ______________ to solve for the parameters of the equation.

The simplest way to fit a parabola to a set of data points from a quadratic model is to pick three of the points and find the equation of the parabola that passes through those three points.

■ **Example 2** Major Motors Corporation is testing a new car designed for in-town driving. The data below show the cost of driving the car at different speeds. The speeds, v, are given in miles per hour, and the cost, C, includes fuel and maintenance for driving the car 100 miles at that speed.

v	30	40	50	60	70
C	6.50	6.00	6.20	7.80	10.60

Find a possible quadratic model,

$$C = av^2 + bv + c$$

that expresses C in terms of v.

Solution When the data are plotted, it is clear that the relationship between v and C is not linear, but it may be quadratic. We will use the last three data points, $(50, 6.20)$, $(60, 7.80)$, and $(70, 10.60)$, to fit a parabola to the data. We would like to find the coefficients a, b, and c of a parabola $C = av^2 + bv + c$ that includes the three data points. This gives us a system of equations:

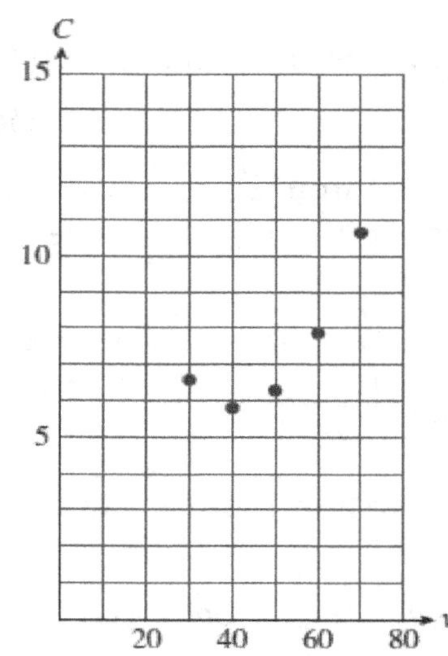

$$\begin{aligned} 2500a + 50b + c &= 6.20 \quad (1) \\ 3600a + 60b + c &= 7.80 \quad (2) \\ 4900a + 70b + c &= 10.60 \quad (3) \end{aligned}$$

Eliminating c from Equations (1) and (2) yields Equation (4), and eliminating c from Equations (2) and (3) yields Equation (5).

$$1100a + 10b = 1.60 \qquad (4)$$
$$1300a + 10b = 2.80 \qquad (5)$$

Eliminating b from Equations (4) and (5) gives us

$$200a = 1.20$$
$$a = 0.006$$

We substitute this value into Equation (4) to find $b = -0.5$, then substitute both values into Equation (1) to find $c = 16.2$. Thus, our quadratic model is

$$C = 0.006v^2 - 0.5v + 16.2$$

The graph of this equation, along with the data points, is shown at right.

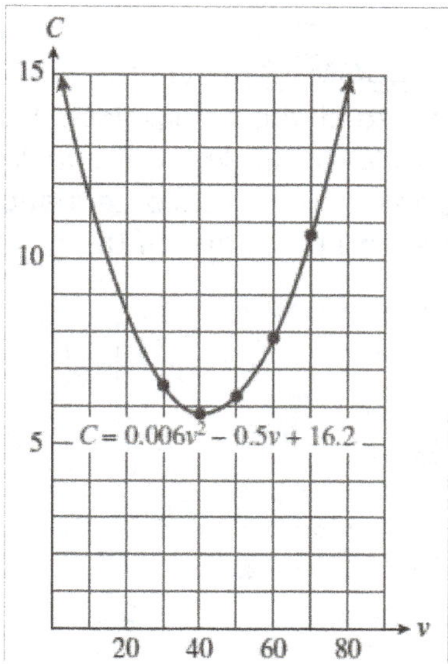

As was the case with linear regression, the graph of the regression equation may not pass through all of the data points, but it should be close to most of them.

> **RQ2** True or false.
> **a.** We can plot the data points to see what type of curve is appropriate as a model.
> **b.** We write a system of equations in which the y-coordinates of the data points are the unknown values.
> **c.** A good regression equation will pass through all of the data points.
> **d.** According to the model in Example 2, higher speeds always result in higher driving costs.

Using a Calculator for Quadratic Regression

We can use a graphing calculator to find an approximate quadratic fit for a set of data. The procedure for quadratic regression follows closely the steps for linear regression.

Example 3a. Use your calculator to find a quadratic fit for the data in Example 2.
b. How many of the given data points actually lie on the graph of the quadratic approximation?

Solutions a. We press [STAT] [ENTER] and enter the data under columns $\mathbf{L_1}$ and $\mathbf{L_2}$, as shown in figure (a) below. We calculate the quadratic regression equation and store it in Y_1 by pressing [STAT] [▷] [5] [VARS] [▷] [1] [1] [ENTER]. The regression equation has the form $y = ax^2 + bx + c$, where $a \approx .0057$, $b \approx -0.47$, and $c \approx 15.56$, as shown in figure (b). Notice that a, b, and c are all close to the values we computed in Example 2.

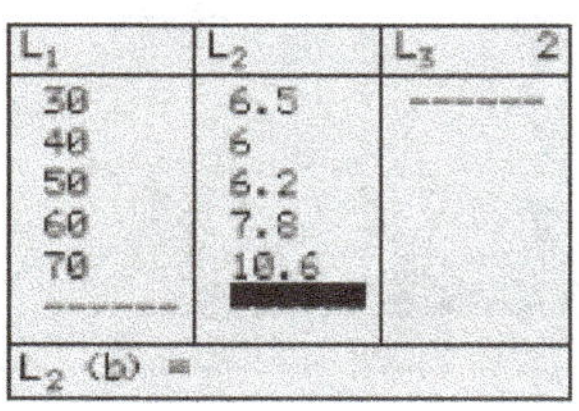

a.

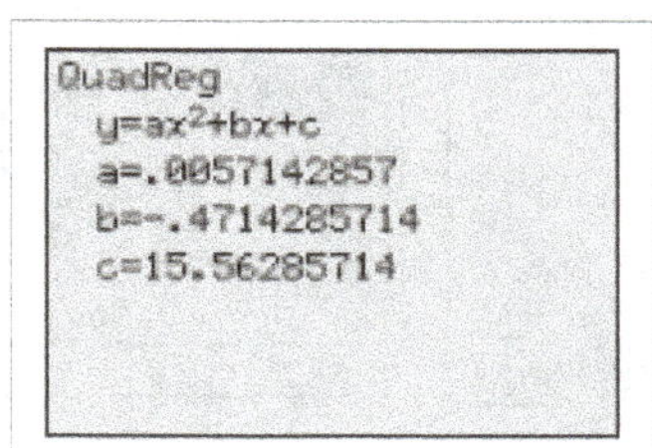

b.

b. Next we'll graph the data and the regression equation. We press [Y=] and select **Plot1**, then press [ZOOM] [9] to see the graph shown in figure (a) below. The parabola seems to pass close to all the data points. However, we can use either the **value** feature or a table with $x = 30,\ 40,\ 50,\ 60, 70$, as shown in figure (b), to find the y-coordinates of points on the graph of the regression equation. By comparing these y-coordinates with the data points, we find that none of the given data points lies precisely on the parabola.

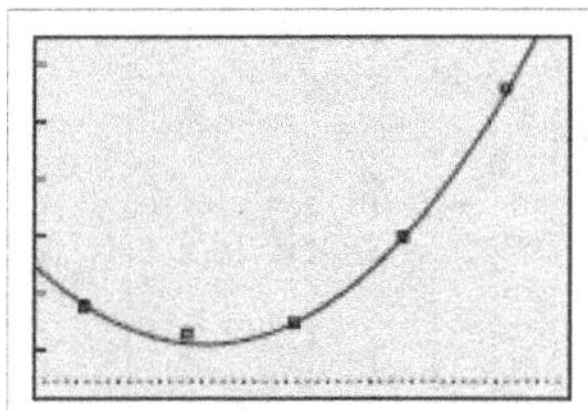

a.

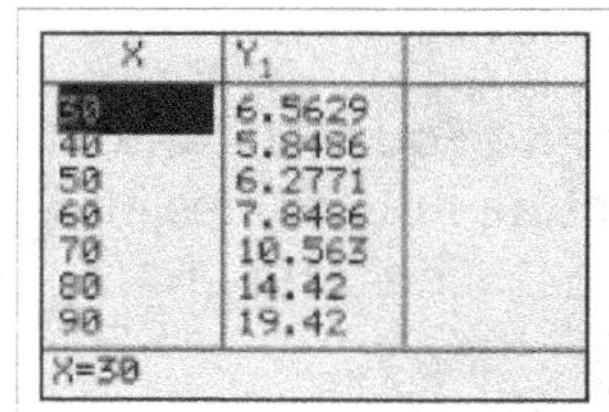

b.

Choosing an Appropriate Model

We must be careful that our data set gives a complete picture of the situation we want to model. A regression equation may fit a particular collection of data and still be a poor model if the rest of the data diverge from the regression graph. In Example 2, suppose Major Motors had collected only the first three data points and fit a line through them, as shown at right. This regression line gives poor predictions for the cost of driving at 60 or 70 miles per hour.

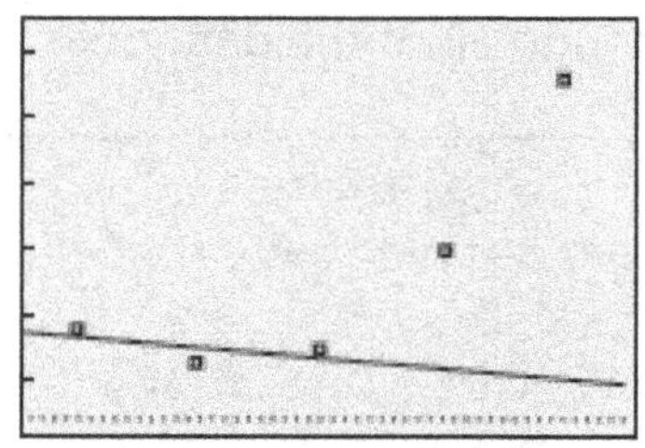

Example 4 Delbert records the height of the tip of the minute hand on the classroom's clock at different times. The data are shown in the table, where time is measured in minutes since noon. (A negative time indicates a number of minutes before noon.) Find a quadratic regression equation for the data and use it to predict the height of the minute hand's tip at 40 minutes past noon. Do you believe this prediction is valid?

Time (minutes)	−25	−20	−15	−10	−5	0	5	10	15	20	25
Height (feet)	7.13	7.50	8.00	8.50	8.87	9.00	8.87	8.50	8.00	7.50	7.13

Solution We enter the time data under L_1 and the height data under L_2. We calculate and store the quadratic regression equation in Y_1 as we did in Example 3.

$$y \approx -0.00297x^2 + 0x + 8.834$$

From either the graph of the regression equation or from the table (shown below), you can see that the fit is not perfect, although the curve certainly fits the data better than any straight line could.

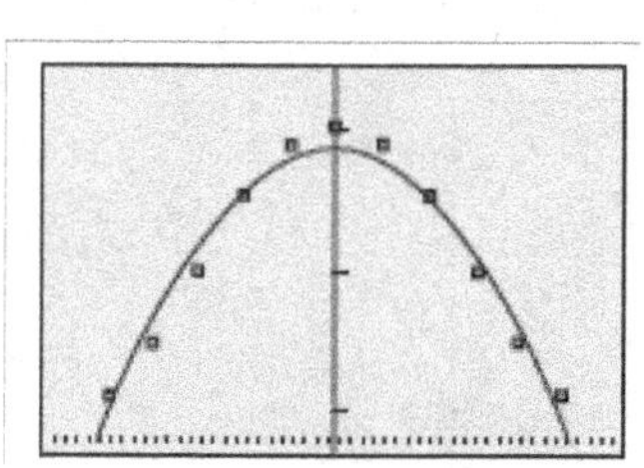

a.

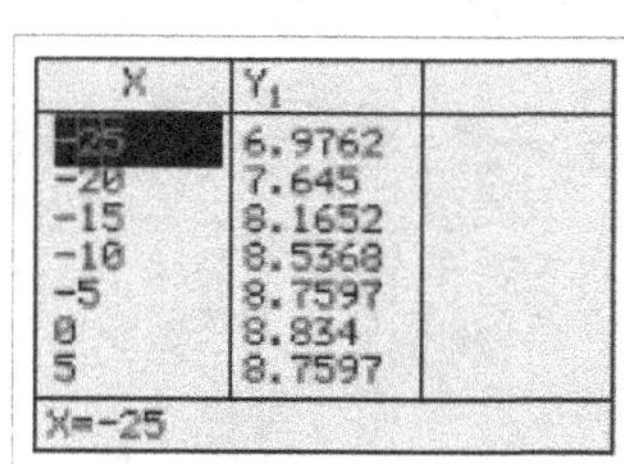

b.

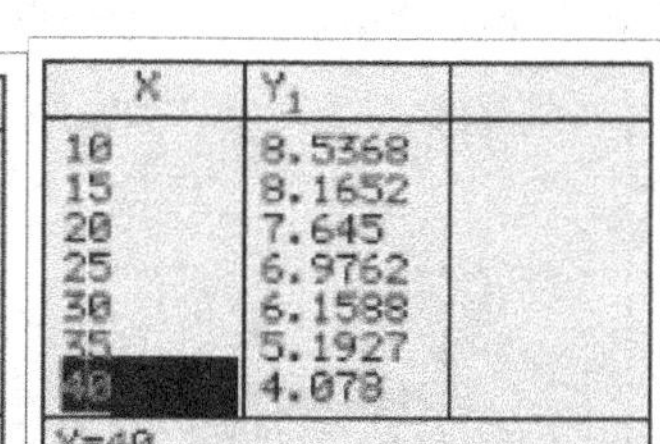

c.

We can scroll down the table to find that this equation predicts a height of approximately 4.08 feet at time 40 minutes. (See figure (c).) This is a preposterous estimate! The position of the minute hand at 40 minutes after noon should be the same as it was exactly one hour earlier (at 20 minutes before noon), when it was at 7.50 feet. ■

Caution! Using the wrong type of equation to fit the data is a common error in making predictions. In Example 4, we know that the minute hand of a clock repeats its position every 60 minutes. The graph of the height of its tip oscillates up and down, repeating the same pattern over and over. We cannot describe such a graph using either a linear or a quadratic equation. The graph of the height is shown below, along with the graph of our quadratic regression equation. You can see that the regression equation fits the actual curve only on a small interval.

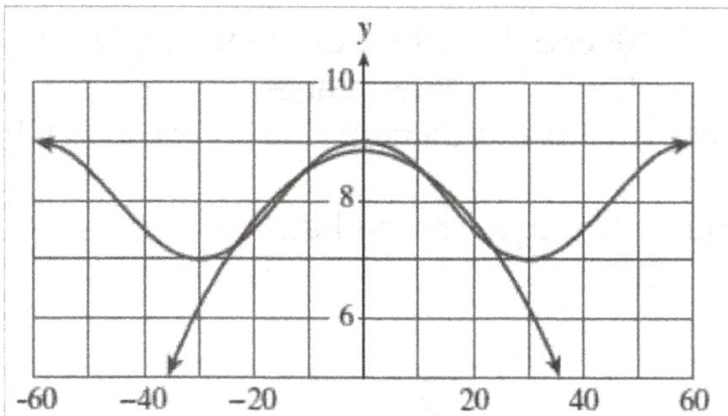

Your calculator can always compute a regression equation, but that equation is not necessarily appropriate for your data. Choosing a reasonable type of regression equation for a particular data set requires knowledge of different kinds of models and the physical or natural laws that govern the situation at hand. ■

Skills Warm-Up

■ Solve the system by elimination. Begin by eliminating c.

1. $$\begin{aligned} a+b+c&=-3\\ a-b+c&=-9\\ 4a+2b+c&=-6 \end{aligned}$$

2. $$\begin{aligned} a+b+c&=10\\ 4a+2b+c&=19\\ 9a+3b+c&=38 \end{aligned}$$

3. $$\begin{aligned} a-b+c&=12\\ 4a-2b+c&=19\\ 9a+3b+c&=4 \end{aligned}$$

4. $$\begin{aligned} 4a+2b+c&=14\\ 9a-3b+c&=-41\\ 16a-4b+c&=-70 \end{aligned}$$

Answers: 1. $(-2, 3, -4)$ **2.** $(5, -6, 11)$ **3.** $(1, -4, 7)$ **4.** $(-3, 8, 10)$

Homework 6.6

For Problems 1-2, find a quadratic equation that fits the data points.

1. Find values for a, b, and c so that the graph of the parabola $y = ax^2 + bx + c$ includes the points $(-1,0)$, $(2,12)$, and $(-2,8)$. Write a quadratic equation that fits the data points.

2. A survey to determine what percent of different age groups regularly use marijuana collected the following data.

Age	15	20	25	30
Percent	4	13	11	7

   a. Use the percents for ages 15, 20, and 30 to fit a quadratic equation $P = ax^2 + bx + c$ to the data, where x represents age.
   b. What does your equation predict for the percent of 25-year-olds who use marijuana?
   c. Sketch the graph of your quadratic equation and the given data on the grid provided.

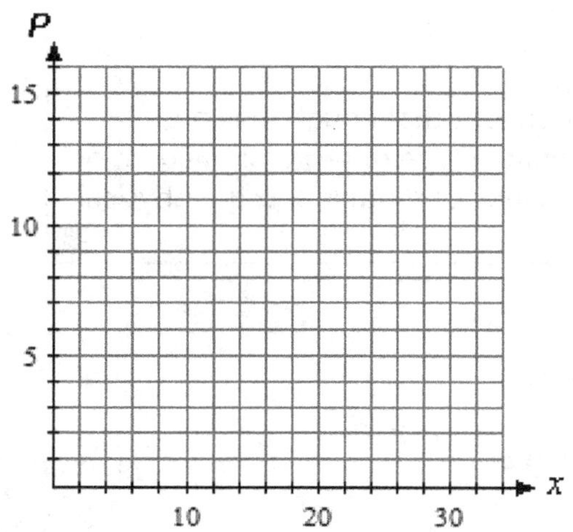

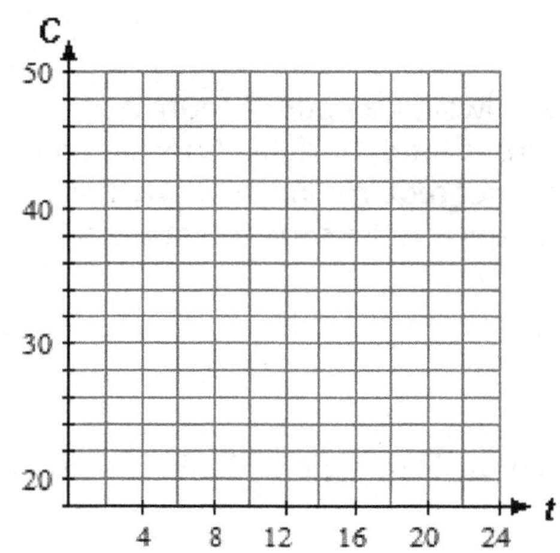

3. The data below show Americans' annual per capita consumption of chicken for several years since 1970.

Year	1970	1975	1980	1985	1989	1990	1991	1992
Pounds of chicken	27.7	26.4	32.5	36.1	40.5	42.1	43.9	45.9

   a. Use the values for 1970, 1985, and 1990 to fit a quadratic equation $C = at^2 + bt + c$ to the data, where t is measured in years since 1970.
   b. What does your equation predict for per capita chicken consumption in 1992?
   c. Sketch the graph of your equation and the given data on the grid provided.

■ For Problems 4-6, use your calculator's statistics features.

4. In a trial in Thailand, scientists applied different amounts of nitorgen fertilizer to a crop of soybeans and recorded the yield for each. The table shows the yield in tons for each amount of fertilizer. (Source: www.arc-avrdc.org)

Nitrogen (kg)	0	15	30	60	120
Yield (tons)	2.12	2.46	2.65	2.80	2.60

a. Fit a quadratic regression equation to the data.
b. Use your model to predict the maximum yield and the amount of nitrogen needed.

5. The table shows the height in meters of a projectile at different times after it was fired.

Time (seconds)	2	4	6	8	10	12	14
Height (meters)	39.2	71.8	98.0	117.8	131.0	137.8	138.0

a. Find the equation of the least-squares regression line for height in terms of time.
b. Use the linear regression equation to predict the height of the projectile 15 seconds after it was fired.
c. Make a scatterplot of the data and draw the regression line on the same axes.
d. Find the quadratic regression equation for height in terms of time.
e. Use quadratic regression equation to predict the height of the projectile 15 seconds after it was fired.
f. Make a scatterplot of the data and draw the regression curve on the same axes.
g. Which model is more appropriate for the height of the projectile, linear or quadratic? Why?

6. The number of daylight hours increases each day from the beginning of winter until the beginning of summer, and then begins to decrease. The table below gives the number of daylight hours in Delbert's home town last year in terms of the number of days since January 1.

Days since January 1	0	50	100	150	200	250	300
Hours of daylight	9.8	10.9	12.7	14.1	13.9	12.5	10.7

a. Find the equation of the least-squares regression line for the number of daylight hours in terms of the number of days since January 1.
b. Use the linear regression equation to predict the number of daylight hours 365 days after January 1.
c. Make a scatterplot of the data and draw the regression line on the same axes.
d. Find the quadratic regression equation for the number of daylight hours in terms of the number of days since January 1.
e. Use quadratic regression equation to predict the number of daylight hours 365 days after January 1.
f. Make a scatterplot of the data and draw the regression curve on the same axes.
g. Predict the number of daylight hours 365 days since January 1 without using any regression equation. What does this tell you about the linear and quadratic models you found?

Chapter 6 Summary and Review

Section 6.1 Factoring Quadratic Trinomials

- If we can factor the quadratic trinomial $ax^2 + bx + c$, we can apply the zero-factor principle to solve the equation $ax^2 + bx + c = 0$.

Sign Patterns for Factoring Quadratic Trinomials

Assume that b, c, p, and q are positive integers. Then

1. $\boldsymbol{x^2 + bx + c = (x + p)(x + q)}$
   If all the coefficients of the trinomial are positive, then both p and q are positive.
2. $\boldsymbol{x^2 - bx + c = (x - p)(x - q)}$
   If the linear term of the trinomial is negative and the other two terms positive, then p and q are both negative.
3. $\boldsymbol{x^2 \pm bx - c = (x + p)(x - q)}$
   If the constant term of the trinomial is negative, then p and q have opposite signs.

Difference of Squares

We can factor a quadratic expression that is the difference of two squares.

$$\boldsymbol{x^2 - p^2 = (x - p)(x + p)}$$

Section 6.2 Quadratic Models

- There is a connection between:
  1. the **factors** of $ax^2 + bx + c$
  2. the **solutions** of the equation $ax^2 + bx + c = 0$, and
  3. the **x-intercepts** of the graph of $y = ax^2 + bx + c$.
- A quadratic equation in one variable always has two solutions (which may be the same.)
- Given any two rational numbers, we can write a quadratic equation that has those solutions.

Section 6.3 Completing the Square

- Not every quadratic equation can be solved by factoring. However, we can solve any quadratic equation by completing the square.

To Solve a Quadratic Equation by Completing the Square:

1. Write the equation in standard form.
2. Divide both sides of the equation by the coefficient of the quadratic term, and subtract the constant term from both sides.
3. Complete the square on the left side:
   a. Multiply the coefficient of the first-degree term by one-half, then square the result.
   b. Add the value obtained in (a) to both sides of the equation.

4. Write the left side of the equation as the square of a binomial. Simplify the right side.
5. Use extraction of roots to finish the solution.

- The maximum or minimum value of a quadratic expression occurs at the vertex of its graph. The x-coordinate of the vertex is given by

$$x_v = \frac{-b}{2a}$$

Section 6.4 Quadratic Formula

- We can express the solutions of a quadratic equation in terms of its coefficients.

The Quadratic Formula

The solutions of the equation $ax^2 + bx + c = 0,\ a \neq 0,$ are

$$x = \frac{-b \pm \sqrt{b^2 - 4ac}}{2a}$$

- We have four methods for solving quadratic equations: extracting roots, factoring, completing the square, and the quadratic formula.

To graph the quadratic equation $y = ax^2 + bx + c$:

1. Determine whether the parabola opens upward (if $a > 0$) or downward (if $a < 0$).
2. Locate the vertex of the parabola.
   a. The x-coordinate of the vertex is $x_v = \dfrac{-b}{2a}$.
   b. Find the y-coordinate of the vertex by substituting x_v into the equation of the parabola.
3. Locate the x-intercepts (if any) by setting $y = 0$ and solving for x.
4. Locate the y-intercept by evaluating y for $x = 0$.
5. Locate the point symmetric to the y-intercept across the axis of symmetry.

- Not all quadratic equations have solutions that are real numbers; their solutions may involve the square roots of negative numbers. These numbers are called **imaginary numbers**.

Imaginary Unit

We define the imaginary unit i by

$$i^2 = -1 \qquad \text{or} \qquad i = \sqrt{-1}$$

Imaginary Numbers

For $a > 0$,

$$\sqrt{-a} = \sqrt{-1}\sqrt{a} = i\sqrt{a}$$

Complex Numbers
A **complex number** can be written in the form $a + bi$, where a and b are real numbers.

- The graph of the quadratic equation $y = ax^2 + bx + c$ may have two, one, or no x-intercepts, according to the number of distinct real-valued solutions of the equation $ax^2 + bx + c = 0$.

Discriminant
The **discriminant** of a quadratic equation $ax^2 + bx + c = 0$ is

$$D = b^2 - 4ac$$

1. If $D > 0$, the equation has two unequal real solutions.
2. If $D = 0$, the equation has one real solution of multiplicity two.
3. If $D < 0$, the equation has two complex solutions.

- Irrational and complex solutions of a quadratic equation always occur in **conjugate pairs.**

Section 6.5 Problem Solving

- Many quadratic models arise as the product of two variables, one of which increases while the other decreases.
- The revenue from sales of an item is given by the formula

Revenue = (price of one item) (number of items sold)

Vertex Form for a Quadratic Equation
A quadratic equation $y = ax^2 + bx + c,\ a \neq 0,$ can be written in the **vertex form**

$$y = a(x - x_v)^2 + y_v$$

where the vertex of the graph is (x_v, y_v).

- Any quadratic equation in vertex form can be written in standard form by expanding, and any quadratic equation in standard form can be put into vertex form by completing the square.

Section 6.6 Quadratic Curve Fitting

- To fit a parabola to a curve we need three data points.
- We can use a graphing calculator to find an approximate quadratic fit for a set of data.
- A regression equation may fit a particular collection of data and still be a poor model if the rest of the data diverge from the regression graph.
- Choosing a reasonable type of regression equation for a particular data set requires knowledge of different kinds of models and the physical or natural laws that govern the situation at hand.

Review Questions

For Problems 1-20, use complete sentences to answer the questions.

1. Give an example of a quadratic equation you could solve by extraction of roots.
2. Explain how to fill in the squares for the box method of factoring a quadratic trinomial.
3. Can the equation $x^2 + 2x - 24 = 12$ be solved by factoring? Why or why not?
4. Do all rectangles with the same perimeter have the same area? Support your answer with examples.
5. The perimeter of a rectangle is 50 meters. Write an expression for the length of the rectangle in terms of its width.
6. A fly ball reaches its maximum height 3 seconds after it was hit. After 2.25 seconds, its height was 42 feet. Find another time when the ball's height was 42 feet.
7. How are the factors of $ax^2 + bx + c$ related to the x-intercepts of the graph of $y = ax^2 + bx + c$?
8. What must you add to $x^2 + bx$ to complete the square?
9. What is the first step for solving $3x^2 - 6x + 8 = 0$ by completing the square?
10. What is the difference between a quadratic equation and the quadratic formula?
11. Sing the quadratic formula to the tune of *Pop Goes the Weasel.*
12. Explain how you could use the quadratic formula to factor a quadratic trinomial.
13. List four important points you should find when sketching the graph of a parabola.
14. Explain why the graph of $y = x^2 + 5$ has no x-intercepts.
15. What is an **imaginary number**? What is a **complex number**?
16. What is the **discriminant**? How is it related to the quadratic formula?
17. If one solution of a quadratic equation is $2 - 3\sqrt{6}$, what is the other solution?
18. State a formula for the revenue from sales of an item.
19. Explain why increasing the price of an item does not always increase your revenue.
20. Match the form of the quadratic equation with the point it helps you find:

I. $y = ax^2 + bx + c$	**a.** x-intercepts
II. $y = a(x - r_1)(x - r_2)$	**b.** vertex
III. $y = a(x - x_v)^2 + y_v$	**c.** y-intercept

Review Problems

For Problems 1-12, factor.

1. $24x^2 - 18x$	**2.** $-32y^2 + 24y$	**3.** $a^2 - 18a + 45$
4. $x^2 - 14x - 51$	**5.** $4y^2 + 16y + 15$	**6.** $8b^2 - 18b + 9$
7. $14w - 5 + 3w^2$	**8.** $-3 + 2p^2 - 5p$	**9.** $z^2 - 121$
10. $81 - 4t^2$	**11.** $6x^2 + 21x + 9$	**12.** $8y^2 - 6y - 2$

For Problems 13-22, solve by factoring.

13. $0 = m^2 + 10m + 25$	**14.** $b^2 - 25 = 0$	**15.** $4p^2 = 16p$
16. $11t = 6t^2 + 3$	**17.** $(x - 5)(x + 1) = -8$	**18.** $2q(3q - 1) = 4$

19. $x(3x + 2) = (x + 2)^2$ **20.** $6y = (y + 1)^2 + 3$

21. $4x - (x + 1)(x + 2) = -8$ **22.** $3(x + 2)^2 = 15 + 12x$

For Problems 23-24, write a quadratic equation with integer coefficients and with the given solutions.

23. $\dfrac{-3}{4}$ and 8

24. $\dfrac{5}{3}$ and $\dfrac{5}{3}$

For Problems 25-26, graph the equation using the **ZDecimal** setting. Locate the x-intercepts, and use them to write the quadratic expression in factored form.

25. $y = x^2 - 0.6x - 7.2$

26. $y = -x^2 + 0.7x + 2.6$

For Problems 27-30, solve by completing the square.

27. $x^2 - 4x - 6 = 0$

28. $x^2 + 3x = 3$

29. $2x^2 + 3 = 6x$

30. $3x^2 = 2x + 3$

For Problems 31-36, solve by using the quadratic formula.

31. $\dfrac{1}{2}x^2 + 1 = \dfrac{3}{2}x$

32. $x^2 - 3x + 1 = 0$

33. $x^2 - 4x + 2 = 0$

34. $2x^2 + 2x = 3$

35. $2t^2 + 6t + 3 = 0$

36. $\dfrac{x^2}{4} + 1 = \dfrac{13}{12}x$

For Problems 37-40, solve the formula for the indicated variable.

37. $K = \frac{1}{2}mv^2$, for v

38. $a^2 + b^2 = c^2$, for b

39. $h = 6t - 3t^2$, for t

40. $D = \dfrac{n^2 - 3n}{2}$, for n

For Problems 41-44, solve the formula for x.

41. $x^2 + a^2 = b^2$

42. $x^2 - b^2 = c^2$

43. $\dfrac{x^2}{4} - y^2 = 1$

44. $y^2 + \dfrac{x^2}{9} = 1$

For Problems 45-46, use the discriminant to determine the nature of the solutions of the equation.

45a. $4x^2 - 12x + 9 = 0$

b. $2y^2 = 3y - 4$

46a. $2t^2 + 6t + 5 = 0$

b. $\dfrac{x^2}{4} = x + \dfrac{5}{4}$

For Problems 47-48, write the expression as a complex number, $a + bi$.

47a. $\sqrt{-25} - 3$

b. $\dfrac{6 - \sqrt{-9}}{6}$

48a. $1 - \sqrt{-4}$

b. $\dfrac{-2 + \sqrt{-36}}{4}$

■ For Problems 49-52,

a. Find the coordinates of the vertex and the intercepts.

b. Sketch the graph.

49. $y = x^2 - x - 12$

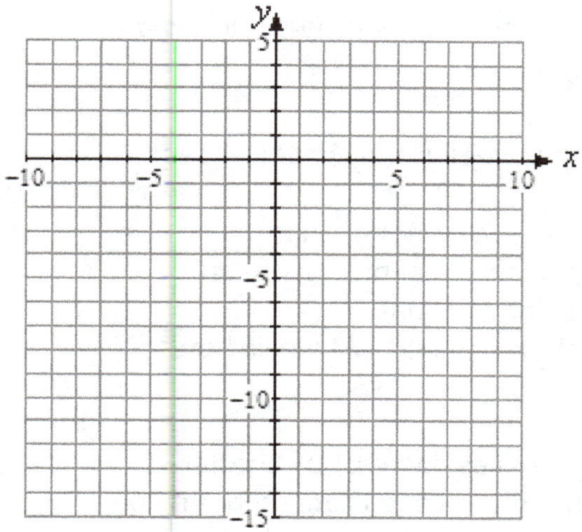

50. $y = -2x^2 + x - 4$

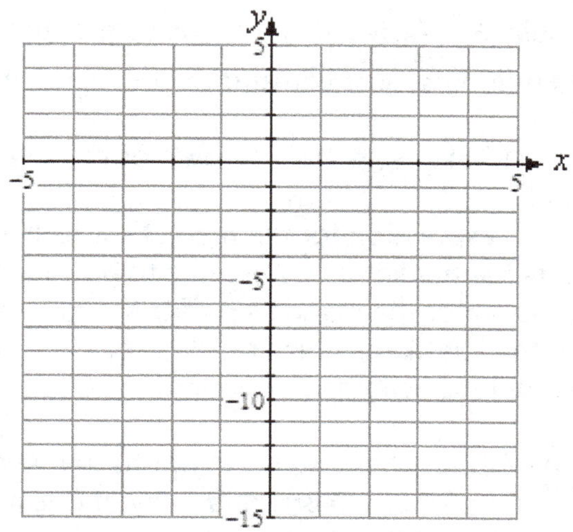

51. $y = -x^2 + 2x + 4$

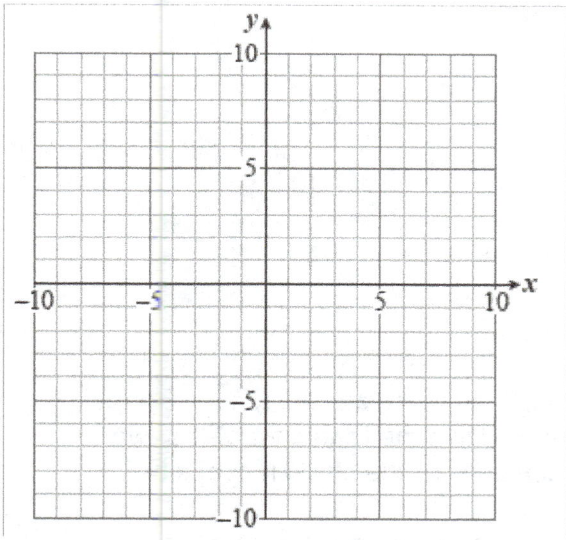

52. $y = x^2 - 3x + 4$

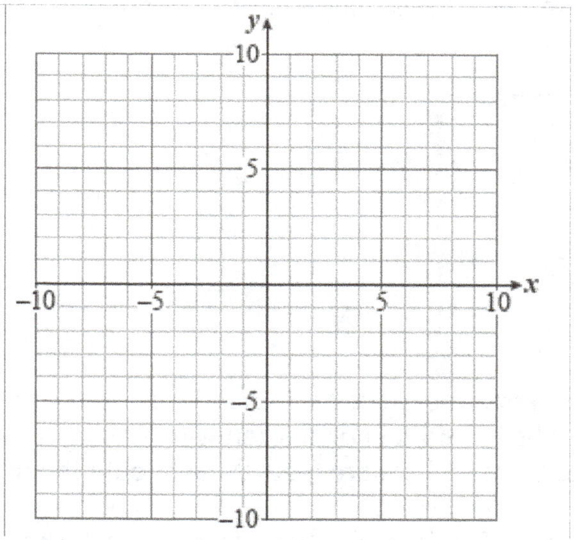

53. a. Find the vertex of the graph of $y = \frac{1}{3}(x+3)^2 - 2$

b. Find the x-intercepts of the graph.

c. Sketch the graph.

d. Write the equation of the parabola in standard form.

54. a. Find the vertex of the graph of $y = -2(x-1)^2 + 5$

b. Find the x-intercepts of the graph.

c. Sketch the graph.

d. Write the equation of the parabola in standard form.

55. Audrey launches her experimental hydraulic rocket from the top of her apartment building. The height of the rocket after t seconds is given in feet by

$$h = -16t^2 + 40t + 80$$

How long is the test flight before the rocket hits the ground?

56. The formula $S = \frac{1}{2}n^2 + \frac{1}{2}n$ gives the sum of the first n counting numbers. How many counting numbers must you add to get a sum of 325?

57. The Corner Market sells $160 - 2p$ pounds of bananas per week if they charge p cents per pound.

a. Write an expression for the market's weekly revenue, R, from bananas.

b. It costs the market $80 + 24p$ cents to buy and display the bananas. Write an expression for the market's profit, M, from selling the bananas.

c. If the market made a profit of $22 (or 2200 cents) on bananas last week, how much did they charge per pound?

58. The town of Amory lies due north of Chester, and Bristol lies due east of Chester. If you drive from Amory to Bristol by way of Chester, the distance is 17 miles, but if you take the back road directly from Amory to Bristol, you save 4 miles. How far is it from Bristol to Chester? (Hint: make a sketch.)

For Problems 59-60, find the sides of the right triangle.

59.

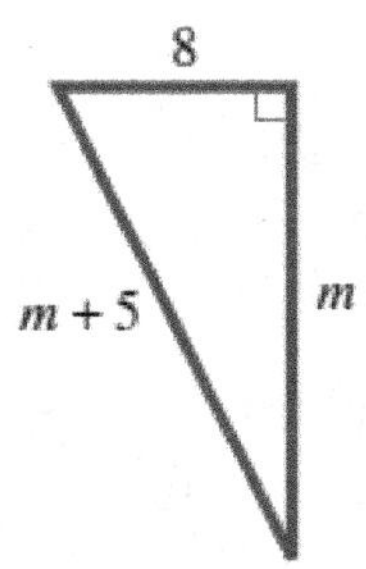

60.

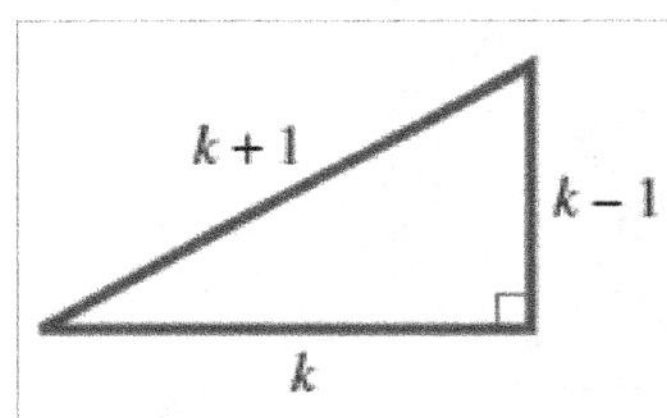

61. Irene wants to enclose two adjacent chicken coops of equal size against the henhouse wall. She has 66 feet of chicken wire fencing and would like the total area of the two coops to be 360 square feet. What should the dimensions of the chicken coops be?

62. The base of an isosceles triangle is one inch shorter than the equal sides, and the altitude of the triangle is two inches shorter than the equal sides. What is the length of the equal sides?

63. Green Valley Nursery sells $120 - 10p$ boxes of rose food per month if they charge p dollars per box. How much should they charge per box to maximize their revenue, and what will the maximum revenue be?

64. The Locker Room finds that it sells $1200 - 30p$ sweatshirts each month when it charges p dollars per sweatshirt. How much should they charge for a sweatshirt to maximize thier revenue, and what will the maximum revenue be?

65. A beekeeper has beehives distributed over 60 square miles of pastureland. When she places four hives per square mile, each hive produces about 32 pints of honey per year. For each additional hive per square mile, honey production drops by 4 pints per hive.

a. Write an equation for the total production of honey, in pints, in terms of the number of additional hives per square mile.

b. How many additional hives per square mile should the beekeeper install in order to maximize honey production?

66. A small company manufactures radios. When it charges \$20 for a radio, it sells 500 radios per month. For each dollar the price is increased, 10 fewer radios are sold per month.

a. Write an equation for the monthly revenue in terms of the price increase over \$20.

b. What should the company charge for a radio in order to maximize its monthly revenue?

67. Find an equation for a parabola whose vertex is $(15, -6)$ and that passes through the point $(3, 22.8)$.

68. Find an equation for a parabola whose vertex is $(-18, -125)$ and that passes through the point $(-4, -27)$.

69. The Tallahatchee bridge is 42 feet above the river below, and the railing on the bridge is 4 feet tall. Billy Joe is standing 4 feet from the railing, and tosses something over in a parabolic path. The object is 3 feet above the bridge when it leaves his hand, and reaches the vertex of its trajectory 3 feet directly above the railing.

a. Make a sketch illustrating the path of the object. Use a coordinate system with the origin at water level directly below the railing, where h is the height of the object and x is its horizontal displacement. Find the coordinates (x, h) of the object when it leaves Billy Joe's hand, and at the vertex of the parabolic path.

b. Find an equation for the height, h, of the object

c. How far from the origin will the object be when it hits the water?

70. Mac bought a restaurant with a yellow parabolic archway. The arch is 20 feet high at the vertex, and 16 feet wide at its base. Mac wants to enlarge the restaurant so that cars can drive under the arch to a take-out window. This will mean cutting one side of the arch so that the cut end rests on the 8.75-foot high roof of the take-out window.

a. Choose a coordinate system and write an equation for the arch.

b. What is the horizontal clearance between the take-out window and the base of the other side of the arch?

c. What is the horizontal clearance between the take-out window and the arch at a height of five feet from the ground?

71. Find values of a, b, and c so that the graph of the parabola $y = ax^2 + bx + c$ contains the points $(-1, -4), (0, -6)$, and $(4, 6)$.

72. a. Find the equation of a parabola $y = ax^2 + bx + c$ that passes through the points $(0, -2)$, $(-6, 1)$ and $(4, 6)$.

b. Plot the data points and sketch the graph on the grid.

73. Find a parabola that fits the following data points.

x	−8	−4	2	4
y	10	18	0	−14

74. Find a parabola that fits the following data points.

x	−3	0	2	4
y	−46	8	−6	−60

75. The height of a cannonball was observed at 0.2-second intervals after the cannon was fired, and the data recorded in the table.

Time (seconds)	0.2	0.4	0.6	0.8	1.0	1.2	1.4	1.6	1.8	2.0
Height (meters)	10.2	19.2	27.8	35.9	43.7	51.1	58.1	64.7	71.0	76.8

a. Find the equation of the least-squares regression line for height in terms of time.
b. Use the linear regression equation to predict the height of the cannonball at 3 seconds and at 4 seconds after it was fired.
c. Make a scatterplot of the data and draw the regression line on the same axes.
d. Find the quadratic regression equation for height in terms of time.
e. Use the quadratic regression equation to predict the height of the cannonball at 3 seconds and at 4 seconds after it was fired.
f. Draw the quadratic regression curve on the graph from part (c).
g. Which model is more appropriate for the height of the cannonball, linear or quadratic? Why?

76. Max took a sequence of photographs of an explosion spaced at equal time intervals. From the photographs, he was able to estimate the height and vertical velocity of some debris from the explosion, as shown in the table. (Negative velocities indicate that the debris is falling back to earth.)

Velocity (meters/second)	67	47	27	8	−12	−31
Height (meters)	8	122	196	232	228	185

a. Enter the data into your calculator and create a scatterplot. Fit a quadratic regression equation to the data, then graph the equation on the scatterplot.
b. Use your regression equation to find the vertex of the parabola. What do the coordinates represent, in terms of the problem? What should the velocity of the debris be at its maximum height?

Answers to Homework Problems, Part 1

Homework 1.1

1. $g = m + 3$

m	g
2	5
3	6
5	8
10	13
12	15
16	19
18	21
m	$m+3$

2. $w = 20 - t$

t	w
0	20
2	18
4	16
5	15
6	14
10	10
12	8
t	$20-t$

3. $x = \frac{b}{2}$

b	x
0	0
2	1
4	2
5	2.5
6	3
8	4
9	4.5
b	$\frac{b}{2}$

4.

z	3	6	8	12	15	18	20	z
r	2	4	$\frac{16}{3}$	8	10	12	$\frac{40}{3}$	$\frac{2}{3}z$

$r = \frac{2}{3}z$

5. Answers will vary.

Sample Answer:

n	0	5	10	15	20
W	0	6	12	18	24

6. Answers will vary.

Sample Answer:

x	0	2	4	6	8
M	0	3	6	9	12

7. a, b, and c.

x	y
4	0.5
8	1
10	1.25
16	2

The three tables have identical values.

8. a, b, and c.

x	y
5	3
10	6
12	7.2
1	0.6

The three tables have identical values.

9.

Ridgecrest	70	75	82	86	90	R
Calculation	$70+15$	$75+15$	$82+15$	$86+15$	$90+15$	$R+15$
Sunnyvale	85	90	97	101	105	$R+15$

a. Add 15° to Ridgecrest's temperature to find Sunnyvale's temperature.

b. Temperature in Sunnyvale = Temperature in Ridgecrest + 15

c. $S = R + 15$

d.

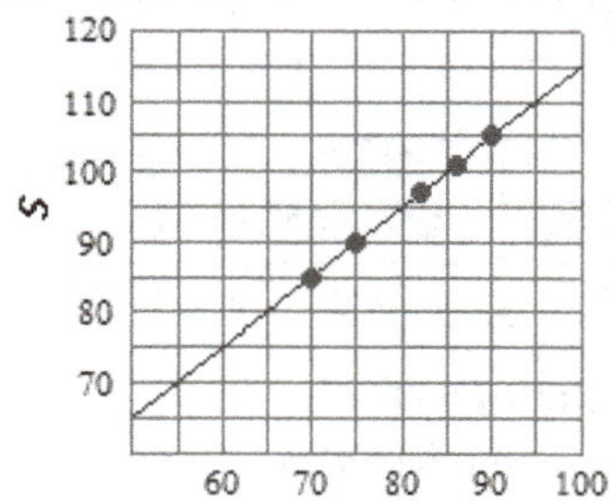

10.

driven	40	60	90	120
Calculation	$200-40$	$200-60$	$200-90$	$200-120$
remaining	160	140	110	80

driven	140	170	d
Calculation	$200-140$	$200-170$	$200-d$
remaining	60	30	$200-d$

a. Subtract the miles driven from 200 to get Jerome's remaining miles.

b. Miles remaining = 200 − miles driven

c. $r = 200 - d$

d.

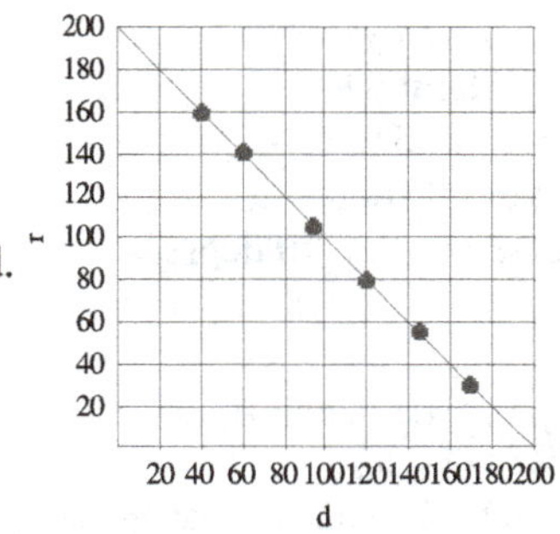

11.

Total	15	30	45	60	75	81	b
Calculation	$\frac{15}{3}$	$\frac{30}{3}$	$\frac{45}{3}$	$\frac{60}{3}$	$\frac{75}{3}$	$\frac{81}{3}$	$\frac{b}{3}$
share	5	10	15	20	25	27	$\frac{b}{3}$

a. Divide the total bill by 3 to get Milton's share.

b. Milton's share = $\frac{\text{total bill}}{3}$

c. $s = \frac{b}{3}$

d. 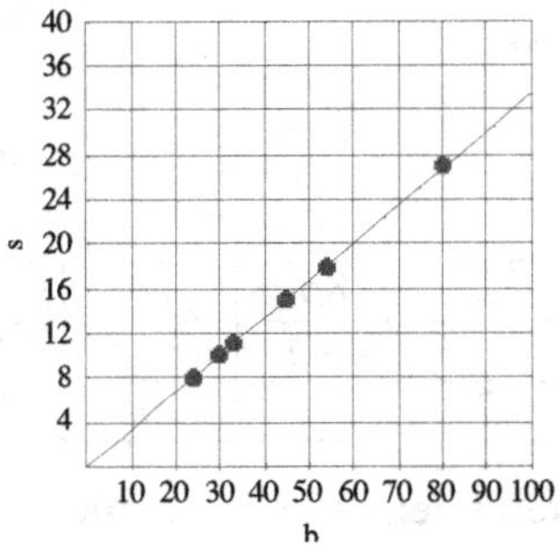

12.

Total Calories	1000	1500	2000
Calculation	0.30 × 1000	0.30 × 1500	0.30 × 2000
Fat Calories	300	450	600

Total Calories	2500	3000	C
Calculation	0.30 × 2500	0.30 × 3000	0.30 × C
remaining	750	900	0.30 × C

a. Multiply the Total Calories by 0.30
b. Fat Calories = 0.30 × Total Calories
c. $F = 0.30 \times C$

d. 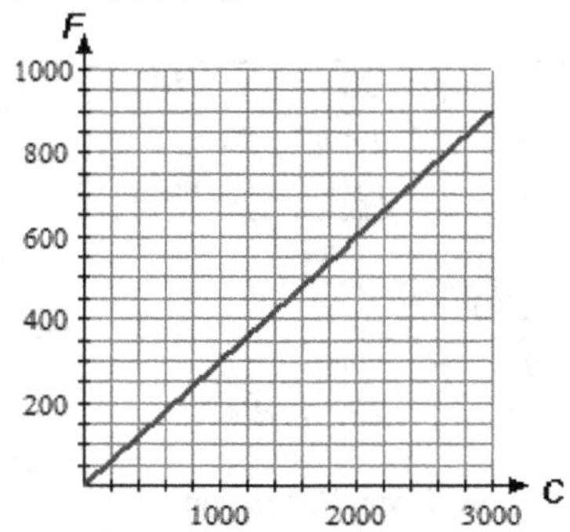

13. a. Horizontal: Number of years, Vertical: Amount in account
    b. About \$3000
    c. Approximately after 13.75 years
    d. About \$200 e. About \$400
14. a. Horizontal: Number of minutes, Vertical: Temperature of soup
    b. After about 3.5 minutes c. 70 degrees
    d. About 55 degrees e. 60 degrees
15. Graph C
16. a. I. Graph C II. Graph D III. Gaph A
    b. Answers will vary. Sample Answer: Delbert walks toward school, about halfway there he realizes that he left a book at home. He finds and retrieves the book, then walks back to school.

17. a. 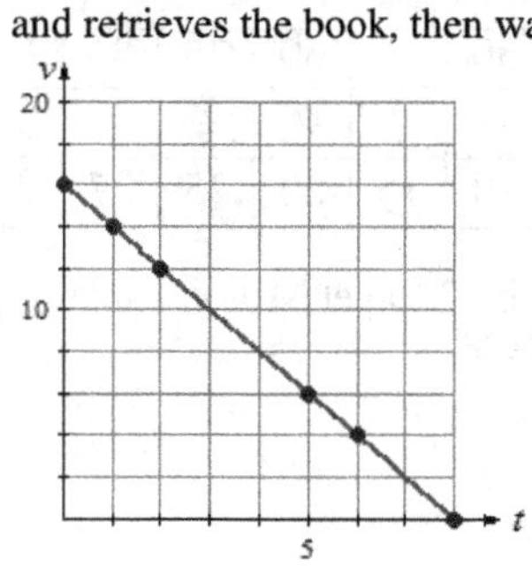

b. 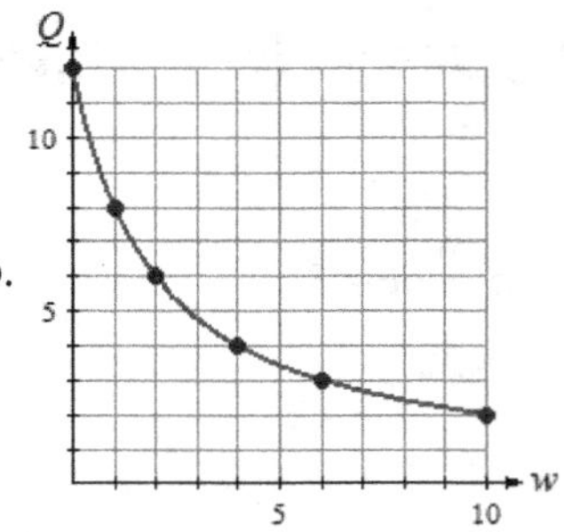

18. a.

x	0	10	30	40	60	70
y	70	60	40	30	10	0

b. $y = 70 - x$

19. a. $y = 2 - x$

b.

x	0	0.2	0.3	0.5	0.6	0.7
y	2	1.8	1.7	1.5	1.4	1.3

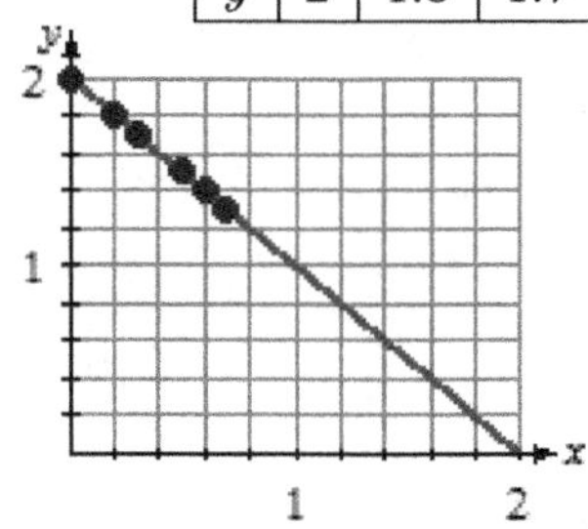

20. a. $y = \frac{120}{x}$

b.

x	1	2	3	4	6	8
y	120	60	40	30	20	15

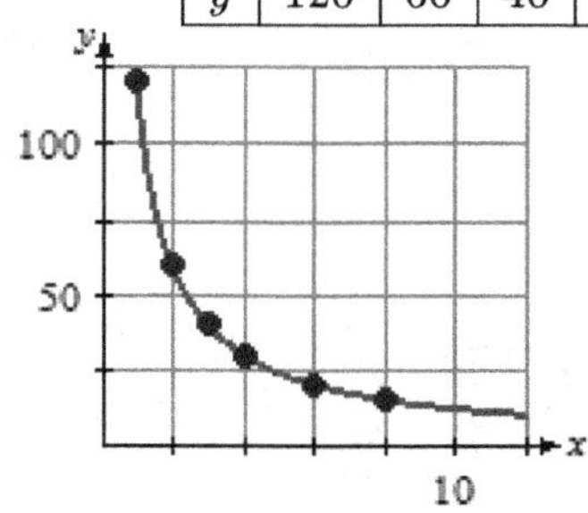

Homework 1.2

1. $4y$ 2. $2b$ 3. $1.15g$
4. $t - 5$ 5. $\frac{7}{w}$ 6. $\frac{B}{4}$
7. $T + 20$ 8. $16 - p$ 9. $\frac{15}{M}$
10. $R - 3.5$ 11. $n - 6$ 12. $6n$
13. $6 - n$ 14. $\frac{6}{n}$ 15. $2x$
16. $4b$ 17. $2v$
18.a. $\frac{x}{6}, \frac{1}{6}x$ b. $0.06, \frac{6}{100}x$
19. a. $\frac{1}{5}m$

b.

Spending Money	m	20	25	60	80
Amount Saved	$\frac{1}{5}m$	4	5	12	16

20. a. $\frac{1}{4}W$

b.

Width of photo	W	6	10	24	30
Height of photo	$\frac{1}{4}W$	1.5	2.5	6	7.5

21. a. $\frac{2}{3}p$

b.

Members present	p	90	96	120	129
Votes Needed	$\frac{2}{3}p$	60	64	80	86

22. a. $\frac{3}{5}D$

b.

Stock Dividend	D	40	85	115	170
Maria's Share	$\frac{3}{5}D$	24	51	69	102

23. Ticket Price: p; $15p$
24. Cost of a light bulb: b; $3b$
25. Savings account balance: s; $\frac{3}{5}s$
26. Price of the pizza : p; $\frac{p}{6}$
27. Weight of copper : w; $\frac{w}{16}$
28. School buses : b, $0.09b$
29. Cost of vaccine : v, $v - 16$
30. Cost of each computer : c, $32c$
31. a. $480 - x$ b. $500 - x$ c. $30 - x$
32. a. $x - 15$ b. $x - 12$ c. $x - 18$
33. Amount of milk : m, $0.70m$
34. Number of days : d, $\frac{1200}{d}$
35. Votes opponent received : v, $v - 432$
36. Budget : B, $B + 2000$
37. a. $d = 180t$
    b. 360 miles; 630 miles; 2160 miles
38. a. $C = 0.0002g$
    b. 0.08 gal, 1 gal, 10 gal
39. a. $B = 0.085R$
    b. \$8500, \$42500, \$170,000
40. a.. $P = R - 4000$
    b. \$6000, \$2500, $-$ \$1500

Homework 1.3

1. a. no b. yes c. no d. no
2. a. yes b. yes c. no d. no
3. a. yes b. no c. no d. yes
4. a. no b. yes c. yes d. no
5. a. no b. no c. yes d. yes
6. a. yes b. yes c. no d. no
7. horizontal: 0.25; vertical: 4
8. horizontal: 50; vertical: 0.5

9. a.

x	0	2	5	6	10	12	16
y	16	14	11	10	6	4	0

b. $y = 16 - x$

10. a.

x	0	1	4	5	7	8	10
y	0	4	16	20	28	32	40

b. $y = 4x$

11. f 12. g 13. c, e 14. b

15. a. \$35

b.

w	12	16	20	30	36	40
p	15	20	25	37.50	45	50

c.

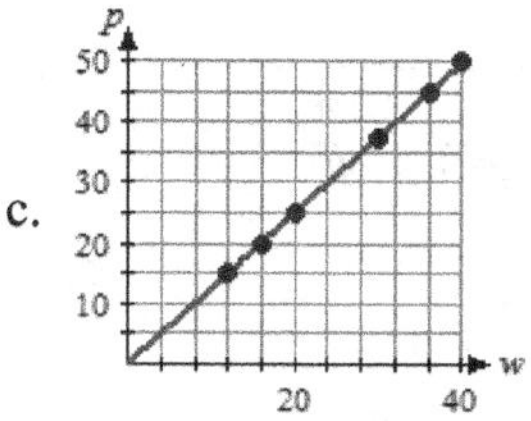

16. a. $D = 1000 - L$

b.

L	250	300	450	600
D	750	700	550	400

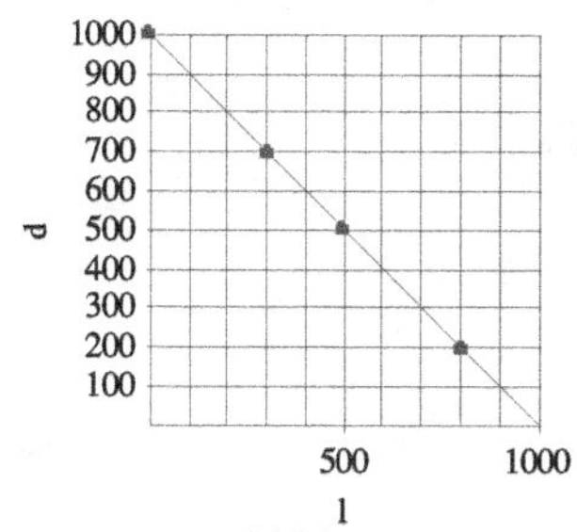

17. a. $S = \dfrac{3600}{m}$

b.

m	20	40	60	100
S	180	90	60	36

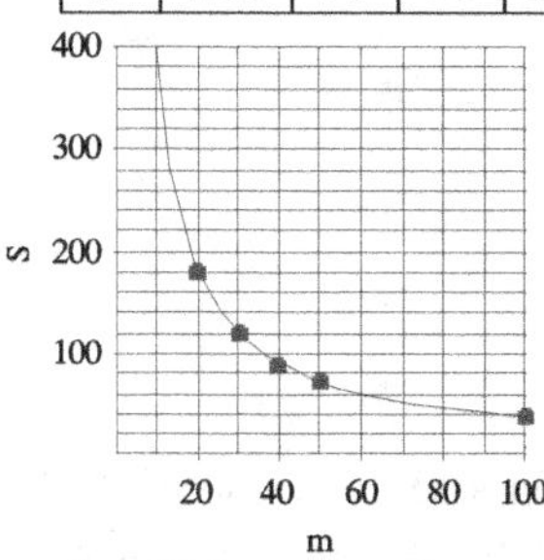

18. a. $p = 110$ b. $p = 60$
19. a. $g \approx 13.9$ b. $g \approx 11.7$
20 a. $(50, 4)$
    b. The tax on a \$50 item is \$4.
21 a. $(4, 16)$
    b. A 4-inch tall bunny weighs 16 ounces.
22. a $I = 80t$

b.

t	2	4	7
I	\$160	\$320	\$560

c.

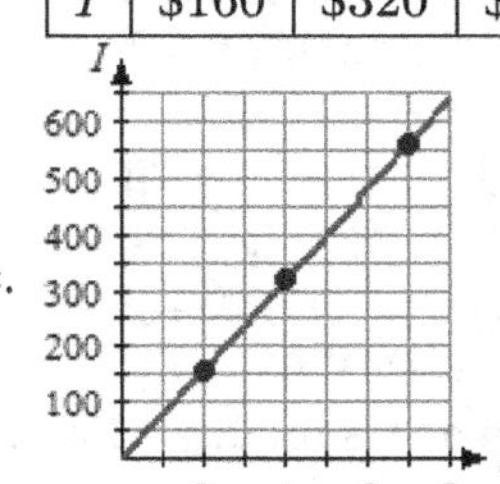

23. a $P = 6000 - C$

b.

C	\$ 800	\$1000	\$2500
P	\$5200	\$5000	\$3500

c.

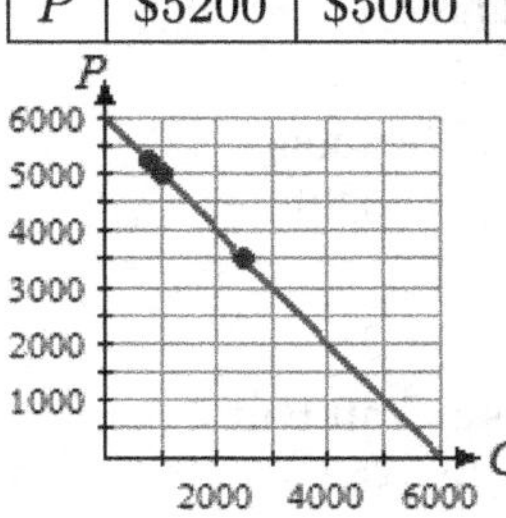

24. a $I = 15000r$

b.

r	8%	$10\frac{1}{4}\%$	12%
I	\$1200	\$1537.50	\$1800

c. 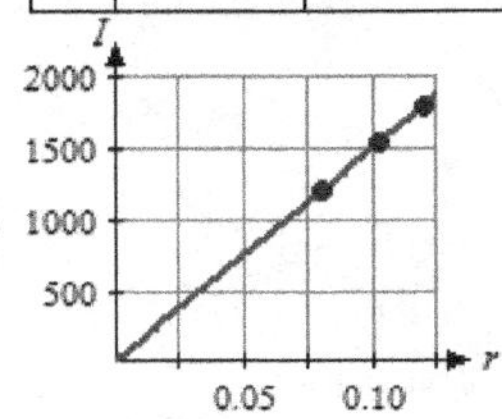

25. a. $S = \frac{540}{n}$

b.

n	20	25	30
S	27	21.6	18

c. 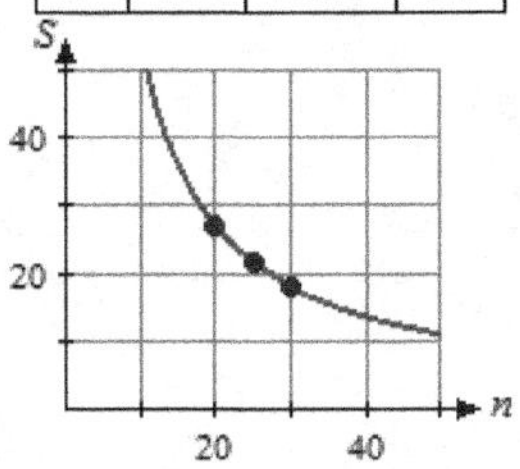

Homework 1.4

1.

t	q
2	11
4	13
6	15
9	18
21	30
30	39

To find a value for q, add 9 to the value of t. To find a value for t, subtract 9 from the value of q.

2.

n	p
0	0
2	10
4	20
5	25
7	35
11	55

To find a value for p, multiply the value of n by 5. To find a value for n, divide the value of p by 5.

3. a. yes b. no c. yes
4. $x = 14$ 5. $y = 2.8$ 6. $y = 36$
7. $b = 12$ 8. $a = 3.9$ 9. $x = 4$
10. $x = \frac{106}{17}$ 11. $z = \frac{10}{3}$ 12. $k = 0$
13. a. $I = Prt$ becomes $75 = P \cdot 0.03 \cdot 1$
 b. $P = 2500$; Clive loaned his brother \$2500.
14. a. $A = \frac{S}{n}$ becomes $38.25 = \frac{S}{8}$
 b. $S = 306$; Andy had 306 points.
15. a. $d = rt$: $234 = 13t$
 b. $t = 18$; It takes 18 hours.
16. a. $A = lw$: $400 = l \cdot 16$
 b. $l = 25$; The roll is 25 feet long.
17. $x + 7 = 26$ 18. $x + 7 = 26$
19. $\frac{x}{7} = 26$ 20. $7x = 26$
21. $\frac{x}{26} = 7$ 22. $x - 7 = 26$
23. a. $B =$ Amount before Craft Fair
 b. $B - 24 = 39$ c. \$63
24. a. $B =$ Brenda's weight
 b. $B + 32 = 157$ c. 125 pounds
25. a. $w =$ Hourly wage
 b. $20w = 136$ c. \$6.80
26. a. $P =$ Total Profit
 b. $\frac{P}{8} = 64$ c. \$512
27. $89 = p - 26$
 Her mother paid \$115.
28. $360 = 0.40I$
 Emily makes \$900 per month.
29. $1978 = B - 2378$
 There wase \$4356 in her account.
30.a. price of new car : n; price of used car : u;
 $u = n - 3400$
 b. \$11,100 c. \$12,600
31. a. number of games played: p;
 number of games won: w; $w = 0.60p$
 b. 72 c. 160
32.a. selling price : s; profit : P;
 $P = 0.18s$
 b. \$10.80 c. \$40

Homework 1.5

1. 14 2. 3 3. 72 4. 18
5. $\frac{3}{2}$ 6. 23 7. 0 8. 9
9. 4 10. 9 11. 3 12. 1
13. 0.859 14. 0.729
15. 42.705 16. 2204.533
17.a. 18 b. 50 18a. 3 b. 18
19a. 2 b. 8 20a. 42 b. 12
21a. 2 b. 18 22a. 125 b. 5

23.

z	$5z$	$5z - 3$
2	10	7
4	20	17
5	25	22

24.

Q	$12 + Q$	$2(12 + Q)$
0	12	24
4	16	32
8	20	40

25. 26 26. 53 27. $\frac{7}{12}$
28. $\frac{19}{18}$ 29. $\frac{15}{2}$ 30. $\frac{13}{10}$
31. a. $(20 - 2) \cdot 8 + 1 = 145$
 b. $20 - 2 \cdot (8 + 1) = 2$
 c. $20 - 2 \cdot 8 + 1 = 5$
32. $4(5 + 7) \div (10 - 8)$
33.a. $\dfrac{20 - 8}{6 + 4 + 10}$ b. $(20 - 8) \div (6 + 4 + 10)$
34. First multiply 32 and 12, then divide the result by 4, then subtract the result from 825, and finally add the result to 2.
35. $\dfrac{12}{m - 3}$ 36. $\dfrac{m}{12} - 3$
37. $\dfrac{m - 3}{12}$ 38. $\dfrac{12}{m} - 3$
39. Subtracting $9 - 3$ is incorrect. Should be

$$9 - 3(2) = 9 - 6 = 3$$

40. $(2.3 + 5.7)6 - (1.2 + 3.3)2 = 39$ sq cm

Chapter 1 Review

1.

Gallons	3	5	8	10	11	12
Miles	66	110	176	220	242	264

b. Multiply 22 by the number of gallons.

c. $m = 22g$

d. 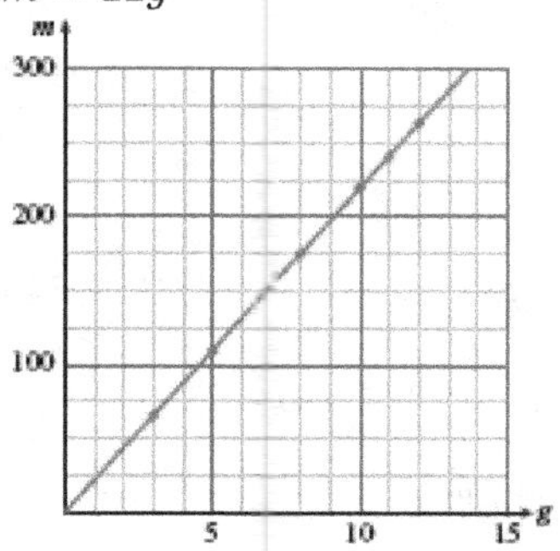

3.

x	.5	1	1.5	2	4	6	7.5
y	.125	.25	.375	.5	1	1.5	1.875

$y = \frac{x}{4}$

5. $z + 5$ 7. $f - 60$

9. $y + 6$ 11. $0.08b$

13a. $d = 88t$ b. 440 ft, 44 ft, 2640 ft

15a. $I = 35t$ b. $35, $70, $175

17a. $n = b - 60$

b.

b	100	120	150	180	200
n	40	60	90	120	140

19a. $n = 2s$

b.

s	2	3	5	8
n	4	6	10	16

21. a, c, d

23a. yes b. no

25. 9 27. 0.4 29. 0

31a. 32 b. 24

33. $171 = 0.095P$; $1800

35. $106,000 = 0.53W$; $200,000$

37. $\frac{w}{85} = 0.7$; 59.5 lb

39. 3 41. 6

43. $\frac{1}{2}d + 4$ 45. $3(l + 5.6)$

47. 4 49. 5

51. $1300

Homework 2.1

1a. 2 b. −2 c. −8 2a. −35 b. −5 c. 5

3a. −25 b. 4.1 c. $\frac{-1}{6}$ 4a. −49 b. 2.4 c. $\frac{-9}{8}$

5a. $4 + (-8) = -4$ b. $3 + 9 = 12$ c. $-8 + 6 = -2$

d. $-6 + (-5) = -11$

6a. 6 b. 10 c. −10 d. 2

7a. −11 b. −9 c. 0 d. −8

8a. 34 b. 5 c. −2 d. −10

9a. 10 b. 10 c. 10

10a. −8 b. −3 c. −8

11a. −5 b. −9 c. 5 d. 9 e. 5

12a. 8 b. −18 c. −8 d. 18 e. −8

13a. −6 b. −18 14a. −13 b. −34

15a. −2 b. 12 c. −39 16a. 4 b. −6 c. −9

17a. 32 b. −3 c. 4

18a. 18 b. undefined c. 0

19a. 0 b. −8 c. undefined

20a. $\frac{-3}{8}$ b. 2.6 c. $\frac{2}{3}$

21a. −16 b. 48 c. −8 d. −16 e. −3

22a. −6 b. 9 c. 0 d. −6 e. 1

23a. 5, −5 b. 5, 5 c. 0, 5 d. 0, 5

24a. 5, $\frac{1}{5}$ b. −5, $\frac{1}{5}$ c. 5, 5 d. 5, −5

25. $x = 5$ 26. $z = \frac{-4}{3}$ 27. $a = -32$

28. $x = 6$ 29. $t = -13$ 30. $b = \frac{9}{2}$

31a. $x = 3$ b. $x = -6$ 32. answers may vary

33. $500 34. 40 feet 35. −6°

36. − $695 37. 14,776 feet 38. −$35.40

39. −48 meters 40. −28.75° per hour

41. $n + 12 = 5$ 42. $-12 + n = -5$

43. $-12 + n = -5$ 44. $-12 + n = 5$

Homework 2.2

1. 29.8 meters 2. 24 square cm 3. $5400

4. 37° C 5. −18 6. 2

7. −15 8. −15

9a. −24 b. 24 c. −24 d. 48 e. −48 f. 48

10a. negative b. positive

11a. 2 b. −2 12a. 7 b. −22

13a. −6 b. −2 14a. 12.96 b. −19.2

15a. −60 b. 7 c. 27 d. 6 e. −23 f. 35

16a. 2 b. −3 c. −6 d. 27 e. 2 f. −8

17. 13 18. −12 19. −18 20. $\frac{1}{13}$

21. $-\frac{1}{9}$ 22. $\frac{7}{6}$ 23a. i. $-\frac{15}{4}$ ii. $-\frac{15}{4}$ iii. −3.75

b. No 24. a, d 25. $2x + 10$ 26. $7x + 42$

27a. $4400 ;$2600 b. Multiply the number of weeks by $200, then subtract from $5000.

c.

weeks	2	4	5	6	10	15	20
left	4600	4200	4000	3800	3000	2000	1000

d. $5000 - 200w$ e. $S = 5000 - 200w$

28.a. $720; $960 b. Subtract $2000 from her income, then multiply the result by 0.12

c.

I	5000	7,000	12,000	15,000	20,000	24,000	30,000
T	360	600	1200	1560	2160	2640	3360

d. $0.12(I - 2000)$ e. $T = 0.12(I - 2000)$

29a. w = width (inches): $2w - 3$ b. 23 inches

30a. P = principle ($): $0.40P + 20$ b. $220

31a. t = number of childrens tickets; $8(150 - t)$

b. $536

32a. p = profit; $\frac{1}{3}(p - 50)$ b. $150

33. $65 + 30t$ 34. $800 + \frac{a}{2}$ 35. $\frac{80 - M}{4}$

36. $50 + \frac{1}{3}I$ 37. $200 - 15w$ 38. $600 + 80t$

39. $\frac{c - 50}{10}$ 40. $35(x + 3)$

41a. $\frac{x}{4 + x}$ b. Answers may vary

42a. $(a + 5)(a - 5)$ b. Answers may vary

Homework 2.3

1a. No b. Yes c. No d. Yes

2a. Yes b. Yes c. No d. No

3a. $(-6, 2)$ b. $(-5, -7)$ c. $(8, -4)$

d. $(1, 7)$ e. $(0, -9)$ f. $(-2, 0)$

4a. $(-10, 0)$ b. $(-2, -4)$ c. $(0, -5)$

d. $(2, -6)$ e. $(10, -10)$

5a. $(0, -600)$ b. $(10, -200)$
c. $(15, 0)$ d. $(20, 200)$
6a. $P = 100 - 6d$
b.

d	0	2	5	10	15
P	100	88	70	40	10

c.
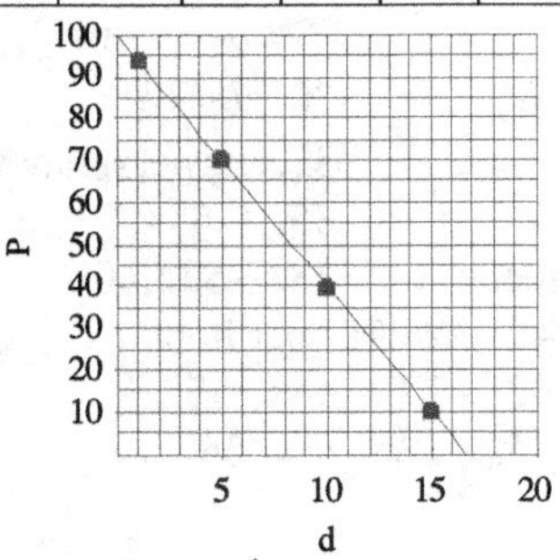

7a. $M = 200 + \frac{1}{5}U$
b.

U	20	40	80	100	200
M	204	208	216	220	240

c.
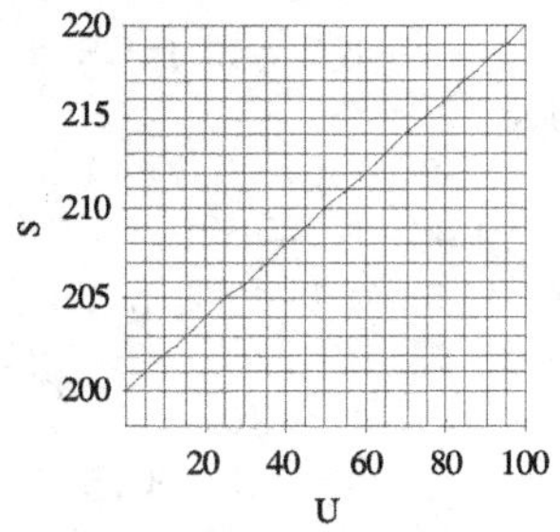

8a. $B = -100d + 2000$
b.

d	B
-15	3500
-5	2500
0	2000
5	1500
15	500
20	0

c.
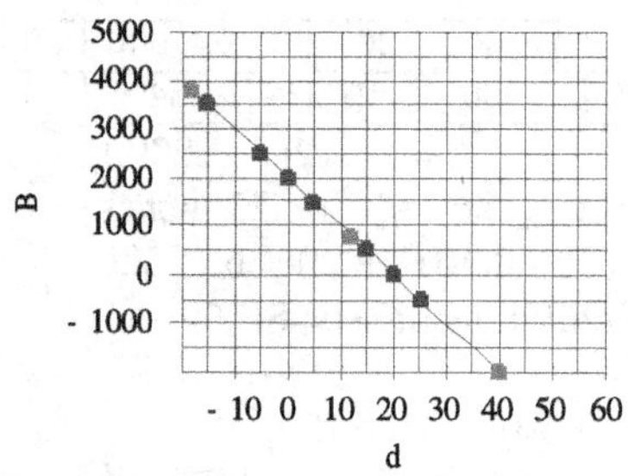

d. -27 feet e. floor 8
9a. \$540 b. \$30 c. \$360 d. 18
e. $B = 540 - 30m$
10a. $s = 100 - 5x$
b.

x	2	5	6	12
s	90	75	70	40

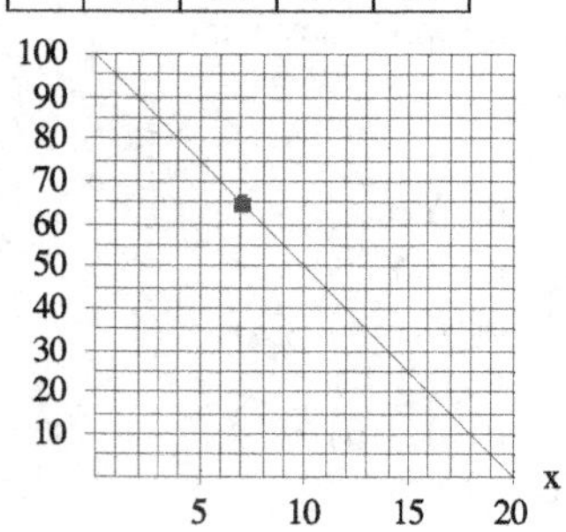

c. $65 = 100 - 5x$; 7
11a. $d = 120 - \frac{1}{2}w$
b.

w	24	36	84	120
d	108	102	78	60

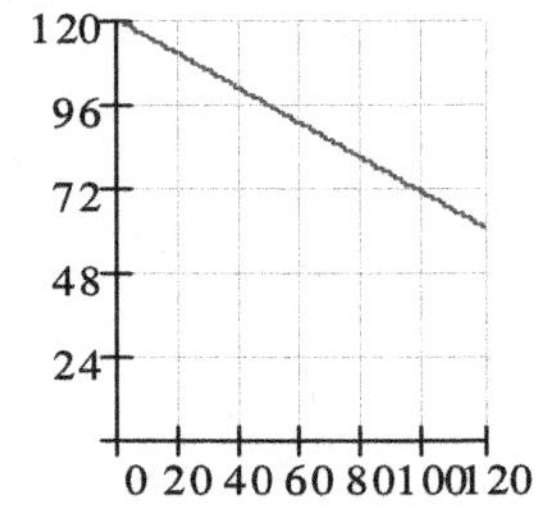

c. 48 weeks; $96 = 120 - \frac{1}{2}w$

12.
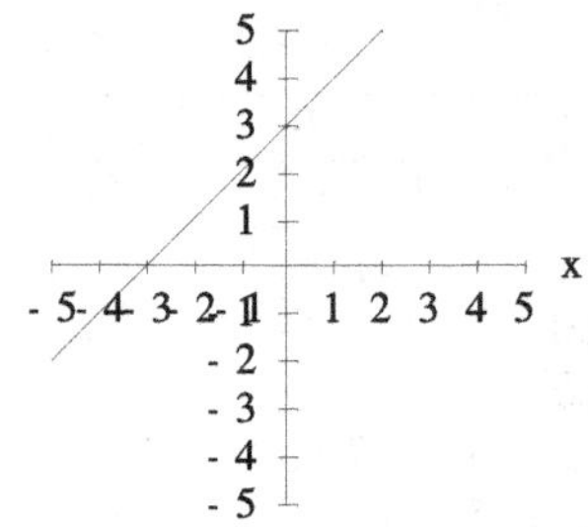

13.
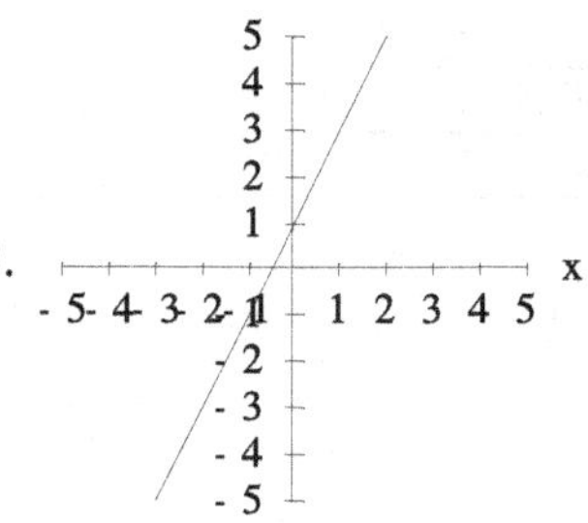

14.
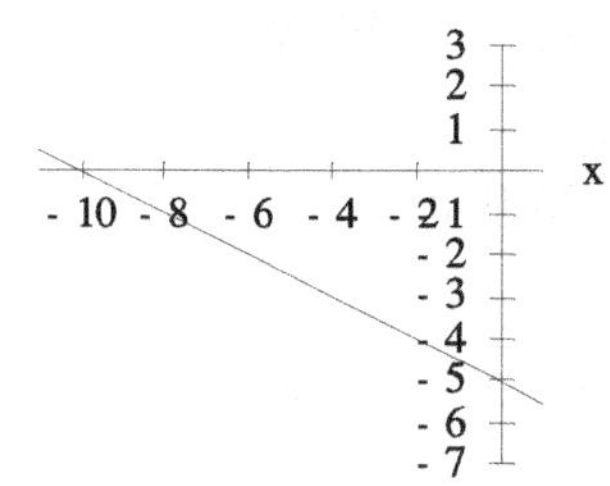

15.
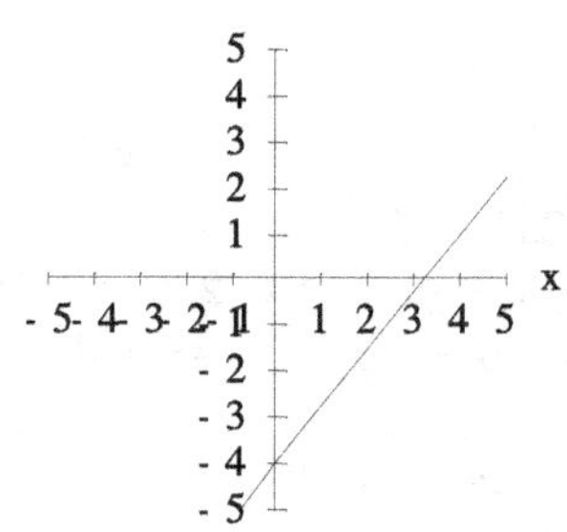

16a. -4 b. -5 c. $-4 = 2(-5) + 6$
d. 1 e. -4.9, -4, etc.
17a. $x = -16$ b. $x = -40$
18a. $x = 6$ b. $x = 12$
19. $t \approx 1.5$ 20. $x \approx -2.5$

Homework 2.4

1.

x	$2x$	$2x-4$
3	**6**	10
6	**12**	16
5	**10**	14
8	**16**	20

2.

q	$q-3$	$5(q-3)$
3	**0**	**0**
5	**2**	10
4	**1**	**5**
7	**4**	20

3. $x=3$ 4. $a=20$ 5. $x=4$
6. $x=0$ 7. $p=13$ 8. $z=0$
9. $k=41$ 10. $x=18$ 11. $b=6$
12. $c=-2$ 13. $t=1$ 14. $h=5$
15. $b=18$ 16. $y=10$ 17. $x=-4$
18. $w=13.4$ 19. $x=3.4$ 20a. $x \leq -15$
b. 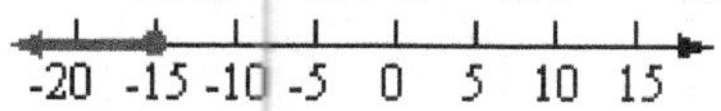

c. $x=-15.1$ is a solution, $x=-14.9$ is not.
21a. $y>-5$
b. 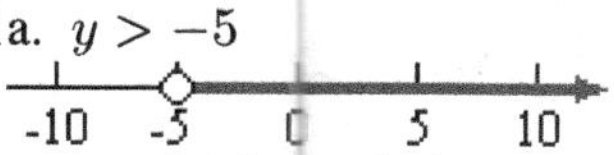

c. $y=-4.9$ is a solution, $y=-5$ is not.
22a. $x \leq 12$
b. 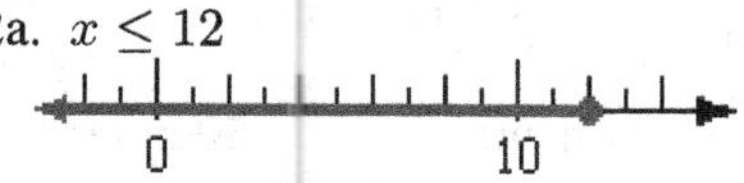

c. $x=12$ is a solution, $x=13$ is not.
23. $x>2$ 24. $x \geq -3$ 25. $x<-6$
26. $-1 \leq x \leq 4$ 27. $-7<b \leq -2$
28. $-7 \leq w < -3$ 29. 7
30. 6 31. -3
32a. 8.6 lbs b. 36 weeks c. $w=3.8+0.6t$
d. $26=3.8+0.6t$; $t=37$ weeks
33a. $P \approx 20$ b. $m \approx 80$ minutes
c. $P=13+0.15m$
d. $25=13+0.15m$; $80=m$
34a. m = each installment ($)
b. $10200=1200+36m$
c. Each installment is $250.
35a. d = assigned distance
b. $5(d+0.5)=22.5$
c. She is assigned 4 miles.
36a. p = number of planted seeds
b. $112=0.6p-38$
c. 250 seeds were planted.
37a. c = average temperature change per day
b. $26=-6+4c$
c. The average change was 8 degrees per day.
38a. w = time it takes (weeks) b. $162=196-4w$
c. It will take 8.5 weeks.
39. 175 yards 40. 6 meters
41. $4x+8=36$; $x=7$
42. $\frac{1}{2}(2)(3+x)=20$; $x=17$

Homework 2.5

1. $-4x$ 2. $-12.8a$ 3. t 4. $-bc$
5. $12y-4$ 6. $-8st+9s-2$ 7. $9x-2$
8. $6y-10$ 9. $2a$ 10. $-5b+2$ 11. 4
12. $\frac{-9}{2}$ 13. 0 14. $x>-2$ 15. $\frac{8}{3}$
16. $y \leq -4$ 17. $w \geq 21$ 18. $p=-1$
19. a. -21 b. $-6y-6$ c. -21 20. -170
21. $x=0$: $2+7x=2+7(0)=2$;
$9x=9(0)=0$
22. $a=0$: $-(a-3)=-(0-3)=3$;
$-a-3=-0-3=-3$
23. $6x+8y$ 24. $13a+7$
25a. $d+8$ b. $2d+8$ c. There are 16 dogs.
26a. $3x$ b. $4x$ c. Nine are smokers.
27a. $9.8t$ b. 98 feet c. 15 seconds
28a. $0.08x$ b. $1.08x$ c. $860
29a. $596x-1355$ b. 30
30. $8b=5b+12$; $b=4$ oz.
31. $3p+33=p+131$; $p=49$ in
32. $12c+1.8=8c+5$; $c=\$0.8$
33. $3g-70=2g+220$; $g=290$ mg
34. $9s+15+14=10s+23$; $s=6$ points

Chapter 2 Review

1. f 3. c 5. b
7. $\frac{5}{3}$ 9. 8 11. $V=2000-200t$

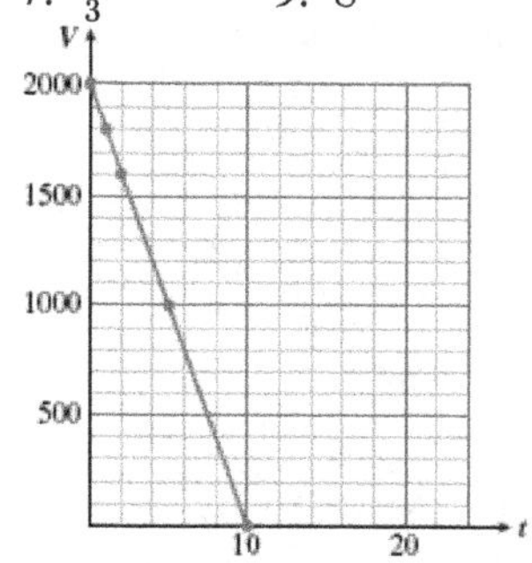

13. $5.75+4.50b=19.25$; 3 bushels
15a. 3 b. -6 c. 6 d. -3
17. $>$ 19. $>$ 21. 20 23. 9
25. 31 27. $\frac{-13}{16}$ 29. -3 31. -1
33. 6°F 35. $460,000 37. -12 yd
39. 6.9, 6, 5 41. 3, 2, 1

43a.

h	-4	-2	0	1	3	5	8
T	30	24	18	15	9	3	-6

b. $T=18-3h$
c. 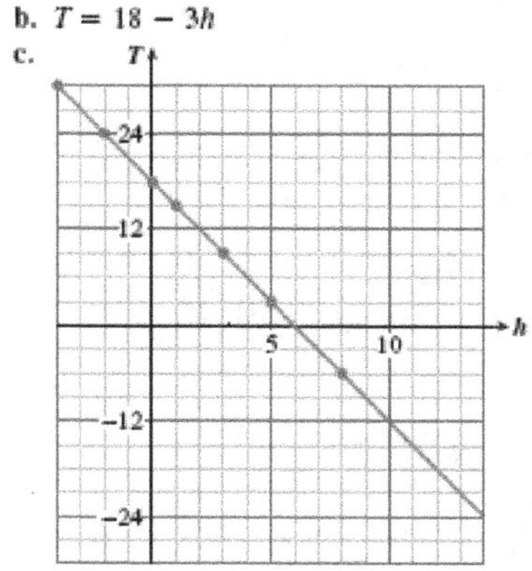

d. 24°F e. 11 p.m. f. -18°F

45. 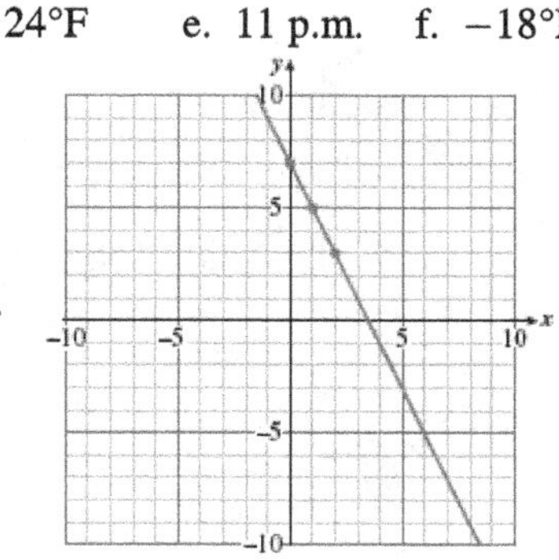

47. $2m + 7n$ 49. $z = -1$
51. $w = -3$ 53. $h = -6$
55. a. -5 b. 2
57. a. $3x$ b. $8x$ c. 6 cm by 18 cm

59. $z \geq 3$ 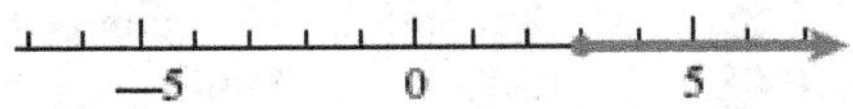

61. $k > -6$ 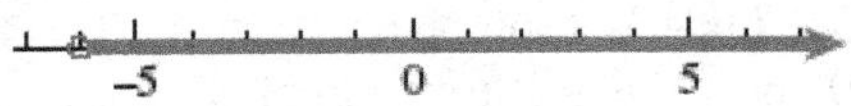

63. $7 > n \geq 3$ 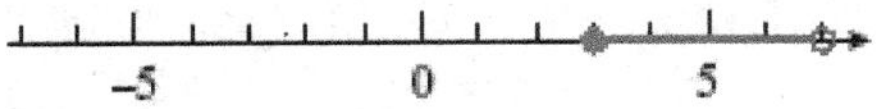

Homework 3.1

1. 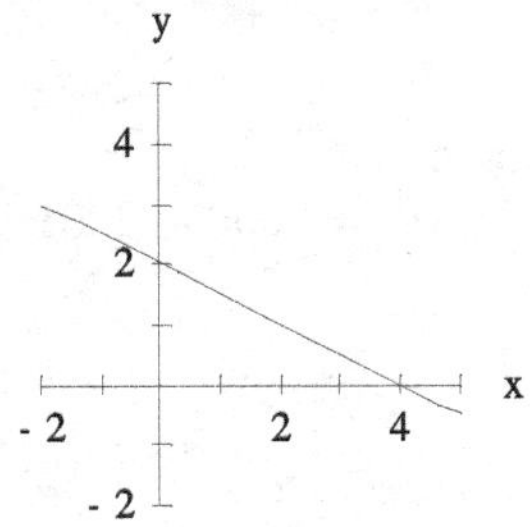 $(0, 2), (4, 0)$

2. 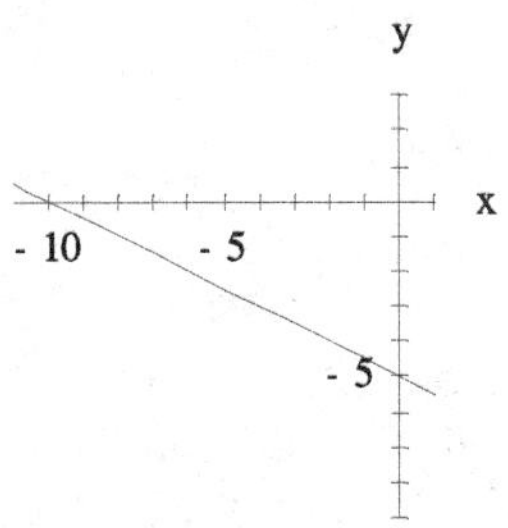 $(0, -5), (-10, 0)$

3. 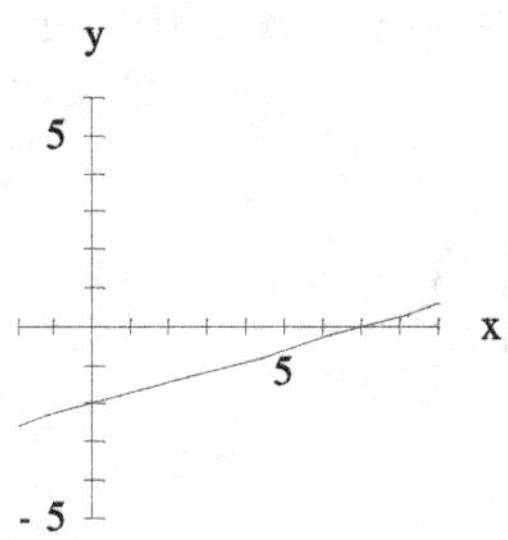 $(7, 0), (0, -2)$

4. 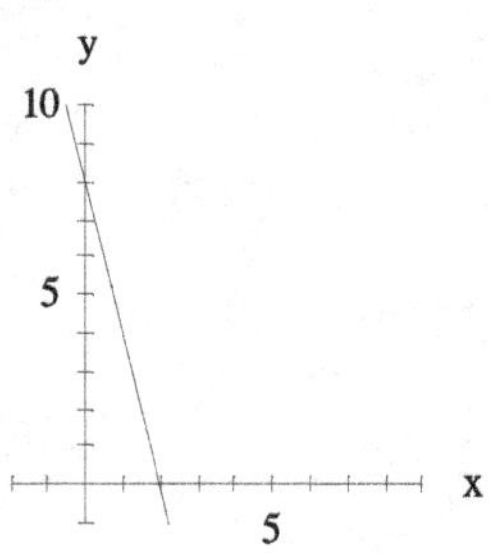 $(0, 8), (2, 0)$

5a. $(20, 0), (0, 30)$ b. 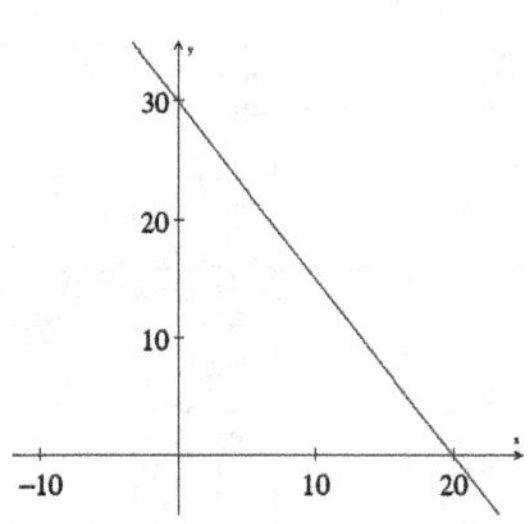

6. 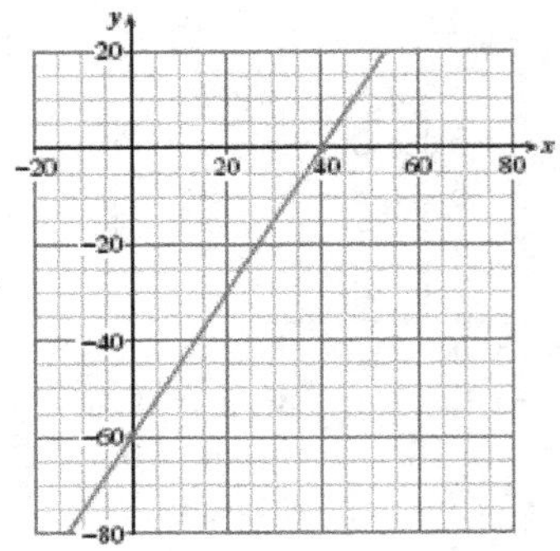 $(0, -60), (40, 0)$

7. d 8. a 9. c 10. b 11. a 12. c

13a. $x = -3$ b. $x = -\frac{k}{2}$

14a. $x = 3$ b. $x = 7 - k$

15a. $x = 14$ b. $x = k + 5$

16a. $x = \dfrac{5}{2}$ b. $x = \dfrac{8-k}{2}$

17a. $x = 3$ b. $x = \dfrac{k-15}{-4}$

18a. $x = \dfrac{-10}{3}$ b. $x = \dfrac{-10}{k}$

19a. 360 miles b. 24 hours c. 150 miles

d. 60 miles 20a. $(13\frac{1}{3}, 0), (0, 200)$

b. 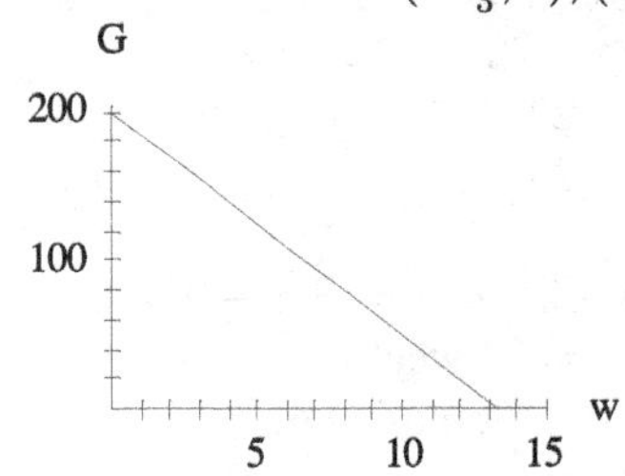

c. The w-intercept at $w = \frac{40}{3}$ shows that the fuel will run out after $\frac{40}{3} \approx 13.3$ weeks. The G-intercept at $G = 200$ shows that there were 200 gallons in the tank when they turned on the furnace.

21a. $(-9, 0), (0, 225)$

b. 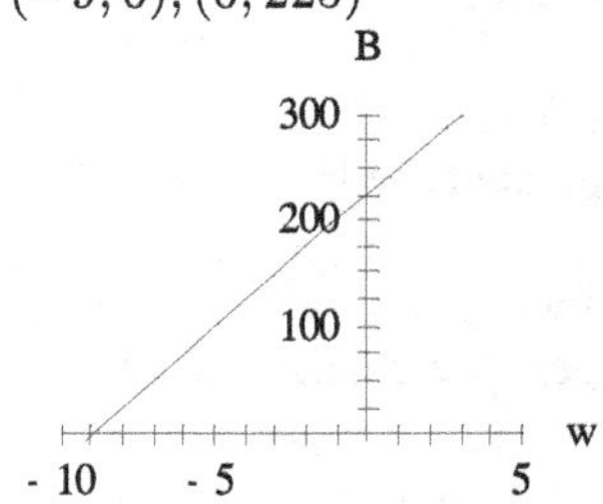

c. The B-intercept at $B = 225$ shows that she has \$225 this week. The w-intercept at $w = -9$ shows that she had no balance 9 weeks ago.

22a. $(15, 0), (0, -600)$

b. 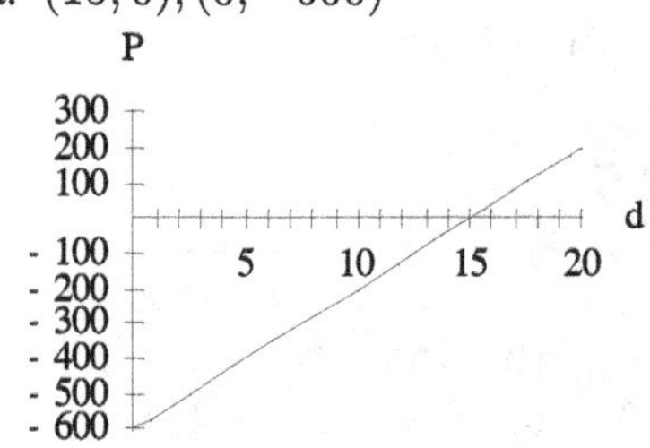

c. The horizontal intercept at $d = 15$ shows that Delbert must groom 15 dogs to break even. The vertical intercept at $P = -600$ shows that Delbert had spent \$600 before grooming the first dog.

23a. $(12.5, 0)$ b. $(0, -5)$

24. Increasing 25a. $A = 250 - 15w$

b. 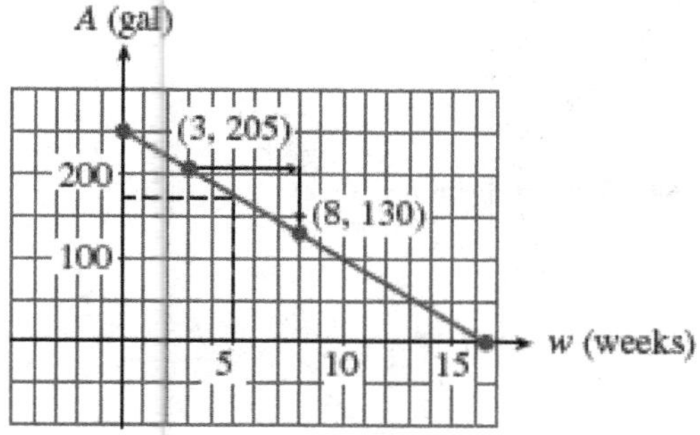

c. $(0, 250)$: They start with 250 gals.
$(16\frac{2}{3}, 0)$: They use all the oil in $16\frac{2}{3}$ weeks.

26. a. $P = -800 + 40t$

b. 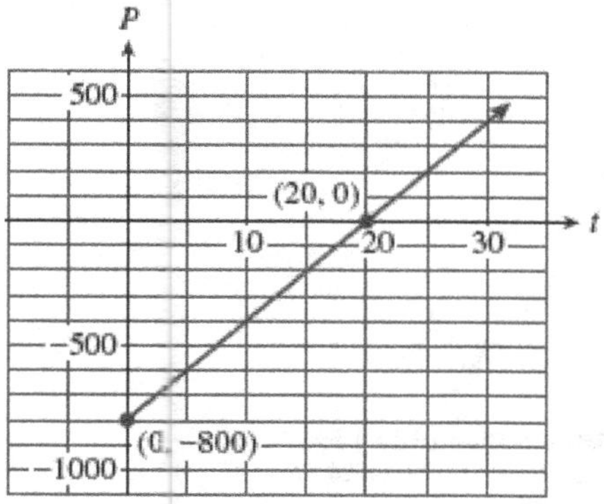

c. (0, −800): Phil and Ernie start out $800 in debt. (20, 0): They break even after 20 hours of copying.

27. a. $\$2.40x$, $\$3.20y$

b. $2.40x + 3.20y = 19,200$

c. 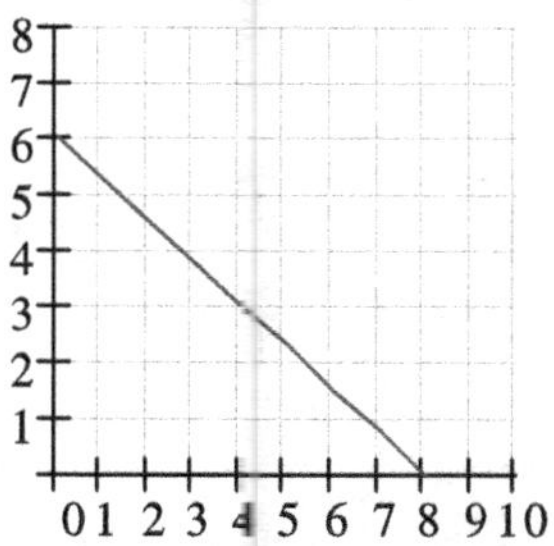

d. $(8000, 0)$ If he buys no premium, he can buy 8000 gals of regular. $(0, 6000)$ If he buys no regular, he can buy 6000 gals of premium.

28a. $9x$; $4y$ b. $9x + 4y = 1800$

c. 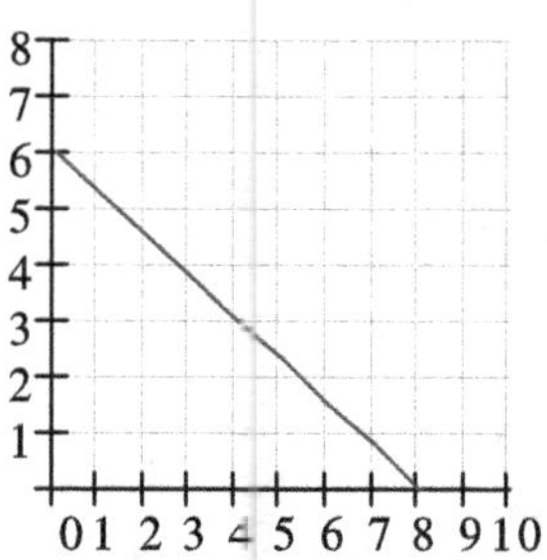

d. $(0, 450)$ If he eats no figs, he should eat 450 gm of banana. (200, 0) If he eats no bananas, he should eat 300 gm of figs.

Homework 3.2

1a. $y = 6$

b. 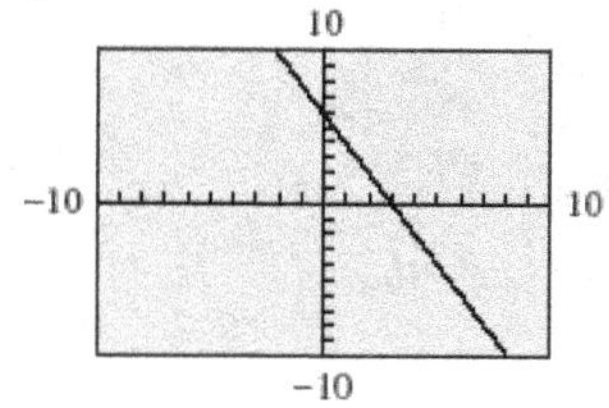

c. 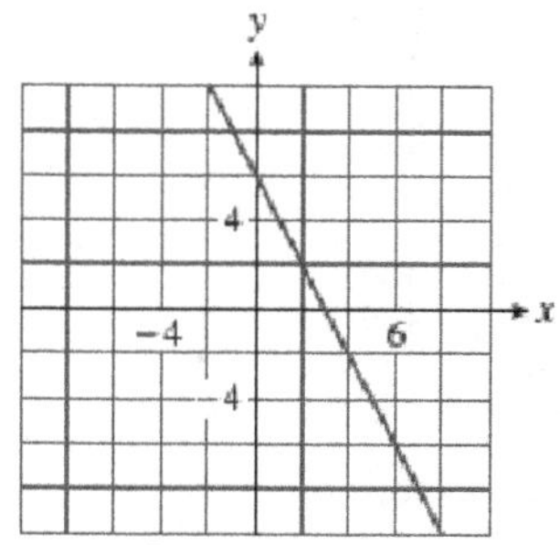

2a. $y = \frac{3}{4}x - 300$

b. 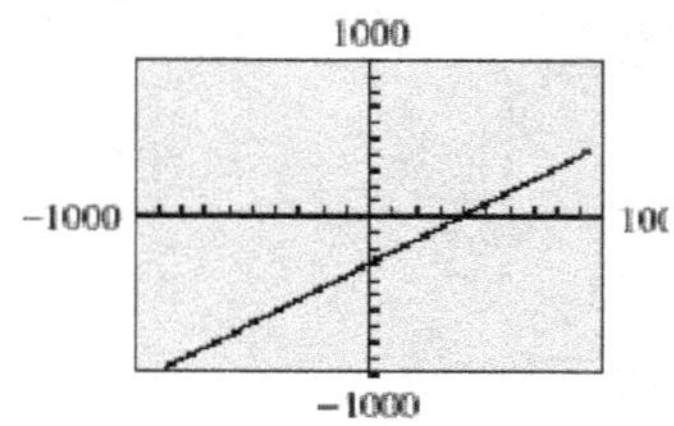

c. 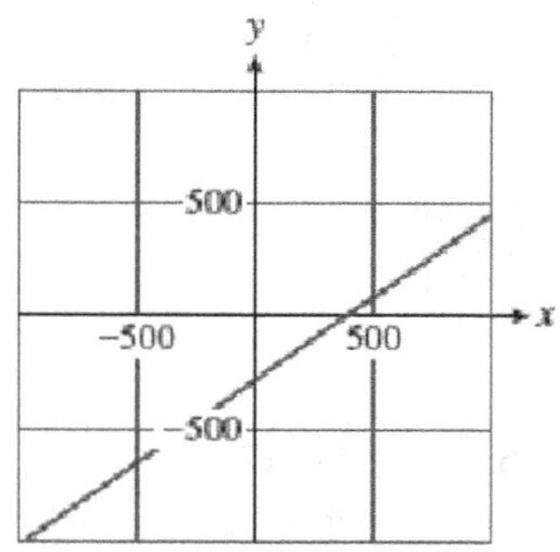

3a. $y = 0.02 - 0.04x$

b. 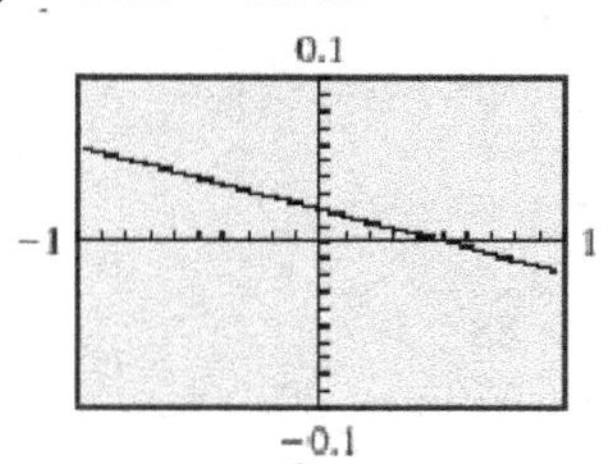

b. 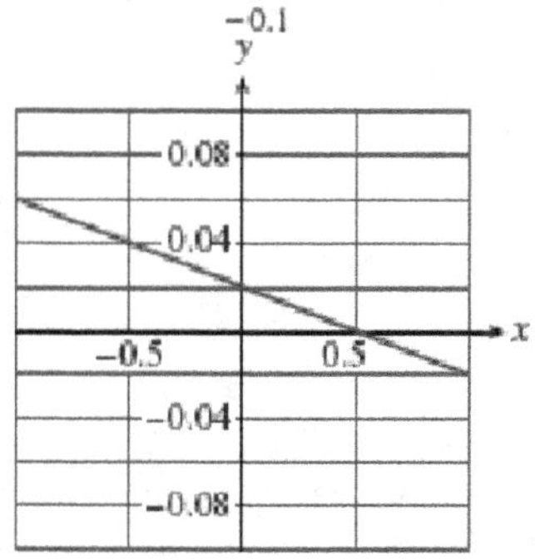

4a. $y = 210 - 35x$

b. 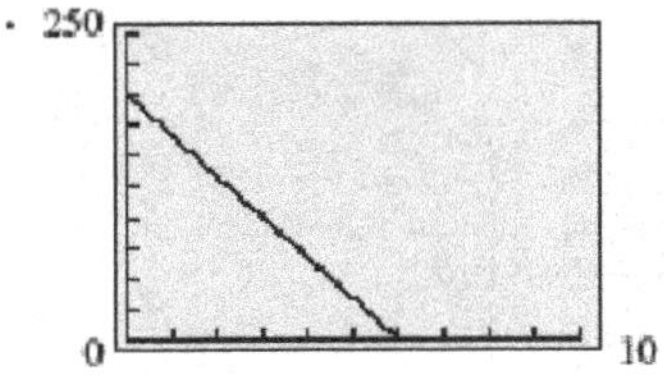

c. 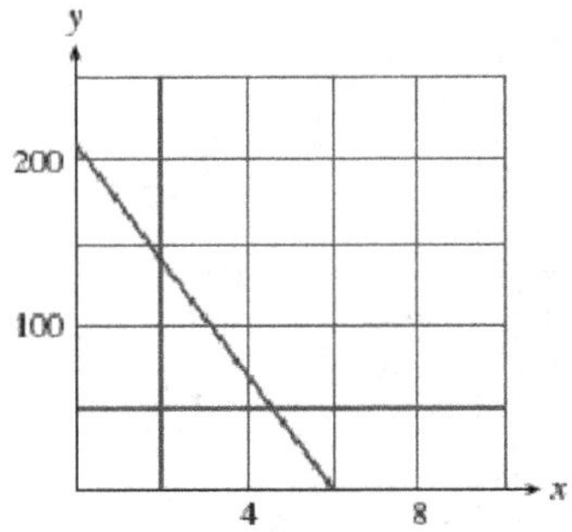

5a. $(100, 0)$, $(0, 100)$ b. $y = -x + 100$
c. Xmin = -20, Xmax = 120, Ymin = -20, Ymax = 120

d. 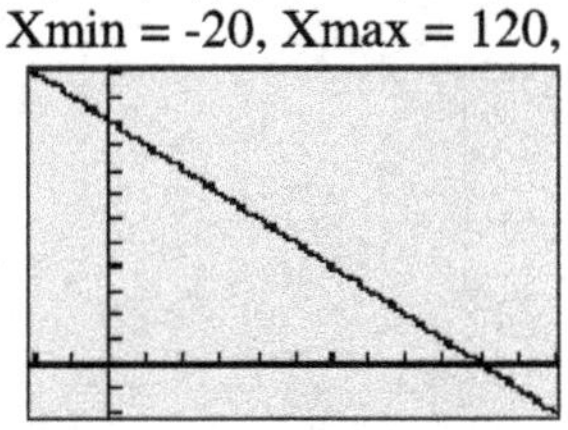

6a. $(0.04, 0)$, $(0, -0.028)$ b. $y = (25x - 1)/36$
c. Xmin = -0.1, Xmax = 0.1, Ymin = -0.1, Ymax = 0.1

d. 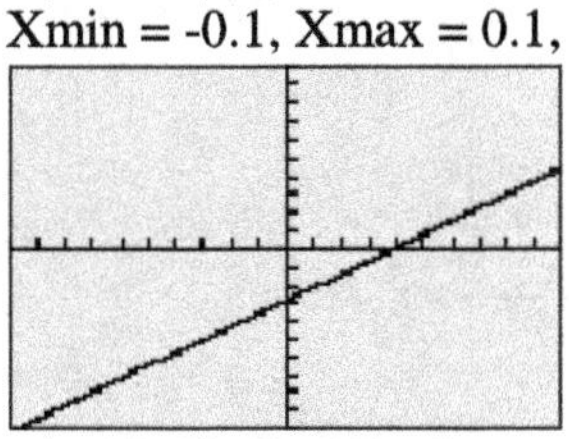

7a. $(-47, 0)$, $(0, 12)$ b. $y = 12(x/47 + 1)$
c. Xmin = -50, Xmax = 10, Ymin = -5, Ymax = 15

d. 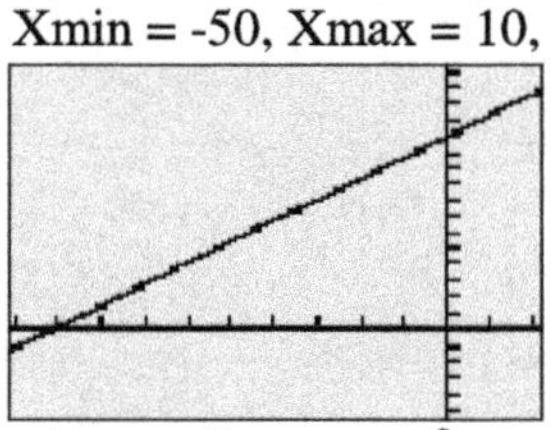

8a. $(-42, 0)$ b. $y = \frac{-2}{3}x - 28$
c. Xmin = -50, Xmax = 10, Ymin = -30, Ymax = 5

d. 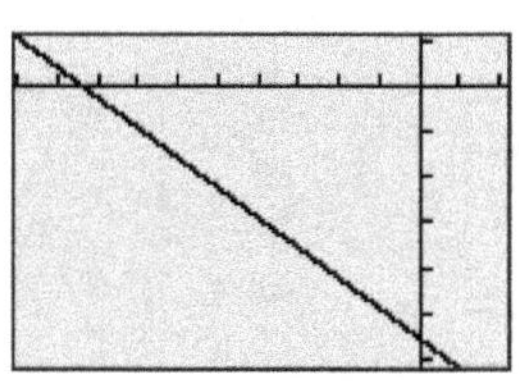

9a. i) $x = -3$ ii) $x < -3$ iii) $x > -3$
b. i) $x = -3$ ii) $x < -3$ iii) $x > -3$
c. The answers to parts (a) and (b) are te same because, for points on the graph of $y = -2x + 6$, the values of the y-coordinates are the same as $-2x + 6$.

10. 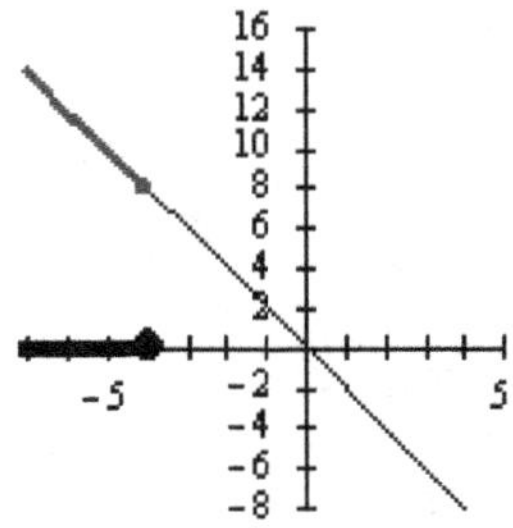 $x \le -4$

11. 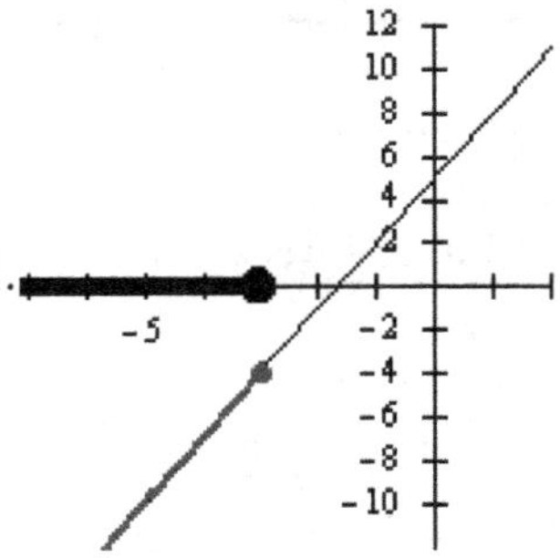 $x \le -3$

12. 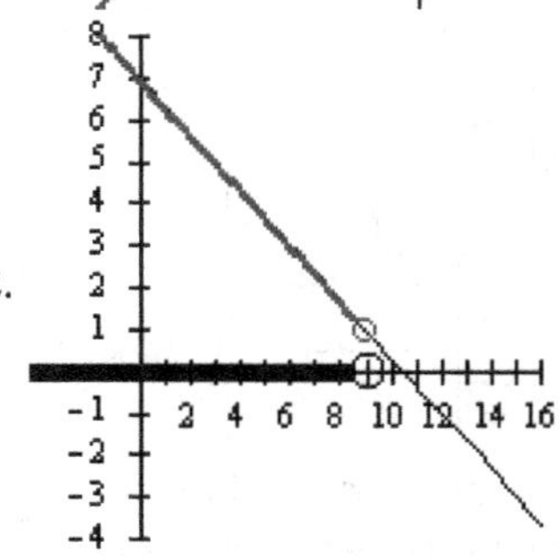 $x < 9$

13a. 0.6 b. -0.4 c. $x > 0.6$ d. $x < -0.4$
14a. 4 b. -5 c. $x > 1$ d. $x < 14$
15a. 11 b. -10 c. $x \ge -5$ d. $x \le 8$
16a. 4 b. $x < 22$
17a. 20 b. $x \le 7$

18a.

Time on treadmill (minutes)	0	1	2	5	10
Blood pressure (mm Hg)	120	126	132	150	180

b. $p = 120 + 6t$

c. 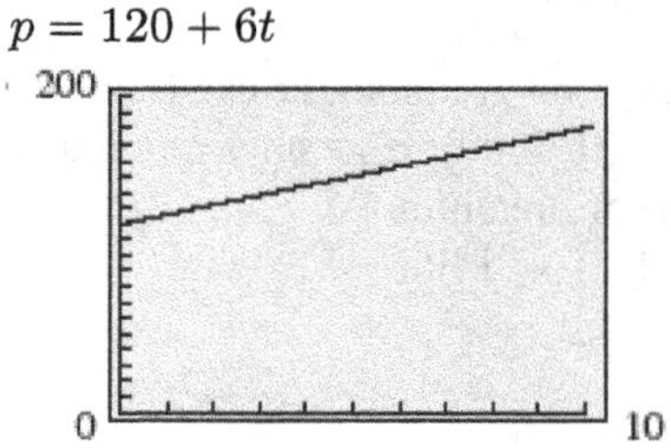

d. 141 e. At 7.5 minutes

Homework 3.3

1. $\frac{5}{7}$ 2. 8.72 3. $\frac{11}{6}$ 4. $\frac{30}{7}$
5. (a) and (d) are proportions.
6. $x = 6$ 7. $w = \frac{15}{2}$ 8. $b = \frac{63}{5}$ 9. $x = \frac{7}{2}$
10. yes 11. no 12. no 13. no
14. yes 15. yes 16. no 17. 40 lbs
18. 100 liters 19. 80 kilometers
20. 5753 21. about 13.78 in. 22. 2 cups

23a. $w = 7.5h$

b. 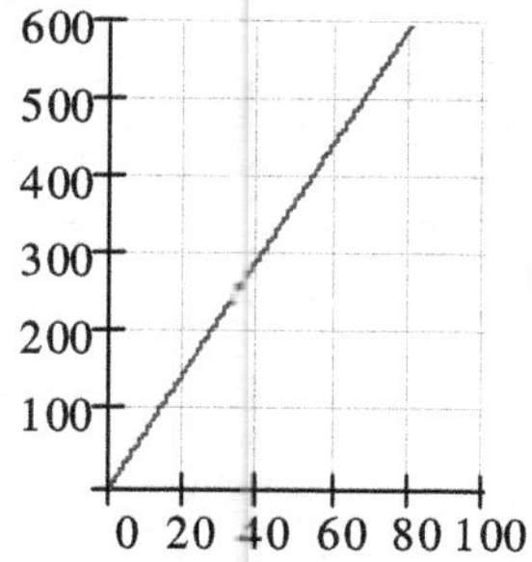 c. 67 hours

24a. $c = 0.03g$

b. 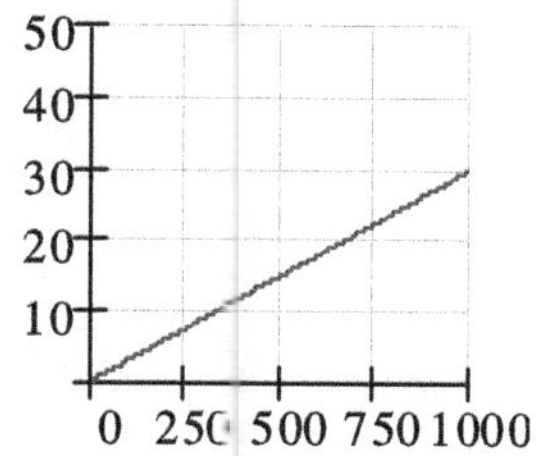 c. 30 grams

25a. $C = 0.04S$ b. $8000

26a. $d = 8t$

b. 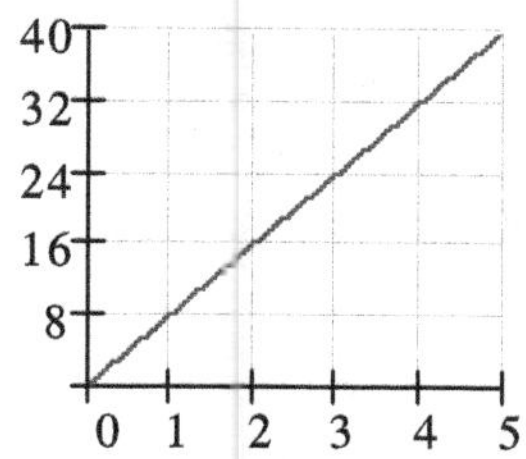

c. The distance doubles also.

27a. $T = 0.06P$

b. 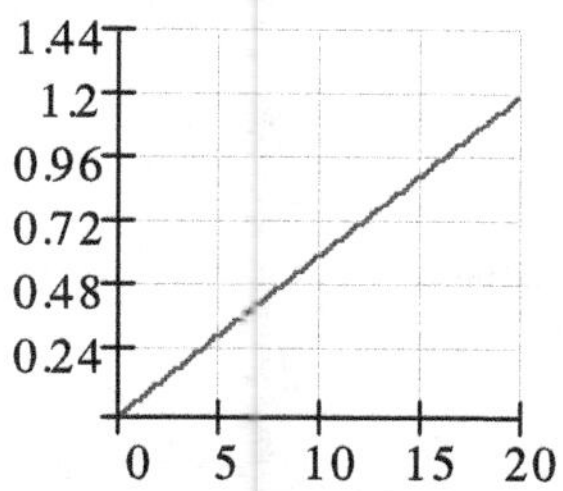

c. The tax is also cut in half.

28a.

Units	Tuition	Units	Tuition
3	620	6	740
5	700	10	900
8	820	16	1140

b. No c. No

29a. $C = 2\pi r$

b.

r	2	5	7	10
C	12.6	31.4	44.0	62.8

c. 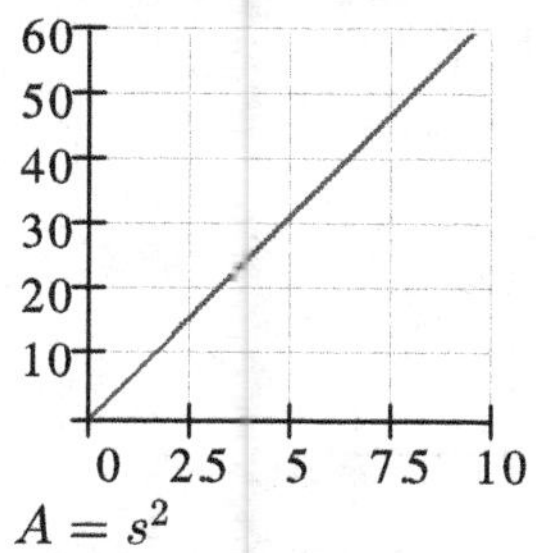 d. Yes

30a. $A = s^2$

b.

s	2	5	6	8
A	4	25	36	64

c. 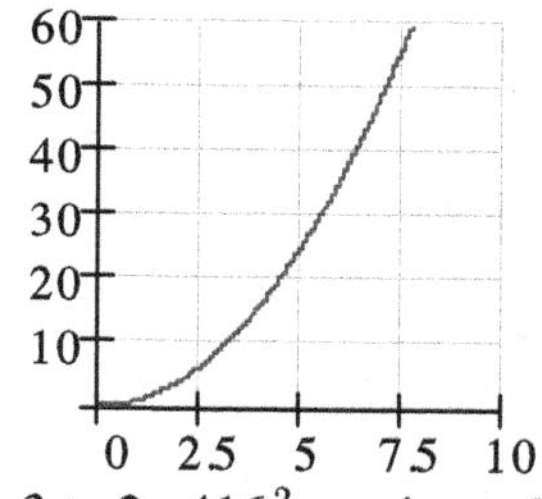 d. No

31a. 3 to 2; $416\frac{2}{3}$ servings; $7\frac{1}{2}$ cans

b. 2.4 oz; 15.5 lb c. 24.75 cans; 90 lb

Homework 3.4

1a. $\frac{3}{2}$ b. -2 c. $\frac{-2}{3}$ d. $\frac{1}{2}$

2a. $\frac{-1}{4}$ b. 3 3. 15

4. -12 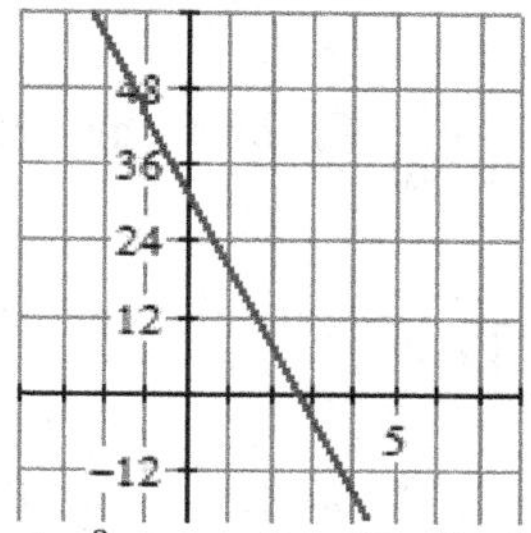

5. $\frac{-1}{2}$ 6. $\frac{3}{2}$ 7a. $(0, 4)$, $(6, 0)$

b. 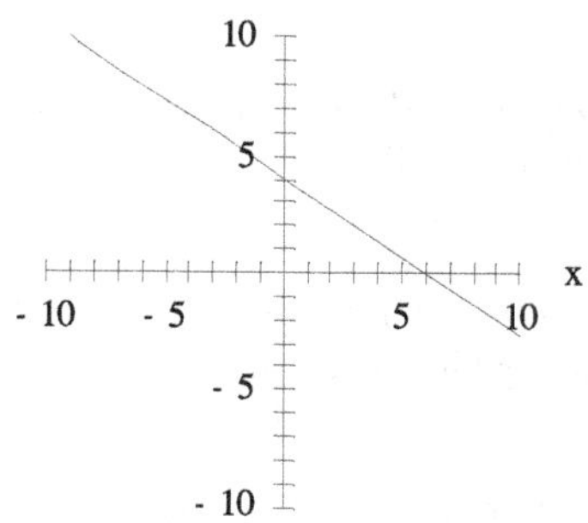 c. $\frac{-2}{3}$

8a. $(0, -5), (2, 0)$

b. 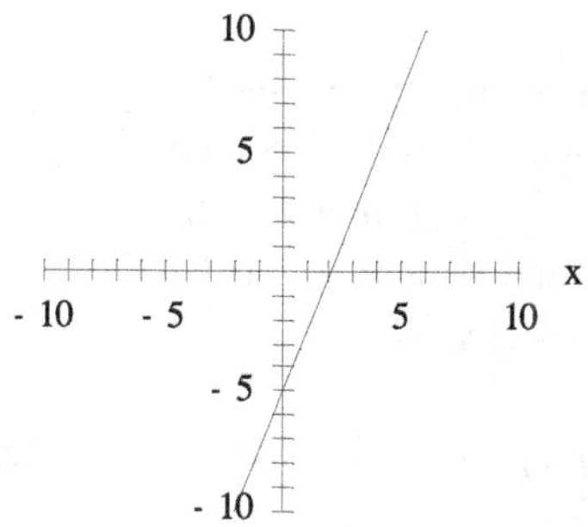 c. $\frac{5}{2}$

9a. $(0, 5), (5, 0)$

b. 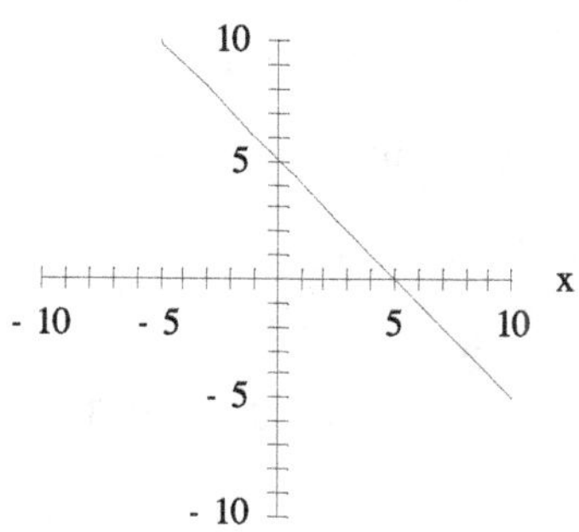 c. -1

10a. $(0,-2),(4,0)$

b. 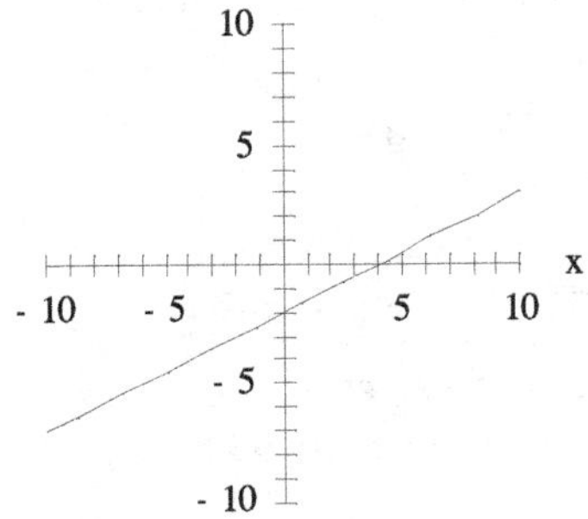

c. $\frac{1}{2}$

11. $\frac{2}{3}$ 12. $\frac{-5}{3}$

13. The line with slope $\frac{5}{3}$ is steeper.

14. The line with slope -2 is decreasing.

15. $\Delta y = 17.5$ 16. $\Delta x = 1.5$

17a. $d = 25g$ b. $\frac{25 \text{ miles}}{1 \text{ gallon}}$; the car gets 25 miles per gallon.

18a. $T = 4p$ b. $\frac{4 \text{ cents}}{1 \text{ dollar}}$; the sales tax is 4 cents per dollar purchased.

19a. \$50 b. $(0,200),(2,300)$, $\frac{\Delta A}{\Delta w} = 50$

c. dollars/week. She is saving \$50 per week.

20a. 4 lbs; 3 lbs b. $\frac{1}{2}$pound per week

21. 950.4 ft 22. 16 ft

23a. 0 b. All the y-coordinates on the line are the same, so the change in y is 0.

c. $\frac{\Delta y}{\Delta x} = \frac{0}{\Delta x} = 0$

24a. undefined b. All the x-coordinates on the line are the same, so the change in x is zero.

c. $\frac{\Delta y}{\Delta x} = \frac{\Delta \text{y}}{0} = 0$

25a. (1) b. (1): $m = 3$; (2): $m = 20$; (2)

26a. $(8, 2.4)$, $(16, 4.8)$ b. 0.3 dollars/oz

Homework 3.5

1. $y = 3x + 4$; slope 3 and y-intercept $(0,4)$.
2. $y = -2x + \frac{5}{3}$; slope -2 and y-intercept $\left(0, \frac{5}{3}\right)$.
3. $y = \frac{2}{3}x - 2$; slope $\frac{2}{3}$ and y-intercept $(0,-2)$.
4. $y = \frac{5}{4}x$; slope $\frac{5}{4}$ and y-intercept $(0,0)$
5. $y = 3x - 4$ 6. $y = \frac{2}{3}x - 3$
7. $y = \frac{-8}{5}x + 1$

8a. slope $-\frac{1}{3}$; y-intercept $(0,-1)$ b. $y = -1 - \frac{1}{3}x$

9a. 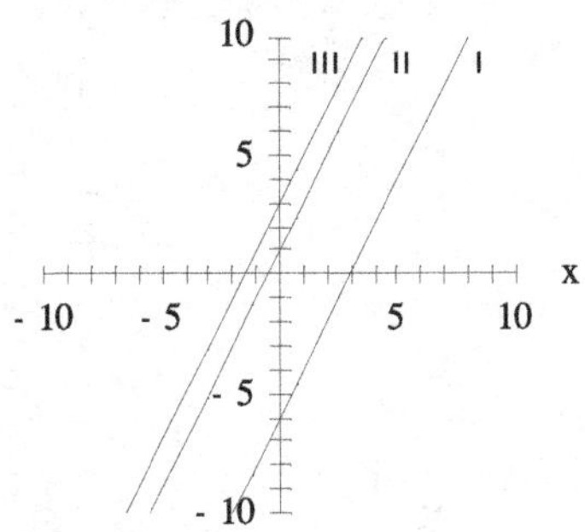

b. The slopes are all 2. c. $(0,-6),(0,1),(0,3)$

10a. 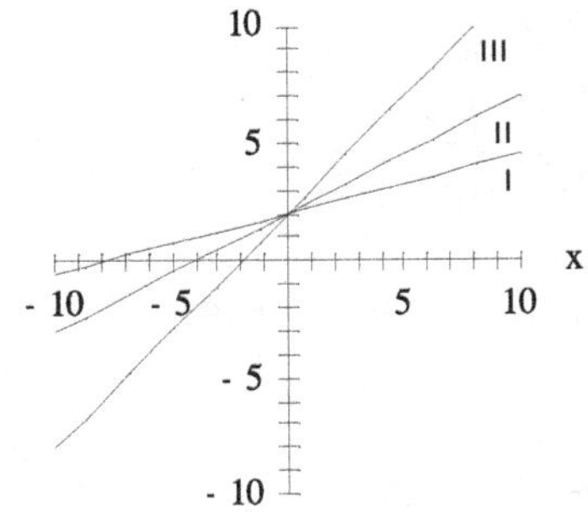

b. I. $\frac{1}{4}$ II. $\frac{1}{2}$ III. 1 c. $(0,2)$

11a. $(0,3)$, $(4,0)$

b. 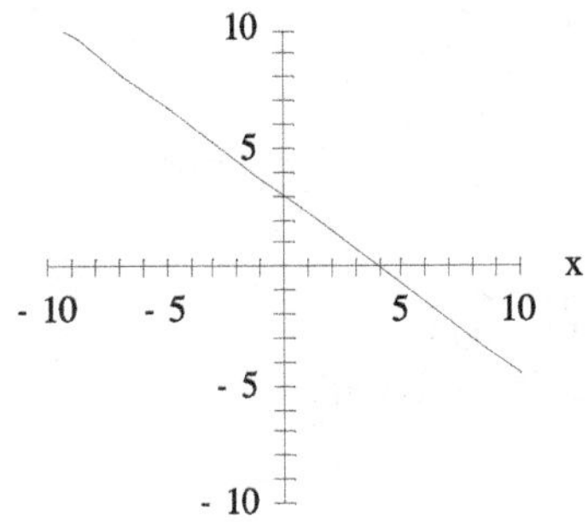

c. $\frac{-3}{4}$ d. $y = \frac{-3}{4}x + 3$

12a. $(0,8)$, $(\frac{8}{3},0)$

b. 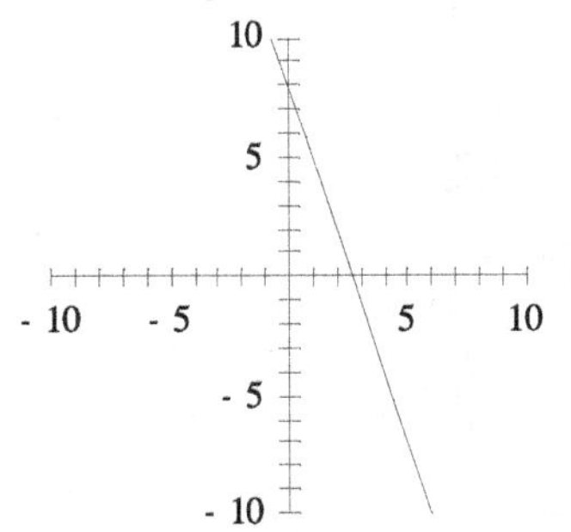

c. -3 d. $y = -3x + 8$

13a. $y = \frac{3}{5}x$ b. $(0,0)$, $\frac{3}{5}$

c. 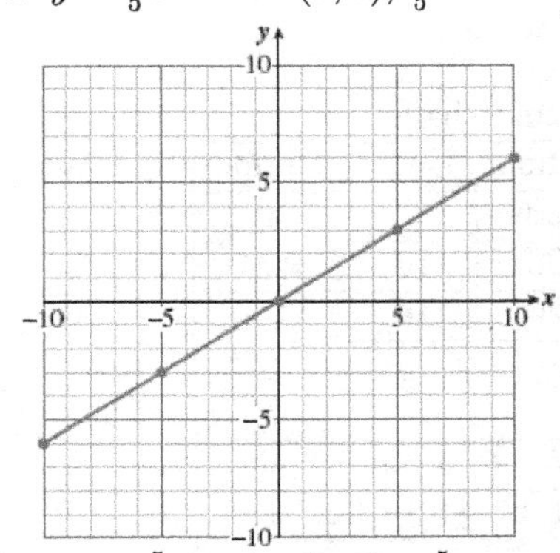

14a. $y = \frac{-5}{4}x$ b. $(0,0)$, $\frac{-5}{4}$

c. 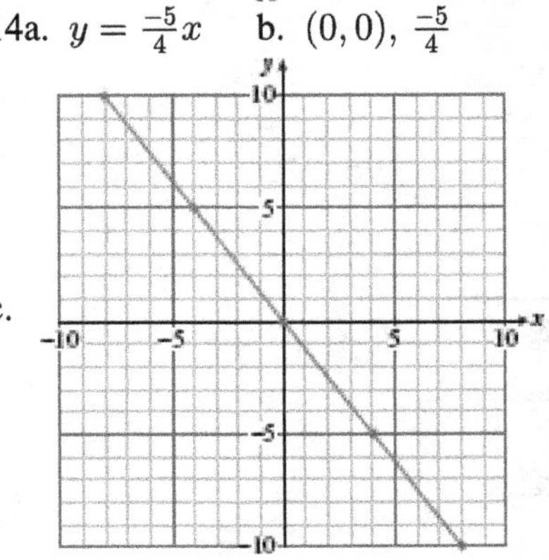

15. 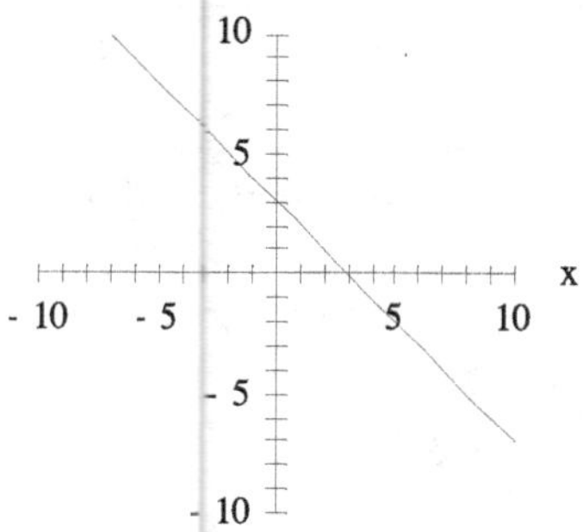

16. 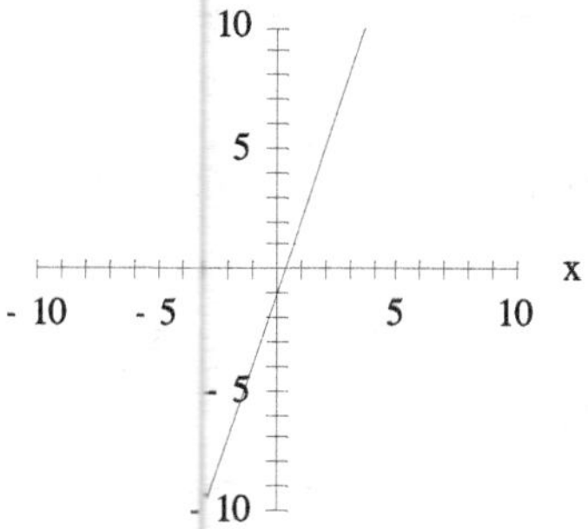

17. 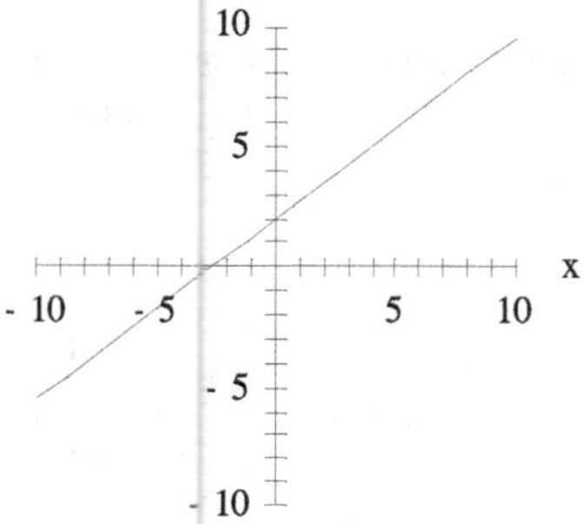

18. 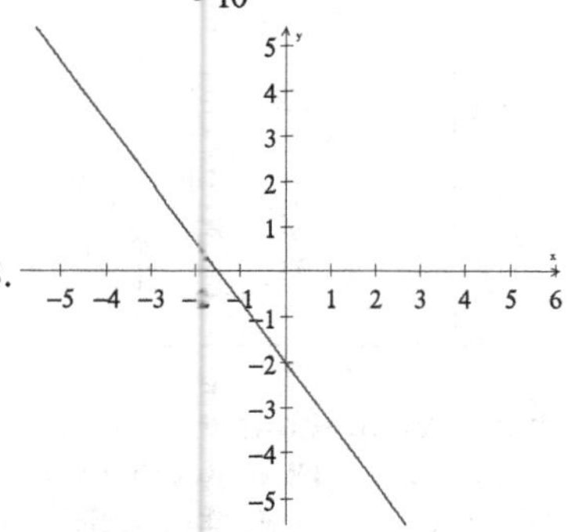

19. 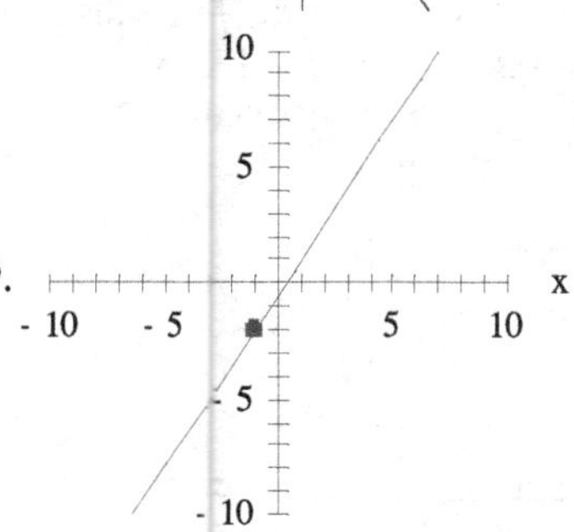

20. 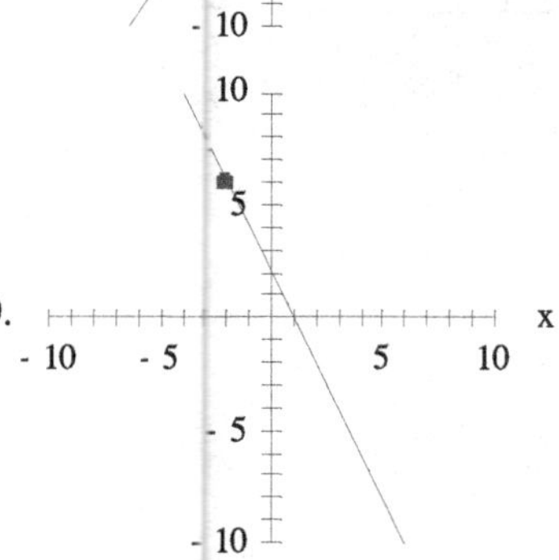

21a. 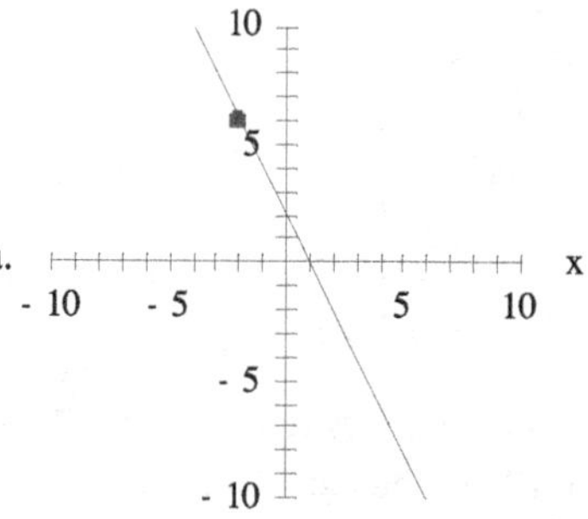

b. The slope means that for each incorrect answer, her score decreases by 4 points. The S-intercept means that if she has zero wrong answers, her score will be 120 points.

22a. 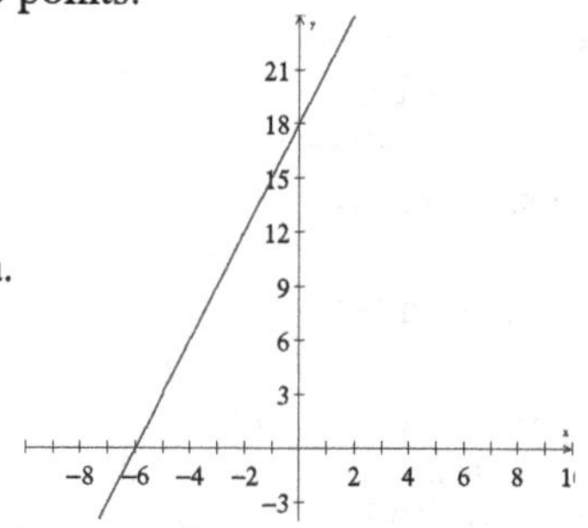

b. The slope gives the daily rate at which the beans are growing: 3 inches per day. The h-intercept, (0,18), means that today the beans are 18 inces tall.

23a. 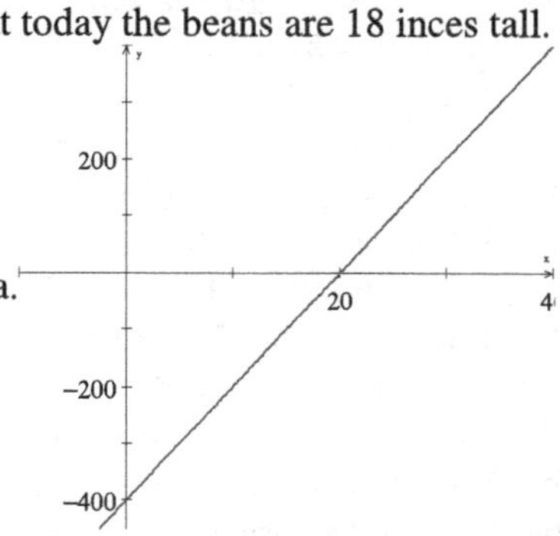

b. The slope, 20 points per question, means that his score increases by 20 points for each question he gets right. The S-intercept, $(0, -400)$, means that his score at the end of the first round is -400.

24a. $(0, -240)$ b. 3 c. $P = -240 + 3x$

d. He spent \$240 to open the shop and makes \$3 of profit per smoothie.

25a. $a = 100 + 150t$ b. $(0, 100)$, $m = 150$. The skier's starting altitude is 100 feet and she rises at 150 feet per minute.

26. a. $G = 25 + 12.5t$ b. $(0, 25)$, $m = 12.5$. There were 25 tons in the dump before the regulations, and the dump is filling at 12.5 tons per year.

27a. $M = 7000 - 400w$ b. $(0, 7000)$, $m = -400$. Tammy had \$7000 saved, and she spends \$400 per week.

28. a. II b. III c. I d. IV

29a. 50°F b. −20°C

c. 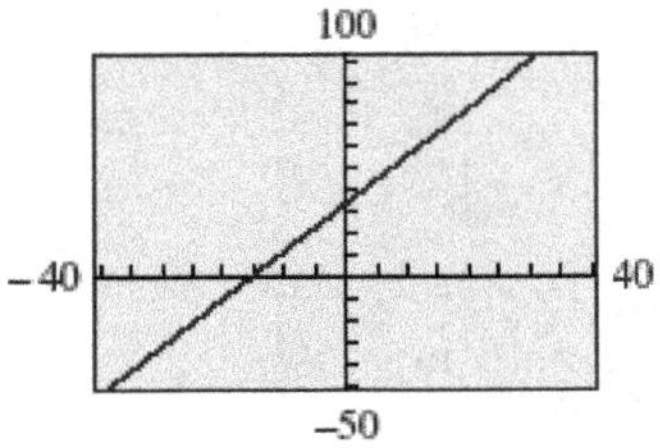

d. The Fahrenheit temperature increases $\frac{9}{5}$ of a degree for every degree increase in the Celsius temperature.
e. The C-intercept gives the Celsius temperature at 0°F, and the F-intercept gives the Fahrenheit temperature at 0°C.

Homework 3.6

1. $10y - 15$ 2. $-8x - 16$ 3. $-5b + 3$
4. $36 - 6t$ 5a. $32c$ b. $32 + 8c$
The distributive law is used in (b).
6a. $-16 - 2t$ b. $-16t$
The distributive law is used in (a).
7. d 8a. $2xy$; 54 b. $2x + 2y$; 24
c. $2 - xy$; -25 d. $-2xy$; -54
9. $-4x - 6$ 10. $x + 23$ 11. $-12z - 15$
12. $-24a + 4$ 13. $y = -2$ 14. $w = 6$
15. $c = 0$ 16. $a \le -4$ 17. $t < 2$
18. $x = 9$ 19. $a = 4$ 20. $b = 3.25$
21a. $x = 0$: $5(x + 3) = 5(0 + 3) = 15$;
$5x + 3 = 5(0) + 3 = 3$
b. $c = 0$ 22a. 48 b. $-12x + 12$ c. 48
23a. $2w - 8$ b. $6w - 16$ c. $2w^2 - 8w$
24a. $108 - c$ b. $216 - 2c$ c. $324 - 2c$
25a. $3s$ b. $3s + 45$ c. 45 d. $6s + 45$
26a. $15 - a$ b. $0.35a$ c. $2.25 - 0.15a$
d. $0.2a + 2.25$ 27a. $8 - x$ b. $0.6x$
c. $2 - 0.25x$ d. $0.35x + 2$ 28a. $10 - x$
b. $0.12x$ c. $5.5 - 0.55x$ d. $5.5 - 0.43x$
29. $5(2x + 3)$; $10x + 15$ 30. $8x(3y + 2)$; $24xy + 16x$
31a. $6w + 12$ b. 5 by 16 yards
32. \$13,714,385.71
33a. $260 - a$ b. $780 - a$ c. 120
34a. $47 - x$ b. $10x$ c. $282 - 6x$ d. $4x + 282$
35. $10x + 6(47 - x) = 330$; She sold 12 reserved tickets.

Chapter 3 Review

1a.

n	100	500	800	1200	1500
C	4000	12,000	18,000	26,000	32,000

b. $C = 20n + 2000$

c. 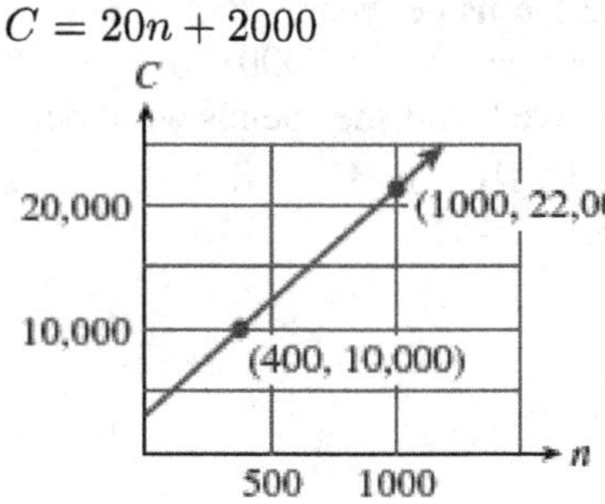

d. \$22,000 e. 400

3a.

t	5	10	15	20	25
R	1560	1460	1360	1260	1160

b. $R = 1660 - 20t$ c. (0, 1660), (83, 0)

c. 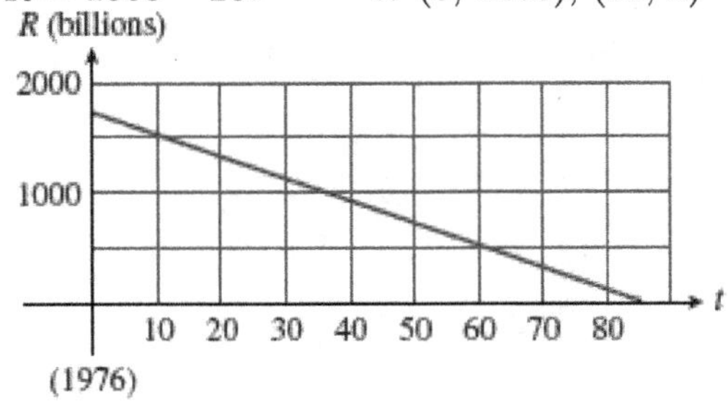

d. In 1976, R was 1660. The world's oil reserves will be gone in 2059.

5a. $5A + 2C = 1000$ b. 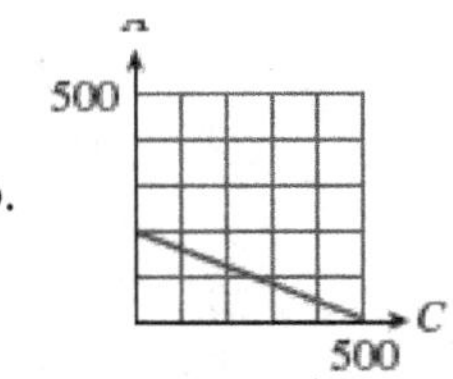

c. 200 children's tickets d. C-intercept (500, 0): only children's tickets are sold. A-intercept (0, 200): only adults' tickets are sold
7a. $S = 7800 - 600m$

b.

m	S
0	7800
13	0

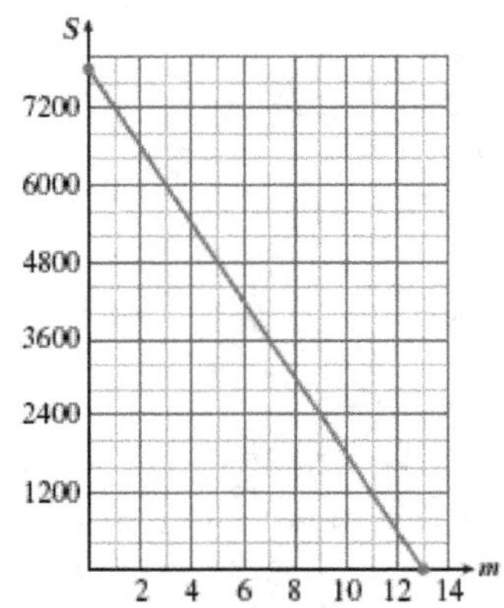

c. The S-intercept at $S = 7800$ shows that she started with \$7800. The m-intercept at $m = 13$ shows that she will use all the money in 13 months. d. 9 months
9a. -5 b. $x \le -2$ 11. 119 13. 50
15. 52 hr 17. $\frac{-3}{2}$ 19. Volleyball 21. Hwy 33
23. 80 ft 25a. $(0, -6)$, $(3, 0)$

b. 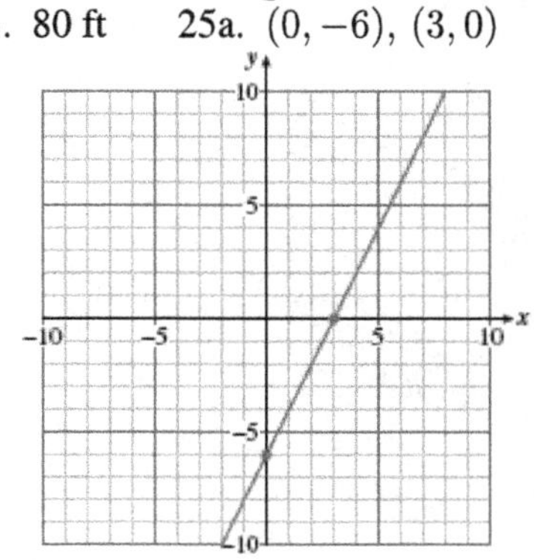

c. $m = 2$

27a. $(2, 0)$, $(0, 3)$

b. 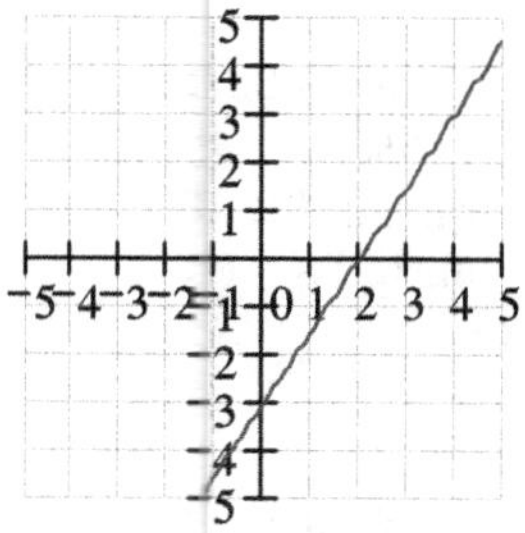

c. $m = \frac{3}{2}$

29. $\frac{3}{5}$

31. $m = \frac{1}{2},\ b = \frac{-5}{4}$ 33. $m = -4,\ b = 3$

35a. $m = -2,\ b = 3$ b. $y = -2x + 3$

37a. $m = \frac{-5}{2},\ b = -5$

b. 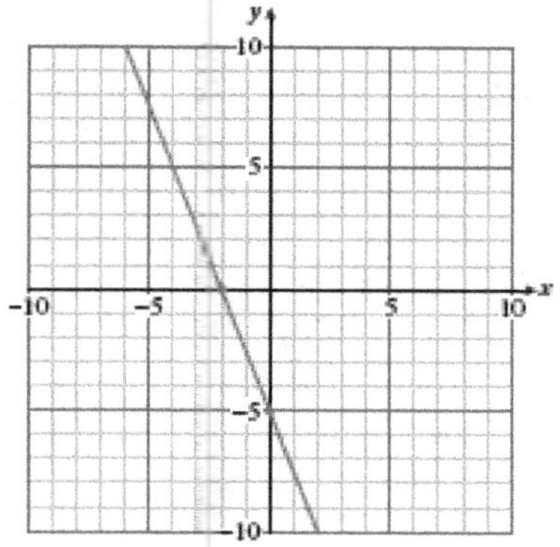

39a. $m = \frac{3}{4},\ b = 2$

b. 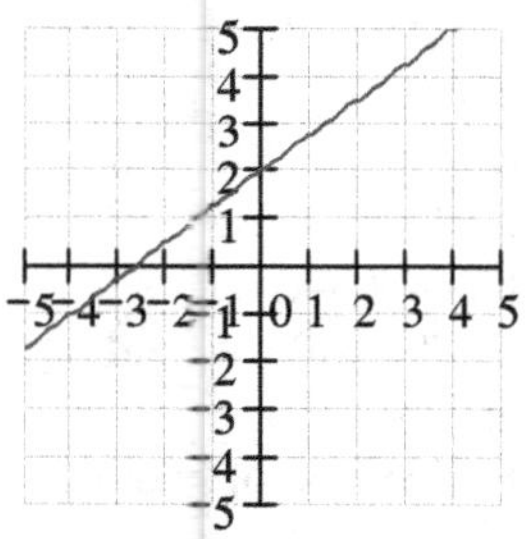

41a. $B = 800 - 5t$

b. 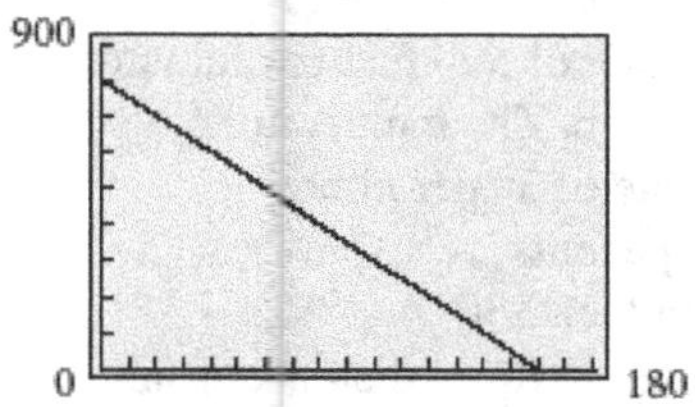

c. $m = -5$ barrels per minute is the rate at which oil is leaking

43. $m = 0.85$ is the price per topping, and $b = 1.85$ is the cost of the yogurt without toppings.

45a. $30m$ b. requires the distributive law: $-30 + 5m$

47. $2m + 7n$ 49. $-15w + 26$

51. $p = 25$ 53. $a < 1$

55a. $3x$ b. $8x$ c. width: 6 cm; length: 18 cm

57a. $30 - x$ b. $1200x$ c. $800(30 - x)$

d. $400x + 24,000$ e. 12 with speakers and 18 without speakers

59. 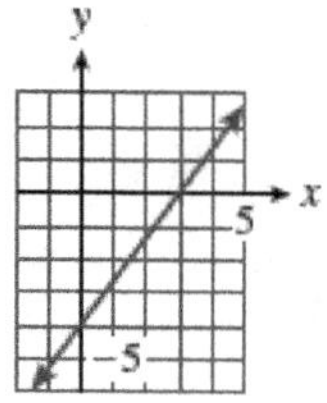

61. 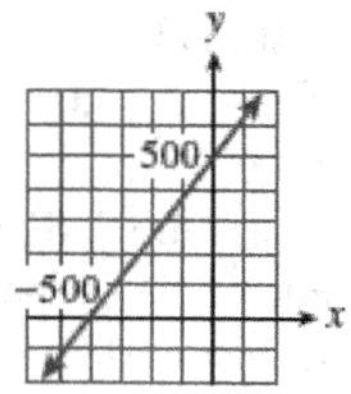

63. 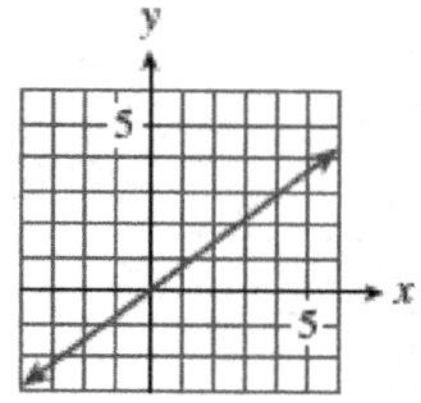

65. 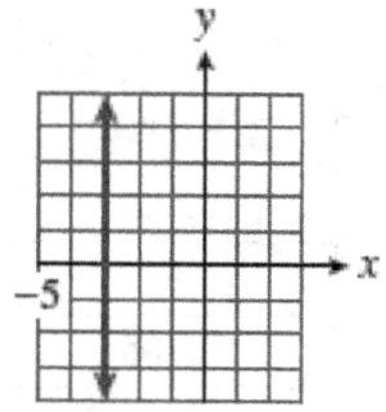

67. a. $T = 62 - 0.0036h$ b. $-46°$F, $108°$F

c. $-71°$F

Homework 4.1

1. No 2. Yes 3. (3,0) 4. $(50, 70)$
5. $(1, -3)$ 6. $(-4, -5)$
7. $(-2, 3)$ 8. $(2, 3)$
9. Inconsistent 10. Dependent
11a. Plan A: $y = 0.03x + 20,000$
Plan B: $y = 0.05x + 15,000$

b.

x	Earnings under Plan A	Earnings under Plan B
0	20,000	15,000
50	21,500	17,500
100	23,000	20,000
150	24,500	22,500
200	26,000	25,000
250	27,500	27,500
300	29,000	30,000
350	30,500	32,500
400	32,000	35,000

c.

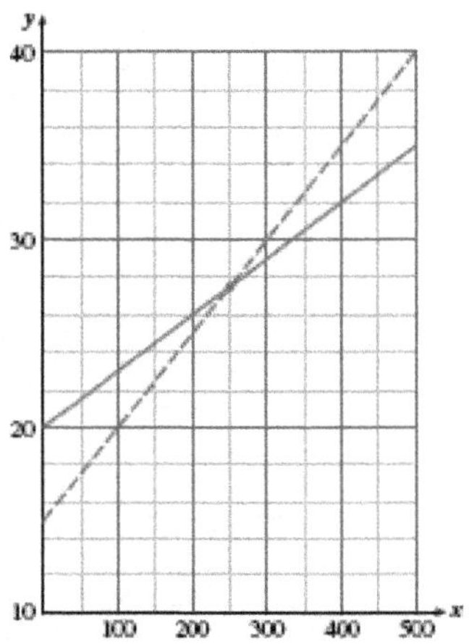

d. $250,000

12a. $C = 200 + 4x$ b. $R = 12x$ c. 25 pendants
13a. $y = 50x$ b. $y = 2100 - 20x$
c. 30 cents per bushel, 1500 bushels
14. rose: $1.50, carnation: $1.00
15. base angles: 65°, vertex angle: 50°
16. From the slope-intercept form, both lines have slope 3, but their y-intercepts are different, so the lines are parallel.

Homework 4.2

1. $(2, 4)$ 2. $(-1, 2)$
3. $(5, 2)$ 4. $(22, -28)$
5. $(2/3, -4/3)$ 6. $(1/3, -2)$
7. $(3, 2)$ 8. $(1, 1)$
9. $(1, 2)$ 10. $(-3, -2)$
11. $(1, -1)$ 12. $\left(\frac{3}{4}, \frac{-1}{4}\right)$
13. inconsistent 14. inconsistent
15. dependent 16. elimination
17. elimination
18. hamburger: 650 cal; shake: 380 cal
19. 12 tables and 48 chairs
20. 17 meters by 4 meters
21. bacon: $2.20; coffee: $5.60
22. a. $S = 35x$, $D = 1700 - 15x$
b. $34 per pair, 1190 pairs

Homework 4.3

1. bonds: $90,000; mutual fund: $60,000
2. $25,000 at 12%; $5000 at 15%
3. student loan: $5000; car loan: $10,000
4c. 26.4% 5c. 40%
6. 15 liters of 12% solution;
30 liters of 30% solution
7. 160 women
8. Delbert: 50 mph; Francine: 45 mph
9. 7 hours; 280 miles 10. 10 hours; 400 miles
11. 40 miles 12a. $g = 0.4t + 43.2$
b. 69.2, 75.2 c,d. 92

Homework 4.4

1. $-4/3$ 2. 0 3. -4
4. $5/3$ 5. -1 6. $1/9$
7. 0
8a. $(5, 0)$, $(0, 7)$ b. $m = -7/5$
9a. $(-2.4, 0)$, $(0, 1.6)$ b. $m = 2/3$
10a. $(1/3, 0)$, $(0, 7/2)$ b. $m = -21/2$
11b. $y = -2x - 2$ c. $(-1, 0)$
12b. $y = \frac{1}{2}x - 5$ c. $(10, 0)$
13b. $y = \frac{-2}{3}x - 2$ c. $(-3, 0)$
14b. $y = 5$ c. no x-intercept
15b. $y = -\frac{5}{2}x + 5$ c. 2
16b. $y = 3x - 4$ c. $\frac{4}{3}$
17. $m = \frac{3}{5}$, $(0, -7)$ 18. $m = 3$, $(-5, 2)$
19a. $m = \frac{-7}{4}$ b. $(1, -8)$
20a. III b. IV c. II d. I
21a. $(0, -3)$ b. $(-2, -7)$
22a. $(6, -8)$ b. $(-4, -3)$
23a. $1200/1 year b. The college fund grows at te rate of $1200 per year.
24a. $0.08/1 page b. The cost of making copies is 8 cents per page.
25a. −500 feet/1min b. The elevation decreases at an average rate of 500 feet per minute.
26a. −3°/1 hour b. The temperature dropped at an average rate of 3° per hour.
27a. cost per pound of peanuts
b. 1.50 dollars per pound c. $30
28a. −75 miles/1 hour b. Roy's distance from home decreases by 75 miles each hour.
c. 800; Roy boarded the train 800 miles away from home.
d. $10\frac{2}{3}$; Roy will arrive home in about ten hours and 40 minutes.

Homework 4.5

1. parallel 2. neither 3. neither

4.a.

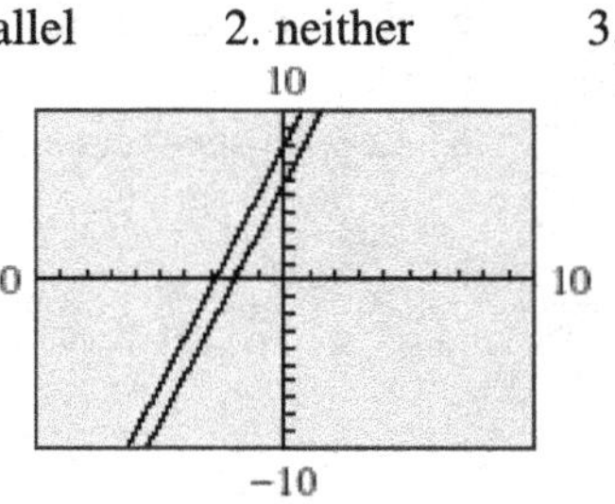

b. $m = 3$, $m = 3.1$. No.

c. $y = 68$ for both lines. The lines intersect at $(20, 698)$.

5. Parallel: a, g, h; parpendicular: c, f

6a. 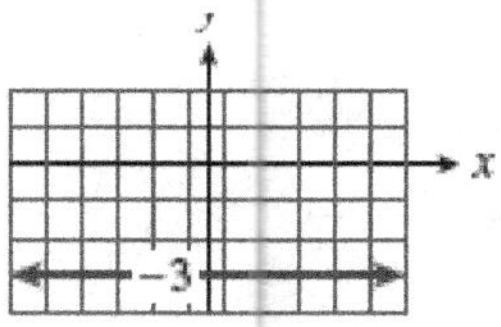

b. $m = 0$

7a. 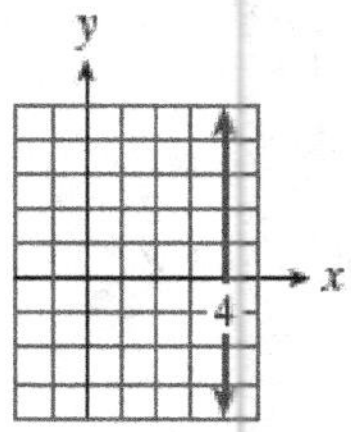

b. m is undefined

8a. 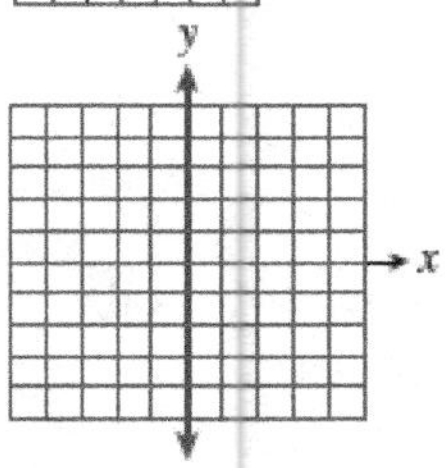

b. m is undefined.

9. $y = -5$ 10. $x = 2$
11. $x = -8$ 12. $y = 0$
13. $y = -12$ 14. $y = x + 6$
15. $y = \frac{5}{3}x$ 16. $y = \frac{-1}{8}x + \frac{19}{4}$
17. $y = \frac{-5}{7}x + \frac{11}{7}$ 18. $y = \frac{1}{4}x - \frac{1}{4}$
19. $y = \frac{-5}{3}x - \frac{1}{3}$ 20. $y = \frac{10}{7}x - \frac{8}{7}$
21a. $y = \frac{3}{2}x + \frac{5}{2}$ b. $\frac{3}{2}$ d. $y = \frac{3}{2}x + \frac{13}{2}$
22a. $y = \frac{1}{2}x - \frac{5}{2}$ b. -2 d. $y = -2x + 5$
23a. 2 b. 2 c. $y = 2x + 5$
24a. -3 b. $\frac{1}{3}$ c. $y = \frac{1}{3}x + \frac{8}{3}$
25a. $y = -0.0025x + 85$
b. $m = -0.0025$ degrees/ft c. 60° d. 85°
26a. $y = 200x + 25,000$
b. $m = 200$ dollars/dryer c. 375 d. \$25,000
27. $C = 65h + 125$ 28. $p = 2.2k$

Homework 4.6

1. 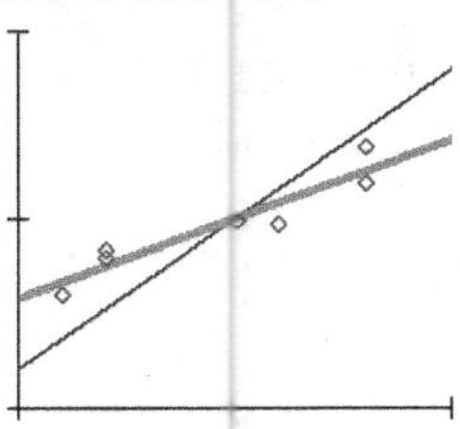

3a. 12 seconds b. 39

c. 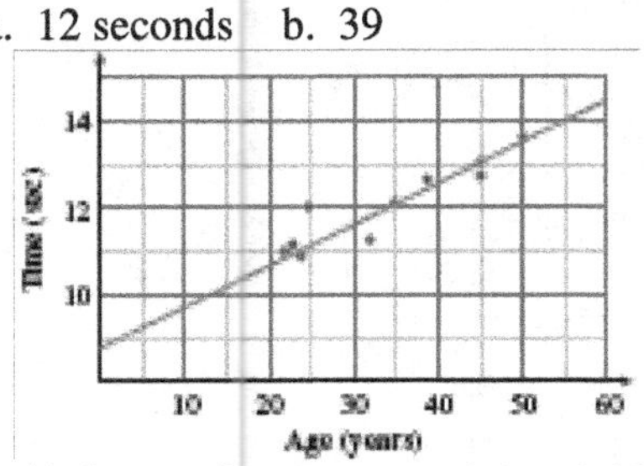

d. 11.6 seconds e. $y = 9.1 + 0.09x$
f. 12.7 seconds; 10.18 seconds; The prediction for the 40-year-old is reasonable, but not the prediction for the 12-year-old.

4a. 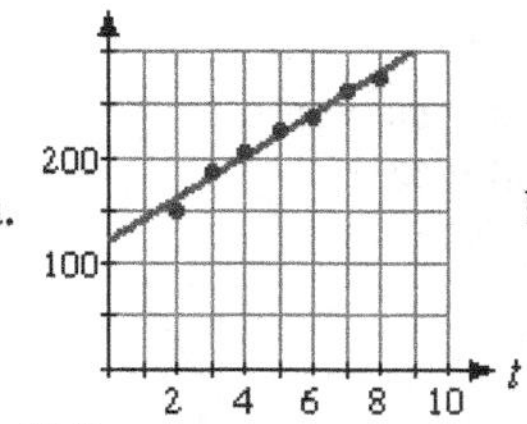

b. $y = 121 + 19.86t$
c. 419

5a. 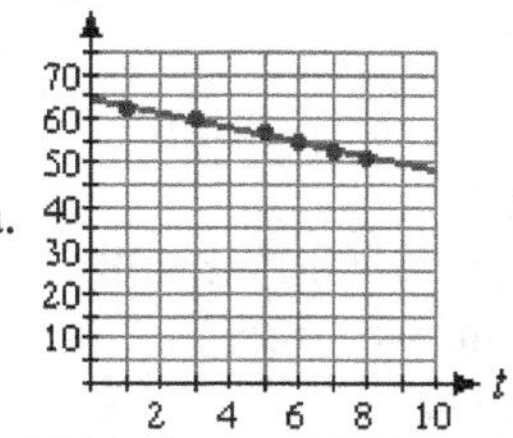

b. $y = 64.2 - 1.63t$
c. 58 births per 1000 women
d. 32 births per 1000 women

6a. 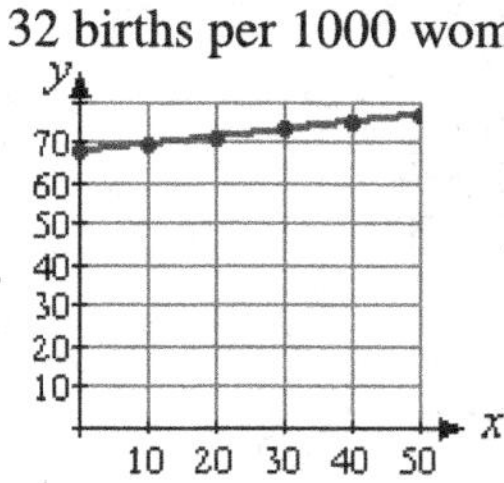

b. $y = 0.18t + 67.9$
c. 74.7 years d. 78.9 years

7a. 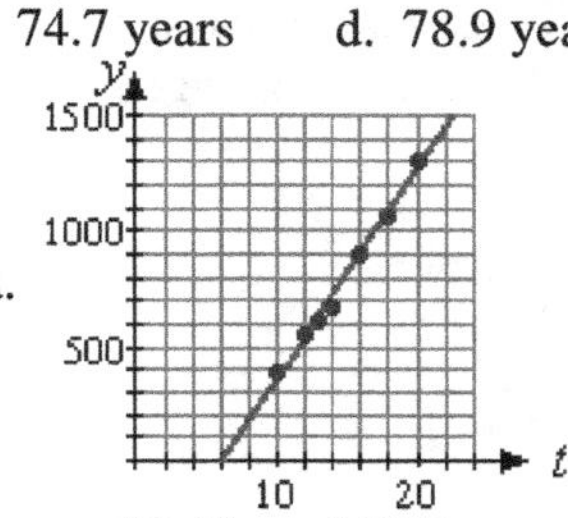

b. $y = 90.49t - 543.7$
c. 90.49 dollars/year: Each additional year of education corresponds to an additional \$90.49 in weekly earnings.
d. No: The degree or diploma attained is more significant than the number of years. So, for example, interpolation for the years of education between a bachelor's and master's degree may be inaccurate because earnings with just the bachelor's degree will not change until the master's degree is attained. And the years after the professional degree will not add significantly to earnings, so extrapolation is inappropriate.

8a. 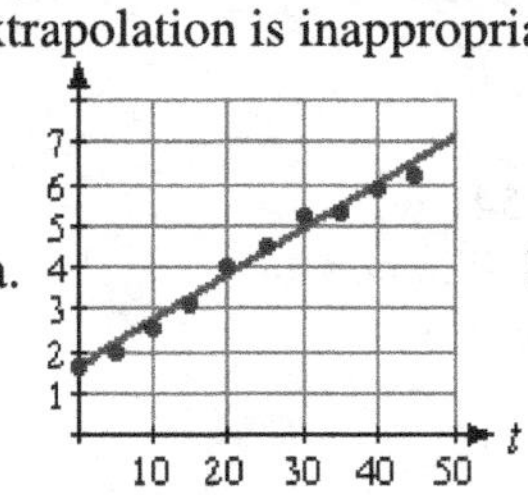

b. $y = 1.6 + 0.11t$
c. 6.2 billion tons

9a. 75°
b. The slope of -2 degrees/hour says that temperatures are dropping at a rate of 2° per hour.
10a. 20 mph b. The slope of 60 mph/second says the car accelerates at a rate of 60 mph per second.
11a. 2 min: 21°C; 2 hr: 729°C; The estimate at 2 minutes is reasonable; the estimate at 2 hours is not reasonable. 12. 128 lb

13a. 0.34 meters per year b. $y = 0.34x$
($b = 0$ because the plant has zero size until it begins.) c. Over 1300 years

14a, b.
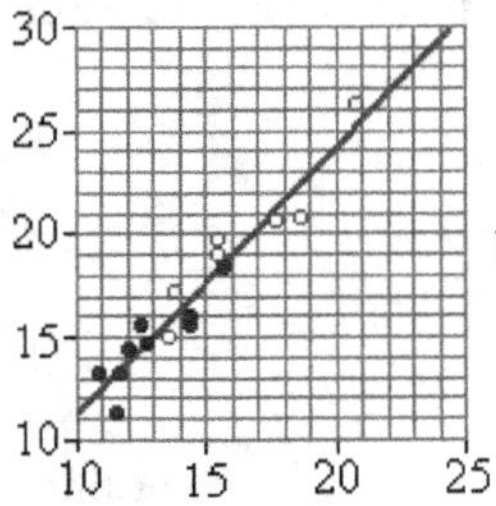

b. $y = 1.29x - 1.62$

c. The slope, 1.29 kg/sq cm, tells us that strength increases by 1.29 kg when the muscle cross-sectional area increases by 1 sq cm.

15a. E b.
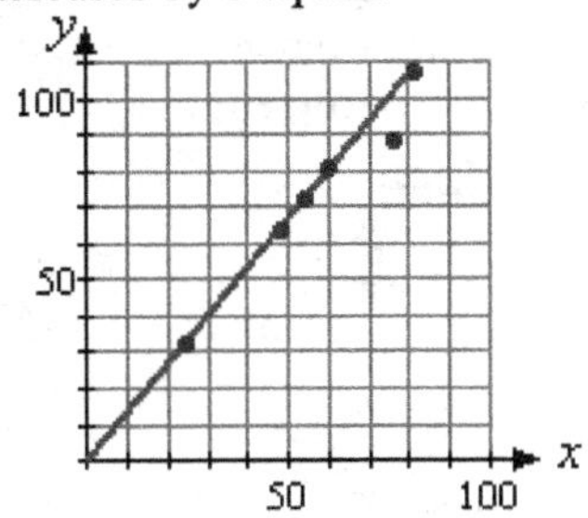

c. $y = 1.33x$; There should be no loss in mass when no gas evaporates. d. 1333 mg e. Oxygen

16a.
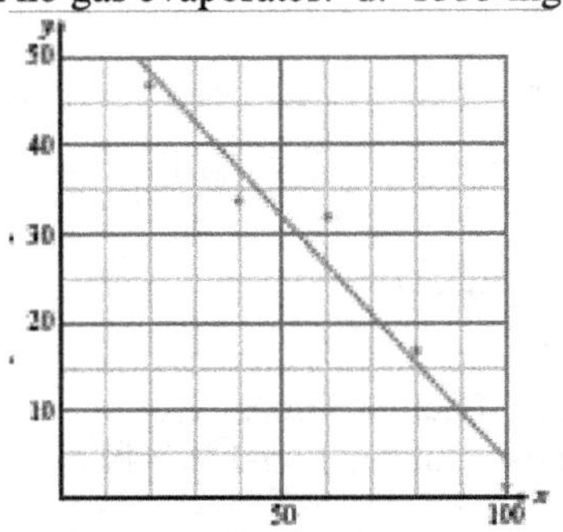

b. 31.7% c. 90 meters
d. The regression line gives a negative probability, which is not reasonable.

17a.
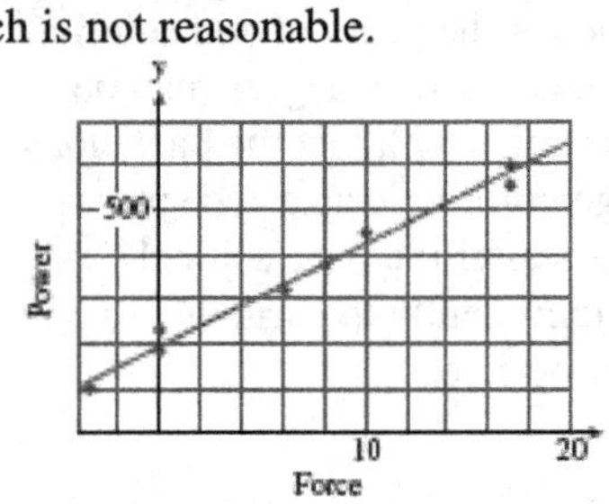

$y \approx 23.048x + 195.3$ b. 541 watts
c. 198.5 watts d. -8.5 newtons
e. 3.4 watts f. 1.7%

Chapter 4 Review

1. No 3. $(3, 2)$ 5. $(-1, 2)$
7. $(-3, -5)$ 9. $(1, -3)$ 11. $(\frac{1}{2}, \frac{7}{2})$
13. $(12, 0)$ 15. Consistent 17. Dependent
19. 12 lb of cereal and 18 lbof dried fruit
21. \$500 in the first account and \$700 in the second
23. 2 lb of the 60% alloy and 6 lb of the 20% alloy
25. Alida: 62 mph, Steve: 31 mph
27. 26 29. \$3181.82 at 8%, \$1818.18 at 13.5% 31. $\frac{-3}{2}$ 33. -0.4

35. Neither 37. $d = 1$, $V = 4.2$

39a.
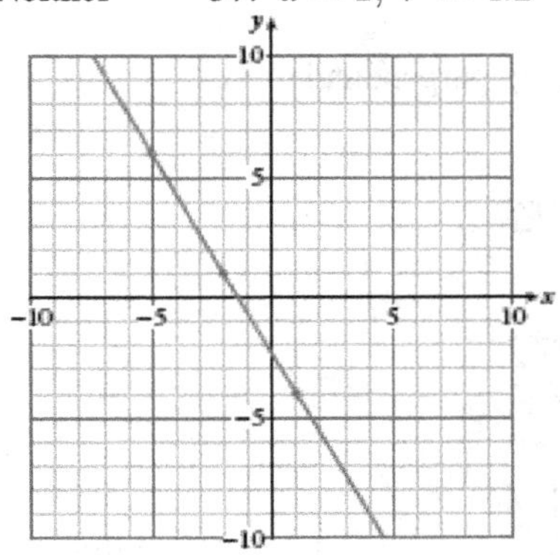

b. $y = \frac{-5}{3}x - \frac{7}{3}$ 41a. $F = 500 + 0.10C$

b.
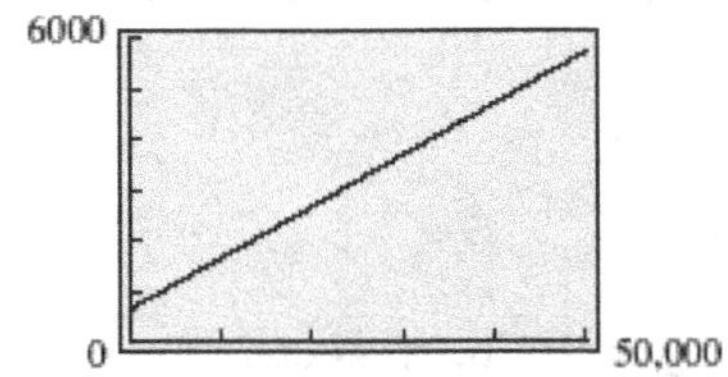

c. $m = 0.10$. The decorator charges 10% of the cost of the job (plus a \$500 flat fee.)
43. $y = \frac{-9}{5}x + \frac{2}{5}$

45.

t	0	15
P	4800	6780

b. $P = 4800 + 132t$

c. $m = 132$ people per year gives the rate of population growth
47. parallel 49a. neither
b. perpendicular c. parallel
51. $y = \frac{-2}{3}x + \frac{14}{3}$ 53. $y = \frac{1}{5}x - 1$
55. $x = -4$ 57. $y = -x - 2$
59. $y = \frac{3}{2}x + 7$ 61. 6

63a.
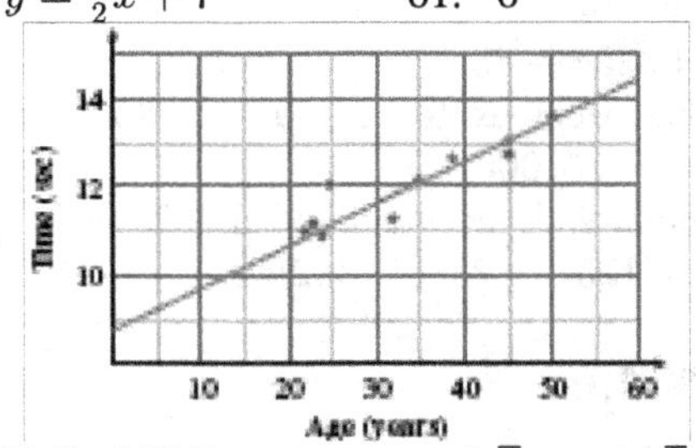

b. 129 lb, 145 lb c. $y = 2.\overline{6}\,x - 44.\overline{3}$
d. 137 lb e. $y = 2.84x - 55.74$
f. 137.33 lb

65a.
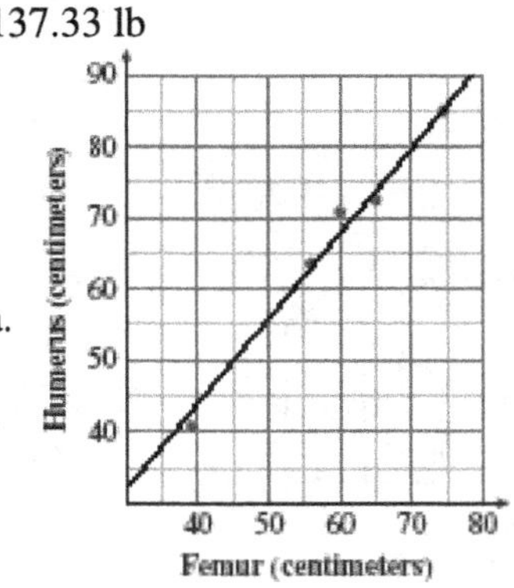

b. 45 cm c. 87 cm d. $y = 1.2x - 3$
e. 69 cm f. $y = 1.197x - 3.660$; 68.16 cm

Homework 5.1

1a. 64 b. 125 c. 625

2a. $\frac{16}{81}$ b. $\frac{64}{125}$ c. $\frac{121}{81}$

3a. 29.79 b. 45.70 c. 0.41

4a. -25 b. -125 c. 25 d. -125

5a. -4 b. 8 c. -12 d. -36

6a. -40 b. 20 c. 1 d. 13

7a. 192 b. 61 c. 361 d. -84

8a. $3x$ b. x^3 9a. $2a^2$ b. $3a$

10a. $-3q$ b. $-q^3$ 11a. $-4m$ b. $-3m^2$

12. $x + x = 2x$ 13. $x \cdot x = x^2$

14. $x^2 + x^2 = 2x^2$ 15. $x \cdot x^2 = x^3$

16. $-2a^2$ 17. $3t - 2t^2$ cannot be simplified.

18. $-2m^2$ 19. $3k^2$ 20. cannot be simplified

21. $7k^2$ 22. $12b^3$ 23. $2y^3 + 2y^2 + y + 1$

24. $3x^3 + 3x^2 + 11$

25. The exponents can't be combined; $14w^3$

26. The terms are not like terms, so they can't be combined; $6 + 3x^2$

27. $7 - (-5) = 12$; $t^2 + 12$

28. The terms are not like terms, so they can't be combined; $5b^2 - 3b$

29a. 350 b. 3500 c. 35,000

30a. 7.4 b. 74 c. 740

31a. 2400 b. 891,000 c. 30

32a. 39 b. 234 c. 3456

33. 49 34. $5 + x^3$ 35. $0.25h^3$ 36. $a^2 + b^2$

37. Area; $49\pi \approx 153.94$ square inches

38. Circumference; 23 inches

39. $\approx 105,904.14$ cubic feet

40. $144\pi \approx 452.40$ square feet

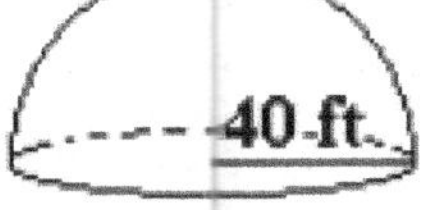

41a. b. $\frac{128,000\pi}{3}$ cubic feet

c. 3200π sq. ft.

42a. 210π sq. cm. b. 90 feet

c.

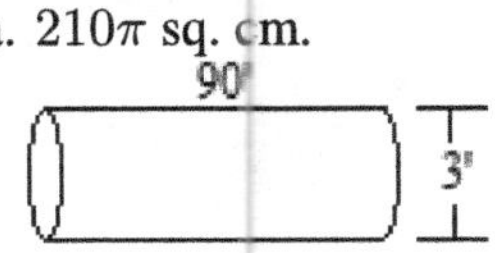

43. $w = \dfrac{v}{lh}$ 44. $m = \dfrac{2E}{v^2}$

45. $h = \dfrac{2A}{b+c}$ 46. $C = \dfrac{5}{9}(F - 32)$

47. $h = \dfrac{A - 2\pi r^2}{\pi r}$ 48. $x = a\left(1 - \dfrac{y}{b}\right)$

49a. $-0.3x^2 + 9x - 50$ b. \$10, \$17.50, \$10

50a. $0.3h$ b. $50 - h$

c. $0.25(50 - h)$ d. $0.3h + 0.25(50 - h)$

51a. For $z = 2$, $3z^2 = 3(2)^2 = 3 \cdot 4 = 12$, but $(3z)^2 = (3 \cdot 2)^2 = 6^2 = 36$

b. $(3z)^2 = (3z)(3z) = 9z^2$, so $(3z)^2 = 9z^2$

52a. $x = 0$, $x = 2$ b. Any values except $x = 0$, $x = 2$ c. No

Homework 5.2

1a. -12 b. -2 2a. 9 b. 3

3a. $19/2$ b. $-2/3$ 4. 8.660 5. -3.055

6. 1.899 7. -15.544 8. 1.293 9. -1.512

10. b 11. c 12. b 13a. 85

b. 168 c. 4082 14a. 3, 5, 8, 12 b. a

c. 4082 15a. 16 b. 29 c. x

16a. 7 b. 20 c. 4 17a. $\sqrt{6}$

b. $-\sqrt{15}$ c. $2\sqrt{m}$ 18a. $2b$ b. $9a$

c. $8b$ 19. $\frac{5}{4}$, 2, 2.3, $\sqrt{8}$

20. $\sqrt{6}$, 3, $2\sqrt{3}$, $\dfrac{23}{6}$ 21. $\sqrt{5} \approx 2.236$; 1; 0

22. -1; $2 - \sqrt{7} \approx -0.646$; $10 - \sqrt{103} \approx -0.149$

23. 30; $4 - 4\sqrt{2} \approx -1.657$; $\frac{-7}{8}$

24. ≈ 193.81 miles 25. ≈ 9.61 sec

26. A rational number is one that can be expressed as a ratio of integers, such as 5, $\frac{3}{4}$, and $\frac{-11}{2}$. An irrational number cannot be expressed as a ratio of integers. $\sqrt{3}$, $-\sqrt{5}$, and π are irrational numbers.

27. $\sqrt{4}$, $\sqrt{9}$, $\sqrt{64}$. $\sqrt{2}$, $\sqrt{6}$, $\sqrt{11}$

28. Rational: b, c, e, f

29. No

a	b	$a+b$	$(a+b)^2$	a^2	b^2	a^2+b^2
2	3	5	25	4	9	13
3	4	7	49	9	16	25
1	5	6	36	1	25	26
-2	6	4	16	4	36	40

30. No

a	b	$a+b$	a^2	b^2	a^2+b^2	$\sqrt{a^2+b^2}$
3	4	7	9	16	25	5
2	5	7	4	25	29	$\sqrt{29} \approx 5.4$
1	6	7	1	36	37	$\sqrt{37} \approx 6.1$
-2	-3	-5	4	9	13	$\sqrt{13} \approx 3.6$

31. No

a	b	$a+b$	$\sqrt{a+b}$	$\sqrt{a}$	$\sqrt{b}$	$\sqrt{a}+\sqrt{b}$
2	7	9	3	$\sqrt{2}$	$\sqrt{7}$	$\sqrt{2}+\sqrt{7} \approx 4.1$
4	9	13	$\sqrt{13} \approx 3.6$	2	3	5
1	5	6	$\sqrt{6} \approx 2.4$	1	$\sqrt{5}$	$1+\sqrt{5} \approx 3.2$
9	16	25	5	3	4	7

32. No

a	b	$a+b$	$\sqrt{a}$	$\sqrt{b}$	$\sqrt{a}+\sqrt{b}$	$\left(\sqrt{a}+\sqrt{b}\right)^2$
4	9	13	2	3	5	25
1	4	5	1	2	3	9
3	5	8	$\sqrt{3}$	$\sqrt{5}$	$\sqrt{3}+\sqrt{5}$	≈ 15.7
6	10	16	$\sqrt{6}$	$\sqrt{10}$	$\sqrt{6}+\sqrt{10}$	≈ 31.5

33a. $6a^3$ b. $22a^2$ 34a. $6m^2$

b. $4\sqrt{3}m^2 + 4\sqrt{3}m + 6m$

Homework 5.3

1. a,c

2.

x	-3	-2	-1	0	1	2	3
y	13	18	5	4	5	8	13

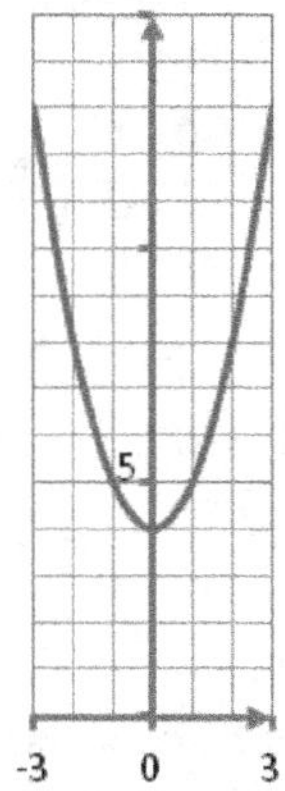

3.

x	-3	-2	-1	0	1	2	3
y	5	0	-3	-4	-3	0	5

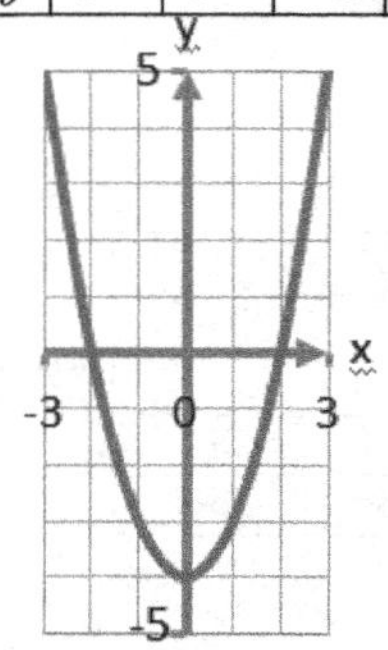

4.

x	-3	-2	-1	0	1	2	3
y	-5	0	3	4	3	0	-5

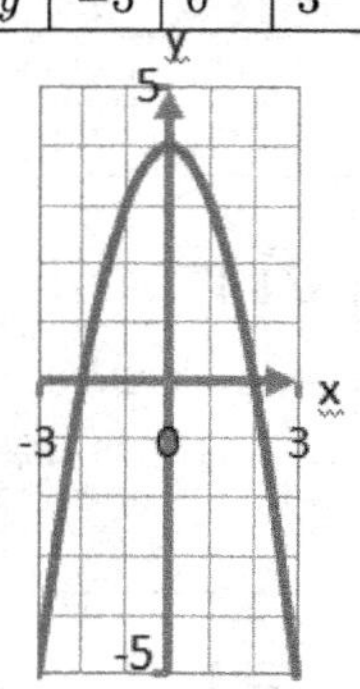

5.

x	-4	-2	-1	0	1	2	4
y	4	1	0.25	0	0.25	1	4

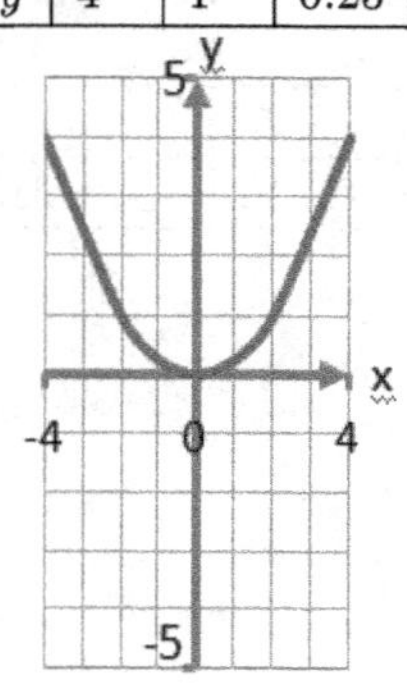

6a. 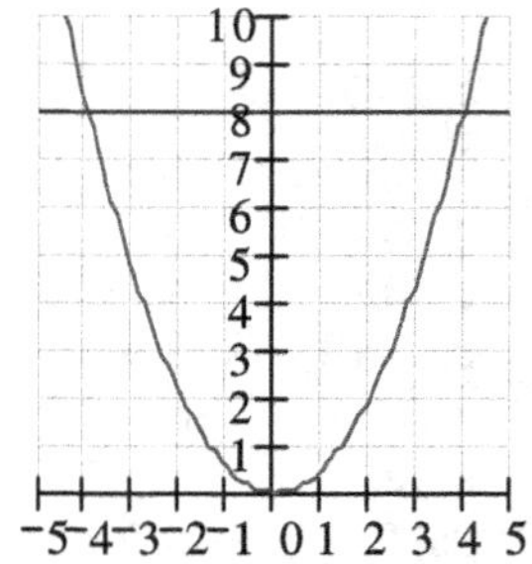 b. -4, 4

7a. 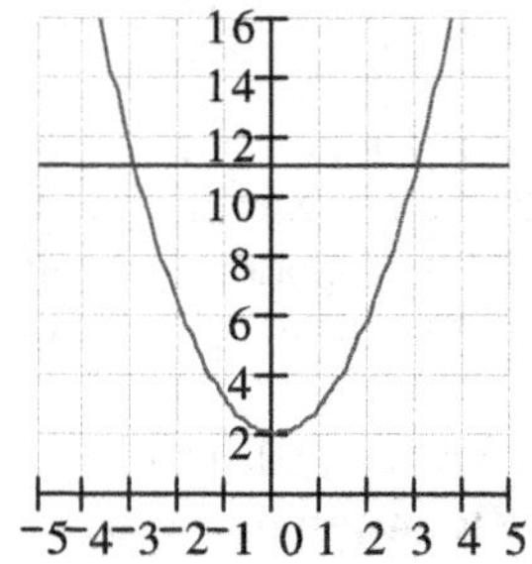 b. -3, 3

8a. 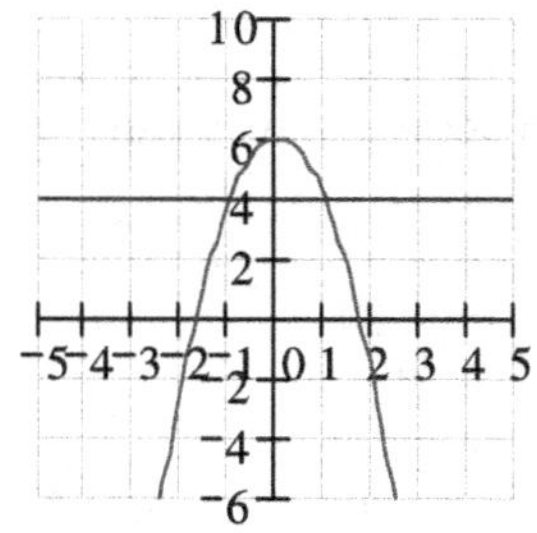 b. -1, 1

9. ± 11 10. ± 7 11. $\pm\frac{5}{3}$ 12. $\pm\sqrt{18}$
13. $\pm\sqrt{6}$ 14. ± 30 15. ± 7.25
16. ± 4 17. ± 15 18. No solution
19. ± 5.72 20. ± 5.73 21. 8 22. 37
23. $\sqrt{68}$ 24. $\sqrt{15}$ 25. $\sqrt{50}$, $3\sqrt{50}$
26. No 27. Yes 28. No
29. It is not a right triangle.
30. $8^2 + 4^2 = x^2$ 31. $x^2 = 2^2 + 11^2$
32. $x^2 + 16^2 = 25^2$ 33. $\pm\sqrt{\frac{A}{4\pi}}$
34. $\pm\sqrt{\dfrac{2F}{m}}$ 35. $\pm\sqrt{\dfrac{d-6}{k}}$
36. $\pm\sqrt{\dfrac{2(100-h)}{g}}$

37. a.

r	C
0	0
1	0.25
2	1
4	4
6	9
9	20.25
10	25
14	49

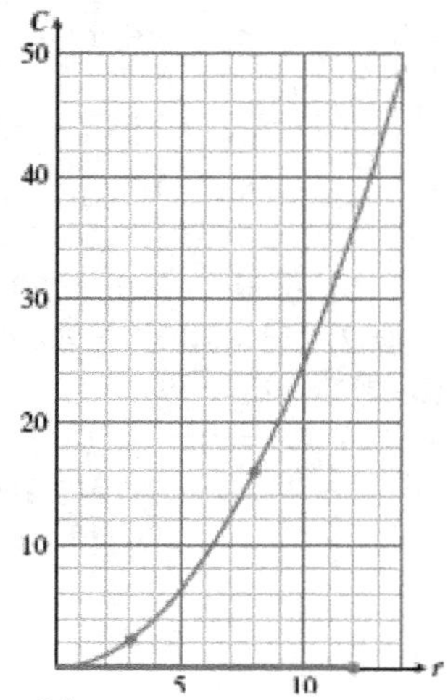

b. \$2.25 c. 8 in.
d. The cost of a radius 6" pizza is \$9.

38a.

Velocity (kph)	5	10	15	20	40	60
Distance (meters)	0.125	0.5	1.125	2	8	18

b. 8 m c. 49 kph d. A distance of 2 m is required to stop a car with a speed of 20 kph.

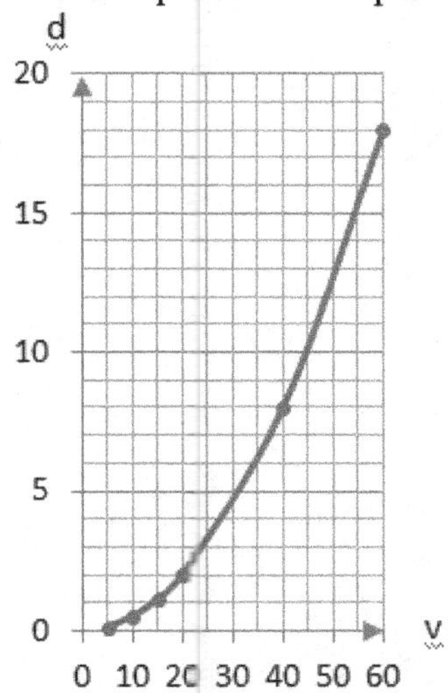

39.

t	h
0	100
0.5	96
1	84
1.25	75
1.5	64
1.75	51
2	36
2.25	19
2.5	0

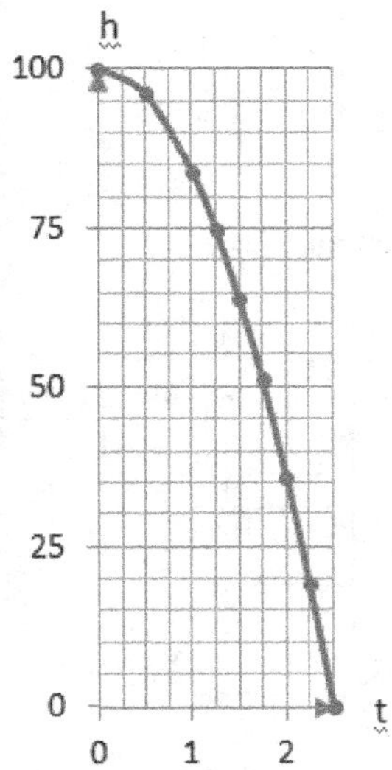

b. 36 ft c. After 1.25 seconds
d. $75 = 100 - 16t^2$ 40a. $\sqrt{27}$ ft
b. $3\sqrt{27}$ sq ft 41a. $\sqrt{8}$ cm b. $\sqrt{14}$ cm
c. $\frac{4\sqrt{14}}{3}$ cm^3 42. 26 ft 43. ≈ 127.28 ft
44. ≈ 8.60 mi 45. ≈ 176.87 ft
46. ≈ 79.20 m 47. $\sqrt{2}$ in, $\sqrt{3}$ in, 2 in, $\sqrt{5}$ in, $\sqrt{6}$ in, $\sqrt{7}$ in, $\sqrt{8}$ in, 3 in, $\sqrt{10}$ in
48. 5.463 cm 49. 1405 ft 50a. $V = 62.8r^2$
b.

r	1	2	3	4
V	62.8	251.2	565.2	1004.8

r	5	6	7	8
V	1570	2260.8	3077.2	4019.2

The volume decreases by a factor of 4.
c. 1.95 cm

Homework 5.4

1a. $12n$ b. $12n^2$ c. $64n^3$ 2a. $-30x^3$
b. $30x^3$ c. $25x^2$ 3a. $-16p^2$ b. $-16p^2$
c. $-4p^2$ 4a. $-40xt$ b. $-40x - 8xt$
c. $40x + 8xt$5a. $-12n^2$ b. $-n$
c. $3n^2 - 4n$ 6a. $-10x^3$ b. $2x - 10x^2$
c. $2x - 25x^2$7. $2a(5a + 3) = 10a^2 + 6a$
8. $3xy(2x - 5 + y) = 6x^2y - 15xy + 3xy^2$
9. $-12b^2 + 4b$ 10. $18a^2 - 15a$
11. $15v^2 - 6v^3$ 12. $-8x^3 - 12x^2y$
13. $2y^4 + 6y^2 - 4y$ 14. $-2x^3y + x^2y^2 - 3xy^3$
15. -3 16. $-3m$ 17. $4 - 3h$
18. $2a(a^2 - 3a + 5)$ 19. $-ax + 15a$
20. $4x + 2$ 21. $-3ax$ 22. $2ab^2 + 3a^2b^2$
23. $x^2 + 4x + 3x + 12 = (x + 3)(x + 4)$
24. $4x^2 - 4x - 2x + 2 = (4x - 2)(x - 1)$
25a.

	a	-3
a	a^2	$-3a$
-5	$-5a$	15

b. $(a - 5)(a - 3) = a^2 - 8a + 15$
26a.

	$3y$	-2
y	$3y^2$	$-2y$
1	$3y$	-2

b. $(y + 1)(3y - 2) = 3y^2 + y - 2$
27a.

	$4x$	3
$5x$	$20x^2$	$15x$
-2	$-8x$	-6

b. $(5x - 2)(4x + 3) = 20x^2 + 7x - 6$
28. $x^2 + xy - 2y^2$ 29. $6s^2 + 11st + 3t^2$
30. $2x^2 - 7ax + 3a^2$ 31. $6x^2 - 20x + 6$
32. $-3x^2 - 9x + 12$ 33. $-4x^2 + 5x + 6$
34. $20x^2 - 5$ 35a. $12x$ b. $60b^2$
36a. $63v^2$ b. $32v$ 37. $16w^2 - 36t^2$
38. $4\pi w^2 - 25$ 39. $\frac{32}{3}\pi h^3$ 40. $42c^3$
41. 1512 42. $800 + 480 + 20 + 12 = 1312$
43a. $-3x$ b.

	x	-9
x	x^2	$-9x$
6	$6x$	-54

44a. $3x$ b.

	x	4
$2x$	$2x^2$	$8x$
-5	$-5x$	-20

45. $x^2 - 9$ 46. $x^2 - 4a^2$
47. $9x^2 - 1$ 48. $w^2 + 8w + 16$
49. $z^2 - 12z + 36$ 50. $9a^2 - 12ac + 4c^2$
51. No

a	b	$a - b$	$(a - b)^2$	a^2	b^2	$a^2 - b^2$
5	3	2	4	25	9	16
2	6	-4	16	4	36	-32
-4	-3	-1	1	16	9	7

52. $(x - y)^2 = (x - y)(x - y) = x^2 - xy - xy + y^2 \neq x^2 - y^2$
53. $x^2 + 7x + 7x + 49 = (x + 7)^2$
54. $x^2 - 5x - 5x + 25 = (x - 5)^2$
55. No: If $x = 1$, then $(x + 4)^2 = (1 + 4)^2 = 5^2 = 25$, but $x^2 + 4^2 = 1^2 + 4^2 = 1 + 4 = 5$.
56. $x^2 - 4x + 4$ 57. $4x^2 + 4x + 1$
58. $9x^2 - 24xy + 16y^2$ 59. $h^2 + \left(\sqrt{7}\right)^2 = (h + 1)^2$
60. $(x - 1)^2 + 12^2 = (2x + 1)^2$

Homework 5.5

1. $-1,\ 4$ 2. $0,\ -7$ 3. $\frac{-3}{2},\ \frac{1}{4}$ 4b. $9,\ -3$
c. $9^2 - 6(9) - 27 = 0;\ (-3)^2 - 6(-3) - 27 = 0$
5b. $9,\ -9$ c. $9^2 - 81 = 0;\ (-9)^2 - 81 = 0$
6. $4x$ 7. $3ab$ 8. $5w$ 9. -4
10. $6t$ 11. $7x - 5$ 12. $3n;\ 9n$ 13. $2a(3a - 4)$
14. $6v(-3v - 1)$ or $-6v(3v + 1)$
15. $h(4 - 9h)$ 16. $d = k(1 - at)$
17. $r(2h - \pi r)$ 18. $V = h\left(\pi r^2 - \frac{1}{3}s^2\right)$
19. $0,\ \frac{3}{2}$ 20. $0,\ \frac{1}{20}$ 21. $0,\ -48$
22. $0,\ \frac{2}{3}$ 23. $0, -1$ 24. $0, 2$
25. $(3,0),\ (-\frac{5}{2}, 0)$ 26. $(0,0),\ (3,0)$
27. $(0,0),\ (\frac{8}{3}, 0)$ 28. $5, -1$
29. $\frac{5}{2}, \frac{-3}{2}$ 30. $-2 \pm \sqrt{3}$ 31. $\frac{1}{2} \pm \frac{\sqrt{3}}{2}$
32. $\frac{-2}{9}, \frac{-4}{9}$ 33. $\frac{7}{8} \pm \frac{\sqrt{8}}{8}$ 34. $\pm\sqrt{\frac{bc}{a}}$
35. $a \pm 4$ 36. $\frac{-b \pm 3}{a}$
37. $4x^2 = 12x,\ 4x^2 - 12x = 0,\ 4x(x - 3) = 0,$
$x = 0, 3$
38. $9x^2 - 4 = 16,\ 9x^2 = 20,\ x^2 = \frac{20}{9},\ x = \pm\frac{\sqrt{20}}{3}$

39a.

w	A
0	0
4	80
8	128
12	144
16	128
20	80
24	0

b.

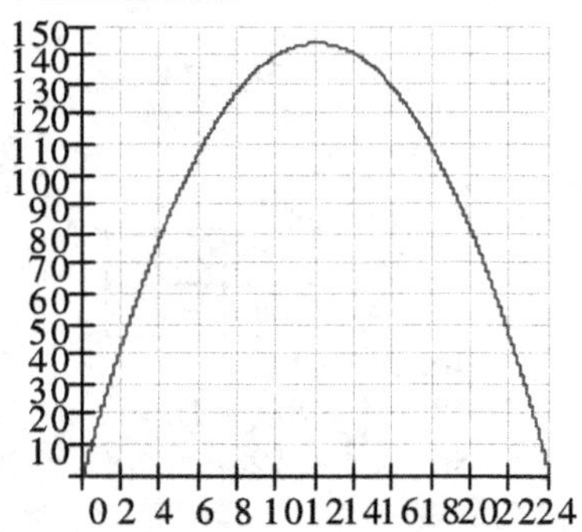

c. $(0,0),\ (24,0)$ d. $144\ \text{ft}^2$, 12 ft by 12 ft
40a. $B = 5000(1 + r)^2$

b.

r	0.02	0.04	0.06	0.08
A	5202	5408	5618	5832

c. 11.8% 41. 8% 42. 7.98 mm

Homework 5.6

1. Narrower than basic parabola

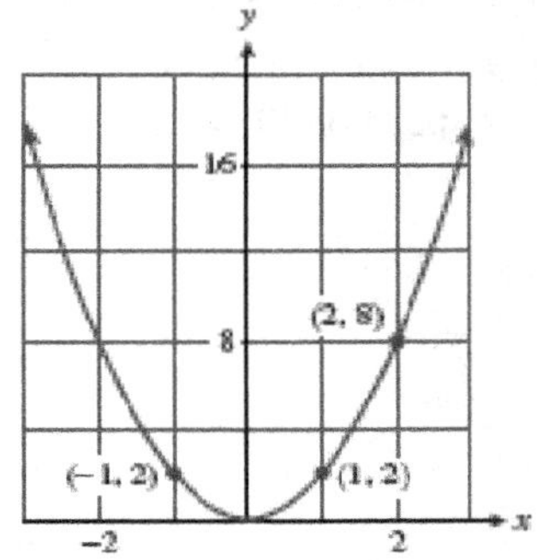

2. Wider than basic parabola

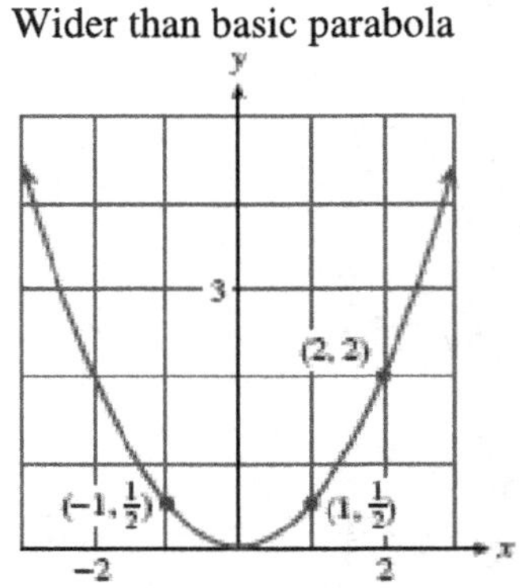

3. Reflected about x-axis

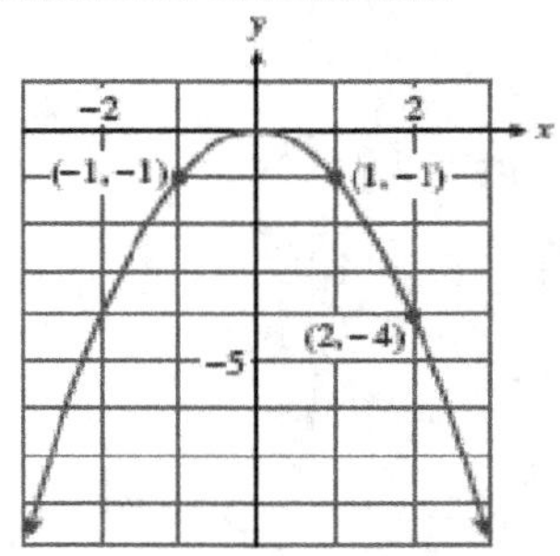

4. Wider and reflected about x-axis

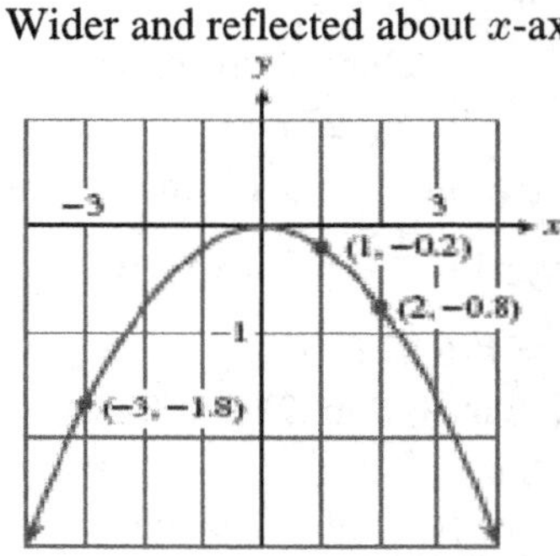

5. Shifted up 2 units

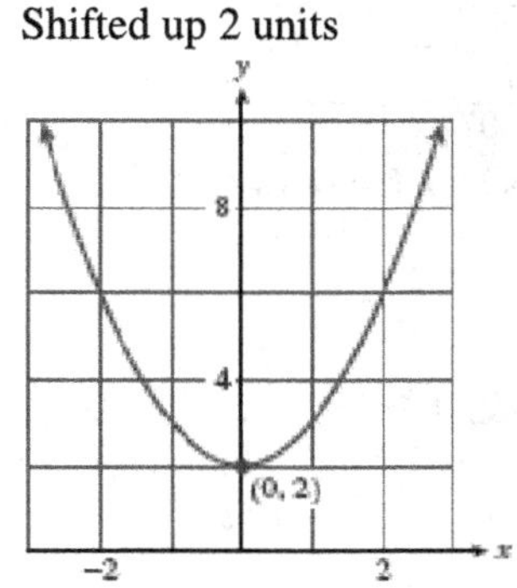

6. Shifted down 1 unit

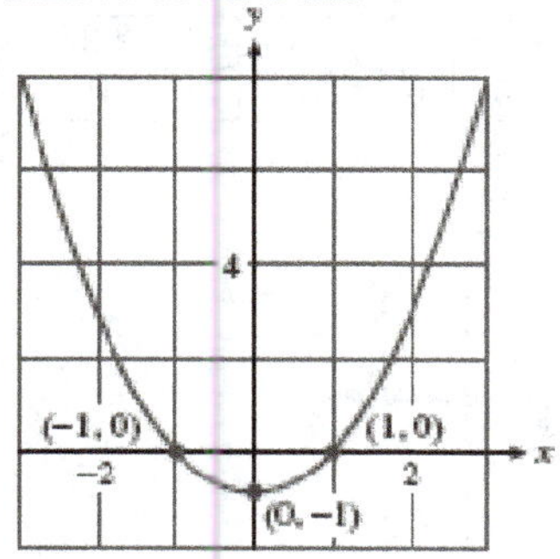

7. Shifted down 5 units

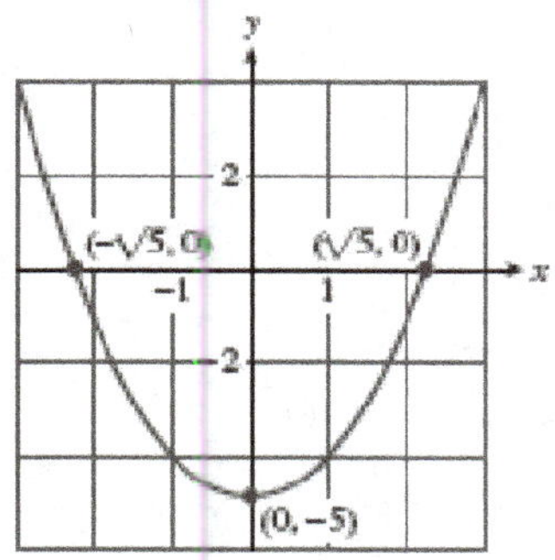

8. Reflected about x-axis, then shifted up 100 units

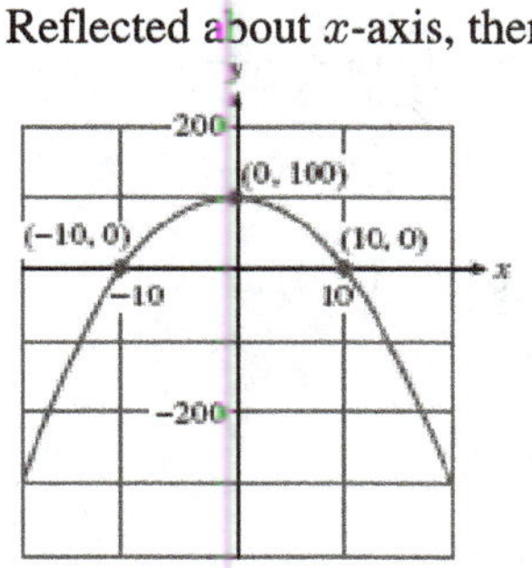

9. $(0,0)$, $(4,0)$; $(2,-4)$

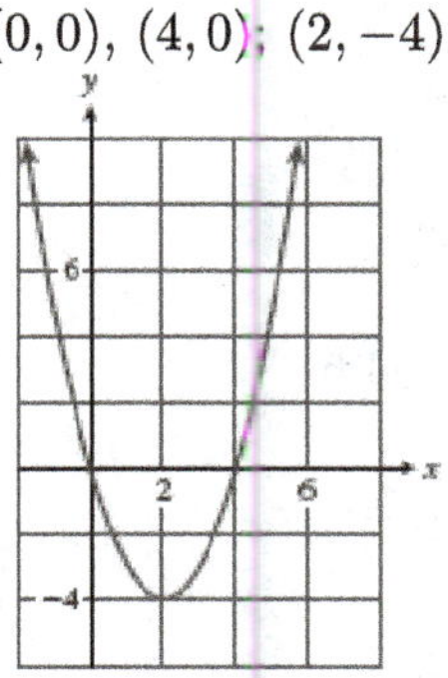

10. $(0,0)$, $(-2,0)$; $(-1,-1)$

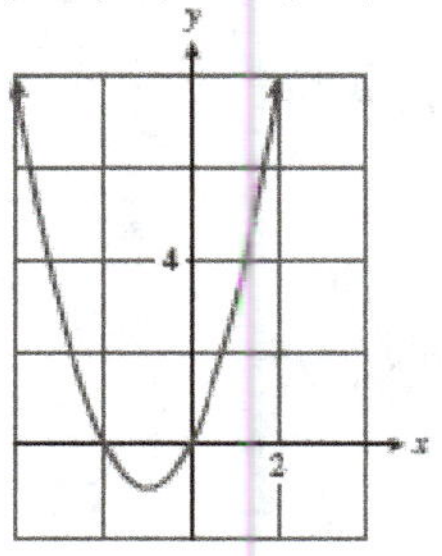

11. $(0,0)$, $(-2,0)$; $(-1,-3)$

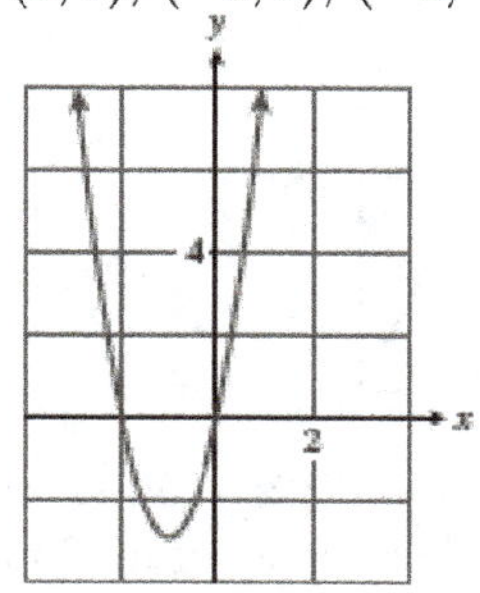

12. $(0,0)$, $\left(\frac{5}{2},0\right)$; $\left(\frac{5}{4},3\frac{1}{8}\right)$

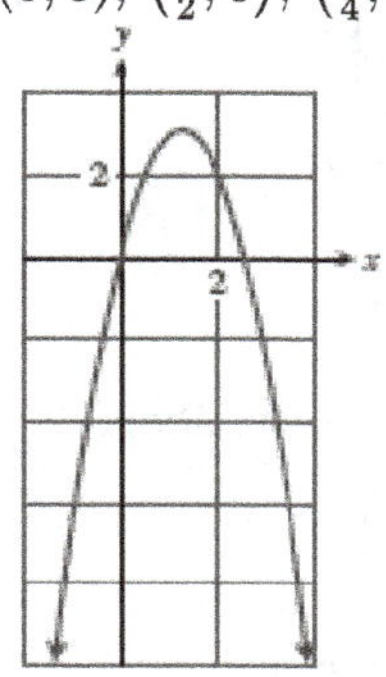

13a. II b. IV c. I d. III e. VI f. V

14a. $-2n(n-40)$

b.

n	R
0	0
5	350
10	600
20	800
30	600
40	0

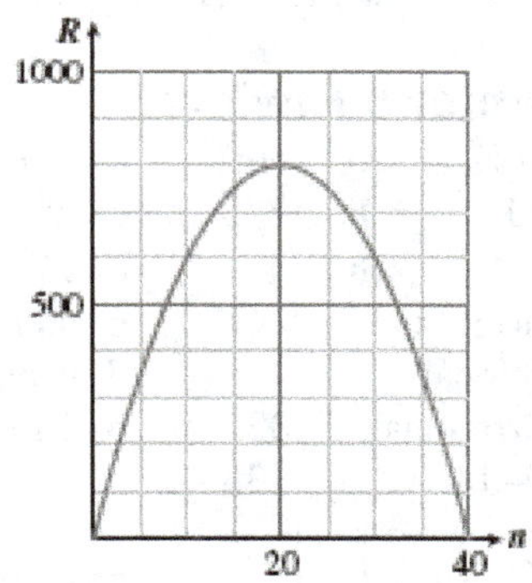

c. \$800; 20 bracelets

15a. $-16t(t-5)$

b.

t	h
0	0
1	64
2	96
2.5	100
3	96
4	64
5	0

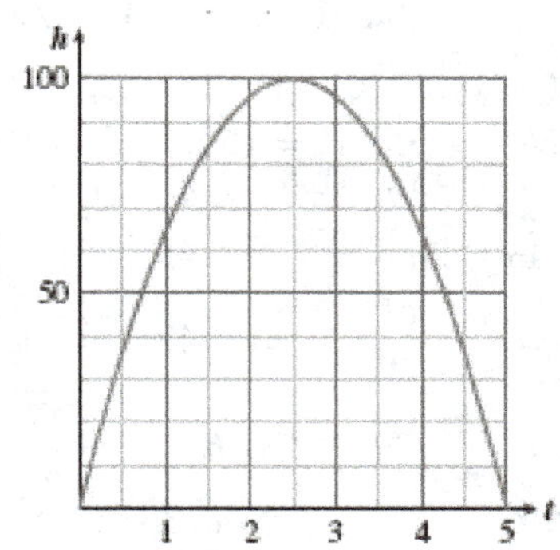

c. 100 ft; 2.5 sec

16a. At $x=0$ and $x=4000$, there is no increase in biomass.

b. $(2000,400)$. The largest annual increase in biomass, 400 tons, occurs when the biomass is 2000 tons.

c. 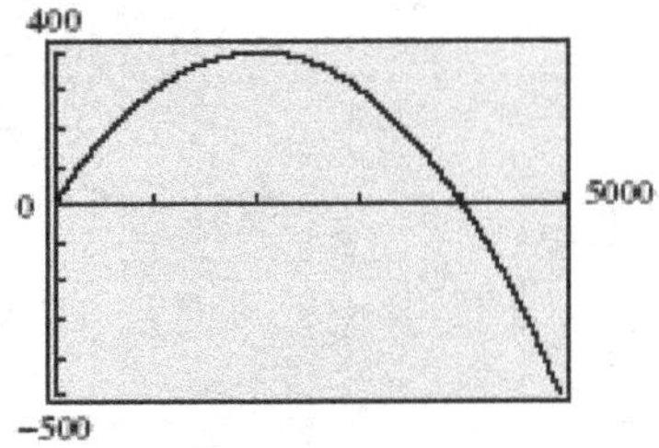

d. The biomass decreases if $4000 < x \le 5000$. If the population becomes too large, its supply of food may be inadequate.

17. 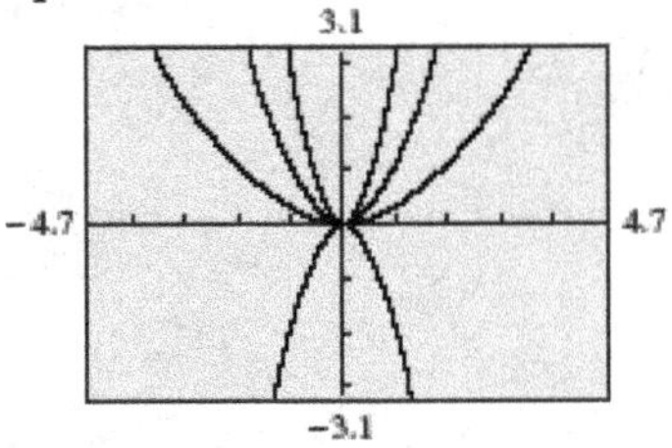

The constants affect the width of the parabola.

18. 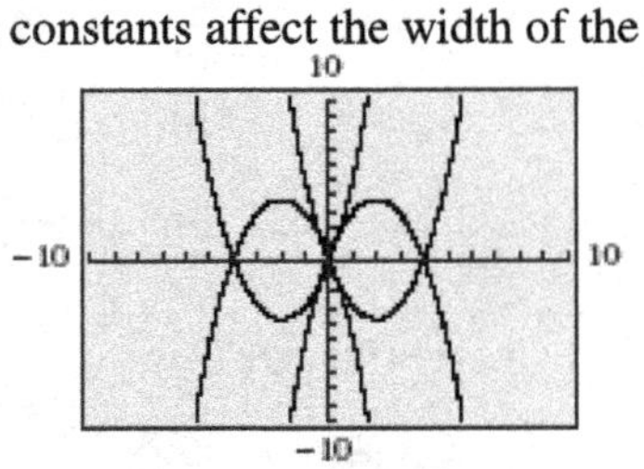

The constants affect the location of the x-intercepts.

Chapter 5 Review

1a. 147 b. −441 c. −40 d. 16

3a. 1 b. 7/5 c. $\frac{12}{25}$ d. $\frac{7}{5}$

5a. 4 b. 4 c. 27 d. 27

7a. irrational b. undefined

c. rational d. irrational

9a. irrational, 17.32 b. irrational, 7.47

11. −14 13. −2 15. −20.247

17a. −4 b. 4 19a. −4 b. 4

21a. −13 b. 48 23a. 12 b. 18

25a. 6 b. 12 27a. 17 b. 40

c. 60 29a. 1 b. $3+\sqrt{3}$ c. $\dfrac{\sqrt{3}}{2}$

31. $V = 126$; $S = 30\sqrt{7} + 84$

33. $16b^3$ 35. $V = 150p^3$, $S = 190p^2$

37. $\pm\frac{2}{3}$ 39. ± 2 41. $\pm\frac{2}{3}$

43. $-4\pm\sqrt{20}$ 45. 4, 1

47. $v = \dfrac{s}{t} - \dfrac{at}{2}$ 49. $s = \pm\sqrt{\dfrac{3V}{h}}$

51. $a = \dfrac{2S}{n} - f$ 53. $h = \pm\sqrt{\dfrac{C}{br}}$

55. 11% 57a. 3 sec b. 36 ft

59. 13.9 yd 61a. 2.5 ft b. 78.54 sq ft

63. 11.40 m 65. 10 m 67. $-20x + 15x^2$

69. $6x^2y - 12xy - 3xy^2$ 71a. $-2t^2$

b. cannot be simplified c. $-48t^3$

73. $9a^2 - 3a$ 75. $-2a^2 - 9a + 8$ 77. 1

79a. $5x^2 - 2x + 5x - 2$ b. $(x+1)(5x-2)$

81. $x^2 + 2x - 8$ 83. $25a^2 - 1$

85. $4q^2 + 20q + 25$ 87. $u^2 - 7u + 10$

89. $a^2 + 12a + 36$ 91. $6a^2 - 11ac - 10c^2$

93a.

t	0	0.5	1	1.25	1.5	1.75	2	2.25	2.5
h	100	96	84	75	64	51	36	19	0

b. 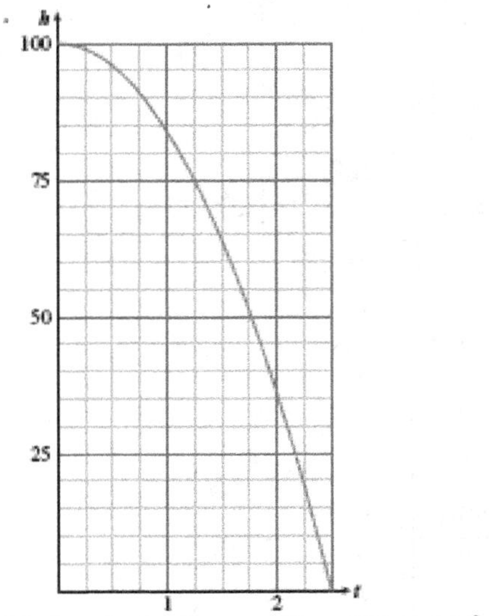

c. 36 ft d. 1.25 sec, $75 = 100 - 16t^2$

95a. $(0,0)$ b. 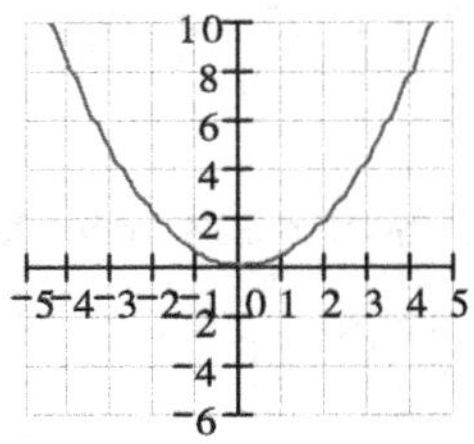

97a. Vertex: $(\frac{9}{2}, \frac{-81}{4})$, y-int: $(0,0)$
x-ints: $(0,0), (9,0)$

b. 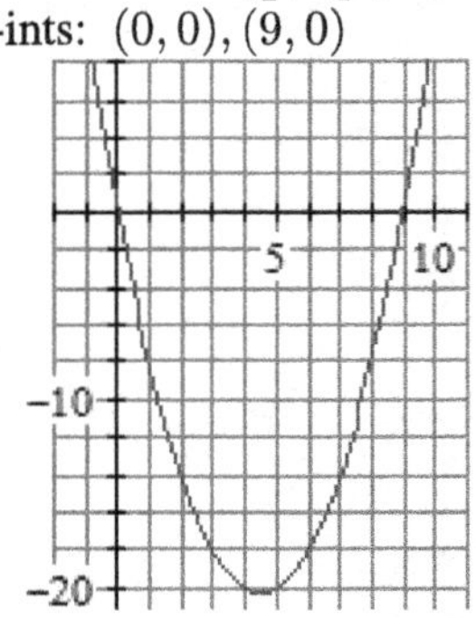

99a. $h = 100t - 2.8t^2$ c. 893 ft

d. $15\frac{5}{7}$sec and 20 sec

101. 64 103. 18

Homework 6.1

1a.

	x	6
x	x^2	$6x$
5	$5x$	30

b. $(x+5)(x+6) = x^2 + 11x + 30$

2a.

	x	−3
x	x^2	$-3x$
−9	$-9x$	27

b. $(x-9)(x-3) = x^2 - 12x + 27$

3a.

	x	2
x	x^2	$2x$
8	$8x$	16

b. $(x+8)(x+2) = x^2 + 10x + 16$

4.

	x	8
x	x^2	$8x$
-4	$-4x^2$	-32

5. $(n+2)(n+8)$ 6. $(h+2)(h+24)$
7. $(a-2)(a-6)$ 8. $(t-3)(t-12)$
9. $(x+2)(x-5)$ 10. $(a-2)(a+10)$
11. $(x-2)(x-15)$ 12. doesn't factor
13. $(y+1)(y-45)$ 14. doesn't factor
15. $(q+1)(q-6)$ 16. $(n-2)(n-3)$
17. $(x-3)(x+3)$ 18. $(2-w)(2+w)$
19. $(b-11)(b+11)$ 20. $(2x+1)(x+5)$
21. $(5t+2)(t+1)$ 22. $(3x-5)(x-1)$
23. $x=2, x=-5$ 24. $t=6, t=-7$
25. $x=-1, x=6$ 26. $n=7$
27. $q=0, q=2$ 28. $z=1, z=\frac{2}{3}$
29. 6 hr 30. $1\frac{1}{2}$ sec

31a. 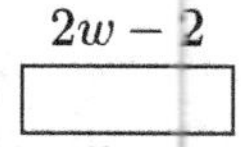

b. $w(2w-2)=60$

c. $w=-5, w=6$: The circuit board should be 6 cm by 10 cm.

32a. 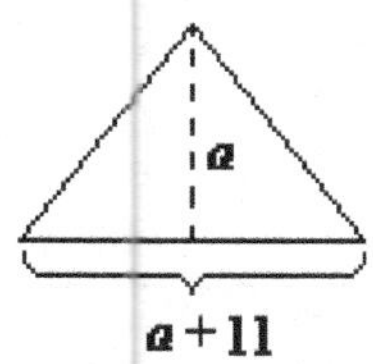

b. $\frac{1}{2}(a+11)a=40$

c. $a=-16, a=5$: The base is 16 inches and the altitude if 5 inches.

33a. 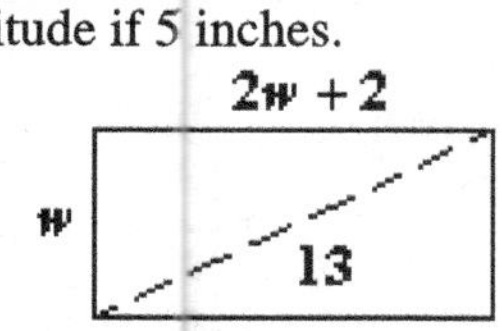

b. $w^2+(2w+2)^2=13^2$

c. $w=\frac{-33}{5}, w=5$: The locker is 5 feet by 12 feet.

34. $5, -4$ 35. $\frac{1}{4}$
36. $x+8$ 37. $b+20$
38. $n+3$ 39. $3u+1$
40. $2t-9$

41. $(2x-9)(x-2)$

	x	-2
$2x$	$2x^2$	$-4x$
-9	$-9x$	18

42. $(5x-4)(x+4)$

	x	4
$5x$	$5x^2$	$20x$
-4	$-4x$	-16

43. $(2h+1)(3h+2)$

	$3h$	2
$2h$	$6h^2$	$4h$
1	$3h$	2

44. $(9n+1)(n-1)$

	n	-1
$9n$	$9n^2$	$-9n$
1	n	-1

45. $(3t+5)(2t-5)$

	$2t$	-5
$3t$	$6t^2$	$-15t$
5	$10t$	-25

46. $(5x+6)(x-4)$

	x	-4
$5x$	$5x^2$	$-20x$
6	$6x$	-24

47. $9x^2-21x-8=(3x-8)(3x+1)$
48. $16z^2-2z-5=(8z-5)(2z+1)$
49. $4a^2+23a-6=(4a-1)(a+6)$
50. $n=\frac{4}{3}, n=-1$ 51. $t=\frac{1}{3}, t=\frac{3}{2}$
52. $y=1/2$ 53. $z=\frac{-5}{2}, z=\frac{1}{3}$
54. $3x+5,\ 2x+3$ 55. $2m+5,\ 6m-5$

Homework 6.2

1a. 10 ft by 20 ft, 5 ft by 25 ft; 200 sq ft, 125 sq ft

b. $30-x$ c. $x(30-x)$

2a.

t	0	1	2	3	4	5
h	300	304	276	216	124	0

b. 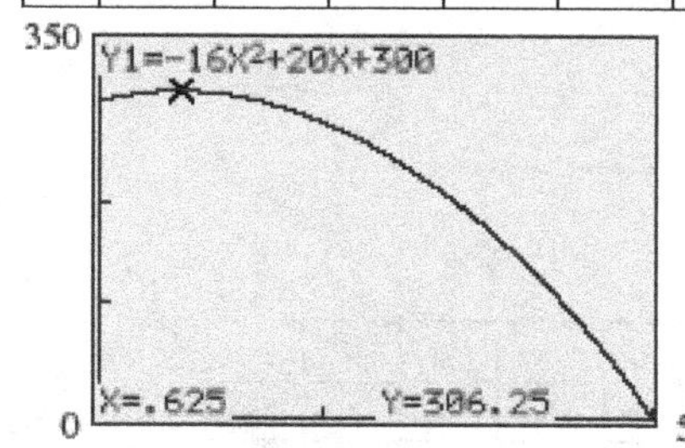

c. 306.25 ft at 0.625 sec d. 1.25 sec

e. 5 sec

3. $x=\frac{1}{2}, 1$ 4. $t=2, 3$
5. $z=-1, 2$ 6. $v=-3, 6$
7, 8. The graphs have the same x-intercepts.
9. $x^2+x-2=0$ 10. $x^2+5x=0$
11. $2x^2+5x-3=0$
12. $8x^2-10x-3=0$
13. $0.1(x-18)(x+15)$
14. $-0.08(x-18)(x+32)$

15a.

Height	Base	Area	Height	Base	Area
1	34	34	10	16	160
2	32	64	11	14	154
3	30	90	12	12	144
4	28	112	13	10	130
5	26	130	14	8	112
6	24	144	15	6	90
7	22	154	16	4	64
8	20	160	17	2	34
9	18	162	18	0	0

b. 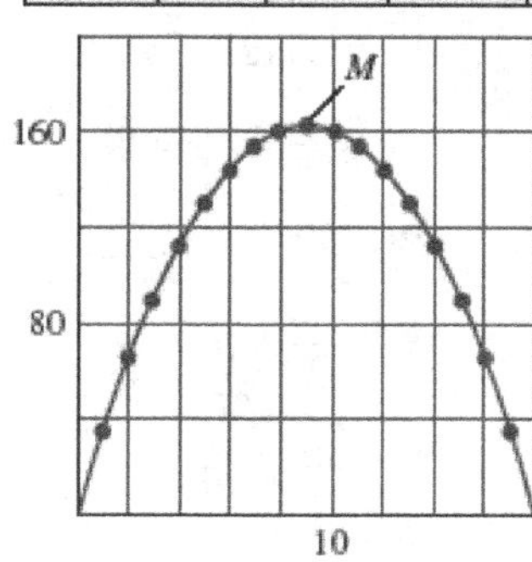

c. 162 sq in, 9 in by 18 in

d. Base: $36-2x$, area: $36x-2x^2$

f. 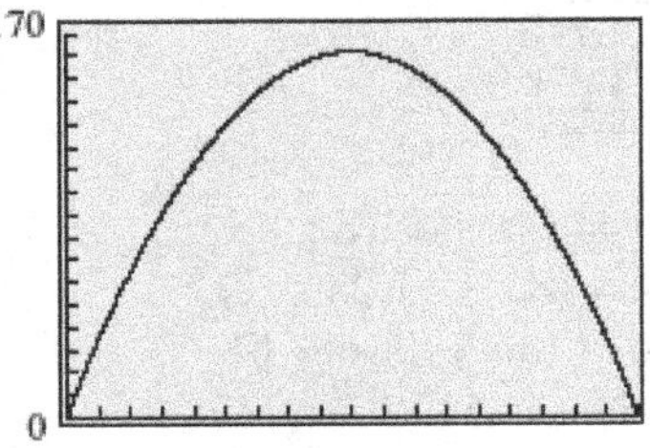

g. Two possible rectangles, one with height 11.5 in. and the other with height 6.5 in.

16a. $h=-16t^2+16t+8$ b. 12 ft, 8 ft

c. $11=-16t^2+16t+8$; at $\frac{1}{4}$sec and $\frac{3}{4}$ sec

d. ΔTbl $=0.25$

e. 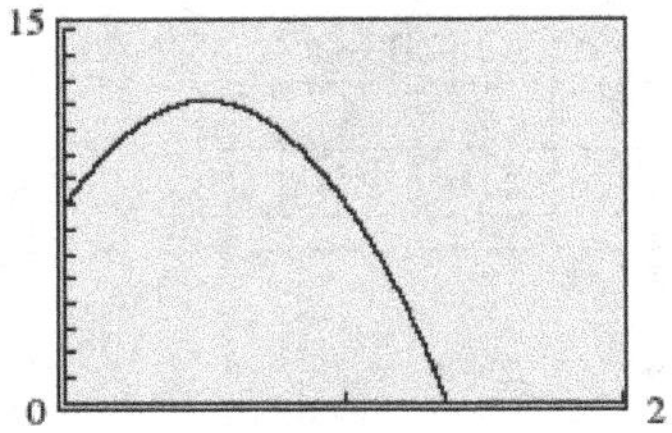

f. 1.37 sec

17a.

Width	10	20	30	40	50
Length	170	160	150	140	130
Area	1700	3200	4500	5600	6500

Width	60	70	80
Length	120	110	100
Area	7200	7700	8000

b. $l=180-x$, $A=180x-x^2$; 80 yd by 100 yd

c. $180x-x^2=8000$, 80 yd by 100yd, or 100 yd by 80 yd. There are two solutions because the pasture can be oriented in two directions.

18a. $l=x-4$, $w=x-4$, $h=2$, $V=2(x-4)^2$

b.

x	4	5	6	7	8	9	10
V	0	2	8	18	32	50	72

c. As x increases, V increases d. 9 in. by 9 in.

e. $2(x-4)^2=50$, $x=9$

Homework 6.3

1. $(x+4)^2$ 2. $(x-\frac{7}{2})^2$ 3. $(x+\frac{3}{4})^2$
4. $(x-\frac{2}{5})^2$ 5. $1,1$ 6. $-4,-5$
7. $\frac{-3}{2}\pm\sqrt{\frac{21}{4}}$ 8. $-1\pm\sqrt{\frac{5}{2}}$ 9. $\frac{-4}{3},1$
10. $\frac{1}{4}\pm\sqrt{\frac{13}{16}}$ 11. $\frac{4}{3},-1$ 12. $\frac{2}{5},-2$
13. $x=-5$, $x=8$ 14. $x=3\pm\sqrt{40}$
15. $x=\frac{-1\pm\sqrt{145}}{6}$ 16. $x=\frac{-2\pm\sqrt{24}}{2}$
17. $-1\pm\sqrt{1-c}$ 18. $\frac{-b}{2}\pm\sqrt{\frac{b^2-4}{4}}$
19. $\frac{-1}{a}\pm\sqrt{\frac{4a+1}{a^2}}$ 20. $3\pm\sqrt{\frac{V}{\pi h}}$
21. $\pm\sqrt{\frac{2(E-mgh)}{m}}$ 22. $\pm\sqrt{\frac{V}{2w}-s^2}$
23. $0, x^2$ 24. $\frac{-3x\pm3}{2}$
25. $\pm\sqrt{\frac{4x^2-36}{9}}$ 26. $\frac{\pm2x}{5}$
27. $(-3,-11)$, minimum 28. $(1.5, 4.25)$, maximum
29. $(\frac{2}{3},\frac{1}{9})$, minimum 30. $(-4.5, 18.5)$, maximum

31a. $A=(x+y)^2$ b. $A=x^2+2xy+y^2$

c. x^2, xy, xy, y^2 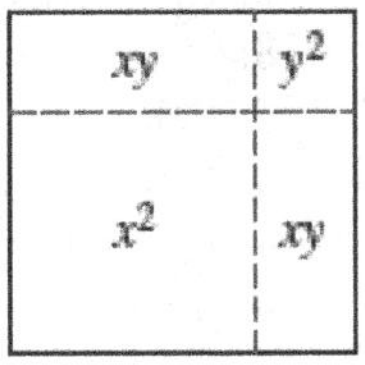

32a. $A=\frac{1}{2}(x-y)^2$ b. $A=\frac{1}{2}(x+y)(x-y)$

c. 18 sq ft 33. $\frac{2}{5}, -2$

34. The result does not give a value for x that solves the equation; it is another form of the original equation.

35a. x-intercepts: $(-0.65,0)$, $(1.15,0)$

b. vertex: $(0.25,-3.25)$ c. 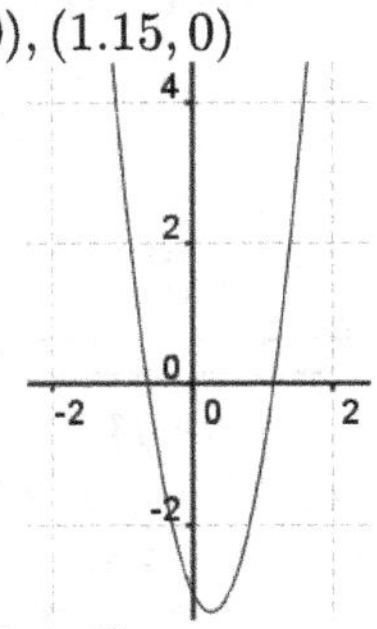

36a. x-intercepts: $(-2,0)$, $(0.4,0)$

b. vertex: $\left(\frac{-4}{5},\frac{-36}{5}\right)$ c. 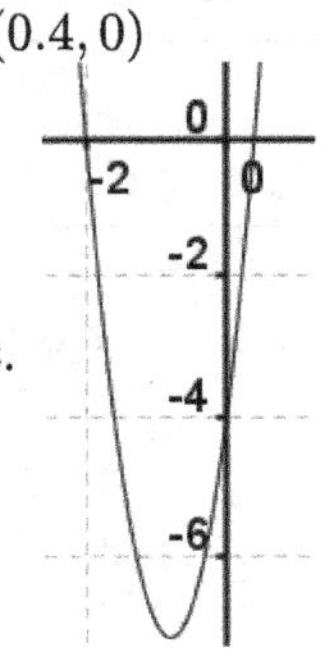

Homework 6.4

1. 1.618, −0.618 2. 1.449, −3.449
3. 1.695, −0.295 4. 1.434, 0.232
5. −5.894, 39.740 6. $x=-18$, $x=2$
7. $x=8$, $2x-1=15$ 8. $k=6$, $2k-5=7$
9. $-4+5i$ 10. $-4+i$
11. $\frac{-5}{6}-\frac{\sqrt{2}}{6}i$ 12. $-3\pm2i$
13. $\frac{1}{6}\pm\frac{\sqrt{11}}{6}i$ 14. Two complex solutions
15. One repeated real solution
16. Two distinct real solutions
17. $w=\frac{-4l\pm\sqrt{16l^2+8A}}{4}$
18. $t=\frac{4\pm\sqrt{16-64h}}{32}$
19. $t=\frac{v\pm\sqrt{v^2-2as}}{a}$

20. $y = \dfrac{-x \pm \sqrt{8 - 11x^2}}{2}$

21a. 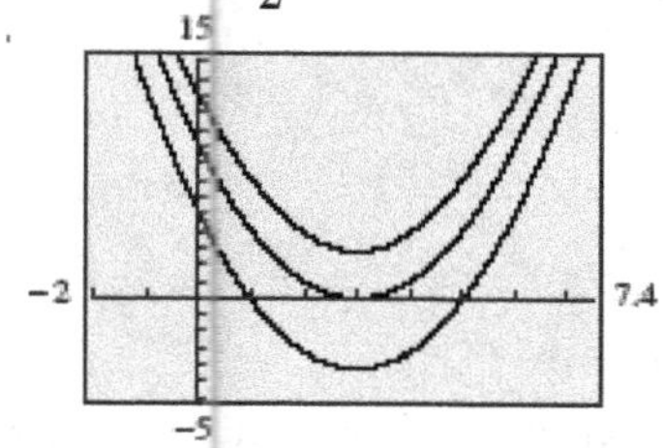

b. $(5, 0)$ and $(1, 0)$; $(3, 0)$; no x-intercepts

22a.

s	10	20	30	40	50	60	70	8	90	100
d	9	27	53	87	129	180	239	307	383	467

b. 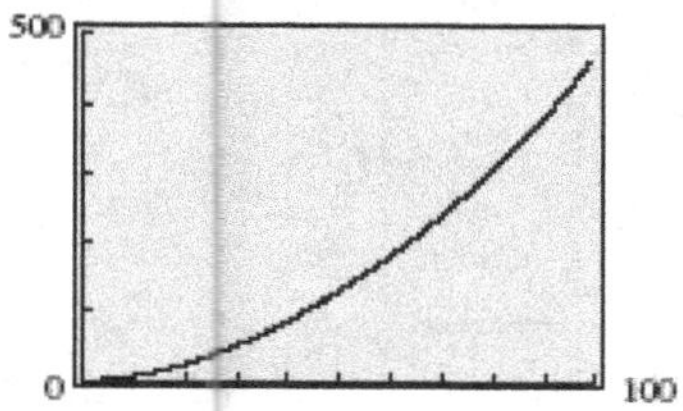

c. $\dfrac{s^2}{24} + \dfrac{s}{2} = 50$, 29.16 mph

23a.

t	0	5	10	15	20	25
h	11,000	10,520	9240	7160	4280	600

b. 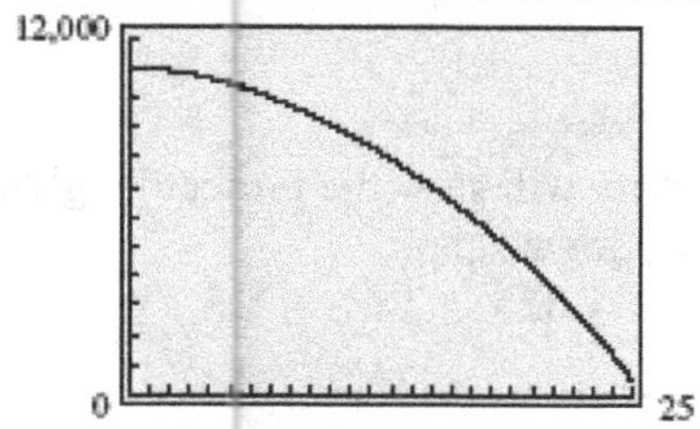

c. $-16t^2 - 16t + 11,000 = 1000$, 24.5 sec

d. 1.2 sec

24a. $2l + 4w = 100$ b. $l = 50 - 2w$

c. $w(50 - 2w) = 250$; $w = 6.91, 18.09$

d. 12.06 m by 6.91 m, or 4.61 m by 18.09 m

25a. 47 mi b. 1.26 mi 26. 15 in × 40 in × 20 in

27. $(\frac{-1}{2}, 0)$, $(4, 0)$, $(0, 4)$; $(\frac{7}{4}, 10\frac{1}{8})$

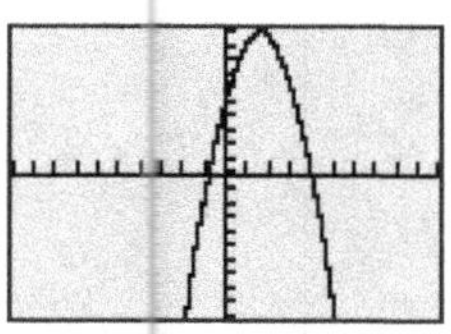

28. $(-2, 0)$, $(1, 0)$, $(0, -1.2)$; $(-0.5, -1.35)$

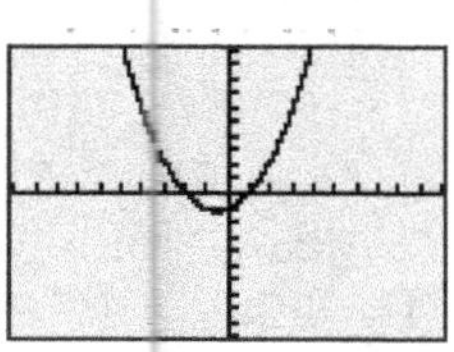

29. $(0, 7)$; $(-2, 3)$ 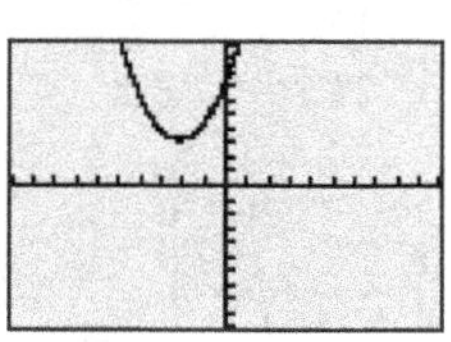

30. $(0.41, 0)$, $(-2.41, 0)$, $(0, -1)$; $(-1, -2)$

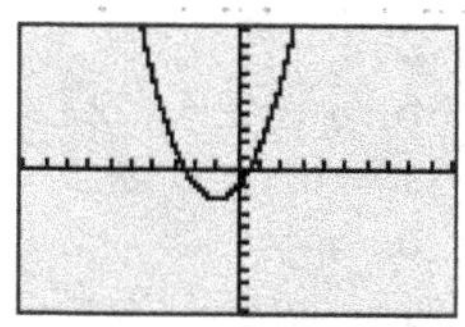

31. $(2.37, 0)$, $(0.63, 0)$, $(0, -3)$; $(1.5, 1.5)$

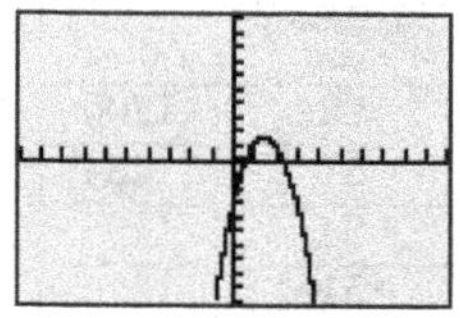

32a. $2 - \sqrt{5}$ b. $x^2 - 4x - 1 = 0$

33a. $4 + 3\sqrt{2}$ b. $x^2 - 8x - 2 = 0$

34. No 35. Yes 36. $\dfrac{-b}{a}$

Homework 6.5

1. 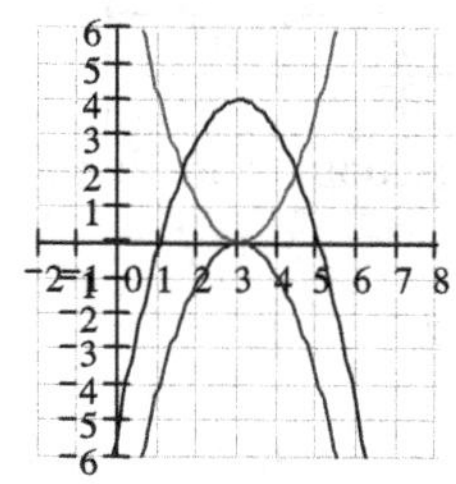

a. $(3, 0)$ b. $(3, 0)$ c. $(3, 4)$

2. 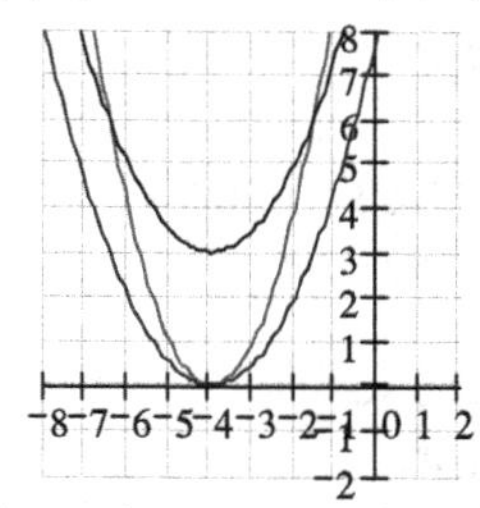

a. $(-4, 0)$ b. $(-4, 0)$ c. $(-4, 3)$

3a. $(3, 4)$ b. $y = 2x^2 - 12x + 22$

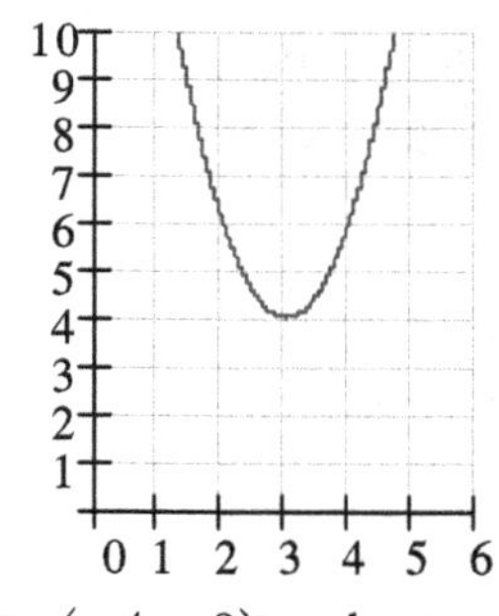

4a. $(-4, -3)$ b. $y = \frac{-1}{2}x^2 - 4x - 11$

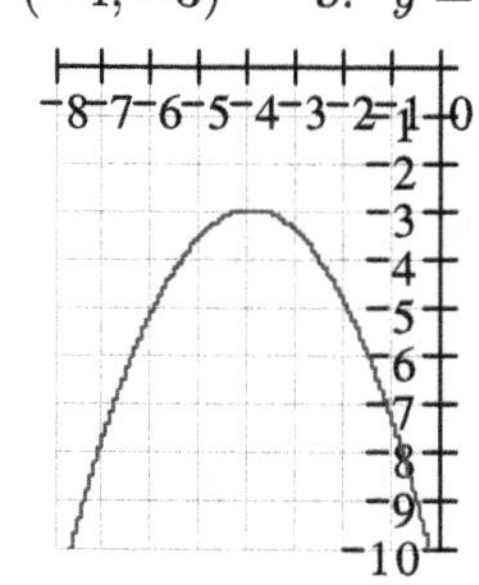

5a. IV b. V c. I d. VII

6. $y = (x - 2)^2 + 3$ 7. $y = 3(x + 1)^2 - 5$

8. $y = -2(x + 2)^2 + 11$

9a. $y = x^2 + x - 6$ b. $y = 2x^2 + 2x - 12$

10a. $y = a(x + 2)^2 + 6$ b. 3

11a. $y = \frac{3}{4}x^2 - 3$ b. $y = -x^2 - 3$

12. $y = x^2 - 9$ 13. $y = -2x^2$

14. $y = x^2 - 2x - 15$ 15. $y = x^2 - 4x + 5$

16. $y = -2(x - 30)^2 + 280$

17a.

No. of price increases	Price of room	No. of rooms rented	Total revenue
0	20	60	1200
1	22	57	1254
2	24	54	1296
3	26	51	1326
4	28	48	1344
5	30	45	1350
6	32	42	1344
7	34	39	1326
8	36	36	1296
10	40	30	1200
12	44	24	1056
16	52	12	624
20	60	0	0

b. Price of a room: $20 + 2x$; rooms rented: $60 - 3x$; revenue: $1200 + 60x - 6x^2$ d. 20

e. 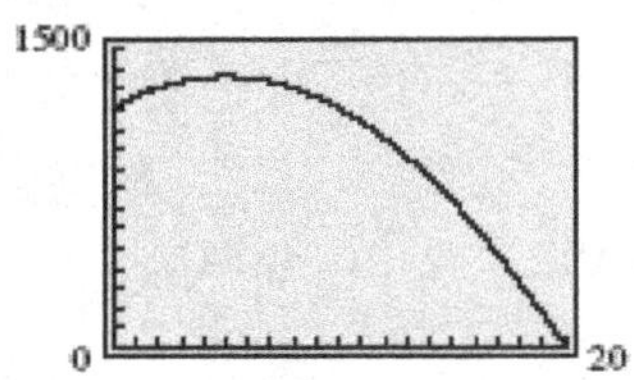

f. \$24; \$36

g. \$1350; \$30; 45 rooms

18. 3 sec, 144 ft 19. 100 baskets, \$2000

20a. length: $50 - w$, area : $50w - w^2$

b. 625 sq in

21a. $300w - 2w^2$ b. 11,250 sq yds

22a. Number of people: $16 + x$, price per person: $2400 - 100x$; total revenue: $38,400 + 800x - 100x^2$

b. 20

23a. $y = \frac{-1}{40}(x - 80)^2 + 164$ b. 160.99 ft

24. $y = 0.00012(x - 2000)^2 + 20$

Homework 6.6

1. $a = 3, b = 1, c = -2$

2a. $P = -0.16x^2 + 7.4x - 71$ b. 14%

c. 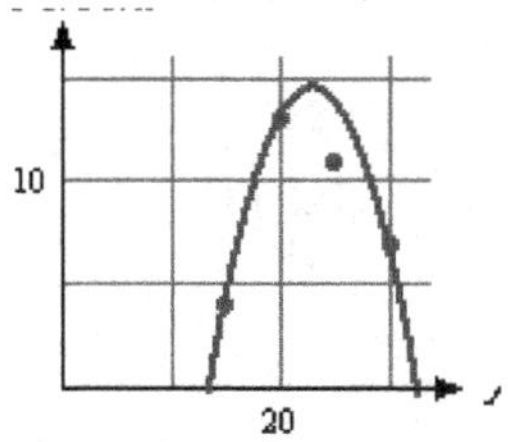

3a. $C = 0.75t^2 - 1.85t + 56.2$ b. 65.7 lb

c. 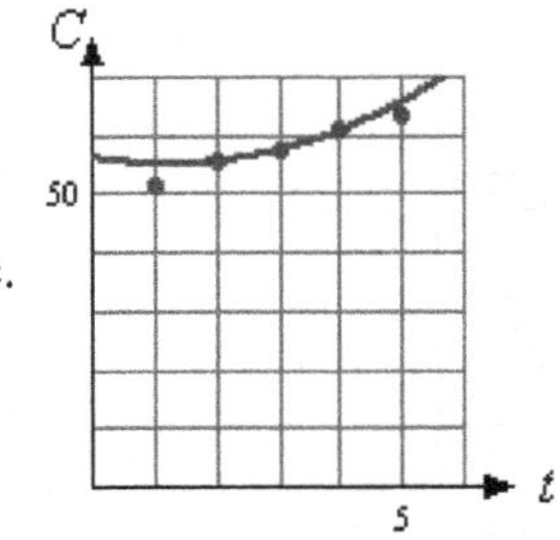

4a. $y = (-1.296 \cdot 10^{-4})x^2 + 0.01916x + 2.1597$

b. For a maximum yeild of about 2.87 tons of soybeans, use 74 kg of nitrogen fertilizer.

5a. $h = 8.24t + 38.89$ b. 162.5 m

c. 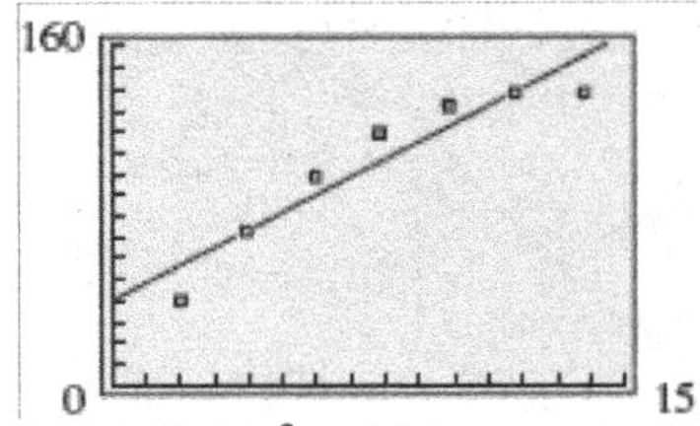

d. $h = -0.81t^2 + 21.2t$ e. 135.7 m

f. 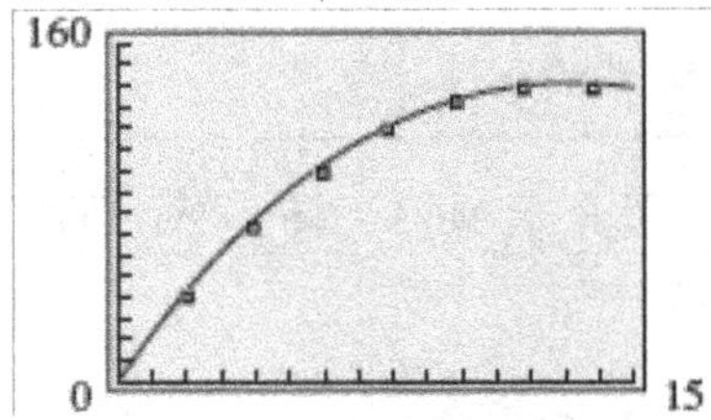

g. Quadratic: gravity will slow the projectile, giving the graph a concave down shape.

6a. $y = 0.0051t + 11.325$ b. 13.2 hr

c. 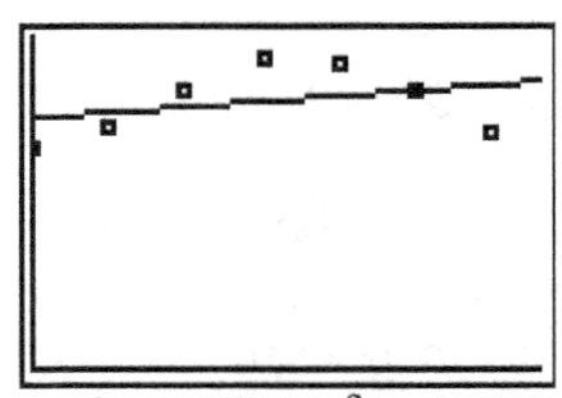 $[0, 330] \times [0, 15]$

d. $y = -0.00016t^2 + 0.053t + 9.319$

e. 7.4 hr

f. 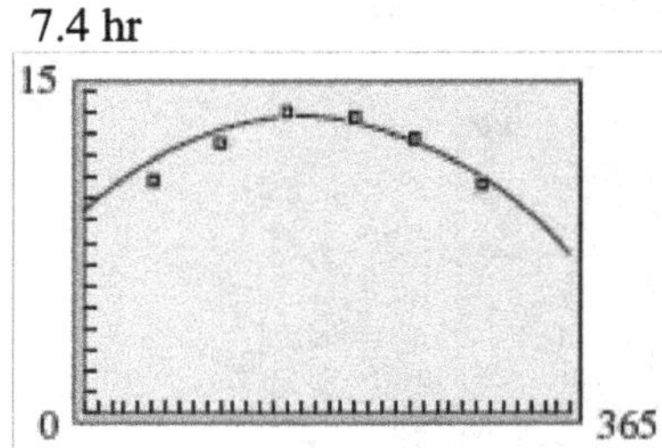

g. 9.8 hr; neither model is appropriate.

Chapter 6 Review

1. $6x(4x - 3)$ 3. $(a - 15)(a - 3)$

5. $(2y + 3)(2y + 5)$ 7. $(3w - 1)(w + 5)$

9. $(z + 11)(z - 11)$ 11. $3(2x + 1)(x + 3)$

13. $-5, -5$ 15. $0, 4$ 17. $1, 3$

19. $-1, 2$ 21. $-2, 3$

23. $4x^2 - 29x - 24 = 0$

25. 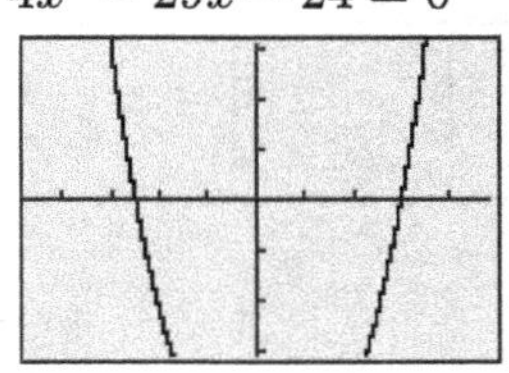 $y = (x - 3)(x + 2.4)$

27. $2\pm\sqrt{10}$ 29. $\frac{3}{2}\pm\sqrt{\frac{3}{4}}$

31. 1, 2 33. $\frac{4\pm\sqrt{8}}{2}$

35. $\frac{-6\pm\sqrt{12}}{4}$ 37. $\pm\sqrt{\frac{2K}{m}}$

39. $\frac{6\pm\sqrt{36-12h}}{6}$ 41. $x=\pm\sqrt{b^2-a^2}$

43. $x=\pm2\sqrt{1+y^2}$

45a. one real solution b. two complex solutions

47a. $-3+5i$ b. $1-\frac{1}{2}i$

49. $(\frac{1}{2}, \frac{-49}{4})$, $(-3,0)$, $(4,0)$, $(0,-12)$

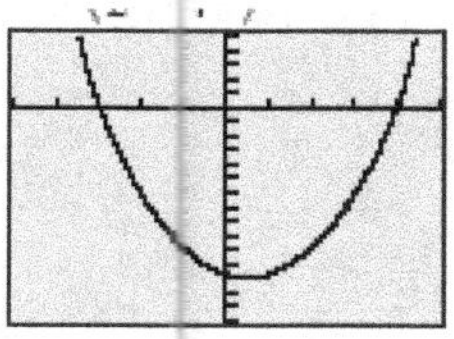

51. $(1,5)$, $(-1.24,0)$, $(3.24,0)$, $(0,4)$

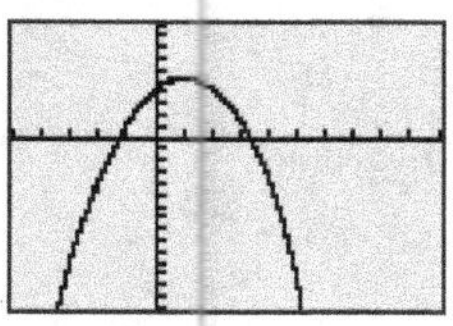

53a. $(-3,-2)$ b. $-3\pm\sqrt{6}$

c. 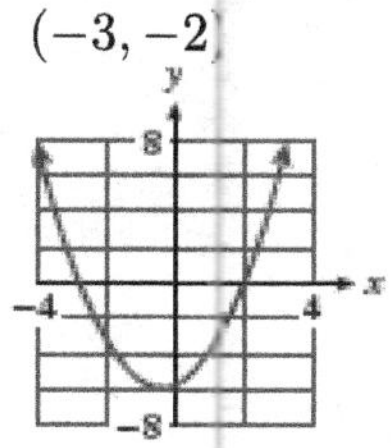

d. $y=\frac{1}{3}x^2+2x+1$

55. 3.81 sec 57a. $R=160p-2p^2$

b. $M=136p-2p^2-80$

c. 30 cents or 38 cents 59. 3.9, 8, 8.9

61. 10 ft by 18 ft, or 12 ft by 15 ft

63. The nursery should charge \$6 per box to obtain a maximum revenue of \$360 per month.

65a. $y=60(4+x)(32-4x)$ b. 2

67. $y=0.2(x-15)^2-6$

69a. (−4, 45), (0, 49)

b. $y=\frac{-1}{4}x^2+49$ c. 14 ft

71. $a=1, b=-1, c=-6$

73. $p(x)=\frac{-1}{2}x^2-4x+10$

75a. $h=36.98t+5.17$ b. 116.1 m, 153.1 m

c. 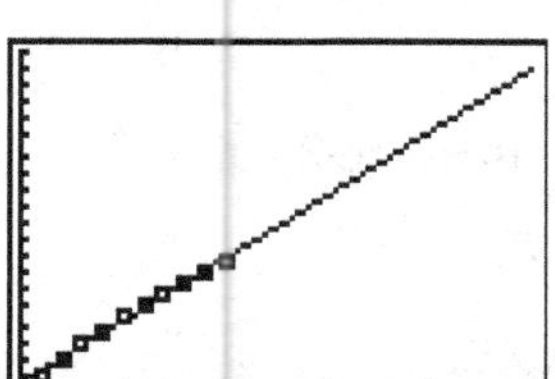

$[0,5]\times[0,200]$

d. $h=-4.858t^2+47.67t+0.89$

e. 100.2 m, 113.9 m

f. 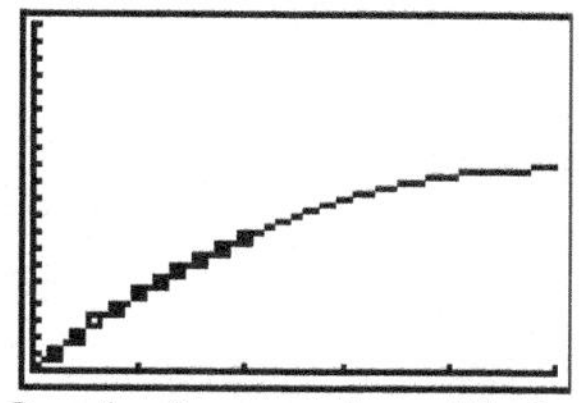

$[0,5]\times[0,200]$

g. Quadratic: gravity will slow the cannonball, giving the graph a concave down shape.

Index